Novel Drug Delivery Approaches

This book presents the evolution, challenges, and opportunities in the field of drug delivery. From foundational concepts to cutting-edge technologies, each chapter of *Novel Drug Delivery Approaches: From Fundamentals to Advanced Carrier Systems* delves into critical aspects that shape the landscape of drug delivery systems. It highlights the crucial role of translational research in bridging the gap between scientific discoveries and clinical applications. Subsequent chapters explore the diverse materials and technologies driving advancements in drug delivery, ranging from polymers and nanoparticles to colonic and intracellular delivery systems. Special emphasis is placed on specific applications and systems, including the role of metallic nanoparticles and nucleic acid-based therapeutic delivery systems. Emerging areas such as nasopulmonary and ocular drug delivery systems are thoroughly examined, uncovering challenges and opportunities in these specialized fields. Additionally, this book addresses the complexities of targeted carrier-based intracellular drug delivery, aiming to enhance treatment efficacy while minimizing toxicity. It provides insights into innovative approaches and challenges in brain drug delivery, transdermal drug therapy, and delivery systems for combating viral infections. Furthermore, this book delves into critical aspects of scaling, efficacy, safety, and regulatory considerations.

This book caters to researchers in biomedical sciences, pharmaceutical sciences, and biotechnology.

Key Features

- Offers an extensive examination of drug delivery, spanning foundational concepts to the latest advancements
- Provides insights into a diverse array of materials and technologies driving drug delivery advancements
- Explores insights into potential future trends and innovations of nasopulmonary and ocular drug delivery systems
- Highlights colonic drug delivery for gastrointestinal treatments and the use of nanoparticles for intracellular delivery
- Discusses quality by design (QbD) approaches in designing novel drug delivery systems and the scale-up technologies for nanomedicine

Novel Drug Delivery Approaches

From Fundamentals to Advanced Carrier Systems

Edited by
Neha Bajwa, Neeraj Mishra,
and N. K. Jain

CRC Press
Taylor & Francis Group
Boca Raton London New York

CRC Press is an imprint of the
Taylor & Francis Group, an **informa** business

Designed cover image: Shutterstock

First edition published 2026
by CRC Press
2385 NW Executive Center Drive, Suite 320, Boca Raton FL 33431

and by CRC Press
4 Park Square, Milton Park, Abingdon, Oxon, OX14 4RN

CRC Press is an imprint of Taylor & Francis Group, LLC

ISBN: 978-1-032-83275-3 (hbk)
ISBN: 978-1-032-85317-8 (pbk)
ISBN: 978-1-003-50860-1 (ebk)

DOI: 10.1201/9781003508601

Typeset in Times
by codeMantra

Contents

Acknowledgments vi
About the Editors vii

Chapter 1 Introduction and Foundation 1
Harpreet Kaur, Amanda Frank Mariki, Mridul Guleria, Arprita Malhan, and Komal Singh

Chapter 2 Evolution of Drug Delivery: A Journey from Bench to Market 21
Meenakshi Dhanawat, Kashish Wilson, Garima, Inderjeet Verma, and Sumeet Gupta

Chapter 3 Bridging the Gap: Translational Research in Drug Delivery 34
Gaurav Kumar, Smita Paul, and Ravindra K. Rawal

Chapter 4 Materials and Technology Involved in Neurological Mechanism Targeting 65
Tejas Girish Agnihotri, Shyam Sudhakar Gomte, Sandeep Kr Maharana, Naveen Singh, and Aakanchha Jain

Chapter 5 Specific Applications and Systems: Protein/Peptide Drug Delivery Systems 78
Virender Kumar, Ashwani Kumar, Shweta Saloni Kakkar, Vandana Garg, Shiv Kumar Yadav, and Davinder Kumar

Chapter 6 Role of Polymers Used in Nanoparticles for Controlled Drug Delivery 94
Sumeyra Sena Dag, Cem Varan, and Erem Bilensoy

Chapter 7 Colonic Drug Delivery Systems as Multiunit Potential: Therapeutic Strategies and Opportunities 108
Mazen Al-Mohaya and Burcu Demiralp

Chapter 8 Nucleic Acid-Based Therapeutic Delivery System 124
Anjali Kumari, Geeta Aggarwal, and Amanpreet Kaur

Chapter 9 Understanding the Concept of Quality by Design (QbD) Strategy: Case Study for Optimization of Lyophilized Solid Lipid Nanoparticles to Boost Oral Bioavailability of Antimalarial Drugs 139
Neha Bajwa, Preet Amol Singh, Narendra Kumar Jain, and Neeraj Mishra

Chapter 10 Nanomedicine Scale-Up Technologies: Feasibilities and Challenges 155
Jayaraman Rajangam, G. S. N. Koteswara Rao, Narahari N. Palei, Haja Bava Bakrudeen, Dinesh Kumar Upadhyay, Sabareesh Mekala, and G. Sabareesh

Chapter 11 Role of Metallic Nanoparticles in Controlled and Targeted Drug Delivery 168
Ahmet Dogan Ergin and Burcu Uner

Chapter 12 Nanoparticle Strategies for Enhanced Protein and Peptide Drug Delivery 184
Burcu Uner, Ahmet Dogan Ergin, and Pankaj Dwivedi

Index 201

Acknowledgments

I would like to express my heartfelt gratitude to the Almighty for his blessings and guidance throughout this journey. I am deeply thankful to my family for their unwavering support, encouragement, and love. I sincerely extend my thanks to Prof. Rajiv Sood Sir, Hon'ble Vice Chancellor, Baba Farid University of Health Sciences, Director, University Centre of Excellence in Research, for giving me the invaluable opportunity and platform to work on this book. His vision and support have been instrumental in making this endeavor possible. I extend my sincere thanks to the CRC team for their support and contribution. Finally, I would like to thank all the authors for their valuable contributions in writing the chapters of this book.

Dr. Neha Bajwa

We would like to express our sincere gratitude to the Honorable Founder-President Dr. Ashok K. Chauhan, Ritnand Balved Education Foundation (RBEF), and Dr. Aseem Chauhan, Chancellor, Amity University Madhya Pradesh. We are thankful to the Honorable Pro-Chancellor Lt. Gen. V. K. Sharma, AVSM (Retd.), Amity University Madhya Pradesh, Gwalior, India; Vice Chancellor Prof. (Dr.) R.S. Tomar, Amity University Madhya Pradesh, Gwalior; and Pro-Vice Chancellor (Research) Prof. (Dr.) M. P. Kaushik, Amity University Madhya Pradesh, Gwalior, India, for their continuous encouragement. We are also grateful to Prof. (Dr.) Manoj Kumar Mishra, Director, Amity Institute of Pharmacy, Amity University Madhya Pradesh, Gwalior, for her support during the writing of this book. Finally, we would like to express our wholehearted appreciation to the faculty and staff members of Amity Institute of Pharmacy, Amity University, Madhya Pradesh, Gwalior.

Prof. Dr. Neeraj Mishra

About the Editors

Dr. Neha Bajwa is a research officer at the University Centre of Excellence in Research and an assistant professor at the University Institute of Pharmaceutical Sciences, Baba Farid University of Health Sciences, Faridkot. Having written over 40 publications in international and national journals, Dr. Bajwa was chosen for a 6-month position at the University of Turin by the Young Investigator Training Program. Additionally, she received a travel grant from the International Pharmaceutical Federation (FIP) to attend the FIP-PSWC 2020 Conference, Montreal, Canada. She is a fellow/life member of the Indian Pharmacy Graduate Association and the Association of Pharmaceutical Teachers of India. Having presented research papers at over 20 national and international conferences, Dr. Bajwa has received various awards, including the Best Poster Research Award at the 6th World Congress on Nanomedical Sciences-ISNSCON-2018 in Vigyan Bhawan, New Delhi. As part of her PhD, she worked on the ICMR-sponsored project, "Molecular Encapsulation Augmented Oral Delivery of Arteether: Formulation Design using Total Quality Management Approach." Furthermore, she has contributed to various government agency-sponsored projects by AICTE, UGC, DST, CYCLONET, and FIP.

Dr. Neeraj Mishra has been professor and head, Department of Pharmaceutics, Amity Institute of Pharmacy, Gwalior since 2019. He has approximately 20 years of teaching and research experience and more than 150 publications of international and national repute on novel drug delivery systems, localized drug delivery, and targeted and controlled drug delivery of nanocarriers/microparticles for the treatment of breast, colon cancer, and neurodegenerative disorders. He has supervised 30 MPharm and 2 PhD students, and has more than 4200 citations with h-index 37. The total cumulative impact factor of his published papers is more than 500 (as per SCOPUS). Dr. Mishra has also written 25 books and 45 book chapters in Elsevier, Bentham, Springer, and Wiley Publication. He has also been granted three international patents and two Indian patents. He is the 2019 recipient of the Distinguished Professor Award from DST-NSTMIS, SPAICS, Indore, and M.P. Recipient of the 2021 Distinguished Professor Award from the Indian Pharmaceutical Association, M.P. State Branch, Indore. Dr Mishra received the 2023 Best Academician Award in the Indian Pharmacy Graduate Association, M.P. State International Conference, Indore, India. He was selected as one of the World Top 2% Scientists (rank 287886) by Stanford University, CA, USA and Elsevier in 2025. In addition, he was placed in the Stanford University Scientists Rankings conducted by AD Scientific Index— World Scientists Rankings in 2022, 2024, and 2025.

Prof. N. K. Jain, a distinguished academician and researcher, served as a senior professor in India for over 40 years before retiring in 2014. Throughout his illustrious career, he held various prestigious positions, including Director, ISF College of Pharmacy in Punjab and emeritus fellow at Rajiv Gandhi Technological University, Bhopal. His contributions to the pharmaceutical sciences are profound, with over two dozen acclaimed books and more than 500 publications in reputable journals. He has supervised numerous PhD and MPharm candidates and received numerous awards, including the IDMA Research Award and the APTI Teacher of the Year Award. His research interests encompass controlled and novel drug delivery systems, nanotechnology, and targeted drug delivery, earning him global recognition. Prof. Jain is actively involved in professional associations and is revered by students and colleagues alike. His recent books include *Introduction to Pharmaceutics* and *Hybrid Nanomaterials for Drug Delivery*, highlighting his continued dedication to advancing pharmaceutical research. In recognition of his outstanding contributions, Prof. Jain was honored with the prestigious APTI Shri Bhojraj Panjamool Lifetime Achievement Award and has been consistently ranked among the Top 2% of Scientists in pharmacology and pharmacy by Stanford University. His legacy in pharmaceutical research, teaching, and administration continues to inspire and shape the future of the field.

1 Introduction and Foundation

Harpreet Kaur[1], Amanda Frank Mariki[2], Mridul Guleria[3], Arprita Malhan[3], and Komal Singh[4]

[1]University Institute of Pharmaceutical Sciences, Panjab University, Chandigarh, India
[2]Tanzania Medicine and Medical Devices Authority, Dar Es Salaam, Tanzania
[3]Department of Pharmaceutical Sciences, Guru Nanak Dev University, Amritsar, Punjab, India
[4]University Institute of Pharmaceutical Sciences, Chandigarh University, Ludhiana Highway, Chandigarh, India

1.1 INTRODUCTION

Drug delivery systems refer to specialized formulations or devices designed to transport active pharmaceutical ingredients directly to their intended site of action. They also promote the efficacy and safety of the ingredients by controlling the rate, timing, and target site [1,2]. Drug delivery systems have been essential in the pharmaceutical sector for many years. They have evolved from basic oral dosage forms to advanced and targeted methods. The concept of modern drug delivery has its roots in the 1950s when the first sustained-release products were introduced. These coated tablets were designed to release medication gradually over an extended period. This period marked the transition from conventional dosage forms to more sophisticated delivery systems. Traditional systems of drug delivery, such as tablets, capsules, syrups, and topical ointments, are known as conventional drug delivery systems. These traditional dosage forms have been commonly used for many years due to their straightforward nature, affordability, and user-friendliness. Nevertheless, conventional drug delivery systems face several drawbacks, like high dosing frequency, poor patient compliance, low bioavailability, premature metabolism, and excretion of drugs from the body, leading to the need to create more sophisticated and targeted alternatives [3,4]. The advent of NDDS marks a significant breakthrough in the search for more effective therapeutic agents. Combining traditional drug delivery approaches along with advanced techniques has given a new direction to the new drug delivery methods [5]. These are systems characterized by smaller particle size, selective site targeting, improved permeability, and enhanced bioavailability, designed to transport therapeutic agents in the body to achieve the desired effect [6]. Usually, a pharmaceutical drug delivery system comprises three main features:

- Ideal dosage formulations responsible for delivering drug components into the body
- Release pattern of the drug components from the dosage form to the cells in the target site
- Ideal pharmaceutical techniques for the manufacturing of the pharmaceutical formulations.

Thereupon, the advancements in novel drug delivery systems (NDDS) can be achieved through:

- SMART nano carrier-based drug delivery systems designed to improve the solubility of drugs for enhanced therapeutic effects.
- SMART extended-release drug delivery systems designed to improve the controlled duration of action.
- Use of new manufacturing techniques such as microfluidics (MF).

1.1.1 Mechanisms Employed in NDDS

In the ever-evolving world of pharmaceutical innovation, NDDS is meticulously engineered. These cutting-edge systems target specific sites in biological systems through two primary mechanisms: physical mechanisms and biochemical mechanisms.

1.1.1.1 Physical Mechanism

These involve the strategic manipulation of drug formulations, such as altering their size, shape, or surface properties, to facilitate targeted delivery [5]. The techniques employed consist of diffusion, osmosis, dissolution, electrotransport, and erosion.

1.1.1.2 Biochemical Mechanism

These involve harnessing the power of biological processes to enhance drug delivery. By leveraging specific molecular interactions or exploiting natural pathways within the body, these systems can selectively transport drugs to their target sites [5]. These mechanisms include gene therapy, vector systems, polymer drug adducts, monoclonal antibodies, and liposomes.

1.2 NEED FOR NDDS

The need for innovative drug delivery systems stems from the shortcomings of the traditional dosage forms. Conventional methods often result in inconsistent drug levels in the bloodstream, unwanted side effects, and

DOI: 10.1201/9781003508601-1

suboptimal therapeutic outcomes. NDDS provide a groundbreaking approach to medication administration [2,3]. NDDS offers a solution by enabling controlled release of drugs, precise dosing, and targeted drug delivery to specific sites in the body. This ensures optimal drug concentrations reach the intended target.

NDDS, particularly in the field of nanomedicine, has demonstrated notable achievements in treating conditions like cancer, genetic disorders, and HIV/AIDS. The capacity of nanoparticles to enhance drug precision, delivery efficiency, and release, along with their potential for integrating diagnostic and therapeutic functions, establishes them as indispensable assets in contemporary healthcare [2,3].

1.3 ADVANTAGES AND LIMITATIONS OF NDDS

The advantages of NDDS can be attributed to various factors like improved stability and solubility, less dosing frequency, site targeting, improved bioavailability, and toxicological protection. However, the limitations associated with NDDS include insufficient localization of the targeted drug delivery systems into the intended target cells or tissues, drug diffusion and re-distribution to unintended regions, and precise dosages of the drug to their target tissues or cells.

1.4 IDEAL CHARACTERISTICS OF NDDS [7]

i. The system should be capable of directing the drug to the intended cells, organs, or tissues and ensure constant distribution in the capillaries.
ii. Drug release rate should be both controllable and predictable.
iii. The system should be non-immunogenic and biochemically inert.
iv. It should be physically and chemically stable both in the body and outside the body.
v. The system should release the therapeutic amount of the drug without affecting its action. It should minimize drug leakage during transit to the target site.
vi. The carriers used ought to be biodegradable or pumped out effectively from the body without any issues.
vii. The NDDS should be easy to prepare in a simple, reproducible, and cost-effective manner (Figure 1.1).

1.5 RECENT DEVELOPMENTS IN NDDS

1.5.1 Nanoparticles

Nanoparticles are tiny, spherical particles made of either natural or synthetic polymers. Their size typically falls within the range of 10–500 nm. With their distinctive shape and the significant surface area relative to their volume, nanoparticles find diverse applications across various fields. Nanotechnology unites physical and biological research by utilizing nanomaterials and nanophases in different scientific domains [8]. Microfluidics, drug delivery, tissue engineering, microarray testing, and biosensors are among the fields they find use in [9].

The unique physical, chemical, and biotic attributes of nanoparticles enable precise delivery and controlled release of the drugs [10,11]. Primarily, the nanoparticle-based therapy employed liposomes and micelles as carriers of drugs and got validation from the federal regulatory agency. Nanostructures shield drugs from deterioration in the adverse conditions of the alimentary canal. By encapsulating drugs within these tiny

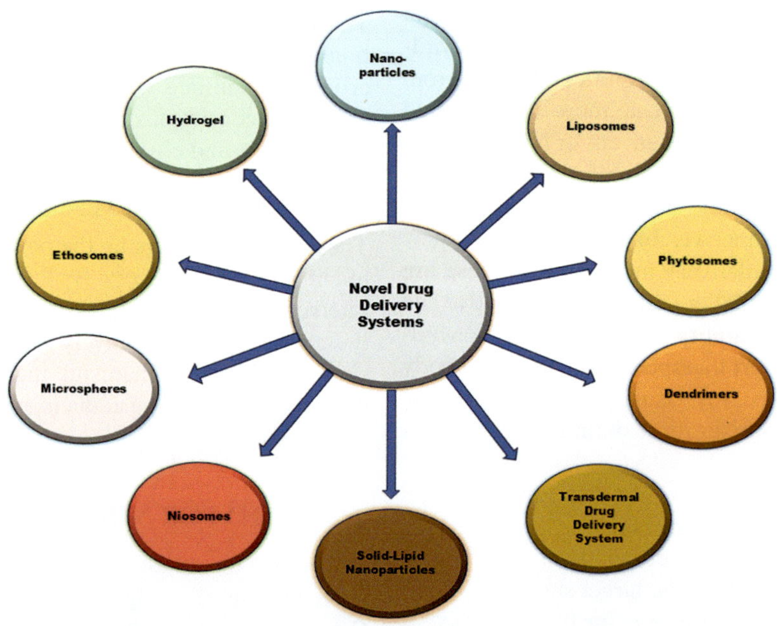

FIGURE 1.1 Types of novel drug delivery systems.

TABLE 1.1
Types of Nanoparticles, Their Characteristics, and Advantages

S. No.	Types of Nanoparticles	Characteristics	Advantages	References
1	Organic nanoparticles	Composed of natural or synthetic polymers Biocompatible	Biodegradability Suitable for effective drug release Versatile applications in medicine and agriculture	[12]
2	Inorganic nanoparticles	Made of inorganic materials (e.g., metals and oxides) High stability and reactivity	Enhanced catalytic activity Optoelectronic application	[12]
3	Carbon-based nanoparticles	Graphite, carbon nanotubes Exceptional mechanical strength Electrical conductivity	Lightweight and flexible Potential in electronics and energy storage	[12]
4	Ceramic nanoparticles	Composed of oxides, carbides, carbonates, and phosphates Exceptional mechanical strength Electrical conductivity	Photocatalysis Biological imaging	[12]
5	Metal-based nanoparticles	Derived from metals (e.g., gold, silver, and iron) Unique optical and electronic properties	Antibacterial agents Cancer therapy	[12,13]
6	Semiconductor nanoparticles	Examples: quantum dots, metal chalcogenides Size-dependent electronic behavior	Fluorescent probes Solar cells	[12,13]
7	Polymeric nanoparticles	Polymer-based carriers (e.g., liposomes and micelles) Controlled drug release	Targeted drug delivery Improved bioavailability	[12,13]
8	Lipid-based nanoparticles	Lipid bilayers or vesicles Biocompatible and amphiphilic	Drug encapsulation Gene therapy Cosmetic formulations	[12,13]

carriers, we prevent their premature breakdown and ensure their safe passage to the intended site. For sparingly water-soluble drugs, nanostructures ferry these hydrophobic compounds to their target locations, overcoming solubility challenges.

Nano drugs exhibit high oral bioavailability due to their unique uptake mechanisms. They are adept at being absorbed through absorptive endocytosis, efficiently entering cells, and reaching their therapeutic targets (Table 1.1).

1.5.1.1 Recent Advancements in Nanoparticle-based Drug Delivery and its Biomedical Applications

Recent advancements in nanoparticle technology have significantly enhanced drug delivery and biomedical applications across various nanoparticle types. Metal-based nanoparticles, like Au NPs (gold nanoparticles), are prevalent in site-specific drug delivery, imaging, and photothermal therapy due to their compatibility with living tissues, stability, and tuneable optical properties [14]. Ag NPs (Silver nanoparticles) are known for their antimicrobial properties. They are applied in wound healing, antibacterial coatings, and cancer therapy. Polymeric nanoparticles, including Poly(lactic-co-glycolic-acid) (PLGA) and chitosan nanoparticles, offer biodegradable and mucoadhesive properties, with advancements in personalized medicine and controlled release of therapeutics [15]. Lipid-based nanoparticles like liposomes have seen progress in targeted delivery and triggered release, enhancing their role in personalized medicine. Solid lipid nanoparticles (SLNs) are notable for their improved stability and controlled drug release [16], particularly in anticancer applications. Iron oxide nanoparticles (Fe_3O_4) are used in magnetic resonance imaging (MRI) and targeted drug delivery systems [17]. Carbon-based nanoparticles [18] are advancing in drug delivery, biosensing, and regenerative medicine, with ongoing research aimed at reducing toxicity and enhancing biocompatibility. Semiconductor nanoparticles like QDs (quantum dots) are used in imaging, biosensing, and cancer diagnosis [19]. These advancements aim to enhance nanoparticle stability, target specificity, and minimize toxicity, thereby facilitating diagnostics and therapeutic interventions for various diseases.

1.5.2 Liposomes

Liposomes are tiny, spherical moieties containing a double layer of lipids (fats) enclosing an inner watery compartment. Liposomes come in various sizes, ranging from a few nanometers to micrometers. Their size can be tailored for specific applications. They may possess one or more concentric lipid layers, making them multilamellar or unilamellar, with a single lipid bilayer. Because unilamellar liposomes are stable, they are frequently employed for medication delivery. The outer surface of liposomes can carry either a positive or negative charge, which influences their interactions with biological tissues and cells. Liposomes can be composed of natural or synthetic lipids. Liposomes derived from natural lipids are generally nontoxic and suitable for clinical use. Researchers utilize liposomes for drug delivery, gene therapy, and cosmetic formulations [20]. Liposomes have emerged as promising carriers for various diseases, from cardiovascular conditions to bacterial infections. Their unique properties enable them to serve multiple purposes: reducing the toxicity of potent drugs, enhancing pharmacokinetics, and improving therapeutic efficacy. Liposomes can accommodate both polar and nonpolar substances. Hydrophobic/nonpolar molecules nestle within the bilayer membrane, while hydrophilic/polar ones find refuge in the central watery compartment [21].

1.5.2.1 Advantages of Liposomes

Liposomes, tiny lipid-based vesicles, offer an innovative approach to drug delivery. Liposomes [22] selectively target tumor tissues, enhancing treatment effectiveness while minimizing side effects. Encapsulating drugs within liposomes increases stability and reduces toxicity. Their improved pharmacokinetics, longer circulation times, and reduced elimination make them valuable allies in precision medicine.

1.5.2.2 Structure and Composition of Liposomes

The formation of liposomes, which are spherical or multilayered vesicles, occurs when diacyl-chain phospholipids self-assemble in aqueous solutions. These lipid bilayers have an amphiphilic structure because of their hydrophilic head and hydrophobic tail [23]. Liposome characteristics, which are impacted by the lipid composition, include particle size, stiffness, stability, and electrical charge. As illustrated in Figure 1.2a, the hydrophilic groups in lipids can

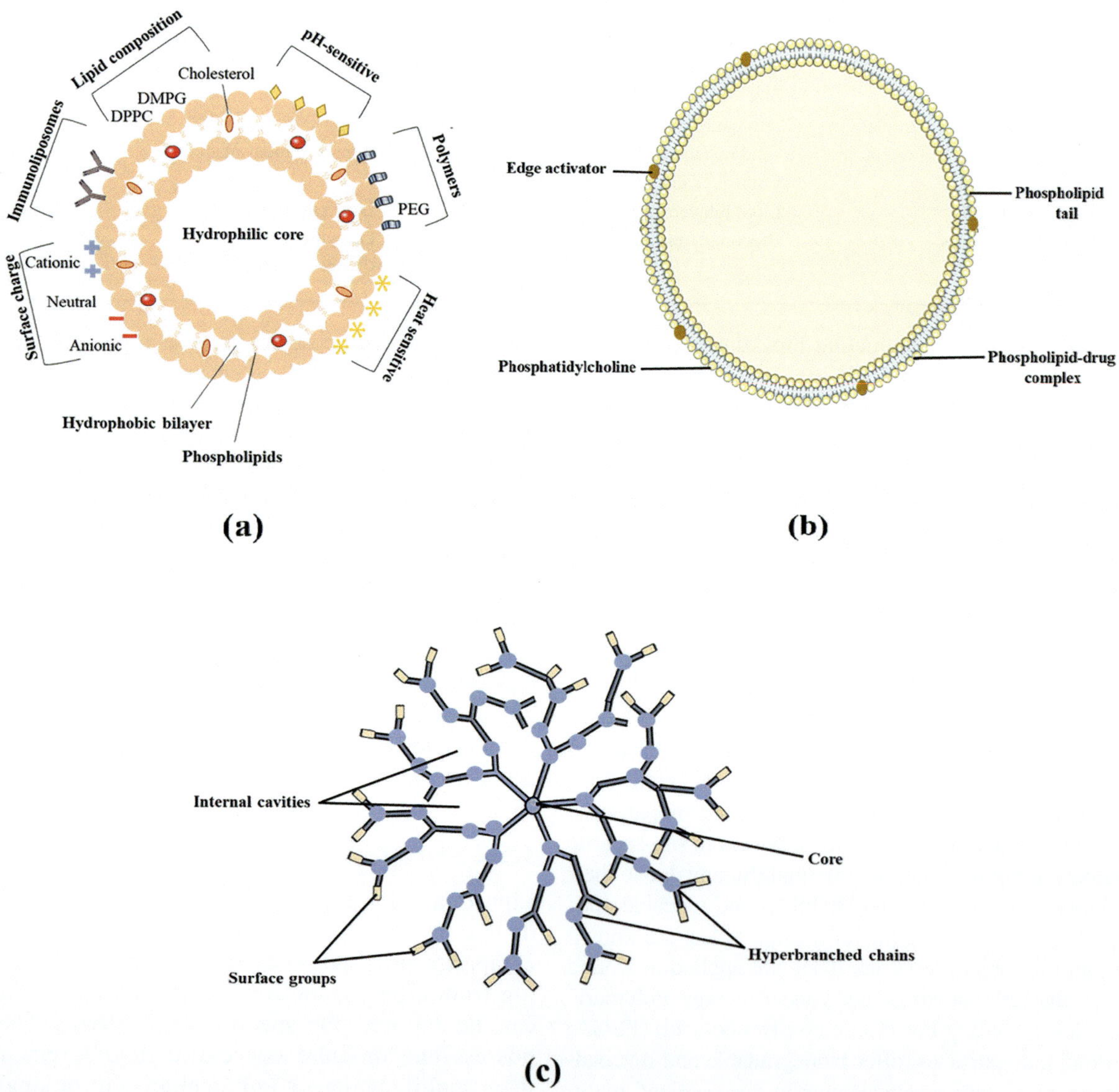

FIGURE 1.2 (a) Liposome, (b) phytosome, and (c) dendrimer.

have either +/- or both charges within the same molecule. This charge fluctuation affects stability through electrostatic resistance. Lipids can be categorized as either natural or synthetic based on their composition.

i. **Natural Lipids**: Glycerophospholipids make up the bilayers of normal cell membranes. They contain two fatty acid molecules and a glycerol unit bound to a phosphate group (PO_4^{2-}). Soy and egg yolks are the primary sources of natural phospholipids. Furthermore, because natural phospholipids include unsaturated hydrocarbon chains, they are less stable than manufactured phospholipids [24].
ii. **Synthetic Lipids**: Synthetic phospholipids are synthesized by chemically modifying natural phospholipids' polar and nonpolar parts. These alterations may produce an unlimited range of synthetic phospholipids [25]. Stearic and/or palmitic fatty acids are the foundation for several saturated synthetic phospholipids.

Cholesterol, a primary steroid, is commonly employed to prepare liposomes. It enhances liposome rigidity and stability by incorporating it into the lipid bilayer of the liposomes [26]. When cholesterol and β-Sitosterol were compared for their effects on liposome membrane properties, it was discovered that both steroids increased absolute zeta potential and decreased membrane fluidity. This resulted in significant alterations to the particle dimension and a drop in the enthalpy and temperature of the phase transition ™ and enthalpy of dipalmitoylphosphatidylcholine (DPPC) [27].

Surfactants alter the characteristics of liposome encapsulation and release [28] by manipulating the tension between non-miscible phases. Commonly used surfactants in liposome formulations include cetyltrimethylammonium bromide, sorbitan esters, and lecithin [29]. These liposomes, which contain surfactants, are used as carriers for active pharmaceutical ingredients. Ultra-deformable liposomes (or transfersomes) stand out among specialized liposomes. Unlike conventional liposomes, ultra-deformable liposomes can rapidly transform their shape in response to osmotic pressure, requiring minimal energy [30].

1.5.2.3 Applications of Liposomes

i. The biocompatibility, low immunogenicity, and ability to encapsulate therapeutic compounds make liposomes ideal for targeted drug delivery of antibiotics, genes, antifungal agents, and anti-inflammatory drugs.
ii. Photothermal or photodynamic materials can be coupled with liposomes to create photodynamic or photothermal therapy (PDT or PTT) at particular wavelengths. These therapies are effective against cancer cells [31].
iii. Liposomes can also serve as carriers for cancer vaccines [31].
iv. Liposomes are essential for the diagnosis process. For instance, precise measurement of low antigen concentrations across a broad dynamic range is now possible because of the invention of the immune-liposome polymerase chain reaction (ILPCR). This innovative approach holds promise for ultrasensitive detection of antigens.
v. Ultra-deformable liposomes (transfersomes) have rendered the administration of transdermal medication [32] more effective. These nano vehicles have shown promise for topical administration of antihypertensive medications.

1.5.3 Phytosomes

Phytosomes are specialized vesicular structures, similar to liposomes, that encapsulate active phytoconstituents within a lipid core, enhancing their permeability across lipid-based cellular membranes and improving bioavailability and efficacy [33]. This technology addresses the challenge many hydrophilic phytoconstituents face in crossing these membranes, such as glycosides and flavonoids. Phytosomes are considered biologically neutral, suggesting their administration is generally safe [34] without triggering immunological responses.

Despite the challenges of scaling up production and maintaining product quality, phytosomes hold significant commercial potential in modern drug delivery systems, promising improved therapeutic outcomes and increased market acceptance.

1.5.3.1 Structure and Composition of Phytosomes

Phytosomes, complex structures designed for enhanced nutrient absorption, consist of three essential components: phospholipids, active phytoconstituents, and solvents, as shown in Figure 1.2b. The phospholipids exhibit a polar head and nonpolar acyl chains linked to alcohol. Varieties of phospholipids such as phosphatidylcholine, phosphatidylethanolamine, and phosphatidylserine contribute to these structures, offering moderate solubility in both aqueous and lipid media. Active phytoconstituents, primarily flavonoids like quercetin and catechins, are selected from standardized plant extracts to form phytosomes, aiming to enhance the bioavailability of poorly absorbed compounds. Lastly, solvents play a crucial role in phytosome preparation, facilitating the mixing of phospholipids and active constituents. While various solvents, including both single and mixed systems like ethanol and chloroform, have been utilized, their choice influences the efficiency and characteristics of the resulting phytosomes, impacting their application in pharmaceutical and nutraceutical formulations.

1.5.3.2 Applications of Phytosomes

i. Phytosomes exhibit enhanced wound healing [35] potential.
ii. Phytosomes efficiently deliver herbal medicines through the skin while maintaining stability and efficacy.

iii. Phytosomes protect key components in herbal extracts from degradation by gut bacteria and digestive secretions, ensuring their integrity and potency.
iv. Phosphatidylcholine, a vital component in the phytosome, is absorbed efficiently in the dermis and epidermis [36], making phytosomes beneficial for skincare applications.

1.5.4 Dendrimers

Dendrimers are highly branched polymeric molecules characterized by a high degree of organization. Typically, dendrimers are symmetric around a central core and often adopt a spherical three-dimensional morphology as illustrated in Figure 1.2c. Dendrimers typically measure 1–10nm in size, although their dimensions can range from 15nm to a few micrometers [37]. Their distinct morphology, which is typified by their mass and modifiable branching units, renders them adaptable instruments across an array of uses. Dendrimers are used in targeting nucleic acids. Their controlled architecture allows precise interactions with genetic material, making them valuable in gene therapy and diagnostics. Dendrimers enhance drug delivery by combating multidrug resistance (MDR). Their tailored properties enable targeted drug delivery and improved therapeutic outcomes [37].

1.5.4.1 Applications of Dendrimers

i. Dendrimers have been used to decrease the amount of contrast agent needed in MRI while enhancing the agent's efficiency and enabling regulated biodistribution [38]. Despite the benefits, metal toxicity remains a concern, prompting research into dendrimer-based contrast agents with functional organic radicals, which show promising potential [39].
ii. The ability of dendrimers to be modified with various biological and chemical substances makes them valuable in tissue engineering for creating artificial extracellular matrixes.
iii. Dendrimers are a useful non-viral vector for delivering genes. Poly(amidoamine) (PAMAM) and lysine-based dendrimers exhibit a strong affinity for DNA and interact electrostatically to form complexes called dendriplexes, which enhances gene delivery efficiency ([40]).
iv. Dendrimers improve drug delivery by enhancing the solubility, stability, and oral bioavailability of different medications (Table 1.2).

1.5.5 Transdermal Drug Delivery System (TDDS)

The transdermal drug delivery system aims to apply medication topically at a regulated and predefined pace. Enhanced patient compliance, fewer adverse effects, longer therapeutic effects, and better bioavailability are just a few benefits that come with TDDS [47].

The epidermis, also known as the stratum corneum, serves as the barrier that limits the pace at which most medication molecules may penetrate it. The three main

TABLE 1.2
Types of Dendrimers, Their Characteristics, and Applications

S. No.	Types of Dendrimers	Characteristics	Applications	References
1	Phosphorus dendrimers	It contains a phosphorous atom with a hydrophobic surface.	Anti-inflammatory and anticancer drug agents. Drugs for neurodegenerative diseases.	[41]
2	Glycodendrimers	It contains naturally occurring glycoproteins, mucin-based dendrimers, and polysaccharides.	Targeted delivery of anticancer drugs.	[42]
3	PAMAM dendrimers	It is hyperbranched and possesses multifunctional terminal surfaces, specified size and shape properties, unmatched molecular homogeneity, and a restricted molecular weight dispersion.	Anticancer potential Delivery of Temozolomide (TMZ) to brain	[43]
4	PPI dendrimers	Strongly branching, easily modifiable terminal end group, well-specified size, and narrow dispersity.	Carrier for chemotherapeutic agents Tumor growth inhibition	[44]
5	Peptide dendrimers	Macromolecules with branches that resemble wedges and surface functional units are covalently linked.	Biomedical diagnostic reagents Protein mimetics Anticancer and antiviral agents Vaccines and gene delivery vehicles	[45]
6	Triazine dendrimers	Structure that allows for surface customization with scalable, orthogonal functional end groups.	Cancer drugs (paclitaxel and camptothecin) Brefeldin A (antiviral and desferrioxamine), Non-viral DNA and RNA delivery systems.	[46]

ways drugs can permeate the skin are appendageal, transcellular, and intercellular pathways [47]. TDDS or transdermal patch delivers the drug from the skin to the bloodstream. Transdermal patches consist of a polymer matrix, membrane, medication, penetration enhancers, and pressure-sensitive adhesive.

There are several benefits of administering drugs transdermally as opposed to orally and via hypodermic injection. The process of administering medication through the skin has significantly aided in handling several illnesses, including diabetes, cancer, heart disease, and disorders of the central nervous system [48]. Various active and passive permeation techniques are employed for the successful diffusion of the drug through the skin barrier. Passive permeation techniques constitute the use of chemical agents, liposomes, and nanoparticles, while active permeation techniques are more robust and advanced nowadays and are summarized in Table 1.3.

1.5.5.1 Factors Affecting the TDDS

Several factors influence the effectiveness of transdermal drug delivery systems. Key physicochemical properties affecting drug penetration [55] include the partition coefficient, pH conditions, penetrant concentration, molecular weight, and solubility. It is necessary to have a lipid/water partition coefficient of at least 1, and a pH typically ranging from 4 to 6. The physicochemical properties of the drug delivery system (DDS), including release characteristics, system composition, and the presence of permeation enhancers, also play significant roles. Additional factors influencing medication absorption are body temperature, skin hydration, age, oxygenation, condition of the local site, and the flora and enzymes of the skin.

1.5.6 Solid Lipid Nanoparticles (SLNs)

SLNs are among the most prominent pharmaceutical nanocarriers for precision drug delivery, known initially as lipospheres [56]. R.H. Muller is regarded as the father of SLNs for introducing SLNs in December 1991. SLNs have a mean diameter that ranges from 50 to 1,000 nm [57]. The ability of SLNs to transport an extensive array of therapies, such as vaccination antigens, genetic material (DNA/siRNA), large biomacromolecules (polysaccharides,) and tiny drug molecules, is a unique characteristic [58]. SLNs are widely recognized as the potential carrier to improve a drug's oral bioavailability despite having a poorer solubility profile. SLNs have distinct characteristics over traditional colloidal carriers such as emulsions and liposomes in terms of drug release and toxicity aspects. As observed in polymeric particles, SLNs are physiologically appropriate carriers [59] from which the lipid matrix maintains and controls the drug release pattern.

1.5.6.1 Various Components of SLNs

Lipids, solid substances by nature at both physiological and normal temperatures, constitute the majority of SLNs shown in Figure 1.3a. They exert a substantial influence on drug loading, release pattern, stability, and encapsulation efficiency. Surfactants control the size and stability of nanoparticles in SLNs. They improve the stability of SLNs by reducing surface tension between the lipid and aqueous phases [59]. The choice of surfactant is dependent on its hydrophilic-lipophilic balance (HLB) value. For steric stabilization of developed SLN, only one or a mixture of surfactants can be employed [60]. Another component, the solvents and co-solvents, is primarily used to dissolve APIs, lipids, surfactants, and co-surfactants. The solvents and co-solvents should primarily be nontoxic and non-irritating.

TABLE 1.3
Active Permeation Techniques Involved in Transdermal DDS, Their Advantages, Disadvantages, and Applications

S. No.	Techniques	Advantages	Disadvantages	Applications	References
1	Iontophoresis	Helps charged and neutral substances to be delivered through the skin more easily.	High current intensity or drug properties can cause skin irritation or erythema.	Delivery of small peptides.	[49]
2	Ultrasound	Improves the administration of medications with high and low molecular weights.	Complex technique, energy-intensive.	Devices like Sonoprep® and SonodermTM.	[50]
3	Microneedles	Reduces microbial penetration compared to hypodermic needles.	It may offer less dosage precision than hypodermic needles.	Systems like MicroCor PTH and Corplex.	[51]
4	Electroporation	Safe and painless method for delivering low molecular weight (LMW) drugs.	Can harm medications that are heat-sensitive as well as induce cell disruption and death.	Transdermal delivery of Piroxicam using surfactants.	[52]
5	Radio-frequency microchannels	Painless and safe alternative to intramuscular or subcutaneous vaccines.	Requires maintenance of high-frequency alternating current [~100 kHz).	Systems like ViaDerm.	[53]
6	Magnetophoresis	Enhances drug permeability through structural changes in the skin.	It can alter skin properties.	Lidocaine transdermal patch.	[54]

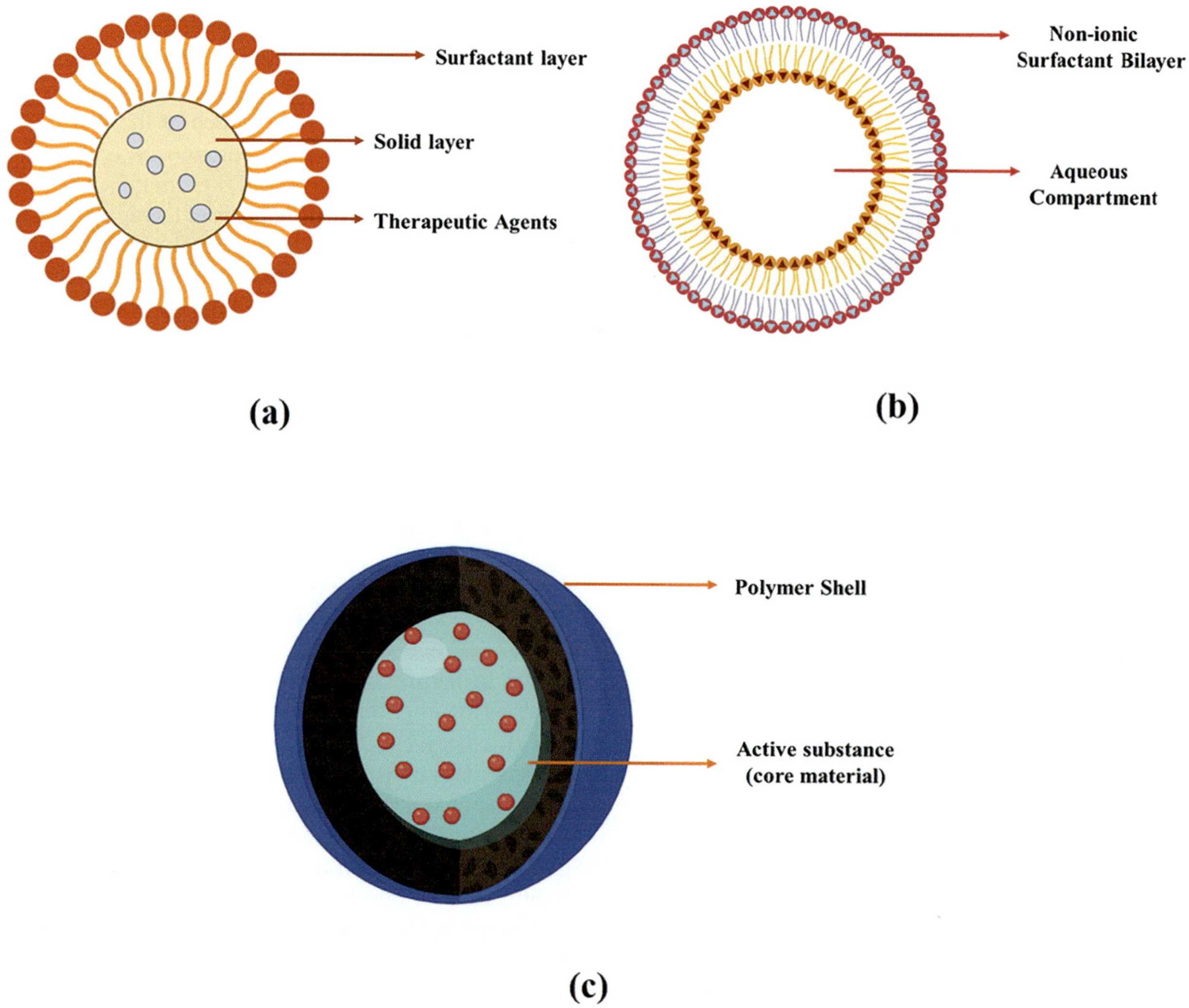

FIGURE 1.3 (a) Solid lipid nanoparticle, (b) niosome, and (c) microsphere.

Additionally, before freezing, the SLNs are protected against freezing and drying stress by adding cryoprotectants to the SLN dispersion. They also retard the aggregation of particles [61] and instability during storage. The primary mechanism contributing to the stabilization of particles is lyoprotectant-induced dehydration. The SLNs can be fabricated [62] through various physical and chemical methodologies like high-shear/pressure homogenization, coacervation, electrospray, solvent injection, microemulsion, and the supercritical fluid method.

1.5.6.2 Advantages and Disadvantages of SLNs

The advantages of SLNs include characteristics like controlled release kinetics, large-scale production, site specificity, and enhanced bioavailability. However, it is equally important to consider their disadvantages, encompassing limited transdermal activity, particle size expansion while storing, and suboptimal drug loading capacity.

1.5.6.3 Applications of SLNs

SLNs act as carriers of nucleic acid while maintaining the stability and efficacy of DNA vaccines [63]. SLNs are prevalent in the treatment of gastrointestinal infections caused by *Campylobacter jejuni* while overcoming antibiotic resistance [64]. SLNs in conjunction with immunosuppressives or glucocorticoids can improve the absorption of poorly water-soluble compounds in ocular medicines [65]. Curcumin-SLN microneedles as a potential preventive measure against Parkinson's disease [66]. SLNs assist in the transport and sustained release of hydrophobic medication simvastatin, which plays a significant role in the management of wound healing [67].

1.5.7 Niosomes (NS)

Niosomes are carriers that deliver drugs in a controlled, targeted, and sustained manner [68]. Liposomes were the original vesicular DDSs, but they have been associated with many shortcomings, like high cost, instability in pH, and toxicity issues. Due to these liposome-related problems, niosomes have become the subject of extensive research. Nonionic surfactants neutralize the harmful effects of niosomes [69]. Hydrophobic drugs enter the NS bilayer structure by segregating, while hydrophilic drugs are trapped in the central aqueous compartment. Niosomes enhance the bioavailability of drugs by allowing APIs to escape the low pH environment of the gut through encapsulation within these vesicles [70]. Many techniques [70] have been developed for preparing niosomes, like the bubble method, heating method, sonication, pronisome method, reverse-phase evaporation, ether injection, and thin-film hydration.

1.5.7.1 Structure of Niosome

Niosomes, as shown in Figure 1.3b, may encapsulate hydrophobic, hydrophilic, and amphiphilic therapeutic molecules with varying solubility ranges due to different HLB values of surfactants consisting of hydrocarbon chains [70]. The term "amphiphile vesicles" refers to the general structure of niosomes, which is composed of nonionic surfactants, such as polyglycerol alkyl ethers or Tweens, cholesterol, and a charge inducer [68]. A crucial element in the formation of niosomes is the molecular composition and HLB values of each surfactant [71]. Niosomes are composed of one or more lamellae, which are bilayers with a hydrophobic and a hydrophilic part joined by amide, ether, or ester linkages [68].

1.5.7.2 Composition of Niosomes

Thermodynamically stable niosomes [72] have a double-layered structure composed of nonionic surfactants that are only synthesized when surfactants and cholesterol are appropriately combined, at temperatures greater than the gel-liquid transition point. The synthesis of niosomes requires three ingredients: cholesterol, a charged molecule, and nonionic surfactants. The concentration of cholesterol in niosomes affects their physical and structural characteristics. Additionally, it enhances vesicle stiffness while simultaneously reducing their permeability for encapsulated molecules, thus avoiding leakage of the entrapped molecules. The charged molecules prevent aggregation of niosomes by increasing the surface charge density of vesicles. The two negatively charged chemicals most frequently used in noisome production are phosphatidic acid and diethyl phosphate. Usually, the niosomal formulation contains 2.5%–5% of the charged molecule [73].

Selection of nonionic surfactants or amphiphilic compounds requires careful consideration of the HLB and critical packing parameter (CPP) values [74]. The range of HLB for nonionic surfactants spans from 0 to 20. However, surfactants with an HLB value of 4–8 are desirable for the synthesis of niosomes. The shape of nanostructures created by amphiphilic molecules self-assembling may be predicted based on the CPP of a surfactant [75]. Niosomes have been distributed in biological environments through several routes [76]. These can take the form of an aerosol, a semisolid such as a gel or cream, a dry powder or suspension, or any combination of these. In vivo, they can be administered parenterally or topically. The reticuloendothelial system (RES) identifies NS as foreign particles when they reach the systemic circulation and ends up with their endocytosis to reach the liver and spleen.

1.5.7.3 Advantages and Disadvantages of Niosomes

There are several advantages and disadvantages of niosomes as carriers in drug delivery systems. The significant advantages are controlled and precise drug administration, cost effectiveness, enhanced oral bioavailability, and convenient storage and transportation. However, the considerable disadvantages worth mentioning are inadequate drug loading capacity, fusion/leaching of the drug, short shelf life, and aggregation.

1.5.7.4 Applications of Niosomes

The niosomal formulations may decrease the adverse effects of anti-leishmanial drugs on other organs. Niosomes can be used as target medications in leishmanial treatments [77]. Niosomes have been recognized as an effective carrier for delivering medication topically. Niosomes have been known as an effective carrier as a drug delivery system for neoplastic drugs. When administered by NS, 5-fluorouracil (5-FU), which is used to treat actinic keratosis and non-melanoma skin cancer, has improved percutaneous penetration [78]. Niosomes are favored for glaucoma treatment over other nanocarriers due to their chemical stability and lower toxicity since they are nonionic.

1.5.8 Microspheres

Microspheres may be described as "a monolithic spherical structure with the drug or as a dispersion of particles throughout the matrix." Proteins or synthetic polymers generate spherical microspheres, with a particle size ranging from 1 to 1,000 μm [79]. Other materials, including glass and ceramics, are effectively utilized in the formulation of microspheres. Drug delivery systems utilize microspheres to enhance the therapeutic efficacy of drugs. The drug is encased in a unique polymeric membrane situated in the center of the microsphere. There is an array of techniques [79] for the synthesis of microspheres. Some of these techniques include the single/double emulsion technique, polymerization, spray drying, solvent evaporation, and the ionic gelation method.

1.5.8.1 Structure and Types of Microspheres

Based on the composition, microspheres are divided into two categories: biodegradable polymeric microspheres and synthetic polymeric microspheres. Alternatively, based on their origin, these are classified as natural and synthetic microspheres exhibited in Figure 1.3c.

Natural polymeric materials can be obtained from various sources, like proteins (gelatin, collagen, and albumin) and carbohydrates (agarose, carrageenan, and starch). Meanwhile, the synthetic polymers are sourced from chemically modified carbohydrates, non-biodegradable (glycidyl methacrylate and acrolein), and biodegradable (polyalkylcyano-acrylates, lactides, and glycosides) materials [79].

1.5.8.2 Advantages and Disadvantages of Microspheres

Microspheres are easily injectable owing to their spherical shape and smaller size, provide controlled and site-specific release of the drug, and possess the ability to alter the distribution and pharmacokinetic profile of the drug. The main disadvantage is the difficulty in maintaining the stability of dosage forms. Additionally, the biological environment is negatively impacted by polymer matrix degradation products that are formed due to hydrolysis, heat, oxidation, biological agents, or solar radiation.

1.5.8.3 Applications of Microspheres

Ophthalmic drug delivery through microspheres is more effective than ointments and solutions. Polymers with bioadhesion and permeability-enhancing qualities are utilized in ophthalmic drug delivery through microspheres [79]. Vaccine-containing microspheres enhance immune response by extending the duration of antigen release for several weeks/months. Controlled administration of vaccines also significantly reduces systemic adverse effects [80]. Microspheres elicit minimal immunological reactions as compared to recombinant adenoviruses [81] and, thus, are preferential in gene delivery. Radioactive microspheres with a β-emitter (such as yttrium-90) are utilized in targeting liver tumors, which die without damaging nearby normal cells. The medication 5-fluorouracil can be administered as a treatment for colon cancer using polymeric microspheres [81].

1.5.9 Ethosome

Ethosomes are recognized as specialized lipid carriers that demonstrate enhanced transdermal delivery capabilities. Water, ethanol, and phospholipid constitute the ethosomal system. Its size varies based on the preparation process and the use of sonication methods, ranging from tens of nanometers to micrometers. In transdermal delivery systems, ethosomes are an effective delivery vehicle and their enhancing impact is well known [82]. Ethanol in lipid bilayers enhances carrier penetration through the stratum corneum layer, enabling both lipophilic and hydrophilic drugs to be delivered locally and systemically. The primary cause of the ethosomal drug's greater penetration and distribution in the skin may be the synergistic interaction between phospholipid and a high ethanol concentration [83].

1.5.9.1 Structure and Composition of Ethosomes

Based on composition, ethosomes are classified into three types: classical, binary, and transethosomes. Classical ethosomes are a variation of conventional liposomes with a 45% w/w alcohol concentration, as shown in Figure 1.4a. In contrast to traditional ethosomes, they exhibit a larger negative zeta potential and improved entrapment efficiencies. The range of molecular weights of drugs entrapped by classical ethosomes is 130.07 Da to 24 kDa. Consequently,

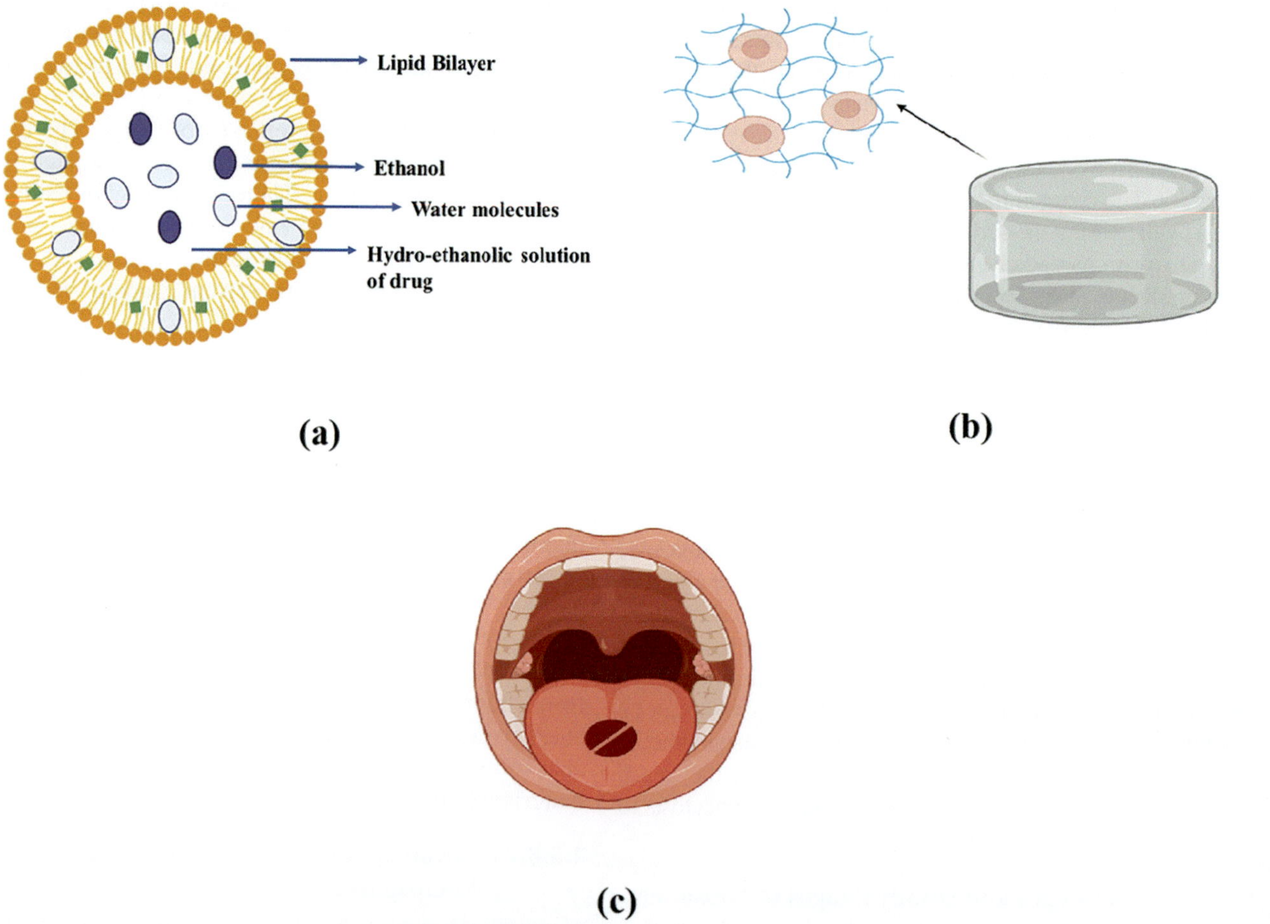

FIGURE 1.4 (a) Ethosome, (b) hydrogel, and (c) mouth-dissolving tablets.

they exhibit enhanced stability and penetration [84]. Binary ethosomes derive their name from the necessity of incorporating an additional alcohol molecule during the synthesis process to optimize their characteristics. Isopropyl alcohol (IPA) and propylene glycol (PG) are frequently used for their synthesis [85]. Transethosomes, ultra-deformable vesicles, are more elastic and flexible. In comparison to conventional formulations, they contain an additional excipient in the form of an edge activator (often a surfactant) and/or penetration enhancer. These can entrap drugs with molecular weights of 200–235 kDa [83].

1.5.9.2 Advantages and Disadvantages of Ethosomes

Ethosomal drug delivery systems are nontoxic and prevalent in pharmaceuticals, nutraceuticals, cosmetics, and veterinary medicines [84]. Additionally, larger molecules like peptides and proteins can be delivered easily, and epidermal absorption of drugs is significantly high. The disadvantages associated with this carrier are the limited size of the drug molecule, the expensive production process, and allergic responses to ethanol or other penetration enhancers used in the preparation of ethosomes. Moreover, unlike other carriers, ethosomal carriers are limited to transdermal delivery only.

1.5.9.3 Applications of Ethosomes

Ethosomal formulations can be used for hormone delivery. For instance, the testosterone-based ethosome, Testosome, in comparison to a standard transdermal patch containing testosterone (Testoderm), showed enhanced skin penetration [85]. Ethosomal formulations can deliver a significant dose of antibiotics across the skin [86] without any harmful outcomes. Encapsulating an anti-arthritis drug in ethosomal formulation is a better approach toward enhanced skin permeability [86] and consequently precise delivery of API.

1.5.10 Hydrogel

Networks of polymer chains distributed in water are called hydrophilic gels or hydrogels. Due to the presence of hydrophilic groups in the polymeric network, the structure transforms into a hydrogel when exposed to water, as illustrated in Figure 1.4b. Crosslinks between network chains provide hydrogels with their resistance to disintegration, while the hydrophilic functional group attached to the polymeric backbone enables them to absorb water. Hydrogels can take a multitude of forms and dimensions and can expand or contract in reaction to variations in temperature, ionic concentration, and pH [87]. They also have a wide range of uses in research for the controlled release of medications and proteins [88].

1.5.10.1 Synthesis of Hydrogels

Polymers, whether synthetic or natural, are used to formulate hydrogels. Hydrogels derived from natural sources are referred to as natural hydrogels, whereas hydrogels made from synthetic polymers are called synthetic hydrogels. A few synthetic polymer examples include ethylene glycol dimethacrylate, N-(2-hydroxypropyl) methacrylamide, acrylic acid, polylactic acid, and vinyl acetate. Owing to their greater compatibility with living tissues and environmentally friendly nature, natural polymers are being employed more frequently than synthetic ones. One more category, known as hybrid hydrogels, originates from the combination of synthetic and natural polymers [89]. The synthetic methods for hydrogel constitute the Photo-Fenton method, ultrafiltration, solvent method, carbodiimide cross-linking, and co-precipitation method.

1.5.10.2 Advantages and Disadvantages of Hydrogel

Hydrogels exhibit several advantages [89] like biocompatibility, easy modification, storage and transportation, low toxicity, and sensitivity to variations in temperature, pH, or metabolite concentration. However, hydrogels [87] are expensive, non-adherent, and difficult to sterilize and load. Several conditions, like dehydration, lens deposition, hypoxia, and red eye reactions, are caused by hydrogels used as contact lenses.

1.5.10.3 Applications of Hydrogel

Hydrogels deliver drugs for treating several oral diseases, viral infections, fungal diseases, oral cavity cancers, and periodontal disease. Hydrogels maintain wound exudates in place by retaining water and medication inside them. Gelatin and sodium alginate-based hydrogel can be applied to a wound to shield it against bacterial infection. Hydrogels having bio-adhesive properties are mainly used in rectal drug delivery. Hydrogels maintain the release of ocular medicines and enhance their efficacy by concentrating them at the site of action.

1.5.11 Mouth-Dissolving Tablets

Mouth-dissolving tablets, Figure 1.4c, are also known as fast-melting, fast-dissolving, orally disintegrating, or orodispersible. Medications for erectile dysfunction, neuroleptics, analgesics, and antiallergics are the most excellent choices for this system. Mouth-dissolving tablets are classified as true-fast dissolving if they dissolve in a matter of seconds after coming into contact with saliva. Other types, which take approximately a minute to dissolve completely, are better known as fast-disintegrating tablets because they contain ingredients that speed up their breakdown in the oral cavity. They are distinct from lozenges, buccal tablets, and traditional sublingual tablets because they require longer to dissolve in the mouth [90].

1.5.11.1 Synthesis of Mouth-Dissolving Tablets

Several techniques are employed for the formulation of mouth-dissolving tablets, such as tablet moulding, sublimation, mass extrusion, lyophylization, direct compression, and spray drying etc.

1.5.11.2 Advantages and Disadvantages of Mouth-Dissolving Tablets

The various advantages [91] associated with mouth-dissolving tablets constitute cost effectiveness, easy administration, easy breakdown in the mouth without chewing, and enhanced bioavailability of hydrophobic and insoluble drugs. The limitations lie in the fact that drugs requiring larger doses for optimal action are unsuitable as candidates for these dose forms. Other drawbacks are extra caution while handling the tablet and an unpleasant mouth feel in case of improper formulation.

1.5.11.3 Application of Mouth-Dissolving Tablets

Mouth-dissolving tablets are helpful for people with dysphasia because they dissolve in the mouth in a couple of minutes and release the active medication and are also cost-effective. The new manufacturing method produces tablets with improved safety, reduced side effects, enhanced bioavailability, and a faster onset of action [91,92].

1.6 CHARACTERIZATION OF NOVEL DRUG DELIVERY SYSTEMS

During the progression of drug delivery systems, the primary objective is to ensure efficacy and the ability of APIs to target specific sites without adverse effects. The introduction of drugs into humans or animals necessitates a method or process for administering the active pharmaceutical ingredient (API). This process is facilitated by the DDS, a technology [6] designed to customize the drug release profile and pharmacokinetic parameters to enhance efficacy at the active site, at the required rate during the treatment period.

Different drug delivery systems entrap the drug molecule and transport it to the precise site of action, surpassing API inactivation by gastric juices and first-pass metabolism in the lung, liver, or intestine. Henceforth, comprehending the attributes of various drug delivery systems is imperative in the advancement of sophisticated drug delivery mechanisms. This involves a comprehensive study of their various features [13] such as size, shape, composition, and physical, electrical, and chemical properties. Additionally, particle size, surface charge, and polydispersity are the foremost properties influencing their potency. A variety of analytical techniques are employed to characterize them and study their surface properties. Two sorts of methodologies can be used for their characterization: spectroscopic methods and microscopic methods, as illustrated in Figure 1.5.

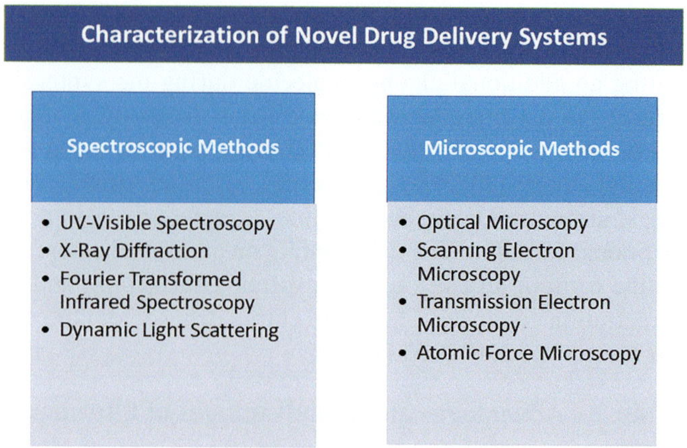

FIGURE 1.5 Techniques for characterization of novel drug delivery systems.

1.6.1 Nanoparticles

Based on their configuration, nanoparticles can be divided into categories, namely: organic (e.g., polymeric nanosphere/nanocapsule, polymeric micelle, liposome, and dendrimers) or inorganic (e.g., gold nanoparticles, carbon nanotubes, and QDs). The complete characterization of these nanoparticles can be conducted through spectroscopic and microscopic methods. Microscopic methods like optical microscopy, scanning electron microscopy (SEM), and transmission electron microscopy (TEM) study the various morphological features [13] of nanoparticles like smoothness, globosity, and aspect ratio/dimension. Meanwhile, spectroscopic methods like X-ray diffraction, and Fourier transform infrared spectroscopy provide information about lattice parameters, crystal structure, and molecular structure of nanoparticles.

1.6.1.1 Morphological Characterization of NPs

The morphological features make the nanoparticles unique and appropriate for various applications [14]. Among various characterization techniques, microscopic techniques such as polarized optical microscopy, SEM, and TEM stand out. The foundation of SEM is that bombarding high-energy electrons on a sample produces signals from the interaction of electrons and the sample being analyzed. This interaction results in the generation of exclusive nanometer-scaled images. This technique highlights their morphology and dispersion in the bulk or matrix. Saeed K et al. [93] have demonstrated morphological, thermal, mechanical, and solvent swelling behavior in carbon nanotube/nylon 6,6 nanocomposites using SEM.

Conversely, TEM operates on the principle of subjecting an ultrathin sample to a high-voltage electron beam, and the chemical interplay results in the formation of an image. It gives precise structural information of NPs at the highest magnification. It enables us to have information about the internal structure of the NP, which is not possible in SEM.

Novel materials with improved magnetic characteristics have applications in drug administration, magnetic resonance imaging, and high-density magnetic storage, according to Upadhyay et al [94]. Fourier Transform Infrared Spectroscopy (FTIR) and Raman Spectroscopy represent sophisticated and extensively validated methodologies for assessing properties like surface energy, charge, and solvation in nanoparticles. The unique identity of nanoparticles is discernible within the fingerprint region of infrared spectra. These techniques quantify the influence of infrared radiation at specific wavelengths on molecular vibration and rotation, facilitating the identification of reactive surface sites. In a recent investigation, León et al. [95] comprehensively characterized a smart drug delivery system for the anti-tumor drug 2-Methoxyestradiol (2ME) using FTIR and Raman Spectroscopy. This involved the complete analysis of vibrational modes for titanium dioxide, polyethylene glycol, and 2-Methoxyestradiol, as well as functionalized nanoparticle conjugate, following the loading of 2ME into polyethylene glycol (PEG) coated titanium dioxide (TiO2) nanoparticles. The FTIR and Raman spectra revealed new bands due to the formation of new hydrogen bridges between polyethylene glycol and 2ME at 1447 cm−1 and 1300 cm−1.

UV-visible spectroscopy is another essential technique for analyzing, studying, and assessing the stability of colloidal solutions containing NPs. This cost-effective, precise, and straightforward method is used to identify NPs based on their optical material properties, which are sensitive to their size, geometry, particle aggregation, and refractive index near the surface.

1.6.2 Liposomes/Solid Lipid Nanoparticles (SLNs)

SLNs [59] and liposomes [23] exhibit similarities in the fact that both nanoformulations are carriers in DDS for APIs that are poorly soluble in water and protect the drug molecule with the outer lipid bilayer. However, SLNs, a successor of liposomes, are more scalable and stable. Morphological and structural characterization of these nanocarriers can be performed using dynamic light scattering (DLS), high-performance liquid chromatography (HPLC), fluorescence spectroscopy, and nuclear magnetic resonance spectroscopy (NMR).

1.6.2.1 Morphological Characterization of Liposomes/SLNs

DLS is utilized in analyzing zeta potential during *in vivo* studies of liposomal/SLN colloidal systems. DLS sensors detect changes in scattered light intensity resulting from liposome/SLN movement due to the interaction of an electric field with nanoparticles. This technique is known as phase analysis light scattering (PALS) or electrophoretic light scattering. [96]

Chiong et al. studied a liposome-based DDS for piroxicam [97], a non-steroidal anti-inflammatory drug. They characterized the morphology of DDS, employing negative staining electron microscopy and TEM. Additionally, the morphological characterization of SLNs included particle size analysis, polydispersity index, and zeta potential using SEM, TEM, and DLS. The SEM analysis showed the spherical and moderately uniform shape of chitosan-carbamazepine SLNs.

1.6.2.2 Structural Characterization of Liposomes/SLNs

SLNs with a diameter of 50–1,000 nm in the colloidal state primarily comprise solid-state lipids and surfactants. These SLNs have broad-spectrum applications in transdermal and oral DDS [76]. Tripathy et al. have exhibited a chitosan-based transport system with the lipophilic drug carbamazepine as an efficient SLN-mediated DDS. The characterization through FTIR revealed extra peaks at 1,732 cm^{-1} and 3444.87 cm^{-1} and the absence of aromatic C-H stretch and methylene C-H asym stretch, confirming the formation of SLN. Other techniques, such as differential scanning calorimetry and in vitro diffusion studies, which can be facilitated by a UV-visible spectrophotometer, are also commonly used for SLNs.

1.6.3 Phytosomes

Phytosomes are herbal delivery systems used to encase organic substances and elixirs for medicinal purposes. Phytosomes are believed to enhance the absorption [36] and utilization of isolated active compounds, whether applied topically or ingested. Many plant-based bioactive constituents are hydrophilic molecules, such as tannins and flavonoids. However, due to their low lipid affinity and large molecular sizes, they have poor bioavailability. This reduced bioavailability can be improved by incorporating soluble phytoconstituents into phospholipids.

1.6.3.1 Morphological Characterization of Phytosomes

Various evaluation techniques for phytosomes include TEM and SEM for surface and size determination, as well as DLS for particle size and zeta potential analysis. The surface tension of the active pharmaceutical ingredient (API) in an aqueous distillate can be determined using the ring method by a Du Nouy ring tensiometer. The quantification of the drug in the phytosome complex is determined via HPLC.

The entrapment efficiency of phytosomal vesicle (PDE) in the DDS of Boswellia serrata distillate was reported [98] by Sahu and Bothara. They studied the in vitro release of boswellic acid from the suspension of phyto-phospholipid complexes using a Franz diffusion cell with pig skin.

1.6.3.2 Structural Characterization of Phytosomes

Structural characterization through ^{1}H-NMR spectra is considered to be essential for evaluating intermolecular interactions between phosphatidylcholine and active phytoconstituents. The phytosomal complex, a novel moiety, results from these interactions. Atoms involved in the creation of the phytosomal complex can provide the 1H-NMR signal, which can be used to determine the structural composition of the complex. Additionally, using the FTIR approach, the stability of the phytosomal complex can be determined by comparing its spectrum with that of the same complex dissolved in water after cryodesiccation at different intervals.

The structural characterization of the Vasaka-loaded phytosome DDS is reported by Nandhni and Ilango [99]. FTIR spectra were acquired from a spectrophotometer, employing the KBr pellet method. In an agate mortar, a blend of phospholipon 90H and Vasaka extract was ground using KBr crystals. A hydraulic press was then used to compress the resultant granules into a disk. The mid-infrared range of the spectra was set at 400–4,000 cm^{-1} with a resolution of 2 cm^{-1}. The characteristic peaks for the Vasaka-loaded phytosome complex were recorded at vibrational frequencies of 3406.69 cm^{-1} (O–H), 2924.89 cm^{-1} (C-H), 2122.28 cm^{-1} (C=N), and 1614.38 cm^{-1} (C=O). The other characterization techniques employed were XRD and in vitro drug release studies. XRD is useful in identifying the various phases found in phytosome and in determining their crystallinity in addition to structural studies. To assess the active ingredient release profile of phytosomes, in vitro drug release studies are necessary. Visible spectral studies and HPLC are used to evaluate the concentration of active ingredients in the sample.

1.6.4 Dendrimer

Dendrimers are highly ordered, branched, three-dimensional, spherical polymeric molecules. They are important in the pharmaceutical and biomedical industries [38] due to their high compatibility with biological systems, intensive branching, and symmetrical spherical structures. Their structures vary from 1–15 nm. Their primary advantage is that they enhance drug bioavailability by modifying pharmacokinetic and pharmacodynamic properties. They have extensive applications in ocular drug delivery [39] and transdermal drug delivery systems.

1.6.4.1 Morphological Characterization of Dendrimers

Transmission spectroscopy and SEM can analyze dendrimer particle diameter and size. Dendrimer-encapsulated Pt nanoparticles [100] have been reported to be characterized using 3D electron tomography, exhibiting the structural attributes of dendrimers.

Superparamagnetic iron oxide nanoparticles (SPION) encapsulated in poly(amidoamine), a class of dendrimers [101] made up of amide and amine subunits, give rise to SPIONPAMAM (ID-NPs), which is characterized extensively using transmission electron spectroscopy and shows the distribution of particle sizes between 5 and 15 nm. Khosroshahi and Tajabadi also evaluated in vitro cytotoxicity using Michigan Cancer Foundation-7 (human breast cancer) and fibroblast-like cells derived from the primary mouse connective tissue (L929 cell lines).

Additionally, energy-dispersive X-ray spectroscopy (SEM-EDS) conjointly with SEM is also insightful in elucidating the morphological characteristics of dendrimers.

1.6.4.2 Structural Characterization of Dendrimers

In research conducted by Khosroshahi and Tajabadi [101], the crystalline stage of dendrimer-encapsulated NPs was identified through X-ray diffraction spectroscopy and Fourier transform infrared (FTIR) spectroscopy. When examined using the XRD method, the findings indicated the development of a highly purified magnetite stage of iron oxide with an inverse spinel structure in superparamagnetic iron oxide nanoparticles.

Another investigation [102] used the FTIR method to study phosphorus dendrimers with a cyclotriphosphazene core and end phenoxy groups. In this instance, the IR spectra of dendrimers were assessed, and the moderate-intensity bands at 1,590 cm^{-1}, 1,488 cm^{-1}, and 1,369 cm^{-1} were recorded. In pursuit of this, the Bruker VERTEX 70 spectrometer can be used to analyze the FTIR spectra of drugs, dendrimers, and complexes. It can record IR spectra from the far-infrared (FIR) to the near-infrared (NIR), usually between 15 and 28,000 cm^{-1} with resolution up to 0.2 cm^{-1}.

Other than mass spectroscopy, fluorescence spectroscopy, small-angle X-ray scattering (SAXS), and laser light scattering (LLS) can also be utilized for the structural analysis of dendrimers.

Mass spectroscopy can be conducted using chemical ionization or fast atom bombardment to characterize small dendrimers with a mass below 3,000 nm.

SAXS determines the average gyration (Rg) radius of dendrimers in solution. The scattering intensity also yields insights into the arrangement of polymer segments. LLS is used to determine the hydrodynamic radius of dendrimers, and dynamic LLS is primarily employed to detect aggregates.

1.6.5 Transdermal Drug Delivery System

The transdermal DDS is a promising method [47] to transport active ingredients across the skin to distribute them throughout the body. These patches or ointments deliver the drug into the bloodstream and make it more effective [103].

Peptides and proteins are examples of large compounds that are not effectively administered orally due to their size and sensitivity to gastric juices. However, transdermal drug administration allows the drug to enter the blood via dermal microcirculation, bypassing the first-pass effect. Nanoformulations, with a smaller size ranging from 10 to 100 nm, are well-suited for transport through the stratum corneum. Some vesicular structures, like liposomes and transferosomes, are potential contenders for transdermal delivery of drugs. Liposomes and transferosomes have enabled efficient skin penetration of some drugs like propranolol and finasteride.

1.6.5.1 In Vitro Permeation Studies

The lipophilic medication pinacidil monohydrate is deemed unsuitable for oral administration. Aqil and Ali synthesized a matrix-type TDDS [104] of pinacidil monohydrate to improve its bioavailability and reduce side effects from overdosing. The in vitro characterization of Pinacidil monohydrate (PM) matrix by drug release studies confirmed its stability while crossing the skin barrier.

Ktoprofen–cyclodextrin complexes encapsulated in liposomes [105] have also been reported by Maestrelli et al. and have been extensively characterized for morphological elucidation using differential scanning calorimetry (DSC), SEM and studying dissolution properties of drug-cyclodextrin matrix.

Artificial membrane and rat skin were used to study the permeability of liposome-encapsulated ketoprofen–cyclodextrin complexes and the free drug.

1.6.6 Niosomes

Niosomes are lipid sacs consisting of nonionic surface-active agents, incorporating cholesterol as an additive. Niosomes share structural similarities with liposomes, as both systems are vesicular in nature, featuring an aqueous core surrounded by a phospholipid bilayer. When loaded, niosomes can deliver the drug [73] more effectively to a specific tissue and provide sustained release to the target.

1.6.6.1 Morphological Characterization of Niosomes

The liposomal vesicles are generally considered to have a spherical shape, and the LLS method can be used to measure their average diameter. Electron microscopy, molecular sieve chromatography, ultracentrifugation, and optical microscopy are equally employed to determine their diameter. Zeta potential determination on niosomes can be carried out with DLS.

1.6.6.2 Structural Characterization of Niosomes

The structural analysis of niosomes involves various analytical techniques like nuclear magnetic resonance (NMR) spectroscopy, SAXS, and electron microscopy. According to Obeid et al., HPLC is well-suited for quantifying the concentrations of both encapsulated and unencapsulated [106] drugs in niosomes.

The cytotoxic effects [107] of a DDS utilizing atenolol niosomes were studied on a breast cancer cell line by Balasubramaniam et al. The investigations were performed with an 3-(4,5-Dimethylthiazol-2-yl)-2,5-Diphenyltetrazolium Bromide (MTT) assay, and absorbance was measured using a UV-visible spectrophotometer.

1.6.7 Microspheres

Small, spherical particles with a diameter of one to a thousand micrometers are known as microspheres. The significance of microspheres lies in improved absorption of traditional medicines and minimizing their side effects [79].

Microspheres are crucial in the treatment of various diseases, particularly in cancer research, as they are introduced into the circulatory system to eliminate cancerous tumors. On the other hand, bioadhesive microspheres are prevalent in controlled release systems, drug targeting, and their administration through various routes like nasal, eye, and gastrointestinal tracts.

1.6.7.1 Morphological Characterization of Microspheres

Microspheres can be identified morphologically [108] via optical microscopy to determine their particle size. SEM enables the determination of substratum morphology.

The entrapment efficiency of microspheres can be found by the difference between theoretical and actual drug content when a mixture of distilled water and an API entrapped in a microsphere is stirred ultrasonically for a few hours, filtered, and analyzed using UV spectroscopy.

Poly (D, L-lactide-co-glycolide) microspheres [109] sheathing aquaphilic bovine serum albumin as prospective drug distributors for proteins designed for controlled release formulation were characterized via zeta potential for particle size determination and SEM for their morphological examination. These studies were carried out for formulations with various organic solvents. Ravi et al. also determined the resulting encapsulation efficiencies.

1.6.8 Ethosome/Transferases

Ethosomes and transferases are cumulatively known as ultra-deformable vesicles (UDV) for skin delivery studies [83] of many molecules, including peptides and proteins. Ethosomes are nanocarriers for drugs injected into biological systems through the stratum corneum and deep skin layers. The novel modified lipid carriers, referred to as ethosomes, comprise ethanol, phospholipids, and water. While transferases are drug delivery carriers that are self-optimized elastic vesicles consisting of a lipid bilayer, phospholipid, and edge activator. These are specifically helpful in delivery of macromolecules like insulin across the stratum corneum layers, which is otherwise difficult.

1.6.8.1 Morphological Characterization of Ethosomes

The morphological characterization of ethosomes/transferases is like other nanoparticles and is conducted through SEM and TEM. The DLS method can study size distribution and zeta potential.

Penetration and permeation studies can be performed by a confocal laser scanning microscope (CLSM). Release studies of the drug in topical delivery [110] of UDV can be performed using vertical Franz diffusion cells with hydrophilic polysulfone membranes. Afterward, HPLC is used to conduct quantitative studies.

The most important and unique parameter for characterization of transfersomes [111], the deformability study, is conducted by passing transferosome solution through different microporous filters in pure water as standard. After each pass, particle size and size distribution are analyzed using DLS.

1.6.8.2 Structural Characterization of Ethosomes

The generation of caffeic acid ethosomes through continuous stirring of water and phosphatidylcholine ethanol solution combined in a mixed form has been documented. Researchers examined the structural properties [112] using various techniques like cryogenic TEM, small-angle X-ray spectroscopy, and photon correlation spectroscopy.

1.6.9 Hydrogels

Hydrogels are widely recognized [89] in diverse domains, such as controlled drug delivery and tissue engineering. Depending upon the type of crosslinking, there are two types of hydrogels: Chemical gels (having covalent interactions) and physical gels (having non-covalent interactions). Several analytical techniques are used to analyze hydrogels.

1.6.9.1 Structural Characterization of Hydrogels

UV-visible spectroscopy is exploited effectively for the structural characterization of hydrogels to identify small-molecule drugs. However, this method is not effective at low concentrations of the sample.

García-Fernández reported [113] on hydrogels of oxidized dextran, gelatin, and hyaluronic acid. They used a UV-visible spectrophotometer to detect absorbance frequencies for the DDS of anti-inflammatory drugs naproxen and dexamethasone. The signals were detected at 230 and 243 nm, respectively. However, this technique is limited to more than one microgram/mL concentration. Another spectrophotometric technique, fluorescence spectroscopy, facilitates the identification of polysaccharides and oligonucleotides [114].

In protein/growth factor delivery, an enzyme-linked immunosorbent assay (ELISA) is utilized to discover and quantitatively assess specific proteins, antibodies, and glycoproteins within the range of picograms per milliliter (pg/mL). ELISA employs a variety of antigen-antibody pairings, consistently integrating an enzyme-labeled antigen or antibody, and quantifies enzyme activity through colorimetric measurement.

HPLC isolates and detects different mixture parts according to their physical and chemical characteristics. It divides the sample into a moving phase (eluent) and a fixed phase (column packing material). The fractions of a sample are separated at distinct intervals because of the intermolecular interactions between the sample molecules and the packing material. Fourniols et al. detailed the quantification of a hydrogel in delivering temozolomide [115] for treating glioblastoma multiforme (GBM) using the HPLC technique.

The Franz cell diffusion assay is a specific technique designed for the *ex vivo* study of how drugs from ointments diffuse through the skin.

Furthermore, this methodology applies to artificially engineered membranes and hydrogel DDSs, both of which are akin to the membranes utilized in transdermal drug delivery.

1.6.9.2 In Vivo Study for Hydrogel

In vivo studies indirectly measure the drug release profile in hydrogel-assisted drug delivery systems [116] by assessing drug efficacy. In tumor inhibition models, key parameters for evaluation include monitoring the dimensions of the tumor, tracking the body mass of mice over a specific period, and recording the cumulative survival probability of the test subjects. Additionally, the datasets are complemented by microanatomy of tissues post-sacrifice, which provides further insights into the effects of the drug delivery system.

1.6.10 Mouth-dissolving Tablets

Mouth-dissolving tablets are a favoured choice for sick persons who have difficulty swallowing, particularly children and the elderly. In contrast to alternative medication delivery methods, these are more stable both chemically and physically, providing a precise dose [91]. They are best suited for drugs with low solubility and high permeability, such as ibuprofen, prednisone, indomethacin, and nimesulide. The disintegration of these tablets in the mouth is significantly influenced [92] by properties like solubility, particle size, compressibility, and bulk density of the drug.

1.6.10.1 In vivo and In vitro studies

The disintegration time of mouth-dissolving tablets can be determined [117] in vivo using various methods, such as the rotary shaft, texture analyser, and charge-couple device method. For in vitro determination, the dissolution test is quite popular. This test assesses the rate at which a medication is discharged from its formulation into the surrounding aqueous media within a specific apparatus. The tablet is placed in a dissolution apparatus with a selective medium. The apparatus is then operated at a specific speed and temperature, similar to that of the human digestive tract. At set time intervals, samples are collected from the dissolution apparatus and analyzed to determine the quantity of dissolved drug in each fraction using HPLC chromatography. The enhancement of a drug's dissolution properties is explained using dissolution efficiency (DE) and relative dissolution rate (RDR).

Ghenge et al. reported in vitro dissolution studies of fast disintegrating tablets of Amlodipine Besylate [118] using a phosphate buffer of pH 7.2 at 37°C ± 0.5°C. The extracted aliquots were analysed for concentration of the API by measuring optical density with a spectrophotometer. The interaction of drug with the excipients was assessed using FTIR analysis.

Thadkala et al. [119] investigated the in vitro disintegration process of fast disintegrating Piroxcam pills. In this experiment, a petri dish with a diameter of 10 cm was filled with 10 ml of water. The tablet was positioned in the middle of a petri plate, and the amount of time needed for it to fully break down into tiny particles was recorded. Another way to measure the in vitro dispersion time in seconds is to suspend a tablet in a measuring cylinder that holds 6 milliliters of pH 6.8 buffer.

1.7 FUTURE PERSPECTIVES

In modern science, novel drug delivery systems have become a very captivating field of research. Numerous clinical trials have been conducted and successfully concluded. Despite certain drawbacks regarding the clinical application, the NDDS hold promising therapeutic potential. Regardless of abundant information on the possibilities of nano-DDSs, the subject is still relatively recent, and numerous crucial basic traits remain unidentified. One important area of research that must be explored is the fundamental indicators of inflamed tissues. More research is needed around selective targeting mechanisms based on biomarkers of inflammation. Future developments in nanomedicine applications will result from a better understanding of the molecular markers of illness. Further research involving animals and human resources needs a substantial investment of time and money. Cutting-edge research will yield valuable data that may find use in pharmacological therapy and diagnostic tests in the future.

1.8 CONCLUSION

The medical industry has derived advantage from advancements in nanoparticles and nanotechnology over the last few decades. Nanoparticles have revolutionized the medical field and opened new pathways for treating diseases. These microscopic particles are extensively used as drug delivery systems. Nanotechnology has made it possible to encapsulate active pharmaceutical ingredients (APIs) in various drug delivery systems, allowing for targeted delivery to specific sites of action while reducing toxicity by preventing the accumulation of active moieties at other sites in biological systems. Nanocarriers as drug delivery systems address these challenges by overcoming issues of bioavailability and solubility. Looking ahead, the focus of nanocarriers research will shift to addressing irregularities in their preparation and related side effects, leading to toxicity in the human body. Despite the competence of nanocarriers as a drug delivery system, numerous challenges in managing drug degradation during systemic circulation and overcoming biological barriers are yet to be addressed.

REFERENCES

1. Jain KK. An overview of drug delivery systems. In: K. K. Jain (Ed.), *Drug Delivery Systems* (Vol. 2059, pp. 1–54). Humana Press, 2020.
2. Ezike TC, Okpala US, Onoja UL, Nwike PC, Ezeako EC, Okpara JO, Okoroafor CC, Eze SC, Kalu OL, Odoh EC, Nwadike U. Advances in drug delivery systems, challenges and future directions. *Heliyon*. 2023;9(6):e17488.
3. Adepu S, Ramakrishna S. Controlled drug delivery systems: Current status and future directions. *Molecules*. 2021;26(19):5905.
4. Saleh Abdullah Mohammed A, Mohammed Saeed Mesfer A, Hadi Aidh Mesfer A, Faisal Sager Bakheet A, Mohammed Ayed Matar A, Fahd Salem Jazaa Al O. Novel drug delivery systems: Advancements and challenges. *Journal of Population Therapeutics and Clinical Pharmacology*. 2022; *29*(4):2573–2576.
5. Prajapati U, Prajapat A, Chaudhary S, Yadav A. An approach to novel drug delivery systems future prospects and opportunities. *Journal of Clinical Advances and Research Reviews*. 2024;25:2060–2077.
6. Faheem AM, Abdelkader DH. Novel drug delivery systems. In: A. Seyfoddin, S. Masoomi Dezfooli, C. A. Greene (Eds.), *Engineering Drug Delivery Systems* (pp. 1–16). Woodhead Publishing, 2020.
7. Alqahtani MS, Kazi M, Alsenaidy MA, Ahmad MZ. Advances in oral drug delivery. *Frontiers in Pharmacology*. 2021;12:618411.
8. Mohanty SK, Swamy MK, Sinniah UR, Anuradha M. Leptadenia reticulata (Retz.) Wight & Arn.(Jivanti): Botanical, agronomical, phytochemical, pharmacological, and biotechnological aspects. *Molecules*. 2017;22(6):1019.
9. Rodrigues T, Reker D, Schneider P, Schneider G. Counting on natural products for drug design. *Nature Chemistry*. 2016;8(6):531–541.
10. Sharma N, Kurmi BD, Singh D, Mehan S, Khanna K, Karwasra R, Kumar S, Chaudhary A, Jakhmola V, Sharma A, Singh SK, Dua K, Kakkar D. Nanoparticles toxicity: An overview of its mechanism and plausible mitigation strategies. *Journal of Drug Targeting*. 2024:1–13. doi:10.1080/1061186X.2024.2316785.
11. Jahangirian H, Lemraski EG, Webster TJ, Rafiee-Moghaddam R, Abdollahi Y. A review of drug delivery systems based on nanotechnology and green chemistry: Green nanomedicine. *International Journal of Nanomedicine*. 2017:2957–2978.
12. Haba Y, Kojima C, Harada A, Ura T, Horinaka H, Kono K. Preparation of poly (ethylene glycol)-modified poly (amido amine) dendrimers encapsulating gold nanoparticles and their heat-generating ability. *Langmuir*. 2007;23(10):5243–5246.
13. Joudeh N, Linke D. Nanoparticle classification, physicochemical properties, characterization, and applications: A comprehensive review for biologists. *Journal of Nanobiotechnology*. 2022;20(1):262.
14. Jindal M, Nagpal M, Singh M, Aggarwal G, Dhingra GA. Gold nanoparticles-boon in cancer theranostics. *Current Pharmaceutical Design*. 2020;26(40):5134–5151.
15. Jeong HH, Choi E, Ellis E, Lee TC. Recent advances in gold nanoparticles for biomedical applications: From hybrid structures to multi-functionality. *Journal of Materials Chemistry B*. 2019;7(22):3480–3496.
16. Zahin N, Anwar R, Tewari D, Kabir MT, Sajid A, Mathew B, Uddin MS, Aleya L, Abdel-Daim MM. Nanoparticles and its biomedical applications in health and diseases: Special focus on drug delivery. *Environmental Science and Pollution Research*. 2020;27(16):19151–19168.
17. Singh KR, Nayak V, Singh J, Singh AK, Singh RP. Potentialities of bioinspired metal and metal oxide nanoparticles in biomedical sciences. *RSC Advances*. 2021;11(40):24722–24746.
18. Singh I, Arora R, Dhiman H, Pahwa R. Carbon quantum dots: Synthesis, characterization and biomedical applications. *Turkish Journal of Pharmaceutical Sciences*. 2018;15(2):219–230. doi:10.4274/tjps.63497.
19. Bajwa N, Mehra NK, Jain K, Jain NK. Pharmaceutical and biomedical applications of quantum dots. *Artificial Cells, Nanomedicine, and Biotechnology*. 2016;44(3):758–768
20. Liu P, Chen G, Zhang J. A review of liposomes as a drug delivery system: Current status of approved products, regulatory environments, and future perspectives. *Molecules*. 2022;27(4):1372.
21. Murthy BRS, Vishakha K, Prakash P, Basha J D, Govardhan TPM. Formulation and in vitro evaluation of liposomes loaded with mupirocin. *International Journal of Research in Pharmaceutical and Nano Sciences*. 2015; 4:162–174.
22. Esfandiari N, Arzanani MKH. The study of toxicity and pathogenicity risk of Potato Virus X/Herceptin nanoparticles as agents for cancer therapy. *Cancer Nanotechnology*. 2018;9(1):1–13.
23. Akbarzadeh A, Rezaei-Sadabady R, Davaran S, Joo SW, Zarghami N, Hanifehpour Y, Samiei M, Kouhi M, Nejati-Koshi K. Liposome: Classification, preparation, and applications. *Nanoscale Research Letters*. 2013;8:102.

24. Wang X, Zhang T, Wang C, Huang Z, Luo X, Deng Y. A review on phospholipids and their main applications in drug delivery systems. *Asian Journal of Pharmaceutical Sciences*. 2015;10:81–98.
25. Van Hoogevest P, Wendel A. The use of natural and synthetic phospholipids as pharmaceutical excipients. *European Journal of Lipid Science and Technology*. 2014;116(9):1088–1107
26. Nakhaei P, Margiana R, Bokov DO, Abdelbasset WK, Jadidi Kouhbanani MA, Varma RS, Marofi F, Jarahian M, Beheshtkhoo N. Liposomes: Structure, biomedical applications, and stability parameters with emphasis on cholesterol. *Frontiers in Bioengineering and Biotechnology*. 2021;9:705886.
27. Jovanović AA, Balanč D, Ota A, Ahlin Grabnar P, Djordjević VB, Šavikin K, Bugarski B, Nedović V. Comparative effects of cholesterol and β-sitosterol on the liposome membrane characteristics. *European Journal of Lipid Science and Technology*. 2018;120:1800039.
28. Cipolla D, Wu H, Gonda S, Eastman S, Redelmeier T, Chan HK. Modifying the release properties of liposomes toward personalized medicine. *Journal of Pharmaceutical Sciences*. 2014;103:1851–1862.
29. Tian Y, Chen L, Zhang W. Influence of ionic surfactants on the properties of nanoemulsions emulsified by nonionic surfactants span 80/tween 80. *Journal of Dispersion Science and Technology*. 2016;37:1511–1517.
30. Chen R, Li R, Liu Q, Bai C, Qin B, Ma Y, Han J. Ultradeformable liposomes: A novel vesicular carrier for enhanced transdermal delivery of procyanidins: Effect of surfactants on the formation, stability, and transdermal delivery. *AAPS PharmSciTech*. 2017;18:1823–1832.
31. Kantoff PW, Higano CS, Shore ND, Berger ER, Small EJ, Penson DF, Redfern CH, Ferrari AC, Dreicer R, Sims RB, Xu Y. Sipuleucel-T immunotherapy for castration-resistant prostate cancer. *New England Journal of Medicine*. 2010;363(5):411–422.
32. Permana AD, Paredes AJ, Volpe-Zanutto F, Anjani QK, Utomo E, Donnelly RF. Dissolving microneedle-mediated dermal delivery of itraconazole nanocrystals for improved treatment of cutaneous candidiasis. *European Journal of Pharmaceutics and Biopharmaceutics*. 2020;154:50–61.
33. Solanki R, Jodha B, Prabina KE, Aggarwal N, Patel S. Recent advances in phytochemical based nano-drug delivery systems to combat breast cancer: A review. *Journal of Drug Delivery Science and Technology*. 2022;77:103832.
34. Barani M, Sangiovanni E, Angarano M, Rajizadeh MA, Mehrabani M, Piazza S, Gangadharappa HV, Pardakhty A, Mehrbani M, Dell'Agli M, Nematollahi MH. Phytosomes as innovative delivery systems for phytochemicals: A comprehensive review of literature. *International Journal of Nanomedicine*. 2021;16:6983–7022.
35. Mazumder A, Dwivedi A, Du Preez JL, Du Plessis J. In vitro wound healing and cytotoxic effects of sinigrin–phytosome complex. *International Journal of Pharmaceutics*. 2016; 498(1–2):283–293.
36. Tripathy S, Patel DK, Barob L, Naira SK. A review on phytosomes, their characterization, advancement & potential for transdermal application. *Journal of Drug Delivery and Therapeutics*. 2013;3(3):147–152.
37. Myc A, Kukowska-Latallo J, Cao P, Swanson B, Battista J, Dunham T, Baker Jr. Targeting the efficacy of a dendrimer-based nanotherapeutic in heterogeneous xenograft tumors in vivo. *Anti-Cancer Drugs*. 2010;21(2):186–192.
38. Longmire M, Choyke PL, Kobayashi H. Dendrimer-based contrast agents for molecular imaging. *Current Topics in Medicinal Chemistry*. 2008;8(14):1180–1186.
39. Zhang S, Lloveras V, Lope-Piedrafita S, Calero-Pérez P, Wu S, Candiota AP, Vidal-Gancedo J. Metal-free radical dendrimers as MRI contrast agents for glioblastoma diagnosis: Ex vivo and in vivo approaches. *Biomacromolecules*. 2022;23(7):2767–2777.
40. Pérez-Ferreiro MM, Abelairas A, Criado A, Gómez IJ, Mosquera J. Dendrimers: Exploring their wide structural variety and applications. *Polymers*. 2023;15(22):4369.
41. Shcharbina N, Shcharbin D, Bryszewska M. Nanomaterials in stroke treatment: Perspectives. *Stroke*. 2013;44(8):2351–2355.
42. Zhang W, Dhumal D, Zhu X, Ralahy B, Ellert-Miklaszewska A, Wu J, Laurini E, Yao YW, Kao CL, Iovanna JL, Pricl S. Bola-amphiphilic glycodendrimers: New carbohydrate-mimicking scaffolds to target carbohydrate-binding proteins. *Chemistry–A European Journal*. 2022;28(58):e202201400.
43. Sharma AK, Gupta L, Sahu H, Qayum A, Singh SK, Nakhate KT, Ajazuddin, Gupta U. Chitosan engineered PAMAM dendrimers as nanoconstructs for the enhanced anti-cancer potential and improved in vivo brain pharmacokinetics of temozolomide. *Pharmaceutical Research*. 2018;35:1–4.
44. Yan X, Yang Y, Sun Y. Dendrimer applications for cancer therapies. In *Journal of Physics*: Conference Series 2021 Jun 1 (Vol. 1948, No. 1, p. 012205). IOP Publishing.
45. Xie F, Li R, Shu W, Zhao L, Wan J. Self-assembly of Peptide dendrimers and their bio-applications in theranostics. *Materials Today Bio*. 2022;14:100239.
46. Haiba NS, Khalil HH, Bergas A, Abu-Serie MM, Khattab SN, Teleb M. First-in-Class star-shaped triazine dendrimers endowed with MMP-9 inhibition and VEGF suppression capacity: Design, synthesis, and anticancer evaluation. *ACS Omega*. 2022;7(24):21131–2144.
47. Ramadon D, McCrudden MT, Courtenay AJ, Donnelly RF. Enhancement strategies for transdermal drug delivery systems: Current trends and applications. *Drug Delivery and Translational Research*. 2021:1–34.
48. Malaiya MK, Jain A, Pooja H, Jain A, Jain D. Controlled delivery of rivastigmine using transdermal patch for effective management of alzheimer's disease. *Journal of Drug Delivery Science and Technology*. 2018;45:408–144.
49. Bakshi P, Vora D, Hemmady K, Banga AK. Iontophoretic skin delivery systems: Success and failures. *International Journal of Pharmaceutics*. 2020;586:119584.
50. Han T, Das DB. Potential of combined ultrasound and microneedles for enhanced transdermal drug permeation: A review. *European Journal of Pharmaceutics and Biopharmaceutics*. 2015;89:312–328.
51. Bariya SH, Gohel MC, Mehta TA, Sharma OP. Microneedles: An emerging transdermal drug delivery system. *Journal of Pharmacy and Pharmacology*. 2012;64(1):11–29.
52. Chen X, Zhu L, Li R, Pang L, Zhu S, Ma J, Du L, Jin Y. Electroporation-enhanced transdermal drug delivery: Effects of logP, pKa, solubility, and penetration time. *European Journal of Pharmaceutical Sciences*. 2020;151:105410.
53. Sintov AC, Krymberk I, Daniel D, Hannan T, Sohn ZE, Levin G. Radiofrequency-driven skin microchanneling as a new way for electrically assisted transdermal delivery of hydrophilic drugs. *Journal of Controlled Release*. 2003;89(2):311–320.

54. Wong TW. Electrical, magnetic, photomechanical, and cavitational waves to overcome the skin barrier for transdermal drug delivery. *Journal of Controlled Release*. 2014;193:257–269.
55. Marwah H, Garg T, Goyal AK, Rath G. Permeation enhancer strategies in transdermal drug delivery. *Drug Delivery*. 2016;23(2):564–578.
56. Salah E, Abouelfetouh MM, Pan Y, Chen D, Xie S. Solid lipid nanoparticles for enhanced oral absorption: A review. *Colloids and Surfaces B: Biointerfaces*. 2020;196:111305.
57. Sastri KT, Radha GV, Pidikiti S, Vajjhala P. Solid lipid nanoparticles: Preparation techniques, their characterization, and an update on recent studies. *Journal of Applied Pharmaceutical Science*. 2020;10(6):126–141.
58. Paliwal R, Paliwal SR, Agrawal GP, Vyas SP. Biomimetic solid lipid nanoparticles for oral bioavailability enhancement of low molecular weight heparin and its lipid conjugates: In vitro and in vivo evaluation. *Molecular Pharmaceutics*. 2011;8(4):1314–1321.
59. Harish V, Mohd S, Tewari D, Pandey NK, Vishwas S, Babu MR, et al. Unravelling the role of solid lipid nanoparticles in drug delivery: Journey from laboratory to clinical trial. *Journal of Drug Delivery Science and Technology*. 2023:104616.
60. Duan Y, Dhar A, Patel C, Khimani M, Neogi S, Sharma P, et al. A brief review on solid lipid nanoparticles: Part and parcel of contemporary drug delivery systems. *RSC Advances*. 2020;10(45):26777–26791.
61. Abdelwahed W, Degobert G, Fessi H. Investigation of nanocapsules stabilization by amorphous excipients during freeze-drying and storage. *European Journal of Pharmaceutics and Biopharmaceutics*. 2006;63(2):87–94.
62. Matta VDR. A concise review on preparation methods used for the development of solid lipid nanoparticles. *Journal of Drug Delivery and Therapeutics*. 2021;11(1-s):162–169.
63. Francis JE, Skakic I, Smooker PM. Design and preparation of solid lipid nanoparticle (SLN)-mediated DNA vaccines. *Vaccine Design: Methods and Protocols, Resources for Vaccine Development*. 2022;3:355–366.
64. Ribeiro LNM, De Paula E, Rossi DA, Martins FA, de Melo RT, Monteiro GP, et al. Nanocarriers from natural lipids with in vitro activity against Campylobacter jejuni. *Frontiers in Cellular and Infection Microbiology*. 2021;10:571040.
65. Abul Kalam M, Sultana Y, Ali A, Aqil M, Mishra AK, Chuttani K, et al. Part II: Enhancement of transcorneal delivery of gatifloxacin by solid lipid nanoparticles in comparison to commercial aqueous eye drops. *Journal of Biomedical Materials Research Part A*. 2013;101(6):1828–1836.
66. Prabhu A, Jose J, Kumar L, Salwa S, Vijay Kumar M, Nabavi SM. Transdermal delivery of curcumin-loaded solid lipid nanoparticles as microneedle patch: An in vitro and in vivo study. *AAPS PharmSciTech*. 2022;23:1–12.
67. Gupta B, Sharma G, Sharma P, Sandhu SK, Kaur IP. Self-gelling solid lipid nanoparticle hydrogel containing simvastatin as a suitable wound dressing: An investigative study. *Gels*. 2022;8(1):58.
68. Izhar MP, Hafeez A, Kushwaha P, Simrah. Drug delivery through niosomes: A comprehensive review with therapeutic applications. *Journal of Cluster Science*. 2023;34(5):2257–2273.
69. Biswas GR, Majee SB. Niosomes in ocular drug delivery. *European Journal of Pharmaceutical and Medical Research*. 2017;4(7):813–819.
70. Mawazi SM, Ann TJ, Widodo RT. Exploring the evolution of niosomes: From past techniques to future advances in preparation methods: A comprehensive review. *BioNanoScience*. 2024:1–22.
71. Maurya K, Singh A, Pandey S, Kumar N, Siddhiqui MA. Niosomes: Classification, preparation and application. *International Journal of Indigenous Herbs and Drugs*. 2021:29–32.
72. Sharma S, Garg A, Agrawal R, Chopra H, Pathak D. A comprehensive review on niosomes as a tool for advanced drug delivery. *Pharmaceutical Nanotechnology*. 2024;12(3):206–228.
73. Ge X, Wei M, He S, Yuan WE. Advances of non-ionic surfactant vesicles (niosomes) and their application in drug delivery. *Pharmaceutics*. 2019;11(2):55.
74. Vaiswade R, Chandrakar S, Roy A, Prasad P, Yadav C. Stability aspects of niosome. *International Journal of Creative Research Thoughts*. 2020;8(12):813–823.
75. Umbarkar MG. Niosome as a novel pharmaceutical drug delivery: A brief review highlighting formulation, types, composition, and application. *Indian Journal of Pharmaceutical Education & Research*. 2021;55:S11–S28.
76. Sharma V, Anandhakumar S, Sasidharan M. Self-degrading niosomes for encapsulation of hydrophilic and hydrophobic drugs: An efficient carrier for cancer multi-drug delivery. *Materials Science and Engineering: C*. 2015;56:393–400.
77. Niroumand U, Motazedian MH, Ahmadi F, Asgari Q, Bahreini MS, Ghasemiyeh P, Mohammadi-Samani S. Preparation and characterization of artemether-loaded niosomes in Leishmania major-induced cutaneous leishmaniasis. *Scientific Reports*. 2024;14(1):10073.
78. Paolino D, Cosco D, Muzzalupo R, Trapasso E, Picci N, Fresta M. Innovative bola-surfactant niosomes as topical delivery systems of 5-fluorouracil for the treatment of skin cancer. *International journal of Pharmaceutics*. 2008;353(1–2):233–242.
79. Dhadde GS, Mali HS, Raut ID, Nitalikar MM, Bhutkar MA. A review on microspheres: Types, method of preparation, characterization, and application. *Asian Journal of Pharmacy and Technology*. 2021;11(2):149–155.
80. Virmani T, Gupta J. Pharmaceutical application of microspheres: An approach for the treatment of various diseases. *International Journal of Pharmaceutical Sciences and Research*. 2017;8(8):3253–3260.
81. Salem R, Mazzaferro V, Sangro B. Yttrium 90 radioembolization for the treatment of hepatocellular carcinoma: Biological lessons, current challenges, and clinical perspectives. *Hepatology*. 2013;58(6):2188–2197.
82. Chauhan N, Vasava P, Khan SL, Siddiqui FA, Islam F, Chopra H, et al. Ethosomes: A novel drug carrier. *Annals of Medicine and Surgery*. 2022;82: 104595.
83. Emanet M, Ciofani G. Ethosomes as promising transdermal delivery systems of natural-derived active compounds. *Advanced NanoBiomed Research*. 2023;3(10):2300020.
84. Devaki J, Pavuluri S, Suma N. Ethosomes: A vesicular carrier as a novel tool for transdermal drug delivery system. *Journal of Drug Delivery and Therapeutics*. 2023;13(4):159–164.
85. Rada SK, Yadav S. Ethosomes: A novel approach in vesicular drug delivery systems. *Journal of Pharmaceutical Research International*. 2022:62–71.
86. Aggarwal D, Nautiyal U. Ethosomes: A review. *International Journal of Pharmaceutics and Medicinal Research*. 2016;4(4):354–363.
87. Thang NH, Chien TB, Cuong DX. Polymer-based hydrogels applied in drug delivery: An overview. *Gels*. 2023;9(7):523.
88. Nasitha I, Krishnakumar K, Dineshkumar B. Hydrogel in pharmaceuticals: A review. *Indo American Journal of Pharmaceutical Sciences*. 2016;3(3):265–270.

89. Kaith BS, Singh A, Sharma AK, Sud D. Hydrogels: Synthesis, classification, properties and potential applications—a brief review. *Journal of Polymers and the Environment*. 2021;29(12):3827–3841.
90. Joshi R, Garud N, Akram W. Fast dissolving tablets: A review. *International Journal of Pharmaceutical Sciences and Research*. 2020;11(4):1562–1570.
91. Jassem NA. Orodispersible tablets: A review on recent trends in drug delivery. *International Journal of Drug Delivery*. 2022;12:432–436.
92. Rahane R, Rachh PR. A review on fast-dissolving tablets. *Journal of Drug Delivery and Therapeutics*. 2018;8(5):50–55
93. Saeed K, Khan I. Preparation and characterization of single-walled carbon nanotube/nylon 6, 6 nanocomposites. *Instrumentation Science & Technology*. 2016;44(4):435–444.
94. Upadhyay S, Parekh K, Pandey B. Influence of crystallite size on the magnetic properties of Fe3O4 nanoparticles. *Journal of Alloys and Compounds*. 2016;678:478–485.
95. León A, Reuquen P, Garín C, Segura R, Vargas P, Zapata P, Orihuela PA. FTIR and Raman characterization of TiO_2 nanoparticles coated with polyethylene glycol as a carrier for 2-methoxyestradiol. *Applied Sciences*. 2017;7(1):49.
96. Giordani S, Marassi V, Zattoni A, Roda B, Reschiglian P. Liposomes characterization for market approval as pharmaceutical products: Analytical methods, guidelines and standardized protocols. *Journal of Pharmaceutical and Biomedical Analysis*. 2023;236:115751.
97. Chiong HS, Hakim MN, Sulaiman MR, Zakaria ZA, Zuraini A, Ong SG, Yuen KH. Development and characterisation study of liposomes-encapsulated piroxicam. *International Journal of Drug Delivery*. 2011;3(1):64–73.
98. Sahu AR, Bothara SB. Formulation and evaluation of a phytosome drug delivery system of boswellia serrata extract. *International Journal of Research in Medicine*. 2015;4(2):94–99.
99. Nandhini S, Ilango K. Development and characterization of a nano-drug delivery system containing vasaka phospholipid complex to improve bioavailability using quality by design approach. *Research in Pharmaceutical Sciences*. 2021;16(1):103–117.
100. Ju Y, Ro HJ, Yi YS, Cho T, Kim SI, Yoon CW, Jun S, Kim J. Three-dimensional TEM study of dendrimer-encapsulated Pt nanoparticles for visualizing structural characteristics of the whole organic–inorganic hybrid nanostructure. *Analytical Chemistry*. 2021;93(5):2871–2878.
101. Khosroshahi ME, Tajabadi M. Characterization and cellular fluorescence microscopy of superparamagnetic nanoparticles functionalized with third-generation nano-molecular dendrimers: In-vitro cytotoxicity and uptake study. *Journal of Nanomaterials & Molecular Nanotechnology*. 2016;11:2.
102. Garea SA, Ghebaur A. FT-IR spectroscopy and Thermogravimetrical Characterization of Prodrugs based on different dendritic polymers and antitumoral drug. *Materiale Plastice*. 2012;49(1):1–4.
103. Wong WF, Ang KP, Sethi G, Looi CY. Recent advancement of medical patches for transdermal drug delivery. *Medicina (Kaunas)*. 2023;59(4):778.
104. Aqil M, Ali A. Monolithic matrix type transdermal drug delivery systems of pinacidil monohydrate: In vitro characterisation. *European Journal of Pharmaceutics and Biopharmaceutics*. 2002;54(2):161–164.
105. Maestrelli F, González-Rodríguez ML, Rabasco AM, Mura P. Preparation and characterisation of liposomes encapsulating ketoprofen–cyclodextrin complexes for transdermal drug delivery. *International Journal of Pharmaceutics*. 2005;298(1):55–67.
106. Obeid MA, Khadra I, Aljabali AA, Amawi H, Ferro VA. Characterisation of niosome nanoparticles prepared by microfluidic mixing for drug delivery. *International Journal of Pharmaceutics: X*. 2022;4:100137.
107. Balasubramaniam A, Anil Kumar V, Sadasivan Pillai K. Formulation and in vivo evaluation of niosome-encapsulated daunorubicin hydrochloride. *Drug Development and Industrial Pharmacy*. 2002;28(10):1181–1193.
108. Galande P, Yadav V, Borkar S. A review on microspheres: Preparation, characterization and applications. *Asian Journal of Pharmaceutical Research and Development*. 2022;10(6):128–133.
109. Ravi S, Peh KK, Darwis Y, Murthy BK, Singh TR, Mallikarjun C. Development and characterization of polymeric microspheres for controlled release protein-loaded drug delivery system. *Indian Journal of Pharmaceutical Sciences*. 2008;70(3):303.
110. Ascenso A, Raposo S, Batista C, Cardoso P, Mendes T, Praça FG, Bentley MV, Simões S. Development, characterization, and skin delivery studies of related ultradeformable vesicles: Transfersomes, ethosomes, and transethosomes. *International Journal of Nanomedicine*. 2015;10:5837–5851.
111. Sharma U, Verma P, Jain NK. A review on novel vesicular drug delivery system: Transfersomes. *International Journal of Pharmacy & Life Sciences*. 2020;11(7):6812–6824.
112. Hallan SS, Sguizzato M, Mariani P, Cortesi R, Huang N, Simelière F, Marchetti N, Drechsler M, Ruzgas T, Esposito E. Design and characterization of ethosomes for transdermal delivery of caffeic acid. *Pharmaceutics*. 2020;12(8):740.
113. García-Fernández L, Olmeda-Lozano M, Benito-Garzón L, Pérez-Caballer A, San Román J, Vázquez-Lasa B. Injectable hydrogel-based drug delivery system for cartilage regeneration. *Materials Science and Engineering: C*. 2020;110:110702.
114. Zhao M, Bozzato E, Joudiou N, Ghiassinejad S, Danhier F, Gallez B, Préat V. Codelivery of paclitaxel and temozolomide through a photopolymerizable hydrogel prevents glioblastoma recurrence after surgical resection. *Journal of Controlled Release*. 2019;309:72–81.
115. Fourniols T, Randolph LD, Staub A, Vanvarenberg K, Leprince JG, Préat V, des Rieux A, Danhier F. Temozolomide-loaded photopolymerizable PEG-DMA-based hydrogel for the treatment of glioblastoma. *Journal of Controlled Release*. 2015;210:95–104.
116. Vigata M, Meinert C, Hutmacher DW, Bock N. Hydrogels as drug delivery systems: A review of current characterization and evaluation techniques. *Pharmaceutics*. 2020;12(12):1188.
117. Dor PJ, Fix JA. In vitro determination of disintegration time of quick-dissolve tablets using a new method. *Pharmaceutical Development and Technology*. 2000;5(4):575–577.
118. Ghenge G, Pande SD, Ahmad A, Jejurkar L, Birari T. Development and characterisation of fast disintegrating tablet of amlodipine besylate using mucilage of plantago ovata as a natural superdisintegrant. *International Journal of PharmTech Research*. 2011;3(2):938–945.
119. Thadkala K, Thadkapally R, Devara R, Kemisetti D. Development and characterisation of fast disintegrating tablets of Piroxicam. *International Journal of PharmTech Research*. 2012;4(2):734–741.

2 Evolution of Drug Delivery
A Journey from Bench to Market

Meenakshi Dhanawat[1], Kashish Wilson[2], Garima[3], Inderjeet Verma[2], and Sumeet Gupta[2]
[1]Amity Institute of Pharmacy, Amity University Haryana, Gurugram, Haryana, India
[2]M.M. College of Pharmacy, Maharishi Markandeshwar (Deemed to be University), Mullana-Ambala, Haryana, India
[3]Chitkara College of Pharmacy, Chitkara University, Punjab, India

2.1 EVOLUTION OF DRUG DELIVERY SYSTEMS

An ointment, pill, capsule, or other drug composition is referred to as a "drug delivery system." "Controlled-release drug delivery system" or "controlled drug delivery system" refers to a formulation that utilizes technology that regulates the drug-releasing kinetics over time. Compared to conventional formulations, which immediately release the majority or all of these loaded medications and lack any authority, controlled-release drug delivery methods are different. As a result, "immediate release" or "IR" formulations are usually referred to as conventional formulations. Another use of the word "controlled" is intended to guarantee an essentially consistent drug concentration in the circulatory system throughout the duration. It was challenging to maintain a constant drug concentration, particularly with oral controlled-release formulations. There are various methods to characterize the formation of drug delivery technologies, like therapeutic classes and delivery paradigms [1]. Here, the Food and Drug Administration (FDA)-approved items are used for the introduction of innovative technologies that illustrate the evolution. The history of controlled drug delivery systems is shaped by significant advancements in drug delivery, as illustrated in Figure 2.1. Even though medication delivery technologies are constantly evolving, a formulation's ability to demonstrate safety and efficacy and receive FDA approval—which allows patients to utilize these novel technologies—is the actual test of its success. Drug release regulated by osmosis, diffusion, dissolution, and ion exchange were four drug release mechanisms that underpinned the early advancements in oral and transdermal formulations. Diffusion- and dissolution-controlled techniques have been applied the most frequently among the four. Although it hasn't been accomplished yet, the primary goal of modulated medication delivery, also known as self-regulated drug delivery, is to regulate the supply of insulin.

As long as the blood drug concentration remains beyond the minimum effective concentration (Cmin) and over the maximum safe concentration (Cmax), a formulation with prolonged release durations is potentially just as effective as an IR formulation. The ratio of Cmax to Cmin is called a therapeutic index. Many drugs have a therapeutic index that is high enough to ensure safety even at high doses; therefore, variations in the blood concentration of the drug usually have the same effect. Furthermore, the use of controlled-release medication delivery systems may result in fewer peaks and troughs in plasma concentrations, reducing the adverse effects and the possibility of ineffectiveness. Numerous formulations, referred to as "sustained" or "extended" release systems, were developed because the blood concentration of a medicine didn't need to stay constant for it to be effective and secure. Different words have been used interchangeably over time. There were other words used as well, like "therapeutic system" and "modified system." However, they are all referring to formulations with controlled release, which is called the drug's optimum release, i.e., at a rate constant of zero or first order. Modified-release solid oral dosage forms are defined by the FDA as enteric-coated, delayed-release, and extended-release medication [2,3]. Delivery strategies have significantly transformed prospective treatments into viable therapies as delivery tactics and technologies have rapidly advanced [4]. As the need for medication administration evolved, the therapeutic landscape changed accordingly. Small-molecule medications were the mainstay of treatment a few decades ago. The small molecule's physicochemical properties play a major part in determining their delivery, and, therefore, delivery efforts of the drugs initially concentrated on increasing the solubility, regulating the release of the drugs, increasing their activities, and modifying their pharmacokinetics (PKs) [5,6,7]. Fresh therapeutic functionalities have been made possible over time by fresh generations of therapies, like live cells, mAbs, nucleic acids, proteins, and peptides [8,9,10].

2.2 CLASSES OF THERAPEUTIC AND DELIVERY CHALLENGES

The objective for drug delivery in every medication is minimizing off-target drug buildup while simultaneously delivering and releasing the drug (either actively or passively) at an intended place inside the body to maximize therapeutic efficacy. Drug PKs can be managed, drug toxicity can be

DOI: 10.1201/9781003508601-2

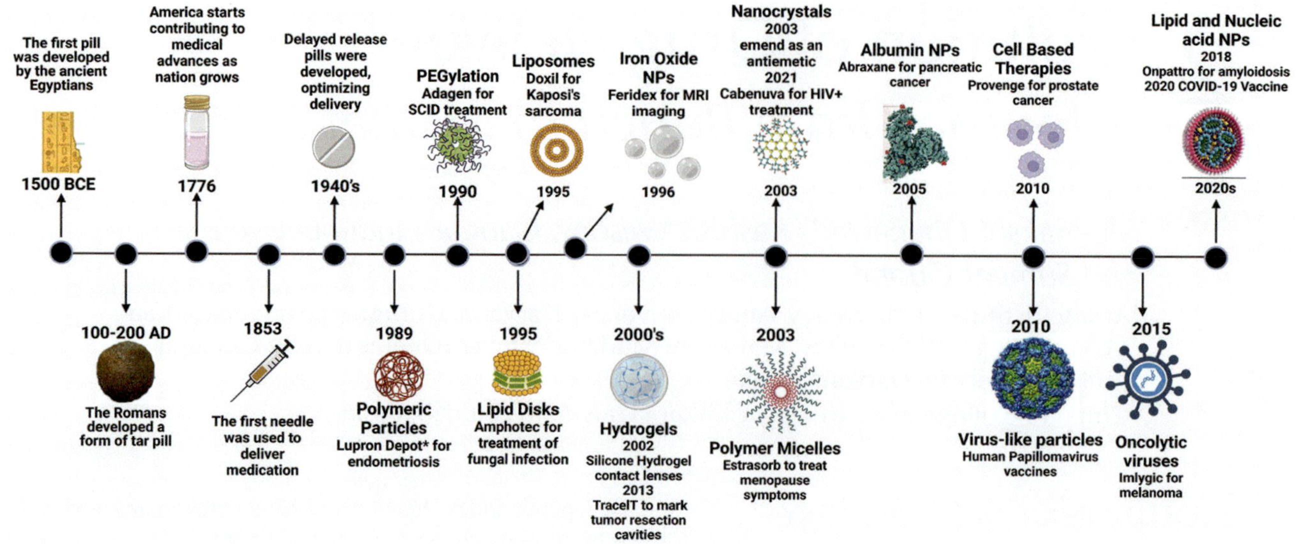

FIGURE 2.1 Evolution of the drug delivery process.

decreased, drug accumulation at the target location can be increased, and patient acceptance and compliance can be enhanced [11–17]. The recognition of distinct delivery issues connected to each treatment class has served as a catalyst for innovation in delivery technology and methods. We highlight the major discoveries in this section, together with their business and clinical achievements [18–20].

2.2.1 Small Molecules

Since the late 1800s, medications containing small molecules (less than 900 Da), such as steroids, antibiotics, and chemotherapeutics, have been discovered, created, and used. Small-molecule medications can quickly permeate biological fluids, multiple biological restrictions, and cell membranes because of their tiny size [21,22]. These advantages allow tiny molecules to interact with almost every bodily tissue and a variety of cells as well as transit the intricate vasculature. However, small compounds need to disintegrate readily in bodily fluids to diffuse rapidly and enter the bloodstream's vasculature; thus, this restricts (or makes difficult) the therapeutic value of weakly soluble compounds [23]. This is still an issue because low-solubility chemicals make up around 90% of preclinical drug candidates. The use of pH modifiers, especially for tiny compounds whose solubility is significantly pH-dependent, was one strategy employed to improve therapeutic solubility and overcome limited bioavailability [24]. Clinical achievements like intravenous ciprofloxacin, generally made using lactic acid to increase its soluble nature through pH regulation, were made possible by this [25]. Alternative approaches centered on modifying the tiny molecules themselves to adjust their physicochemical characteristics for enhanced diffusion, absorption, and solubilization. For instance, alkyl ester prodrugs are used in the commercial formulation of angiotensin-converting enzyme inhibitors, such as enalapril (marketed under the Vasotec brand) and benazepril (marketed under the Lotensin brand), to conceal ionizable groups and enhance the drug's absorption and bioavailability [26]. Meanwhile, thiazole modification has been introduced to the commercialized formulation of ritonavir (Norvir), the primary protease inhibitor utilized for treating HIV, to improve its water solubility and stability in metabolism [27]. PK was founded because of basic research into how drugs are transported into tissues or cells via the vasculature.

Early efforts to enhance drug efficacy centered around regulating dosage and dosing regimen (i.e., quantity and frequency of infusion), because the connection between pharmacokinetics and pharmacodynamics attributes and the duration of the activity, toxicity, and effectiveness of tiny molecules grew more evident [28–30]. Significantly, authorities have approved a minimum of 16 oral delivery methods that utilize osmotic release; the application of osmotic pumps minimizes adverse effects related to substantial changes in the drug's concentration generated by conventional dosage [31]. Furthermore, topical non-invasive controlled-release delivery techniques, such as transdermal strips, have improved patient compliance while simplifying the lengthy treatment of analgesics and medications for stopping smoking [32]. Moreover, systems built on microparticles and nanoparticles have been utilized to minimize side effects that are not intended, make it simpler for tiny molecules to reach their intended areas of action, and overcome solubility problems.

Approved applications for nanoparticle therapies include immunization and cancer treatment [33]. Notably, early research revealed that applying poly(ethylene glycol) (PEG) coatings was a useful method for increasing particle retention at tumor sites and prolonging the particles' half-life in circulation. The US FDA permitted Doxil, the first medical nanoparticle, in 1995 as a result of the discovery of PEGylated liposomal doxorubicin. Ever since, numerous preclinical studies have been conducted on nanoparticles to tackle the problematic concern. Nowadays, delivery systems are frequently employed to regulate the solubility, dosage, and other delivery characteristics of small molecules [34,35].

2.2.1.1 Red Blood Cell Membrane-Camouflaged Nanoparticles Drug Delivery System

Because they are little immunogenic and have a 120-day lifespan in systemic circulation, drug delivery techniques utilizing camouflaged nanoparticles made of red blood cell membranes are extremely promising and provide several benefits. Moreover, due to a great deal of cell membranes, RBC spheres were biodegradable and readily gained substantial weight capacities, leading to aggregation at the target location. Significantly, membrane-coated erythrocyte nano-formulations are being enormously used to achieve excellent results in anticancer research [36], cardiovascular disorders [37], and encephalopathy [38].

2.2.1.2 Hyaluronic Acid-Based Drug Nanocarriers Drug Delivery Systems

Gel, cationic polymer, and nanoparticle Drug Delivery Systems (DDS) are regarded as a few of the DDSs that are enhanced by hyaluronic acid and its derivatives for actively targeting CD44 receptors in cancer cells. Studies show that following delivery, the tumor site is where the accumulation of HA (Hyaluronic acid) and drug conjugates occurs, allowing long-term drug release because HA-based nanocarriers frequently have a negative surface, and the reticuloendothelial system is unable to eliminate the nanocarriers from the body. The hyaluronic acid-based drug nanocarriers, or Novel drug delivery systems (NDDs), target CD44 receptors actively and use the Enhanced permeability and retention (EPR) effect to specifically infiltrate cancer cells [39].

2.2.1.3 Hexagonal Boron Nitride Nanosheet Drug Delivery System

Research and medicine administration have shown the benefits of hexagonal boron nitride. According to a study by Silicka Jedrzejczak and her associates, MCF-7 cell line cultures proliferated less in exposure to H-BN enriched through gold particles than when compared to normal L929 cell lines. Utilizing a modified Hummers' process and sonication treatment, H-BN was chemically exfoliated. Afterward, gold particles were used to functionalize it for the research and subjected to an analysis utilizing the Neutral Red (NR) uptake assay [40–42].

2.2.1.4 Polymer-Lipid Hybrid Nanoparticles Drug Delivery System

To get beyond these drawbacks and create a useful nanomaterial, scientists looked for and created hybrid polymer-lipid nanoparticles—a hybrid system that combines the special qualities of the two types of nanoparticles (Figure 2.2). This hybrid system fulfilled the demands for biological compatibility, substantial consistency of storage, extended medication release, minimized drug spillage, and tiny size of particles, along with effective encapsulating capability [43].

This technique is currently being utilized for an array of therapeutic and diagnostic applications because of its efficacy. Polymer-lipid hybrid nanoparticles consist of three main parts, which are as follows: a polymeric core that efficiently encases medications that are both hydrophobic and hydrophilic. According to research, they can be applied to imaging, alternate magnetic fields (AMF), immunological stimulation, and vaccine administration. For this reason, it is widely used in the rapidly expanding medical field [44].

2.2.1.5 Self-Microemulsifying Drug Delivery System

Recently, there seems to be an abundance of focus on lipid-based drug preparations, particularly on DDS that Self-Microemulsifying Drug Delivery System (SMEDDS). Insufficient bioavailability is one of the toughest elements when developing oral dosage forms for drugs [23]. It has previously been noted that hydrophobic medicines can be included in SEDDS. Additionally, SMEDDS and SNEDDS (self-nanoemulsifying DDS) undergo alteration from SEDDS to SMEDDS and SNEDDS, respectively. Conversely, emulsions are made by combining a liquid phase that contains macroscopic particles with a surfactant-based liquid phase [45]. Additionally, they are semitransparent (sometimes hazy), unstable thermodynamic solutions with characteristics similar to viscous liquids [46]. Three distinct types of emulsions exist: multiple emulsions, oil/water, and water/oil. Additionally, the self-emulsification of conventional micro- or nanoemulsions upon oral ingestion sets them apart from SMEDDS [47].

2.2.1.6 *In situ* Gel Drug Delivery System

The unique capacity of the in situ gel drug delivery system to switch from sol to gel serves to improve patient compliance and comfort while also prolonging and controlling the release of medications [48]. Most of the time, formulations that are a solution turn into gel form before entering the body due to specific physiological circumstances. A solution can become a gel by combining numerous stimuli, including pH alterations, modification of temperature, and solvent interaction [49]. Various research investigations have employed oral, nasal, injectable, vaginal, rectal, ocular, intraperitoneal, and parenteral approaches. Numerous polymeric techniques have been developed for the delivery of medications [50]. These polymers undergo a sol-gel transition when they come

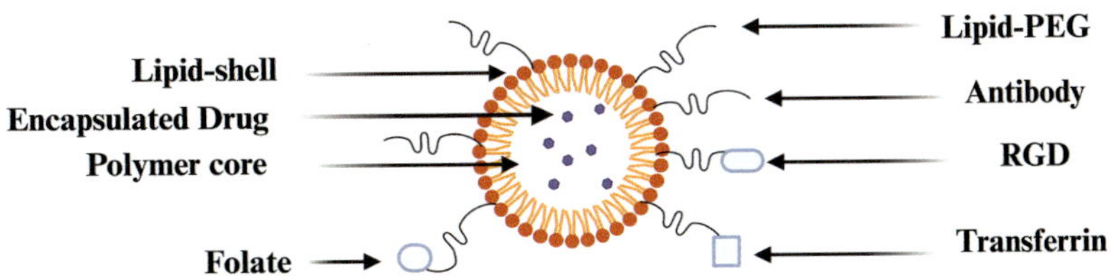

FIGURE 2.2 Polymeric-lipid hybrid nanoparticles.

into interaction with physiologic stimuli. For the creation of in situ gel DDS, several synthetic and natural polymers are utilized. Several procedures were recently recognized for the creation of in-situ gel biomaterials: [1] the effect of pH and temperature alterations; [2] the biomaterials' physical characteristics, such as swelling and solvent transfer; [3] biological alterations, including chemical and enzymatic reactions; and [4] photo-polymerization [51].

2.2.1.7 Micro Electromechanical Systems [MEMS] for Drug Delivery

The applications of MEMS techniques are numerous and include inkjet printing, motion detection, accelerometers, actuators, and medication delivery [52]. Utilizing microfabrication processes, the devices made possible by this technology create mechanical and electromechanical implants that are micro- or nano-sized. It's noteworthy to point out that these strategies enhance the efficacy of these electronics by offering users substantial authority over the size, geography, and microstructure of the finished devices [53]. The most popular methods for creating these MEMS-based devices involve inventive combinations of different micromachining processes, including etching, which is a subtractive process; deposition, which is an additive process; and lithography, which includes micromolding, oxidation, ion implantation, ink jetting, and patterning [54]. Micropumps, detectors, microvalves, polymer compounds, metals, glass, nitrides, and powerful computers are just some of the materials that MEMS technology utilizes for building miniature DDS [55].

2.2.1.8 Combined Drug Delivery Approach

Drug susceptibility is typically an area of worry for medical practitioners. Combination treatment has grown more and more effective because of its wider specificity of target and complementarity to increase therapy's efficacy, as well as enhance clinical results.

The combination delivery of medicines technique is currently extensively used in cancer studies and treatment to tackle resistance to multiple drugs. Combination drug delivery has been shown to retain effectiveness and reduce drug resistance while lowering therapeutic dosage and side responses [56]. Nanoparticles inserted into cells were revealed to be more successful than the nanoparticle drug delivery system, based on a study by Pang et al. on a particular combination of drug delivery technology that concentrated on utilizing cells in conjunction with nanoparticles.

Increased therapeutic effectiveness, prolonged half-lives, enhanced drug release, and lowered immunogenicity and cytotoxicity have all been demonstrated by cell-based therapy. Its capacity to migrate, or chemotaxis, was unaffected by the combination of nanoparticles and exploit cells. Therefore, the combined medication delivery approach is thought to be a viable strategy for medical care and pharmacological research [57].

2.2.1.9 Targeted Drug Delivery System

To reduce adverse effects, the drug's dosage is decreased without compromising its effectiveness or strength. This method makes use of additional drug carriers, including artificial cells, liposomes, micelles, soluble polymers, neutrophils, and biodegradable microsphere polymers [58]. This method is becoming more and more common because it operates particularly in the battle against cancer.

Research by Murugun shows that this method of medication administration works well when applied. Target-negative breast cancer cells (TNBC) (MDA-MB-231) and multidrug-resistant breast cancer cells (MCF-7) were treated using topotecan (TPT) and quercetin (QT) with polyacrylic acid chitosan surface-modified mesoporous silica nanoparticles (MSN).

Arg-Gly-Asp sequences include RGD-peptides, which were grafted onto the nanoparticle surface. The cellular nucleus, endoplasmic reticulum, and mitochondria displayed structural, molecular, and cell death alterations in both cell lines. Additionally, a synergistic antiproliferative impact was noted [59].

2.2.2 Proteins and Peptides

While the structural specifications for tiny molecules established the foundation for therapeutic delivery, the intended recipients of these compounds comprise just 2%–3% of the genome of an individual [60,61] Peptides and proteins are increasingly being used in clinical settings; however, this has led to the emergence of special delivery problems.

Peptides and proteins have better potency and selectivity due to their complex structures; however, this is also a factor in their poor stability when compared to small molecules. They break down easily in ambient storage conditions and become unstable in vivo when exposed to common proteases, physiological temperatures, and pH changes [62]. The quick diffusion of stabilizing excipients in physiological fluids exacerbates this. Additionally, the antigens' immunogenicity on the protein structure, as well as the breakdown, aggregation, or post-translational protein and peptide alteration, might stimulate the immune system [63]. This usually results in immunogenicity-driven adverse effects and quick medication clearance. Synthetic or humanized peptide analogs are designed to overcome structural difficulties by incorporating artificial amino acids or conjugating with chemical moieties that enhance peptides' or proteins' half-lives, stability, receptor affinities, or toxicity (generally known as drug-like characteristics) [64].

Peptides have also been treated with controlled-release methods, which were originally developed for small-molecule medicines. This is especially crucial when the treatment takes the place of normal biological functions, similar to how hormones are released with pulsatility through the entire day or how insulin breaks down according to changes in blood sugar [65].

2.2.3 Antibodies

In the United States, there are now over 500 active clinical trials and over 70 approved clinical trials with antibodies, making them the most common type of therapy [66]. Antibodies' structure, which is very different from other biologic classes' structures, permits interactions between the immune system and therapeutic targets. By attaching to cellular targets, antibodies communicate with the immune system. Antibodies can neutralize a target antigen by attaching to it and preventing the binding of signaling molecules to it, which would start undesired cell processes [67]. Furthermore, antibodies can directly engage in host immune cell interactions that result in the death of unwanted cell populations through phagocytosis and cellular cytotoxicity that depends on complement or antibodies [68].

But the same characteristics that make antibodies special can also result in the production of anti-antibodies, which can have negative effects such as injection site rashes, flu-like symptoms, and the onset of autoimmune disorders [69]. An example of this is muromonab-CD3 (Orthoclone OKT3), the first murine-derived monoclonal antibody (mAb) to be clinically approved in 1986. This mAb generated side effects related to both its mode of action and the immune system's detection of it as a foreign antigen. 2010 saw the drug's discontinuation due to the introduction of more beneficial treatments [70].

Antibody-drug conjugates (ADCs) were developed as a result of antibody application as a treatment approach and constant progress in small-molecule and antibody engineering. ADCs provide synergistic immunomodulatory effects and facilitate highly targeted medication delivery by combining antibodies with cytotoxic small molecules [71]. For instance, the ADC ado-trastuzumab emtansine (Kadcyla) integrates the cytotoxic chemical DM1 (also known as mertansine) with the humanized mAb trastuzumab, a treatment for stomach and breast cancer that targets HER2, also known as human epidermal growth factor receptor. Antibodies have made it easier for tiny compounds (particularly calicheamicins, auristatins, and maytansinoids) that would be extremely dangerous to employ in medicine to be translated into the clinical setting by guiding their transport to target tissues [72].

2.2.4 Nucleic Acids

While the number of druggable targets has increased significantly due to protein and peptide therapies, nucleic acids allow for precise regulation of gene expression, which makes them useful for both driving the expression of therapeutically important genes and silencing or repairing defective genes [73]. It is reasonable to construct nucleic acids and more contemporary gene-editing technologies, like Clustered regularly interspaced short palindromic repeats (CRISPR), to specifically bind to certain regions of the human genome to modify it for therapeutic purposes [74]. The market approval of fomivirsen (Vitravene), an antisense oligonucleotide (ASO) medicine utilized to treat cytomegalovirus retinitis in patients with weakened immune systems, in 1998 served as an example of the potential of nucleic acids. However, due to gene suppression, inadequate levels, and results from the challenges that nucleic acid transport carries by nature, first-generation ASO treatments have had limited clinical effectiveness [75].

The invention of liposomal small-molecule formulations (in addition to the lipid-based nanoparticles' size, charge, and chemistry being optimized for human usage) and the PEGylation application, which was first applied in clinical settings for liposomal doxorubicin, for enhancing drug PKs [76]. Recent advances in nucleic acid delivery are now coming to light with the recent approval of COVID-19 vaccines for emergency use. These vaccines are given via lipid nanoparticles maintained with PEG and depend on chemically modified mRNA [77].

2.2.5 Live-Cell Therapy

The newest class of medicines consists of living cells utilizing some cell types' innate therapeutic properties; live-cell treatments control or facilitate essential biological operations. For example, bacteria may work together with the microbiota to control chronic inflammatory processes, metabolic processes, and mucosal immunity, while vaccinations against diseases and the immune system's capacity to combat cancer may be employed by reprogrammed immune cells. Pluripotent stem cells can regenerate and repair tissues [78]. It's also possible to engineer living cells. T cells with chimeric antigen receptors (CARs) are the most well-known example. These genetically altered cytotoxic T cells, which are directed against particular antigens linked to cancer, were granted clinical approval in 2017 [79–83].

2.3 THREE PARADIGMS OF DRUG DELIVERY

Three fundamental drug delivery principles that pertain to every therapeutic classification were identified: drug modifications, microenvironment modifications, and DDS.

2.3.1 Drug Modifications

A drug's modification includes conjugating it to recognized moieties or targeting ligands, as well as altering its chemical structure (for example, altering nucleic acid backbones, amino acids, or functional groups). Examples include lidocaine, an isoguanine derivative with enhanced strength, toxicity, and duration of action, and fentanyl, a potent opioid with a therapeutic index greater than that of morphine and its parent molecule, meperidine [84]. This set the stage for medicinal chemistry, which in turn led to the development of high-throughput screening technologies that made it possible to rationally design and select molecules with the ideal properties. It also produced general structure-function guidelines, like Lipinski's rule of five, which can be used to identify potential small-molecule pharmaceuticals [85]. New tactics were developed, and old ones were modified in response to the introduction of more recent medications with unique molecular architectures.

High-throughput screening, for instance, has also been used to find therapeutic peptides and proteins. Typically, desirable features are found by the chemical synthesis of huge recombinant libraries employing display methods [86]. Additionally, new methods of modification were needed for peptides and proteins. One such method was the conjugation of PEG to increase immunogenicity and circulation time [87]. Insulin administration has been one area where drug adjustments have been especially significant. For instance, logical changes to the structure of insulin allowed for the development of long-acting formulations such as the modified human insulin glargine (Lantus), which forms micro precipitates in the subcutaneous region, and degludec (Tresiba), an acylated insulin analog that forms high-molecular-weight complexes [88]. These formulations imitate some elements of naturally occurring host-mediated insulin secretion by gradually releasing insulin monomers from a subcutaneous depot. This lowers the frequency of injections, lowers the risk of hypoglycemia, and enhances glycemic management.

The identification of biocompatible therapies has been made possible by the evolution of antibody therapeutic production strategies from mouse immunization to phage display. The development of mAb technology has enabled recent successes of ADCs in resolving the cytotoxicity, off-target toxicity, and lack of tissue-specific distribution of many small-molecule therapies [89]. The development of drug-modification techniques for small molecules and antibodies has been shown to be advantageous in the production of ADCs. Previous conjugation tactics have also benefited nucleic acid medicines, as demonstrated by the recent approval of Givlaari, a GalNAc–siRNA conjugate. Conjugate chemistries for nucleic acids have been based on a combination of conjugate compounds being included as non-nucleoside monomers during oligonucleotide synthesis and previously documented bio-orthogonal click-chemistry processes [90]. Control over the action of nucleic acid medications was made possible, in particular, by chemical changes discovered for the 2′ hydroxyl group of ribose, polyadenylation of the 3′ end, and replacement of the phosphodiester bond [91] (Table 2.1).

TABLE 2.1
Types of Modifications and Selected Examples

Type	Class of Theraputics	Modification	Examples
Drug modification	Small molecules	Modification of functional groups	Ritonavir
		Masking undesirable chemical groups	Benazepril
		High-throughput and combinatorial chemistry libraries	Ezetimibe
		Conjugation of targeting ligands	Vintafolide
		PEGylation	Naloxegol
	Peptides and proteins	Alterations to the amino acid sequence	Belatacept
		Addition of non-natural amino acids	Desmopressin
		Alteration of intramolecular bonding	Cyclosporin A
		Chemical backbone modifications	Cyclosporin A
		PEGylation	Pegademase
	Antobodies	Affinity maturation and variable-region engineering	Remacirumab
		Antibody humanization	Daclizumab
		Modification of Fc binding to the neonatal Fc receptor	MEDI4893
		Modification of antigen binding for pH responsiveness	Ravulizumab
		Conjugation of PEG	Certolizumab pegol
		Conjugation to small molecules	Brentuximab vedotin
	Nucleic acids	Codon optimization and chemical nucleotide modifications	Patisiran, Givosiran, Nusinersen
		Chemical backbone modifications	Fomivirsen, Givosiran
		Conjugation to targeting ligands	Givosiran
	Live-cell therapies	Surface conjugation	PRIME
		In situ delivery of nucleic acids for host-cell reprogramming	Janseen COVID-19 vaccine
		Exposure to external stimuli	Sipuleucel-T
		Genetic engineering	Tisagenlecleucel

(*Continued*)

TABLE 2.1 (*Continued*)
Types of Modifications and Selected Examples

Type	Class of Theraputics	Modification	Examples
Environmental modification	Small molecules	Addition of solubilizing excipients	Ciprofloxacin
		Inhibition of clearance pathways	Co-administration of penicillin and probenecid
		Inhibition of drug metabolism	Not yet in clinical trials
	Peptides and proteins	Use of protease inhibitors	Calcitonin-salmon
		Use of pH modifiers	Semaglutide, calcitonin-salmon
		Use of permeation enhancers	Semaglutide, octreotide
		Use of immunomodulators	Enzyme replacement therapy co-administered with methotrexate in Pompe disease
		Use of hyaluronidases	Hylenex administered with insulin
	Antibodies	Use of enzyme inhibitors	Not yet in clinical trials
		Use of pH modifiers	Adalimumab
		Use of permeation enhancers	Not yet in clinical trials
		Use of immunomodulators	Infliximab co-administered with methotrexate
		Use of dispersion enhancers	Trastuzumab and hyaluronic-oysk
	Nucleic acid	Use of immunomodulators	Patisiran, Givosiran
		Use of endosomal-pH modifiers	Not yet in clinical trials
		Use of endosomal-release agents	Patisiran
	Live-cell therapies	Use of signaling molecules to support in vivo cellular action	Consistent IL-12-secreting CAR T cells
		Modification of the tumor microenvironment	EGFR-IL-12 CART cells
		Use of immunomodulators	TGF-β-insensitive CAR T cells, RTX-240
Drug delivery modification	Small molecules	Controlled-release capsules	Methylated HCl, fexofenadine, and pseudoephedrine
		Controlled-release implants	Etonogestrel implant, fluocinolone
		Inhalable devices	Albuterol inhalation, budesonide/formoterol
		Transdermal patched	Fentanyl transdermal system, transdermal nicotine
		Stimuli-responsive drug release	Verteporfin for injection
		Microparticles	Risperidone
		Nanoparticles	Doxorubicin, protein-bound paclitaxel
		Targeted delivery	Ado-trastuzumab emtansine
	Peptides and proteins	Controlled-release microparticles depot	Leuprolide acetate, somatropin
		Targeted delivery system	No clinical products yet
		In situ forming polymer matrix	Leuprolide acetate
		Implants	Leuprolide acetate implants
		Non-invasive delivery systems	Inhalation powder of human insulin, semaglutide
	Antibodies	ADCs	Brentuximab vedotin, ado-trastuzumab emtansine
		Minimally invasive subcutaneous injections combining humanized antibodies with dispersion enhancers	Trastuzumab and hyaluronidase-oysk
		Non-invasive delivery systems	Inhalation powder
	Nucleic acid	Lipid-based nanoparticles	Patisiran, mRNA COVID-19
		Viral vectors	Recombinant human p53 adenovirus, Janssen COVID-19 vaccine
		Polymer conjugates	Pegabtanib
		Scaffolds to sustain cell populations near the site of action	Matrix-induced autologous chondrocyte implantation
	Liver-cell-therapies	Micro particles and micro particle implants	SIG-001
		Drug-secreting therapeutic bacteria	AG019
		Cellular hitchhiking	PRIME^{IL-15}/RPTR-147

Overall, by stabilizing drugs in physiological fluids, regulating their interactions on specific cells and within them, and altering their interactions with their medicinal targets, drug modifications have made new functions for all classes of therapy possible [92,93].

To supplement the systemic distribution of small molecules, proteins, and peptides, genetically modified cells offer a viable alternative to conjugation techniques in terms of both delivery vehicles and in situ drug-secreting factories. Furthermore, a natural phenotypic shift can be induced in cells by ex vivo stimulation, negating the requirement for genome engineering. For instance, stem cell differentiation can cause the biologic to be secreted, while immune cell stimulation can encourage the presentation of antigens. Cell modifications can also solve persistent issues with small-molecule distribution that is tissue- and site-specific but exhibits significant off-target toxicity [94].

2.3.2 Microenvironmental Modifications

Environmental modifications take several forms, ranging from extremely specific modifications performed at the action site for adjuvants that are given comprehensively and influence the host environment. When taken as a whole, environmental alterations offer an all-encompassing drug delivery strategy that can help biological barriers pass. Excipients were created for tiny molecules to improve medicine's soluble nature in bodily fluids through local pH adjustments [95]. Because nucleic acids' ability to interact with gene targets is limited by their entry into intracellular compartments, prior approaches were modified to include cell-permeating peptides and cationic lipids added for improving intracellular absorption, endosome escape, and targeting of nuclei, as well as to change the pH of endosomes. These methods helped with the current clinical use of Onpattro, an siRNA therapy that helps the siRNA escape into the cytosol and the endosome reorganize following endocytosis by using ionizable cationic lipids [96]. It is also possible to modify physiological liquids or mechanical operations that impede the effects of drugs by using microenvironment modulators. A collection of specified excipients has been created specifically for small-molecule medications to enhance the drugs' absorption in the small intestine. This is achieved either by prolonging the intestinal transit time or by lengthening the time the drugs are in the bloodstream by actively inhibiting particular metabolic pathways [95].

pH modifiers, like citric acid, were first developed as small-molecule solubilizers for proteins and peptides. This allowed them to be more stable in physiological fluids and to prevent proteolysis. Furthermore, systemically given steroids undergo utilization to control the immune system and avoid negative responses to nucleic acid and protein therapy [97]. To increase cell treatments' activity at the target location and, thus, their efficiency, it is necessary to understand how to precisely adjust the microenvironment. Additionally, in response to cues from the environment, a cell's gene-expression profile—and thus, its phenotype—can change. Microenvironmental tactics have been used in clinical settings using CAR T-cell treatment to overcome obstacles related to solid tumors comprising an unfavorable microenvironment with high extracellular pressure, abnormal vasculature, and inhibition of immune signaling [98].

The early years of contemporary drug delivery established that a drug's PK properties, specifically its highest concentration in blood, half-life, overall drug exposure throughout time, and biodistribution, may be changed by varying the drug's rate of release in the body. The introduction of controlled-release delivery systems offers greater authority over these factors, though changes in dosage, frequency, and infusion rates can still have an impact [99]. These systems included polymeric implants and hydrogels focused on four controlled-release mechanisms (dissolution, diffusion, osmosis, and ion exchange), as well as microparticles and nanoparticles, which enable specific interactions with the microenvironment to modify the particle surface and improve drug half-life [100].

The basis of small-molecule controlled-release technologies was crucial to the development of therapeutic peptide and protein delivery techniques, as their short half-lives required frequent, intrusive injections. Leuprolide acetate, which is prepared in Lupron Depot microspheres that release the drug for 6 months, was one of the first examples of such a system. This provided a sustained-release approach that was subsequently used for several other peptide treatments [101].

More research is currently being done on lipid nanoparticles as a means of delivering nucleic acids, such as mRNA. Lipid nanoparticles and nucleoside modifications are employed in the mRNA-based COVID-19 vaccines to enhance the translation, stability, and efficacy of the supplied nucleic acid. Delivery systems can generally enhance the medication's stability, targeting, and PKs. They can also package customized treatments with microenvironmental modifiers to more effectively control how the drug interacts with the tissues, cells, and chemicals in the body. The clinical translation of cell therapies will require novel and repurposed delivery systems, especially those that combine microenvironmental modifications (such as the release, distribution, or simultaneous administration of antibodies, cytokines, or small molecules) with cell modifications (such as surface conjugation, genetic engineering, or stimulus exposure for cellular differentiation or antigen presentation). Numerous drug delivery methods currently in use have been refined for cell therapy protection, targeting, and activation. In a particular instance, β-cells that secrete insulin and are stimulated for treating type-1 diabetes were enclosed in implants made of polymers to prolong their lifespan and sustain blood glucose levels [102].

2.3.3 Drug Delivery

Alongside the creation of medicines, drug delivery has changed over time. Nucleic acids, proteins, peptides, small compounds, and most recently, live-cell treatments have all been used. Drug modifications and microenvironmental alterations have been used to overcome delivery problems for each of these treatment classes.

This is demonstrated through the widespread use of sustained-release and controlled-release systems, which were initially designed for small molecules, across various therapeutic domains. On the other hand, technology and delivery methods created for new therapy methods have been adjusted to improve the way existing therapies are administered. As an illustration, immune-modulating polymers were used in clinical settings for the SIG-001a longer cell therapy deposit, which treats hemophilia A by releasing human factor VIII. In a similar vein, high-throughput screening methods that initially emerged for maximizing small-molecule treatments have assisted in the identification of materials that alter the microenvironment to transport living cells.

A variety of therapeutic classes have also benefited from the repurposing of environmental modifiers. These include the use of cytokines to alter the tumor microenvironment, antibiotics to engraft living biological treatments, chemically generated cell-permeating peptides for improving intracellular transport of small molecules and nucleic acids, and cell-based immunotherapies. Furthermore, developments within one therapeutic class have aided in the practical translation of treatments from other classes. Therapeutic antibodies, for instance, have been employed as small-molecule drug modifiers (ADCs). They allowed cytotoxic drugs with intolerably high levels of toxicity to be translated into the clinic by utilizing the specificity of the antibody (monomethyl auristatin E, a synthetic antineoplastic small chemical, is one example). The three primary unmet challenges in the present state of therapies, as well as delivery strategies (alterations to the environment and drugs), are biological obstacles that impede the introduction of complex therapeutic molecules, precise administration with the conclusion of a single cell or cell compartment, and the development of medication delivery methods which, in response to outside stimuli, rapidly distribute biomolecules into specific tissues at particular times and quantities. While most medicines will still be implemented despite these obstacles, we think that cell therapies can solve both of them at the same time and result in efficient single-dose DDS. Cell treatments can overcome biological obstacles, offer a consistent supply of complicated biologics, and react to host stimuli in a manner that resembles normal biological processes. Thus, cell treatments have the potential to function as a therapeutic as well as a dynamic delivery mechanism. Cell treatments are therefore especially well adapted for the management of weakly responding malignancies, metabolic genetic disorders, and uncommon blood illnesses (such as sickle cell disease and hemophilia). Additionally, because modern cell treatments mimic important biological processes (such as host-responsive insulin secretion), they may lessen the frequency of doses and the quantity or necessity of specific medical interventions. If past experiences are any indication, cell therapies will leverage well-established methods to manipulate medications and their surrounding milieu to regulate drug activity, effectiveness, and toxicity; on the other hand, particular enhancements to the methods will help the other therapeutic classes.

2.4 CHALLENGES ASSOCIATED WITH CURRENT DRUG DELIVERY SYSTEMS

The objective of getting medications from different plant sources to the correct places in the human body to treat diseases has gone quite a distance in the past few decades, and an array of ways to deliver them have proven effective. However, these processes face multiple limitations and challenges when it pertains to therapeutic outcomes; a few of them are covered in the sections that follow. A major barrier to the development of DDS is the restricted amount and range of present documentation. The accessibility of essential data in the literature is an essential requirement for the progress of any research, particularly this study on nanomedicine treatment techniques. Differences in published studies on the documented description of reported results from experiments appear to be another major barrier to the development of medical applications of nanotechnology [103]. Future advances in nanomedicines may be hampered by the incomplete and inconsistent data that ought to serve as a guide for industry, delaying the transition from study and experimentation to clinical application [104]. It is widely acknowledged by researchers that nanoparticles can have positive or negative effects. However, there is a lack of information regarding the safety of these particles, their degree of interaction with non-target proteins, and how they move and interact with organs other than their intended targets [105]. The issue of employing much larger particles can be addressed by using tiny particles for delivering the drug to the individual body. Large particles, which may cause obstacles like low absorption as well as dissolution, *in vivo* instabilities, low bioavailability, target-specific delivery issues, and multiple adverse reactions toward management, are used as carriers in a number of these delivery systems and thus are not very helpful for treatment.

A challenge that all delivery methods face is target-specific delivery complications. Target-specific delivery is currently demonstrated to be more effective and less hazardous; nevertheless, its efficiency is not guaranteed until the drug exists at the intended place at adequate levels.

This occurs when siRNA is administered systematically; when given in huge amounts, its negative charge grows a barrier toward cells' capacity to utilize the drug, leading to very little absorption by the human body. Moreover, siRNA frequently reaches the intended cell or organ because they are quickly broken down by body enzymes. Targeted drug delivery is being investigated utilizing lipid nanoparticles such as liposomes and micelles; nonetheless, the drawback of this strategy is that the way the body reacts to the nanoparticles, such as phagocyte ingestion and hepatic filtration, could reduce their effectiveness and result in toxicity effects [106]. Targeted delivery is challenging because an unconscious patient cannot accept a dose, the target site has limited solubility and permeability, the drug may interact with food, and the gut microbiota may degrade the medication. The toxic nature of the particles utilized for distribution is an additional significant issue for DDS as a group; certain nanomaterials are detrimental to both human health and the environment [107]. Experiments conducted both in vivo and in vitro have demonstrated deleterious effects using silver, titanium, silica, and gold as drug-delivering and coupling nanoparticles. The applications of carbon nanotubes (CNTs) in medication administration, gene therapy, and bioimaging have grown significantly. Researchers are concerned about the properties of carbon nanotubes because of their capacity to damage DNA, the immune system, the liver, the heart, and neurons, as well as embryos. This is especially true when using them as carriers of biomolecules, as experiments have shown that they can cross cell membranes [108]. Even if carbon nanotubes have demonstrated promising outcomes in their application, critical toxicity testing must be conducted to guarantee their safety prior to a broad implementation in treatment [109]. Their side effects have made it difficult to employ them as a cancer treatment. DDS face several difficulties, including acceptability (the drug's biocompatibility—its capacity to interact with an individual under certain conditions—and its ability to be absorbed by an individual without stimulating the immune system). This is because individuals respond to organic materials differently than they do to synthetic ones. Researchers have successfully created medications that have both functions as carriers [107]. A selective permeation of the blood-brain barrier (BBB) makes it challenging to obtain therapeutic medication concentrations in the cerebral tissues. Additionally, the BBB inhibits carrier particles from penetrating the brain and the central nervous system in general. As a result, therapeutic agents fail to efficiently carry and sustain targeted drugs within the brain, rendering it ineffective in addressing cerebral illnesses. The complex framework of the human body could be a contributing cause to these intrinsic limitations of these delivery systems' capabilities. Additionally, monoclonal antibodies (mAb) are among the body's most prevalent carriers because they bind to liposome surfaces to form immunoliposomes. However, the restricted uses of these liposomes—which have reduced contents in body absorption, distribution, metabolism, and elimination—in addition to their ability to cause an immune response, render it challenging to use liposomes as site-specific drug carriers [110]. The body's natural detoxification processes include the kidneys and liver, which may handle nanoparticles as possible waste products. Their actions may hinder the delivery of drugs and cause nanoparticles to accumulate in these organs. Nanomaterials mostly gather in macrophages, sinusoidal endothelial cells, hepatic stellate cells, and Kupffer cells inside the liver. Hepatocytes also contain a small amount of nanomaterials. The shape, charge, and structure of the kidneys determine the outcome of nanomaterials that reach the renal system [111].

2.5 FUTURE DIRECTIONS AND CONCLUSION

Scientists, investigators, and clinical studies all have concentrated a lot on drug delivery and nanomedicine over the past few decades. It has become one of the most stimulating fields of research in modern science. Despite the challenges that have prevented the most recent medication delivery technology out of clinical applications, it still has an abundance of potential. We have to work together across educational theory, laboratory experiments, medical expertise, pharmaceutical understanding, and superior research to achieve this efficiency. According to Vargason et al., using cell treatments might substantially lessen the issues associated with DDS' bio-acceptability [1]. Furthermore, Vargason et al. claim that this method will deliver a successful single dosage while preventing a significant accumulation of drugs in the body of an individual. Cell treatments promise to dismantle inherent biological defenses, generate habits within the system that look natural, and provide a seemingly persistent stream of complicated biologics. Adepu has suggested that inorganic mesoporous nanoparticles, microfluids, and molecular imprinting polymers be used as tactics to deal with some of the issues related to medicine delivery. Khalid et al. state that using priming agents that possess the ability to change ecological environments where they are delivered—especially those with the ability to alter the form and functioning of tissues to render the therapy offered better without placing the person receiving it at risk—is one way to increase the efficacy of drug delivery. Furthermore, as cells are a vital part of the body of an individual, the nanomaterials industry takes into consideration cell-based pharmaceutical systems, which combine nanomaterials with living things. This innovative strategy, which remains in the theoretical stage, seems to be very inventive and encourages the use of drug delivery techniques in an attempt to achieve the ideal drug distribution arrangement. Additional research and clinical investigations remain required to verify the efficacy of such modern medication delivery techniques and the challenges presented by their application.

REFERENCES

1. Vargason AM, Anselmo AC, Mitragotri S. The evolution of commercial drug delivery technologies. *Nat Biomed Eng*. 2021;5(9):951–67. https://www.nature.com/articles/s41551-021-00698-w.
2. SUPAC-MR. *Modified Release Solid Oral Dosage Forms Scale-Up and Post-approval Changes: Chemistry, Manufacturing, and Controls; In Vitro Dissolution Testing and In Vivo Bioequivalence Documentation: Guidance for Industry*. Guidance Portal. [cited 2024 May 12] https://www.hhs.gov/guidance/document/supac-mr-modified-release-solid-oral-dosage-forms-scale-and-postapproval-changes-chemistry.
3. USP-NF. 2021. USP-NF. [cited 2024 May 12] https://www.uspnf.com/official-text/proposal-statuscommentary/uspnf-2021.
4. Allen TM, Cullis PR. Drug delivery systems: entering the mainstream. *Science*. 2004;303(5665):1818–22.

5. Yasaei R, Katta S, Saadabadi A. Gabapentin. *Small Anim Crit Care Med*. 2022;919:21. https://www.ncbi.nlm.nih.gov/books/NBK493228/
6. Serajuddln ATM. Solid dispersion of poorly water-soluble drugs: Early promises, subsequent problems, and recent breakthroughs. *J Pharm Sci*. 1999;88(10):1058–66.
7. Akala EO. Oral controlled-release solid dosage forms. *Theory Pract Contemp Pharm*. 2021;333–66. doi:10.1201/9780203644478-14.
8. Neurath AR, Strick N, Li YY, Debnath AK. Cellulose acetate phthalate, a common pharmaceutical excipient, inactivates HIV-1 and blocks the coreceptor binding site on the virus envelope glycoprotein gp 120. *BMC Infect Dis*. 2001;1(1):1–12. doi:10.1186/1471-2334-1-17
9. Hillery AM, Park K. *Drug delivery: Fundamentals and Applications*, Second Edn. CRC Press. 2016, pp. 1–614. doi:10.1201/9781315382579.
10. Lee PI, Li JX. Evolution of oral controlled release dosage forms. In *Oral Controlled Release Formulation Design: Drug Delivery from Theory to Practice*; Rathbone, M. J., Hadgraft, J., and Roberts, M. S., Eds.; Wiley: Hoboken, NJ, 2010; pp. 21–31. doi:10.1002/9780470640487.ch2.
11. Yun Y, Lee BK, Park K. Controlled drug delivery systems: The next 30 years. *Front Chem Sci Eng*. 2014;8(3):276–9. doi:10.1007/s11705-014-1426-x.
12. Zhang W, Zhao Q, Deng J, Hu Y, Wang Y, Ouyang D. Big data analysis of global advances in pharmaceutics and drug delivery 1980–2014. *Drug Discovery Today*. 2017; 22(8):1201–8.
13. Bangham AD, Horne RW. Negative staining of phospholipids and their structural modification by surface-active agents as observed in the electron microscope. *J Mol Biol*. 1964;8(5):660–8.
14. Bangham AD, Standish MM, Watkins JC. Diffusion of univalent ions across the lamellae of swollen phospholipids. *J Mol Biol*. 1965;13(1):238–52.
15. Shen J, Burgess DJ. Drugs for long acting injections and implants. *Long -Act Inject Implant*. 2012;73–91. doi:10.1007/978-1-4614-0554-2_5
16. Birrenbach G, Speiser PP. Polymerized micelles and their use as adjuvants in immunology. *J Pharm Sci*. 1976; 65(12):1763–6.
17. Couvreur P, Tulkenst P, Roland M, Trouet A, Speiser P. Nanocapsules: a new type of lysosomotropic carrier. *FEBS Lett*. 1977;84(2):323–6.
18. Park K. Drug delivery of the future: Chasing the invisible gorilla. *J Control Release*. 2016;240:2–8.
19. Ye M, Kim S, Park K. Issues in long-term protein delivery using biodegradable microparticles. *J Control Release*. 2010;146(2):241–60.
20. Park K. Controlled drug delivery systems: Past forward and future back. *J Control Release*. 2014;190:3–8.
21. Yun YH, Lee BK, Park K. Controlled drug delivery: Historical perspective for the next generation. *J Control Release*. 2015;219:2–7.
22. Washington N, Washington C, Wilson C. Physiological pharmaceutics: Barriers to drug absorption. *Physiol Pharm*. 2000. doi:10.1201/9780203483701.
23. Savjani KT, Gajjar AK, Savjani JK. Drug solubility: importance and enhancement techniques. *ISRN Pharm*. 2012;2012:1–10.
24. Kalepu S, Nekkanti V. Insoluble drug delivery strategies: review of recent advances and business prospects. *Acta Pharm Sin B*. 2015;5(5):442–53.
25. Sharma PC, Jain A, Jain S, Pahwa R, Yar MS. Ciprofloxacin: review on developments in synthetic, analytical, and medicinal aspects. *J Enzyme Inhib Med Chem*. 2010;25(4):577–89.
26. Beaumont K, Webster R, Gardner I, Dack K. Design of ester prodrugs to enhance oral absorption of poorly permeable compounds: challenges to the discovery scientist. *Curr Drug Metab*. 2003;4(6):461–85.
27. Kempf DJ, Sham HL, Marsh KC, Flentge CA, Betebenner D, Green BE, McDonald E, Vasavanonda S, Saldivar A, Wideburg NE, Kati WM, Ruiz L, Zhao C, Fino L, Patterson J, Molla A, Plattner JJ, Norbeck DW. Discovery of ritonavir, a potent inhibitor of HIV protease with high oral bioavailability and clinical efficacy. *J Med Chem*. 1998;*41*(4):602–14.
28. Kubitza D, Becka M, Voith B, Zuehlsdorf M, Wensing G. Safety, pharmacodynamics, and pharmacokinetics of single doses of BAY 59-7939, an oral, direct factor Xa inhibitor. *Clin Pharmacol Ther*. 2005;78(4):412–21.
29. Chien SC, Rogge MC, Gisclon LG, Curtin C, Wong F, Natarajan J et al. Pharmacokinetic profile of levofloxacin following once-daily 500-milligram oral or intravenous doses. *Antimicrob Agents Chemother*. 1997;41(10):2256.
30. Park K. Controlled drug delivery systems: past forward and future back. *J Control Release*. 2014;190:3–8.
31. Keraliya RA, Patel C, Patel P, Keraliya V, Soni TG, Patel RC et al. Osmotic drug delivery system as a part of modified release dosage form. *ISRN Pharm*. 2012;2012:1–9.
32. Jeong WY, Kwon M, Choi HE, Kim KS. Recent advances in transdermal drug delivery systems: a review. *Biomater Res*. 2021;25(1):1–15. doi:10.1186/s40824-021-00226-6
33. Anselmo AC, Mitragotri S. Nanoparticles in the clinic: An update. *Bioeng Transl Med*. 2019;4(3): e10143.
34. Papahadjopoulos D, Allen TM, Gabizon A, Mayhew E, Matthay K, Huang SK et al. Sterically stabilized liposomes: improvements in pharmacokinetics and antitumor therapeutic efficacy. *Proc Natl Acad Sci U S A*. 1991;88(24):11460.
35. Barenholz Y. Doxil®—the first FDA-approved nano-drug: lessons learned. *J Control Release*. 2012;160(2):117–34.
36. Chen J, Hu L. Nanoscale delivery system for nutraceuticals: preparation, application, characterization, safety, and future trends. *Food Eng Rev*. 2020;12(1):14–31. doi:10.1007/s12393-019-09208-w.
37. Zinger A, Cooke JP, Taraballi F. Biomimetic nano drug delivery carriers for treating cardiovascular diseases. *Nanomedicine*. 2021;33:102360.
38. Nguyen TT, Nguyen TD, Nguyen TKO, Vo TK, Vo VG. Advances in developing therapeutic strategies for Alzheimer's disease. *Biomed Pharmacother*. 2021;139:111623.
39. Jia Y, Chen S, Wang C, Sun T, Yang L. Hyaluronic acid-based nano drug delivery systems for breast cancer treatment: Recent advances. *Front Bioeng Biotechnol*. 2022;10:990145.
40. Jedrzejczak-Silicka M, Trukawka M, Dudziak M, Piotrowska K, Mijowska E. Hexagonal boron nitride functionalized with au nanoparticles—properties and potential biological applications. *Nanomaterials*. 2018;8(8):605.
41. Liu J, Zheng T, Tian Y. Functionalized h-BN nanosheets as a theranostic platform for SERS real-time monitoring of MicroRNA and photodynamic therapy. *Angew Chem Int Ed*. 2019;58(23):7757–61. doi:10.1002/anie.201902776
42. Sharker SM. Hexagonal boron nitrides (white graphene): a promising method for cancer drug delivery. *Int J Nanomedicine*. 2019;14:9983.
43. Rizwanullah M, Alam M, Harshita, Mir SR, Rizvi MMA, Amin S. Polymer-lipid hybrid nanoparticles: a next-generation nanocarrier for targeted treatment of solid tumors. *Curr Pharm Des*. 2020;26(11):1206–15.

44. Mohanty A, Uthaman S, Park IK. Utilization of polymer-lipid hybrid nanoparticles for targeted anti-cancer therapy. *Molecules*. 2020;25(19). doi:10.3390/molecules2519437.
45. Baloch J, Sohail MF, Sarwar HS, Kiani MH, Khan GM, Jahan S et al. Self-nanoemulsifying drug delivery system (SNEDDS) for improved oral bioavailability of chlorpromazine: in vitro and in vivo evaluation. *Medicina (Kaunas)*. 2019;55(5):210.
46. Dokania S, Joshi AK. Self-microemulsifying drug delivery system (SMEDDS)—challenges and road ahead. *Drug Deliv*. 2015;22(6):675–90.
47. Silva BFB, Rodríguez-Abreu C, Vilanova N. Recent advances in multiple emulsions and their application as templates. *Curr Opin Colloid Interface Sci*. 2016;25:98–108.
48. Vigani B, Rossi S, Sandri G, Bonferoni MC, Caramella CM, Ferrari F. Recent advances in the development of in situ gelling drug delivery systems for non-parenteral administration routes. *Pharmaceutics*. 2020;12(9):1–29.
49. Singh B, Khurana RK, Garg B, Saini S, Kaur R. Stimuli-responsive systems with diverse drug delivery and biomedical applications: recent updates and mechanistic pathways. *Crit Rev Ther Drug Carrier Syst*. 2017;34(3):209–55.
50. (4) (PDF) *A Review: In-Situ Gel Drug Delivery System*. [cited 2024 May 12] https://www.researchgate.net/publication/353667314_A_Review_in-situ_Gel_Drug_Delivery_System.
51. (4) (PDF) *An In sight into Novel Drug Delivery System: In Situ Gels*. [cited 2024 May 12] https://www.researchgate.net/publication/349621711_An_In_sight_into_Novel_Drug_Delivery_System_In_Situ_Gels.
52. Khudiyev T, Clayton J, Levy E, Chocat N, Gumennik A, Stolyarov AM et al. Electrostrictive microelectromechanical fibres and textiles. *Nat Commun*. 2017;8(1):1–7. https://www.nature.com/articles/s41467-017-01558-5
53. Koch B, Rubino I, Quan FS, Yoo B, Choi HJ. Microfabrication for drug delivery. *Materials (Basel)*. 2016;9(8):646.
54. Ziaie B, Baldi A, Lei M, Gu Y, Siegel RA. Hard and soft micromachining for BioMEMS: Review of techniques and examples of applications in microfluidics and drug delivery. *Adv Drug Deliv Rev*. 2004;56(2):145–72.
55. Cobo A, Sheybani R, Meng E. MEMS: Enabled drug delivery systems. Adv Healthc Mater. 2015;4(7):969–82. doi:10.1002/adhm.201400772.
56. Wang H, Huang Y. Combination therapy based on nano codelivery for overcoming cancer drug resistance. *Med Drug Discov*. 2020;6:100024.
57. Pang L, Zhang C, Qin J, Han L, Li R, Hong C et al. A novel strategy to achieve effective drug delivery: exploit cells as carrier combined with nanoparticles. *Drug Deliv*. 2017;24(1):83.
58. (4) (PDF) *A Review on Targeted Drug Delivery: Its Entire Focus on Advanced Therapeutics and Diagnostics*. [cited 2024 May 12]. https://www.researchgate.net/publication/265862863_A_Review_on_Targeted_Drug_Delivery_Its_Entire_Focus_on_Advanced_Therapeutics_and_Diagnostics.
59. Murugan C, Rayappan K, Thangam R, Bhanumathi R, Shanthi K, Vivek R et al. Combinatorial nanocarrier-based drug delivery approach for amalgamation of anti-tumor agents in breast cancer cells: an improved nanomedicine strategy. *Sci Rep*. 2016;6.
60. Rask-Andersen M, Masuram S, Schiöth HB. The druggable genome: Evaluation of drug targets in clinical trials suggests major shifts in molecular class and indication. *Annu Rev Pharmacol Toxicol*. 2014;54:9–26. doi:10.1146/annurev-pharmtox-011613-135943.
61. Craik DJ, Fairlie DP, Liras S, Price D. The future of peptide-based drugs. *Chem Biol Drug Des*. 2013 Jan;81(1):136–47.
62. Pisal DS, Kosloski MP, Balu-Iyer S V. Delivery of therapeutic proteins. *J Pharm Sci*. 2010;99(6):2557–75.
63. Baker MP, Reynolds HM, Lumicisi B, Bryson CJ. Immunogenicity of protein therapeutics: The key causes, consequences, and challenges. *Self Nonself*. 2010 Oct;1(4):314–22.
64. Ovadia O, Greenberg S, Laufer B, Gilon C, Hoffman A, Kessler H. Improvement of drug-like properties of peptides: the somatostatin paradigm. *Expert Opin Drug Discov*. 2010 Jul;5(7):655–71.
65. Jain D, Raturi R, Jain V, Bansal P, Singh R. Recent technologies in pulsatile drug delivery systems. *Biomatter*. 2011;1(1):57.
66. Lu RM, Hwang YC, Liu IJ, Lee CC, Tsai HZ, Li HJ et al. Development of therapeutic antibodies for the treatment of diseases. *J Biomed Sci*. 2020;27(1):1–30. doi:10.1186/s12929-019-0592-z.
67. Shih T, Lindley C. Bevacizumab: an angiogenesis inhibitor for the treatment of solid malignancies. *Clin Ther*. 2006;28(11):1779–802.
68. Smith MR. Rituximab (monoclonal anti-CD20 antibody): mechanisms of action and resistance. *Oncogene*. 2003;22(47):7359–68.
69. Atzeni F, Talotta R, Salaffi F, Cassinotti A, Varisco V, Battellino M et al. Immunogenicity and autoimmunity during anti-TNF therapy. *Autoimmun Rev*. 2013 May;12(7):703–8.
70. Reichert JM. Marketed therapeutic antibodies compendium. *MAbs*. 2012;4(3):413. /pmc/articles/PMC3355480/
71. Alley SC, Okeley NM, Senter PD. Antibody-drug conjugates: targeted drug delivery for cancer. *Curr Opin Chem Biol*. 2010 Aug;14(4):529–37.
72. Beck A, Goetsch L, Dumontet C, Corvaïa N. Strategies and challenges for the next generation of antibody–drug conjugates. *Nat Rev Drug Discov*. 2017;16(5):315–37. https://www.nature.com/articles/nrd.2016.268
73. Opalinska JB, Gewirtz AM. Nucleic-acid therapeutics: basic principles and recent applications. *Nat Rev Drug Discov*. 2002;1(7):503–14.
74. Mali P, Yang L, Esvelt KM, Aach J, Guell M, DiCarlo JE et al. RNA-guided human genome engineering via Cas9. *Science*. 2013;339(6121):823–6.
75. Rinaldi C, Wood MJA. Antisense oligonucleotides: the next frontier for treatment of neurological disorders. *Nat Rev Neurol*. 2018;14(1):9–22.
76. Allen TM, Cullis PR. Liposomal drug delivery systems: from concept to clinical applications. *Adv Drug Deliv Rev*. 2013;65(1):36–48.
77. Polack FP, Thomas SJ, Kitchin N, Absalon J, Gurtman A, Lockhart S et al. Safety and efficacy of the BNT162b2 mRNA Covid-19 vaccine. *N Engl J Med*. 2020;383(27):2603–15.
78. Vargason AM, Anselmo AC. Clinical translation of microbe-based therapies: Current clinical landscape and preclinical outlook. *Bioeng Transl Med*. 2018;3(2):124–37.
79. Prasad V. Immunotherapy: Tisagenlecleucel—the first approved CAR-T-cell therapy: implications for payers and policy makers. *Nat Rev Clin Oncol*. 2018;15(1):11–2.
80. Guzman G, Reed MR, Bielamowicz K, Koss B, Rodriguez A. CAR-T therapies in solid tumors: opportunities and challenges. *Curr Oncol Rep*. 2023;25(5):479.
81. Hourd P, Ginty P, Chandra A, Williams DJ. Manufacturing models permitting roll out/scale out of clinically led autologous cell therapies: regulatory and scientific challenges for comparability. *Cytotherapy*. 2014;16(8):1033–47.

82. Jarosławski S, Toumi M. Sipuleucel-T (Provenge(®))-autopsy of an innovative paradigm change in cancer treatment: why a single-product biotech company failed to capitalize on its breakthrough invention. *BioDrugs*. 2015;29(5):301–7.
83. Kauer TM, Figueiredo JL, Hingtgen S, Shah K. Encapsulated therapeutic stem cells implanted in the tumor resection cavity induce cell death in gliomas. *Nat Neurosci*. 2011;15(2):197–204.
84. Nadendla R, Yemineni G. Molecular modification: a strategy in drug discovery and drug design. *Biomed J Sci Tech Res. 2023;52*(2):43511–22.
85. Pereira DA, Williams JA. Origin and evolution of high throughput screening. *Br J Pharmacol*. 2007;152(1):53–61.
86. Hewitt WM, Leung SSF, Pye CR, Ponkey AR, Bednarek M, Jacobson MP et al. Cell-permeable cyclic peptides from synthetic libraries inspired by natural products. *J Am Chem Soc*. 2015;137(2):715–21. doi:10.1021/ja508766b.
87. Milton Harris J, Martin NE, Modi M. Pegylation: a novel process for modifying pharmacokinetics. *Clin Pharmacokinet*. 2001;40(7):539–51.
88. Jonassen I, Havelund S, Hoeg-Jensen T, Steensgaard DB, Wahlund PO, Ribel U. Design of the novel protraction mechanism of insulin degludec, an ultra-long-acting basal insulin. *Pharm Res*. 2012;29(8):2104–14.
89. Sievers EL, Senter PD. Antibody-drug conjugates in cancer therapy. *Annu Rev Med*. 2013;64:15–29.
90. Benizri S, Gissot A, Martin A, Vialet B, Grinstaff MW, Barthélémy P. Bioconjugated oligonucleotides: recent developments and therapeutic applications. *Bioconjug Chem*. 2019;30(2):366–83. doi:10.1021/acs.bioconjchem.8b00761.
91. Khvorova A, Watts JK. The chemical evolution of oligonucleotide therapies of clinical utility. *Nat Biotechnol*. 2017;35(3):238–48.
92. Anselmo AC, Mitragotri S. Cell-mediated delivery of nanoparticles: taking advantage of circulatory cells to target nanoparticles. *J Control Release*. 2014;190:531–541.
93. Song W, Anselmo AC, Huang L. Nanotechnology intervention of the microbiome for cancer therapy. *Nat Nanotechnol*. 2019;14(12):1093–103. https://www.nature.com/articles/s41565-019-0589-5.
94. Xu X, Ho W, Zhang X, Bertrand N, Farokhzad O. Cancer nanomedicine: from targeted delivery to combination therapy. *Trends Mol Med*. 2015;21(4):223–32.
95. Taniguchi C, Kawabata Y, Wada K, Yamada S, Onoue S. Microenvironmental pH-modification to improve dissolution behavior and oral absorption for drugs with pH-dependent solubility. *Expert Opin Drug Deliv*. 2014;11(4):505–16.
96. Wan C, Allen TM, Cullis PR. Lipid nanoparticle delivery systems for siRNA-based therapeutics. *Drug Deliv Transl Res*. 2014;4(1):74–83.
97. Chen S, Zaifman J, Kulkarni JA, Zhigaltsev I V., Tam YK, Ciufolini MA et al. Dexamethasone prodrugs as potent suppressors of the immunostimulatory effects of lipid nanoparticle formulations of nucleic acids. *J Control Release*. 2018;286:46–54.
98. Scarfò I, Maus MV. Current approaches to increase CAR T cell potency in solid tumors: targeting the tumor microenvironment. *J Immunother Cancer*. 2017;5(1):28.
99. Tiwari G, Tiwari R, Bannerjee S, Bhati L, Pandey S, Pandey P et al. Drug delivery systems: An updated review. *Int J Pharm Investig*. 2012;2(1):2.
100. Kamaly N, Yameen B, Wu J, Farokhzad OC. Degradable controlled-release polymers and polymeric nanoparticles: Mechanisms of controlling drug release. *Chem Rev*. 2016;116(4):2602–63. doi:10.1021/acs.chemrev.5b00346.
101. Awwad S, Angkawinitwong U. Overview of antibody drug delivery. *Pharmaceutics*. 2018;10(3):83.
102. Vegas AJ, Veiseh O, Gürtler M, Millman JR, Pagliuca FW, Bader AR et al. Long-term glycemic control using polymer-encapsulated human stem cell–derived beta cells in immune-competent mice. *Nat Med*. 2016;22(3):306–11. https://www.nature.com/articles/nm.4030.
103. Youn YS, Bae YH. Perspectives on the past, present, and future of cancer nanomedicine. *Adv Drug Deliv Rev*. 2018;130:3–11.
104. Pasut G. Grand challenges in nano-based drug delivery. *Front Med Technol*. 2019;1:501667.
105. Sharma S, Parveen R, Chatterji BP. Toxicology of nanoparticles in drug delivery. *Curr Pathobiol Rep*. 2021;9(4):133.
106. Ashfaq R, Rasul A, Asghar S, Kovács A, Berkó S, Budai-Szűcs M. Lipid nanoparticles: an effective tool to improve the bioavailability of nutraceuticals. *International journal of molecular sciences*. 2023;24(21):15764.
107. Singh N, Joshi A, Toor AP, Verma G. Drug delivery: advancements and challenges. *Nanostructures Drug Deliv*. 2017;865–86.
108. Perepelytsina OM, Ugnivenko AP, Dobrydnev AV, Bakalinska ON, Marynin AI, Sydorenko MV. Influence of carbon nanotubes and their derivatives on tumor cells in vitro and biochemical parameters, cellular blood composition in vivo. *Nanoscale Res Lett*. 2018;13(1):286.
109. Zare H, Ahmadi S, Ghasemi A, Ghanbari M, Rabiee N, Bagherzadeh M et al. Carbon nanotubes: smart drug/gene delivery carriers. *Int J Nanomedicine*. 2021;16:1681–706. https://www.dovepress.com/carbon-nanotubes-smart-drug-gene-delivery-carriers-peer-reviewed-fulltext-article-IJN.
110. Spitler R, Zanganeh S, Jafari T, Khakpash N, Erfanzadeh M, Ho JQ et al. Drug delivery systems: Possibilities and challenges. *Drug Deliv Syst*. 2017;1–52. www.worldscientific.com.
111. Bartucci R, Paramanandana A, Boersma YL, Olinga P, Salvati A. Comparative study of nanoparticle uptake and impact in murine lung, liver, and kidney tissue slices. *Nanotoxicology*. 2020;14(6):847–65.

3 Bridging the Gap
Translational Research in Drug Delivery

Gaurav Kumar[1,2], *Smita Paul*[1], *and Ravindra K. Rawal*[1,2]
[1]Natural Product Chemistry Group, Chemical Science and Technology Division, CSIR-North East Institute of Science and Technology, Jorhat, Assam, India
[2]Academy of Scientific and Innovative Research (AcSIR), Ghaziabad, Uttar Pradesh, India

3.1 INTRODUCTION

In the modern healthcare sector, drug delivery is considered a critical aspect. A drug delivery system (DDS) is the process by which a therapeutic substance is transferred to the body to enhance its efficacy by regulating the rate, timing, and drug-release site. DDS acts as a mediator between the patient and the drug, which involves both therapeutic drug formulations and devices employed for drug delivery [1]. DDS are valuable tools for improving therapeutic agent uptake at the target site, especially when free medicines have substantial off-site toxicities or poor pharmacokinetics/biodistribution (PK/BD) [2]. DDS has become more complex with the development of technology, which has complicated the processes involved in drug absorption and disposal after administration. To achieve effective drug delivery, it is essential to comprehend the pharmacokinetic and pharmacodynamic aspects of drugs. Pharmacodynamics focuses on the physiological and biochemical effects of drugs, whereas pharmacokinetics deals with the ADMET properties of drugs (absorption, distribution, metabolism, excretion, and toxicity) within the body. Variability in these parameters among individuals can significantly impact drug efficacy and safety. Due to the significant differences in the processes that influence the pharmacokinetics of conventional medicines and DDSs, a thorough understanding of the in vivo behavior of the carrier is essential for the successful clinical translation of DDSs. Developing methods to improve drug absorption and disposition for the best possible therapeutic outcomes is the main aim of the DDS [3]. Furthermore, DDS materials' biocompatibility and toxicity require comprehensive biocompatibility evaluations. Despite their potential, active targeting techniques encounter challenges such as immunogenicity, ligand selectivity, and lysosomal enzyme degradation. Considering these hurdles, there is a critical need for translational research (TR) to bridge the gap between laboratory findings and clinical use.

3.1.1 Overview of Drug Delivery System

DDS goes over several drug delivery methods, formulations, and devices, including tailored delivery. Delivering biotechnology-based treatments, like proteins, peptides, and cell and gene therapies, presents both obstacles and opportunities. Drug administration is conducted through various anatomical routes for systemic and targeted drug delivery, including oral, transdermal, parenteral, transmucosal, nasal, pulmonary, and rectal routes. Each route has its advantages and limitations. Various routes include oral, intravenous, intramuscular, transnasal, transdermal, and pulmonary. Each route is compared based on factors like onset of action, bioavailability, dose control, patient convenience, and potential adverse effects. The commercial aspects, regulatory controls, and prospects of DDSs, aiming for more refined and effective drug delivery methods, are also significant [4].

The technology for controlled drug administration has advanced over the course of six decades, beginning with the first sustained-release formulation created in 1952. Oral and transdermal systems were the primary focus of the first generation (1950–1980), but the second generation (1980–2010) incorporated self-regulated delivery, zero-order release systems, and systems based on nanotechnology. The Journal of Controlled Release (JCR) founded in 1984, is a premier forum for scientists working in drug delivery, focusing on high-quality research and a significant contribution to the advancement of DDS. The third generation of drug delivery technologies is focused on creating future formulations that address current requirements in a creative manner [5].

Long-term formulations and new in vitro testing methods are being explored to improve targeted delivery. Some examples of nanotechnology-based DDS are nanoparticles used for targeted drug delivery, such as anticancer drugs and siRNA delivery to target cells. Polymeric micelles are being used for drug targeting, especially in cancer treatment, such as PEGylated liposomes. For example, the formulation of doxorubicin is designed to reduce side effects. Dendrimers are utilized in drug delivery due to their unique structure and ability to transport multiple drug molecules simultaneously. These technologies are intended to increase drug targeting and efficiency, especially in cancer treatment [5]. As an alternative to oral and hypodermic injection techniques, transdermal drug delivery comes with advantages like non-invasiveness and improved patient compliance. Current limitations include difficulty delivering hydrophilic drugs and macromolecules through the skin. Emerging technologies like microneedles and thermal ablation show

DOI: 10.1201/9781003508601-3

promise for delivering a wider range of drugs, including vaccines and macromolecules [6]. Anatomically, systematic drug delivery has several roots: oral, gastrointestinal, rectal, and parenteral.

3.2 CHALLENGES IN DRUG DELIVERY SYSTEMS

Emphasis on training clinician-scientists and basic scientists, with mentoring programs to support early-career researchers. Translational research (TR) plays an important role in closing the knowledge gap between fundamental science and clinical practice, ultimately improving patient health. However, several barriers hinder practical TR efforts like cultural differences. Basic scientists and clinicians often have different perspectives, goals, and communication styles. For bridging this gap, fostering collaboration, understanding, and shared objectives is required. Resource limitations are observed due to an inadequate workforce and infrastructure hindering TR. Lack of interdisciplinary staff trained in TR reduces the pace of progress. Clinicians and researchers need protected time for research alongside their clinical and other duties. Regulatory hurdles, such as ethics approvals, intellectual property rights, and manufacturing regulations, are daunting tasks. Cell and gene therapies add further complexity to TR. There is a critical gap between preclinical studies and clinical trials that delays translating promising discoveries into effective treatments.

The UK's National Institute for Health Research (NIHR) created the Integrated Research Application System (IRAS) in 2008. IRAS is an online system that gathers data from a variety of review bodies, such as NHS R&D offices, research ethics committees, and regulatory bodies like the Medicines and Healthcare Products Regulatory Agency (MHRA), that are needed for applications requesting research approval. Additionally, the UK developed the Research Passport system, which makes it easier for researchers who aren't working for the NHS to get honorary research contracts. The NHS established the Health Research Authority (HRA) in December 2011, and it is a crucial component of research governance, significantly streamlining the approval process. The goal of the National Centre for Advancing Translational Sciences (NCATS), based in the USA, founded in 2012, is to improve the creation, evaluation, and use of treatments and diagnostics for a range of human illnesses. It aims to address regulatory bottlenecks and make translational research more efficient, cost-effective, and less risky. These initiatives aim to facilitate research collaboration, integrate scientific ideas into clinical practice, and promote the translation of scientific discoveries into effective treatments.

Nanotechnology creates and deploys nanoscale structures with novel properties, impacting various scientific disciplines globally. Advances in optics, photonics, plasmonics, and electronics, including new biophysics techniques and neural stimulation prosthetics. Innovations in wireless electronic nano-biosensors, tissue growth, infection prevention without antibiotics, and smart sensors for medical implants. Nanotechnology faces significant hurdles in transitioning from research to commercial products. The safety of nanomaterials and the potential health and environmental risks they pose are significant challenges, and there is no standard paradigm for evaluating their safety. While regulations are evolving to address these concerns, striking a balance between safety and innovation is crucial. While organizations such as the FDA and EPA are assessing nanomaterials for environmental sustainability and safety, obstacles remain in creating and implementing scientific data to ensure their safety. Developing effective regulatory oversight that does not hinder research and development is essential. The Rule of Five, which is used to predict the safety of low-molecular-weight therapeutics, does not apply to nanomaterials. Most nanomaterials violate this rule, posing challenges to their development and commercialization. Additionally, the unique properties of nanomaterials often don't align with traditional drug development models, necessitating new approaches for evaluating their safety and efficacy. Regulations have the power to encourage responsible technology use or to hinder new ideas. Effective regulation aims to identify potential risks without causing unnecessary delays or costs. Ultimately, a collaborative effort involving researchers, manufacturers, and regulators is required to establish guidelines for responsible nanotechnology development and ensure its benefits are realized while minimizing risks.

3.3 NANOTECHNOLOGY IN DRUG DELIVERY SYSTEMS

Nanotechnology transforms various scientific disciplines, creating nanoscale structures with novel properties and functions [7]. The field of nanomedicine oversees the creation of focused pharmaceutical treatments for illness prevention and therapy [8]. The sizes of nanoparticles range from 10 to 100 nm. According to Mohanraj and Chen [9], the active medicinal components are dissolved, encapsulated, connected, or trapped within the nanocarriers. The objectives of nanodrug delivery systems are to ensure targeted drug activity and dose efficacy by regulating particle size, enhancing surface features, drug transport, and drug release. The use of nanodrugs offers several advantages, including their small size and frequent use of biodegradable components [10,11]. Since drug nanoparticles have a large surface area and a smaller particle size, they are readily soluble and bioavailable [11]. They are easily absorbed by the tight connection in skin endothelial cells, allowing them to pass the blood-brain barrier. These particles' nanoscale allows maximum absorption by different cell types and the medication's accumulation in the intended location [12,13]. They are resistant to the degradation of the drug brought on by localized endogenous enzymes [14]. To increase effectiveness and decrease instability, nanodrug delivery systems are needed. Drug formulation and delivery methods based on nanotechnology are extensively used (Table 3.1).

TABLE 3.1
Nanotechnology-Based Drug Delivery Systems Utilized in Various Diseases

Disease	Type of Nanocarrier	Nanomedicines/ Formulation	Study/Outcome	References
HIV	PLGA nanoparticle	Water-oil-water emulsion of PLGA nanoparticles	Extended release of Lopinavir (LPV) in Balb/c mice.	[15]
Tuberculosis	ZnO nanoparticles	ZnO nanoparticles from canthiumdicoccum	ZnO-NPs suppressed *M. tuberculosis.*	[16]
	Selenium nanoparticles	Isoniazid–selenium nanoparticles	Mannosylated selenium nanoparticles related to isoniazid-induced Mtb autophagy sequestration, which developed into lysosome-associated autophagosomal Mtb destruction coupled with PI3K/Akt/mTOR and ROS-mitochondrial signaling pathways.	[17]
	TiO_2 nanoparticle	Titanium dioxide (TiO_2) nanoparticles in aqueous form	Mycobacterial metabolic activity was reduced by fourfold, and impacts biofilm formation.	[18]
Cancer	Gold nanoparticles	DNA-stabilized GNP loaded with doxorubicin	The formulation showed enhanced reduction in cell viability in uveal melanoma and breast cancer cell lines. As a result, this formulation can be used to enhance the effectiveness of cancer treatment.	[19]
	SPION coated with PEG	Iron oxide NPs loaded with paclitaxel	In live mice, this formulation demonstrated excellent stability in biological media. Human HER2/neu-positive breast cancer cells demonstrated efficient cellular targeting and selective death.	[20]
	Iron oxide magnetic nanoparticles	Gemcitabine	This formulation is used in the treatment of pancreatic cancer. Selective release in an intracellular environment.	[21]
Cardiovascular disease	Carbon nanotubes		Improves the function of cardiomyocytes and is under development for cardiac tissue engineering.	[22]
	Carbon nanocapsules	Heparin-conjugated carbon Nano capsules	Demonstrate increased antithrombotic activity and prolong the thrombus's development period.	[23]
	Nano patch	hiPSC-derived cm sheets	Improves cardiac functions like left ventricle ejection fraction and volumes in a porcine ischemic cardiomyopathy model.	[24,25]
Covid-19	Magnetic nanoparticles	Ferromagnetic Fe_3O_4 nanoparticles	Weakening the integrity of viral surface proteins, such as HA, neuraminidase, and matrix protein, causes lipid peroxidation in the viral envelope.	[26]
	AgNP	Curcumin-modified silver nanoparticles	Impede the respiratory virus's ability to enter cells by possessing abundant and potent cell numbers.	[27]
Lung disease	Polymeric micelles	Genexol-PM, paclitaxel is encapsulated into nanosized polymeric micelles	Patients with non-small-cell lung cancer (NSCLC) can receive higher doses of paclitaxel without experiencing increased toxicity from their treatment.	[28]
	Nanovesicle	Oncoprex, a cholesterol-based nanovesicle encapsulating the TUSC2 gene	Expresses tumor suppressor candidate.	[29]
	Liposomal nanoparticles	Amikacin-loaded liposome	Combats a wide range of gram-negative bacteria, treats chronic severe lung infections with multidrug-resistant (MDR), extends therapeutic half-life, and minimizes side effects.	[30]

(Continued)

TABLE 3.1 (*Continued*)
Nanotechnology-Based Drug Delivery Systems Utilized in Various Diseases

Disease	Type of Nanocarrier	Nanomedicines/ Formulation	Study/Outcome	References
Urology	Polymeric micelles	SN38-incorporated polymeric micelles ($NK0_{12}$)	Treatment of renal cell carcinoma in human renal cancer cells SKP-9 model and murine Renca cells.	[31]
	Gelatin nanoparticles	Paclitaxel-loaded gelatin nanoparticles	In bladder cancer, it shows higher efficacy, targeted delivery, and higher drug concentration.	[32]
Osteoporosis	Nano-hydroxyapatite (HA)	Crystals of nano-hydroxyapatite	Enhance the bioavailability of calcium supplements, reducing osteoporosis risk and improving bone formation.	[33]
	ZrO_2 & $BaSO_4$ nanoparticle	ZrO_2 (zirconium dioxide) and $BaSO_4$ (barium sulfate) into PMMA (polymethylmethacrylate)	Enhance osteoblast adhesion, which is crucial for the integration of the cement with bone tissue, the material's stiffness and mechanical strength positively.	[34,35]
OrthopEDIC	Nanofiber	Nanoscopic drug delivery of nanofiber poly-L-lactic acid (PLLA)	Bone healing.	[36]
Obesity	Copolymer nanoparticle	PLGA-b-PEG copolymer nanoparticles	Deliver rosiglitazone and prostaglandin E2 analog to promote the browning of white adipose tissue.	[37]
	DBZ-nanoparticle	DBZ-loaded PLGA nanoparticles	Stimulate browning in white adipose tissue with localized sustained release.	[38]
	Rosiglitazone nanoparticle	Microneedle patches with rosiglitazone nanoparticles	Enhance insulin sensitivity, promote energy expenditure, and permit local browning of white adipose tissue.	[37]
Reproductive health	Gold nanoparticles	Gold nanoparticles	An in vivo investigation using a male rat model showed that female impregnation was inhibited for over 2 months while near-infrared radiation made it possible for the rats to become pregnant.	[39]
	Iron oxide nanoparticles	Human mesenchymal stem cell-labeled superparamagnetic iron oxide nanoparticles (SPION-MSCs)	Efficient recovery of erectile dysfunction in rats.	[40]
Neurological disorders	Solid lipid nanoparticles	BEO loaded in a nanotechnological odorless cream	Efficient transdermal release of the Bergamot essential oil. Effective in inflammatory and neuropathic pain.	[41]
Pain control	PLGA nanoparticles	Tramadol hydrochloride (TrHC) loaded PLGA nanoparticles	Enhance pharmacological effects, sustain release, and improve antinociceptive performance.	[42]
	Gold-coated Fe_3O_4 nanoparticles	Intrathecal gold-coated Fe_3O_4 nanoparticles	The formulation used as targeted delivery in the spinal canal showed promising results for treating chronic pain and targeting tumors.	[43]
Dry eye disease	Nanomicells	Aqueous nanomicellar solution of cyclosporine A	The formulation showed enhanced stability and sustained release in in vitro and in vivo studies.	[44]
	Nanoparticles	Nanoparticles coated with poloxamer 407 containing loteprednol etabonate	Mucus-penetrating nanoparticles.	[45]

3.3.1 Cancer

Histone alterations and other epigenetic changes, like DNA methylation, play important roles in regulating gene expression and the tumor microenvironment (TME). TME is made of different cells and molecules that promote tumor progression, metastasis, and recurrence. Epigenetic reprogramming of immune cells can affect tumor immunity. Combining epigenetic modulators with other cancer therapies (e.g., immunotherapy, chemotherapy) can improve antitumor efficiency and overcome drug resistance. Because tumor tissues have vasculature that leaks and poor lymphatic drainage, nanoparticles can aggregate there, thereby benefiting from the increased permeability and retention effect. They are modified with ligands or antibodies that specifically bind to overexpressed receptors on tumor cells, thereby enhancing their uptake. The therapeutic payload of nanoparticles is engineered to be released in response to stimuli, which include enzymes, pH, or redox conditions, within the tumor microenvironment. Using nanoparticles coated with cell-improving membranes that mimic the properties of tumor cells or immune cells makes it possible to improve the targeting and evasion of the immune system. Ligands associated with the active targeting of cancer cells are hyaluronic acid, which targets CD44 receptors in tumor cells; trastuzumab for HER2-positive cancer cells, substance P Peptide for NK-1R receptors in glioma, and lactoferrin, which targets LRP1 and secreted protein acidic and rich in cysteine (SPARC) on tumor cells and tumor-associated macrophages [46]. Glutathione [47] is overexpressed in tumor cells, causing resistance to radiotherapy. Various GSH-responsive nanomedicines and radiosensitizers have been developed, including NIR discrete metallic cycles and Bi_2Se_3 hollow nano-cubes. Precision radiotherapy, along with recent advancements in nanotechnology, has provided strategies for precise and efficient tumor eradication by image-guided precision radiotherapy and tumor-targeted delivery. Red blood cell (RBC) membrane coating, cancer cell membrane camouflage, and hybrid membrane coating improve the nanoreactor's targeting abilities, leading to a higher accumulation of tumors and improved treatment efficiency. Prolonged blood circulation enhances targeting, immune evasion, and blood circulation, leading to better treatment outcomes [48].

Nanoplatforms approved for breast cancer treatment primarily utilize organic nanoparticles due to their advantages like biodegradability, biocompatibility, and low polydispersity, which enhance gene and drug delivery over a prolonged period. These include natural or synthetic polymers, dendrimers, and lipid-based materials. Inorganic nanoplatforms, like silica, gold, and iron oxide nanoparticles, offer additional advantages with versatile properties and functions. Gold nanoparticles provide optical and thermal properties, silica nanoparticles feature nanometer-scale pores for high molecule encapsulation, and iron oxide nanoparticles have superparamagnetic properties. These inorganic platforms are explored for gene therapy, drug delivery, tumor ablation, and imaging.

Nanoparticles like liposomes, polymer-derived nanomaterials, etc, are being explored as delivery vehicles for chemotherapy drugs and RNA interference [49] inhibitors in pancreatic cancer. They enhance drug solubility, stability, and tumor bioavailability, while reducing off-target toxicity. Examples include Doxil, gemcitabine-loaded nanoparticles, and Nab-paclitaxel. Nanoparticles also promise to enhance RNAi delivery and targeting of tumor stroma [50].

3.3.2 Lung Disease

Globally, lung disorders are considered one of the significant causes of death. As estimated, it is responsible for one in five fatalities worldwide by 2030. Nanomedicine is utilized for drug delivery, vaccine development, and disease diagnosis, with promising outcomes in conditions such as chronic obstructive pulmonary disease (COPD), asthma, lung cancer, cystic fibrosis, and bacterial infections. Nanoparticles enhance drug stability, prevent premature degradation, and enable targeted administration to specific disease locations, thereby improving treatment effectiveness and reducing adverse effects. Nanotechnology improves the delivery and effectiveness of asthma medications like salbutamol, enhancing lung deposition and controlled release utilizing liposomes, niosomes, nanoemulsions, etc. Recent advancements in transnational respiratory research (TRR) have identified new therapeutic agents such as mi-RNAs, monoclonal antibodies, and si-RNAs. These developments aim to improve the effectiveness of treatments for respiratory conditions by addressing issues like biofilms. Nanotechnology plays a significant role in these advancements, enhancing patient compliance and offering possible applications in tissue engineering, tumor killing, drug delivery, regenerative medicine, and gene delivery. There aren't many studies on the effects of biocompatible nanoparticles (NPs) with anti-IL4Rα antibodies to diminish asthmatic inflammation or antigen-conjugated polystyrene NPs to suppress Th2 responses. In COPD, the development of microparticulate/nanoparticulate powders for aerosol delivery targets the Nrf2/Keap-1 pathway to reduce inflammation. These are a few highlights of nanomedicines in respiratory diseases.

3.3.3 COVID-19

The rapid development and approval of nanotechnology-based mRNA vaccines against SARS-CoV-2 have significantly advanced the field of nanomedicine. The success of these vaccines has streamlined the path to clinical translation for medication delivery systems in nanomedicine, demonstrating their necessity and effectiveness. Nanotechnology has a vital role in the prevention, diagnosis, treatment, and vaccination against COVID-19. It offers benefits like improved drug delivery and nanosensors [51]. Quantum Dot antigen-presenting cells (APCs) can significantly enhance T-cell proliferation and have applications in diagnosing and detecting viruses like influenza A and coronaviruses. Nanoparticles can enhance vaccine efficacy by improving antigen presentation and stability and carrying adjuvants to boost immune responses. Different nanoparticles (e.g., gold, silver, and selenium) could block viral attachment, replication, and inactivate viruses, showing promise against various viral infections [52]. Targeting COVID-19 binding, entrance, replication, and budding is possible with nanoparticles [51].

3.3.4 Cardiovascular

Globally, cardiovascular illnesses are the primary cause of both fatality and morbidity, with inadequate proliferation of mature cardiomyocytes leading to long-term heart damage. Cardiac fibrosis involves excessive buildup of heart-related extracellular matrix components, which results in diminished heart function and heart failure. Developing nanotechnology-based treatments presents challenges, including ensuring safety, biocompatibility, and adherence to regulations. Dendrimers are perfect for targeted drug administration because they provide exact control over size, surface functionality, and drug-loading capacity, especially in cardiovascular diseases like cardiac fibrosis. Micelles, used in cardiovascular treatment and composed of amphiphilic molecules, are effective for encapsulating and delivering hydrophobic drugs, showing promise in treating conditions like atherosclerosis and cardiac fibrosis. Efforts to improve nanoparticle biocompatibility include surface modifications, material selection, and size optimization, ensuring their safe use in biomedical applications [53]. Stem cell therapy has generated significant interest in cardiovascular disease. Current methods for delivering stem cells to the heart are inefficient, with low cell retention and survival rates. Nanotechnology offers new opportunities to improve stem cell delivery, tracking, and engraftment. Engineered nanostructures can mimic the natural environment, promoting tissue regeneration and cell differentiation. Cardiac myocytes accomplish the role of local signaling in nano-domains. Stem cells cover the applications of micro and nanoscale three-dimensional cardiac stem cell culturing systems [54].

3.3.5 Dementia

Neurodegenerative diseases are most commonly found in the aging population. Fifty million people are suffering from dementia. It is frequently observed in people aged 65 or older. There are several comorbidities also associated with it, like osteoarthritis and diabetes-associated neurological pain. Research is being conducted on the effectiveness of bergamot essential oil (BEO) free of furocoumarins in treating agitation in elderly individuals with severe dementia, utilizing a nanotechnology-based delivery method. A clinical investigation has been conducted to thoroughly assess the safety and effectiveness of an essential oil with potent analgesic properties for treating agitation in patients with advanced dementia. A Bergamot essential oil-based formulation for transdermal applications has been developed as an orderly cream using a nanotechnology-based delivery method [41]. Neuroinflammation is a natural response to injury taken by the brain. If it remains untreated, the neuroinflammation may result in loss of function and aging-related brain cell damage. Potential applications of nanotechnology include the identification and treatment of neurological diseases affecting the central nervous system (CNS). Where microglial targeting is important. With the use of nanotechnology, diseases like Parkinson's and Alzheimer's can be addressed. Nanotechnology-mediated cell therapy involves the use of nanodiamonds, which can be delivered to tumor cells through autophagy and subsequently induced by dendritic cells and macrophages, potentially enhancing GBM immunotherapy. In nanoparticle-based gene delivery, PBAE-BPN nanocomplexes have been reported to result in significant lifespan increases in rat brain cancer models, highlighting the potential of convection-enhanced delivery (CED) for effective gene delivery. Nano-miR treatments targeting cancer stem cells (CSCs) can significantly reduce tumor size and increase lifespan in GBM models. Nanopatterned scaffolds and nanobioglass mats show promise in promoting nerve regeneration and controlled degradation for spinal cord injuries [55].

3.3.6 Human Immunodeficiency Virus

Although a cure for HIV is unavailable, anti-retroviral therapy has proved its role in containing HIV or AIDS-related morbidities. There are a few medicines used to treat this adversity, like protease inhibitors, reverse transcriptase inhibitors, antagonists for CCR5, integrase, and fusion inhibitors. They have significantly cut down on mortality and morbidity caused by HIV. Several anti-retroviral drugs metabolized by the cytochrome P450 pathway possess drawbacks such as poor absorption and issues with drug-drug interactions. Conventional anti-retroviral drugs also face short half-lives, necessitating alternative delivery methods. Nanotechnology offers a promising alternative for delivering antiretroviral drugs (ARVs), potentially improving bioavailability, reducing toxicity, and providing targeted delivery. Inhibitors of Non-nucleoside reverse transcriptase like nevirapine, efavirenz, and etravirine are metabolized by enzymes such as CYP3A4, CYP2B6, and CYP2C191. Efavirenz and etravirine can both induce and inhibit these enzymes. Various non-nucleoside reverse transcriptase inhibitors have been formulated into nanoparticles to enhance drug delivery and efficacy, including efavirenz, etravirine, Nevirapine, Delavirdine, and Rilpivirine. Nano formulations improve drug levels, reduce toxicity, and strengthen protection against HIV infection [56]. HIV-1 can infect macrophages (MΦ) and monocytes, resulting in significant changes to their function and phenotype, which in turn affect innate immunity. HIV-1 replication is more frequent in differentiated macrophages than monocytes, with various factors influencing infection rates and viral replication. HIV-1 infection alters cytokine responses, phagocytosis, and apoptotic responses in macrophages, demonstrating their role in immune defense. HIV-1-infected macrophages often exhibit resistance to apoptosis, which may contribute to the persistence of the virus [57].

3.3.7 Eye Disease

Dry eye disease is a growing health issue with a 5%–50% worldwide prevalence. It is characterized by instability of the tear film, inflammation of the ocular region, and neurosensory abnormalities. Current treatments have low drug bioavailability due to ocular barriers, rapid elimination, and physiological barriers, resulting in the target tissues receiving less than 5% of the administered dosage. Research focuses on developing multifunctional nanosystems, such as nanoemulsions, liposomes, dendrimers, and hydrogels, to enhance drug bioavailability and efficacy. Ocular DDSs together with nanoparticles, liposomes, & micelles have been developed [58].

3.3.8 Orthopedic and Prosthetic

DDS based on nanotechnology have demonstrated decreased side effects and increased treatment efficacy, proving them extremely advantageous in the field of orthopedics. Due to the high ratio of surface area to volume and nanoscale dimension of these materials, DDS based on nanotechnology provides better targeting, decreased toxicity, increased bioavailability, and effective drug delivery. One prominent reason why total joint replacements (TJRs) fail is aseptic loosening. The improved fusion of the implant and the bone, or osseointegration, can reduce the likelihood of implant failure. Although the surfaces of contemporary orthopedic implants are smooth on a nanometric scale, surfaces with nanotexturing may improve osteoblast function and decrease fibroblast activity. A profound side effect of total joint arthroplasty (TJA) with significant morbidity and financial implications is periprosthetic joint infection (PJI). Selenium, silk fibroin, and silver nanoparticles all show promise in stopping bacterial colonization on implants. Drug-loaded nanomolecules and various carrier materials (such as titanium and chitosan) enable long-lasting antibiotic release and targeted medication delivery. Although research is still needed to address potential toxicities, nanotechnology has the potential to transform orthopedic disease prevention and therapy, even though it is still in its early phases [59].

3.3.9 Reproductive Disorders

Human reproductive tract-related functions, as well as associated concerns, are the focus of reproductive medicine. Promising applications of nanotechnology include controlling infertility and other reproductive health disorders, preserving fertility, and diagnosing cancer. Most nanotechnology-related clinical trials are still in the early phases, meaning that improved collaboration across different scientific domains is necessary for successful clinical translation. Nanotechnology breakthroughs have led to the development of vaccinations and biodegradable polymers, offering novel forms of contraception. Techniques such as vaccinations based on nanoparticles and male contraceptive effects through photothermal effects. Solutions for enhancing sperm quality and identifying anomalies in semen can be found in nanotechnology [60].

3.3.10 Obesity

The American Medical Association has identified obesity as a chronic condition with severe comorbidities, including type-II diabetes and cardiac issues. Obesity is a global health concern. Biological barriers, vascular permeability, and regulatory obstacles are among the issues caused by obesity. Current therapies consist of medication, bariatric surgery, and lifestyle modifications; however, all have drawbacks and adverse consequences. By improving medication transport, encapsulation, and targeting, nanotechnology offers intriguing solutions that may help overcome the limitations of existing therapies. Because polyethylene glycol–poly lactic acid-co-glycolic acid (PEG-PLGA) nanoparticles escape the reticuloendothelial system, they have potential use in medication administration [61].

3.3.11 Urology

Nanoparticles enhance non-invasive tumor imaging, increasing sensitivity and specificity in detecting cancers like renal and prostate cancer. Nanotechnology allows rapid and accurate detection of single nucleotide polymorphisms (SNPs) in genes related to diseases like polycystic kidney disease. Nanotechnology-based sensors can quickly and sensitively detect bacteria in kidney infections. Lipid-based nanoparticles that are linked with folate are explored for targeted gene delivery. Nanotechnology promises advancements in continuously functioning artificial nephron systems. Nano-structured polymers and PLGA scaffolds are evaluated for bladder tissue replacement applications. The artificial nephron system, also known as the human nephron filter, is a promising development in nanotechnology for treating end-stage renal disease [62].

3.3.12 Pain Control

The advancement of nanomedical applications in pain management is enhancing the treatment provided to patients experiencing pain. Biomarkers in pain medicine involve increased plasma. IL-6 is associated with painful osteoarthritis. High-sensitivity C-reactive protein and the degradation of C-reactive protein by matrix metalloproteinase are further indicators. Serum biomarkers that may be useful have been discovered for individuals with lower back pain. Increased levels of the cytokine IL-6 and the neuropeptide substance P in cerebrospinal fluid are associated with fibromyalgia. These biomarkers aid in diagnosing and understanding pain conditions, potentially leading to more targeted treatments. Nanoparticles can deliver drugs precisely to targeted sites, reducing side effects and improving efficacy. Examples include TNF nanoplexes and Tramadol-loaded PLGA nanoparticles. Nanomedicines, such as Exparel (liposomal bupivacaine) and nanoparticle formulations of fentanyl citrate, have demonstrated improved performance in clinical trials [63] (Figure 3.1).

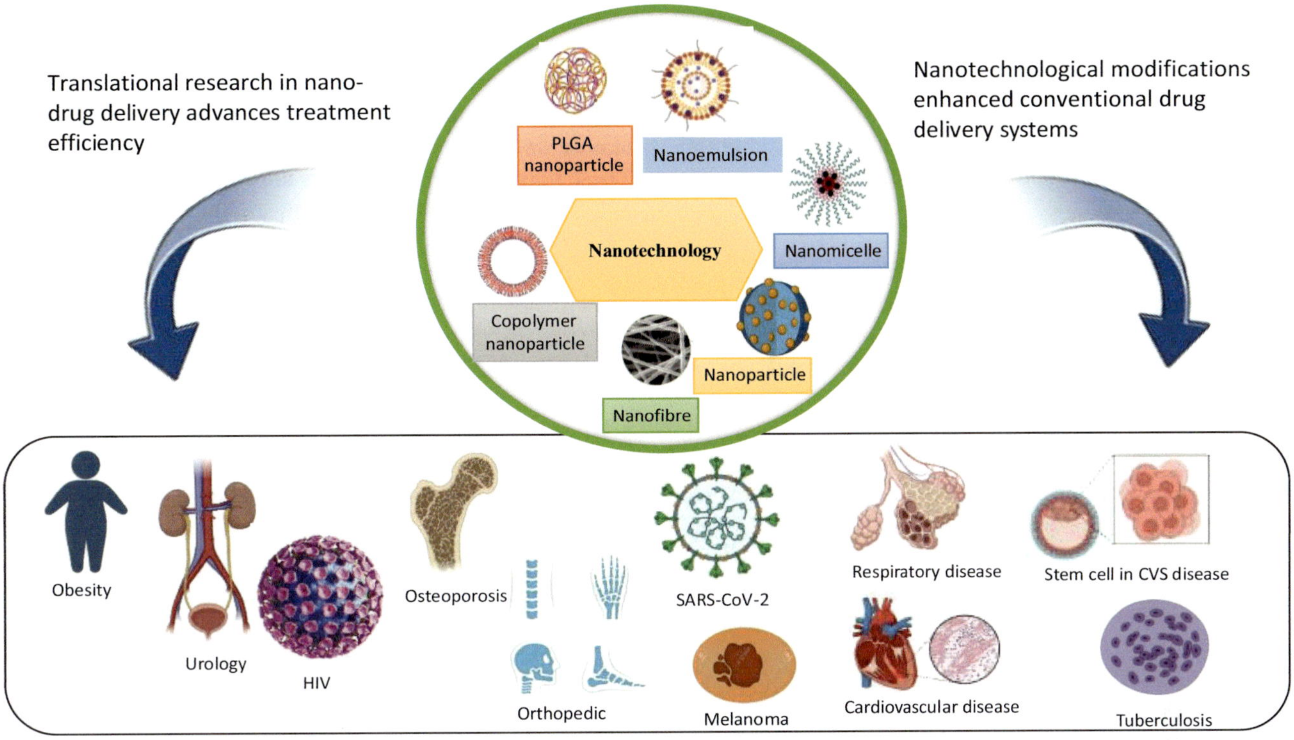

FIGURE 3.1 Nanotechnology drug delivery system targeting different diseases.

3.4 BIOMATERIALS FOR CONTROL RELEASE

Effective DDSs must be developed to improve patient outcomes in a range of medical situations, from acute infections to long-term illnesses. These results are strongly influenced by access to high-quality healthcare and preventive medicine, with differences observed across developed and resource-constrained settings [64]. The application of novel materials in medical technology, including polymers, nanomaterials, and metamaterials, is transforming healthcare. These materials enable breakthroughs in tissue engineering, drug delivery, and smart clinical devices, providing solutions such as biocompatible and biodegradable structures, targeted DDSs, and bioactive coatings for implants. A crucial factor in the design of biomaterials is particle size; nanoparticles are preferred due to their enhanced bioavailability, substantial drug-loading surface area, biocompatibility, and biodegradability, which reduce the need for surgical removal [65]. Carriage mechanisms, such as entrapment, attachment, and encapsulation, play essential roles in promoting drug stability, aiding localization at the target site, and ensuring efficient and targeted drug delivery. These mechanisms offer a complex approach to drug administration, as they are designed with biocompatibility, biodegradability, and responsiveness to environmental stimuli. Beyond the field of medicine, biomaterials have a significant impact on the development of biotechnology and nanotechnology, which makes them essential to contemporary research and invention [66].

Biomaterials play a pivotal role in designing DDSs, as they provide a regulated and targeted release of pharmaceutical compounds. This technique enables prolonged drug release, reduces adverse effects, and increases treatment efficacy, all of which are very advantageous when treating chronic illnesses. For instance, using biodegradable polymer nanoparticles, drugs can be delivered directly to the body's site, ensuring a sustained release. These systems require materials like polymers, lipids, and other biomaterials as they offer the needed structure and functionality to encapsulate and regulate the release of therapeutic substances [67]. For the development of regenerative medicine and tissue engineering, biomaterials play a crucial role, as they provide the scaffolds necessary for cell development and tissue regeneration. The scaffolds are made of biocompatible materials like polymers, hydrogels, and ceramics that provide a structural framework to promote cell adhesion, proliferation, and differentiation, resulting in the development of functional tissues and organs like bones [68]. Biomaterials are also crucial for the development of implantable medical devices, including pacemakers, stents, and artificial joints. To prevent immunological reactions and ensure a smooth integration into the body's tissues, these materials need to be biocompatible. Titanium metals and biodegradable polymers have significantly improved the quality of life for patients requiring such devices [69]. Additionally, biomaterials play a pivotal role in diagnostic tools; for example, nanoparticles coated with specific protein molecules are employed in the early identification of disease. For instance, antibody-loaded magnetic nanoparticles can attach to cancer cells, facilitating their separation and identification [70]. In regenerative medicine, biomaterials are frequently utilized in stem cell therapy. These scaffolds direct stem cell differentiation into specific types of cells, potentially treating conditions such as neurodegenerative diseases and spinal cord injuries [31]. Moreover, novel biomaterials are used to generate three-dimensional cell cultures, including spheroids or organoids, which closely resemble the in vivo environment, thereby enhancing drug screening and testing accuracy [71]. Additionally, biomaterials are employed as contrast agents in ultrasound imaging and magnetic resonance imaging (MRI) procedures [72]. In vaccine development, particularly for mRNA vaccines like those for COVID-19, lipid nanoparticles act as carriers, protecting the fragile mRNA molecules and enabling their entry into cells to elicit an immune response [73].

Biobased materials obtained from biological sources, such as microbes, plants, or animals, are eco-friendly compounds that are vital to medical science and material engineering. They are divided into non-biodegradable and biodegradable categories. Biodegradable materials such as natural biopolymers like chitosan and collagen, and magnesium-based alloys are frequently utilized in orthopedic implants, which decompose over time, minimizing long-term bodily damage. Medical devices made of non-biodegradable materials, such as synthetic polymers and certain metals, offer long-term stability. Polybutylene succinate (PBS), poly-3-hydroxybutyrate-co-3-hydroxyvalerate (PHBV), and polyglycolic acid (PGA) are some biodegradable polymers utilized in packaging, medical instruments, and devices [74]. Biobased polymers that are not biodegradable, such as polylactic acid (PLA) and polyethylene (PE), are used in industrial settings where they do not break down easily in the natural environment [75]. Non-biobased polymers that are biodegradable, like polycaprolactone (PCL) and polybutylene adipate terephthalate (PBAT), are utilized in DDSs. Conversely, non-biodegradable polymers, such as polyethylene (PE), polyvinyl chloride (PVC), and polypropylene (PP), are widely used in various industries that do not naturally break down [76]. Biomaterials are used in medicine and biomedical engineering because of their unique features. Metals like titanium are utilized in implants, while natural polymers like collagen are employed in tissue engineering. Polymers such as PU, PE, and PEG are used in medical devices and for drug delivery [77]. Modern biomaterials, such as gold nanoparticles and bioresorbable magnesium alloys, are revolutionizing tissue engineering, imaging, and drug delivery by offering novel solutions to a range of healthcare issues.

3.4.1 Types of Biomaterials for Controlled Release

3.4.1.1 Polymer

Polymers are the most popular biomaterials used in controlled DDSs due to their versatility in mechanical and chemical composition. Polymers, both synthetic and natural, such as PLA, chitosan, and polyethylene glycol [78], are essential components of DDSs because they provide flexible platforms for encapsulation and regulated release. Because of PEG's well-known hydrophilicity and biocompatibility, various therapeutic medicines with variable release kinetics can be precisely customized. Through regulated rates of degradation, PLA, a biodegradable synthetic polymer, offers continuous medication release, limiting side effects and lowering the frequency of administration [67]. A naturally occurring polymer derived from chitin, chitosan offers a mucoadhesive, biodegradable alternative that enhances the controlled release and absorption of drugs across biological barriers. These polymer's characteristics are essential for maximizing drug-release profiles and ensuring efficient delivery of bioactive compounds with the least side effects. Applications like targeted anticancer drug delivery and chronic pain management greatly benefit from this. Additionally, by lowering dosage frequency, these polymers increase patient compliance and lessen the burden of treatment regimens, particularly for chronic diseases. The versatility of PEG, PLA, and chitosan has made it possible to precisely manage drug encapsulation and release, which has improved treatment efficacy, decreased side effects, and encouraged better adherence to the drug schedule, representing a revolutionary change in patient care [79]. These polymers will remain at the forefront of innovation in pharmaceutical science, offering adaptable solutions to challenging drug delivery problems and expanding the frontiers of contemporary medicine.

3.4.1.2 Lipids

In the pharmaceutical sector, lipid nanoparticles and liposomes are lipid-based nanocarriers that play an essential role in drug delivery. These carriers are designed to encapsulate hydrophobic drugs, overcoming challenges such as the solubility and stability of these therapeutic agents. Liposomes are utilized as drug delivery vehicles because of their versatility. Because of the lipid bilayer's similarity to biological membranes, liposomes are innately biocompatible, which lowers the possibility of adverse reactions and makes them an ideal choice for encapsulating drugs. Their versatility is further enhanced by the ability to modify the lipid content to attain the desired drug-release profile, which further complements their biocompatibility [80]. Liposomes incorporate hydrophobic drugs into their lipid bilayer, thereby increasing their solubility. This is an important benefit because many potent therapeutic compounds are hydrophobic by nature, which makes formulation and administration difficult. Furthermore, liposomes prolong the shelf life of the encapsulated medicine by preventing degradation and ensuring that the therapeutic amount is preserved until it reaches its intended location within the body [81].

Another important lipid-based nanocarrier is lipid nanoparticles. Usually made of lipids, they are smaller than liposomes and can improve the ability of medications to be loaded, especially those that are highly hydrophobic. These nanocarriers can be distributed and stored for a longer period due to their stability. Lipid nanoparticles offer several benefits, one of which is their capacity to provide controlled drug release. This is accomplished by adjusting the lipid content to optimize release kinetics. This characteristic is crucial for maintaining sustained drug concentrations in the bloodstream, thereby ensuring successful therapy. By functionalizing the surface with ligands or antibodies, the surface characteristics of a lipid nanoparticle can also be altered to enable targeted drug administration, thereby improving efficiency and precision while reducing off-target effects [82].

Lipid-based nanocarriers play a crucial role across many medical fields. As such, in oncology, lipid nanoparticles and liposomes enhance drug solubility, reduce systemic toxicity, and increase selectivity for cancer cells by optimizing the delivery of chemotherapeutic agents. They are also essential for the delivery of hydrophobic antimicrobial drugs used in the treatment of infectious disorders [83]. Additionally, these carriers are being employed to encapsulate and deliver nucleic acid-based medicines and genes, addressing specific challenges associated with these novel treatments [84]. These lipid-based nanocarriers have transformed the administration of hydrophobic drugs by improving solubility, stability, and therapeutic efficacy. They are crucial tools in the pharmaceutical sector because of their versatility, biocompatibility, and customizable qualities [85]. Lipid-based nanocarriers will continue to be the focus of drug delivery research, offering novel approaches for the better treatment of a range of diseases and enhancing the overall patient experience.

3.4.1.3 Hydrogels

Hydrogels are 3-D networks of hydrophilic polymers, which have enabled tremendous advancements in drug delivery and tissue engineering. These materials serve as versatile scaffolds in regenerative medicine, providing ideal conditions for the regulated release of drugs. The high-water content of hydrogels mimics the aquatic environment of living organisms, making them extremely biocompatible because they can reduce negative responses and inflammation when introduced into the body. For long-term applications, such as sustained drug delivery and tissue engineering, biocompatibility is crucial. Drug-release profiles can be precisely customized as hydrogels can be modified both in their chemical composition and physical attributes. This flexibility is essential for enhancing patient adherence, managing long-term illnesses, and minimizing adverse effects [86]. DDSs based on hydrogels can be designed for extended-release or single-dose administration to reduce systemic exposure and increase efficacy. This includes the targeted administration of anticancer drugs into tumor tissues. Hydrogels are essential scaffolds in tissue engineering because they replicate the extracellular matrix found in nature and aid in developing functional replacement tissues [68]. These biomaterials have been used by researchers to develop complex organs such as the liver and heart, as well as cartilage, bone, and skin. With the possibility of lab-grown, personalized organs and tissues, this breakthrough has the potential to revolutionize organ transplantation by addressing the lack of donors and lowering the risk of organ rejection [87]. Hydrogels represent a significant advancement in biomedical applications with their remarkable characteristics, biocompatibility, and controlled release capabilities. They will play a vital role in the future of the healthcare sector by offering novel approaches to complex medical issues and developing more efficient, personalized treatments through continuous research and innovation.

3.4.1.4 Nanoparticle

Polymeric nanoparticles are a crucial component of current DDSs due to their adaptability and flexibility. By choosing suitable polymers and formulations, a range of drugs, from hydrophobic to hydrophilic compounds, can be encapsulated. This flexibility is crucial for meeting the distinct needs of various therapeutic agents and applications. A significant advantage of polymeric nanoparticles is their ability to provide controlled drug release, ensuring sustained therapeutic levels over time and thereby improving efficacy while reducing side effects. This precise control over drug release is advantageous for customized medicine, which allows treatments according to specific patient needs. Moreover, imaging agents can be loaded with these nanoparticles, making them powerful tools in medical imaging. They enhance the sensitivity and accuracy of techniques like MRI and fluorescence imaging, enabling the detailed visualization of biological structures and pathological conditions. This ability is essential for early disease detection and monitoring, enabling healthcare professionals to gain clearer insights into patient health [88].

Inorganic nanoparticles, such as silver and gold nanoparticles, exhibit remarkable properties in various clinical applications. Gold nanoparticles are renowned for their surface plasmon resonance properties due to their remarkable tunability in light absorption and scattering. Due to these properties, these nanoparticles can enhance medical imaging, making them effective contrast agents in computed tomography (CT) scans, which provide high-resolution imaging essential for precise pathological and anatomical insights. Additionally, gold nanoparticles also stand out in targeted drug administration by functionalizing their surfaces with antibodies or ligands. This ensures precise distribution to target locations, maximizes therapeutic efficacy, and minimizes harm to healthy tissue. Silver nanoparticles are known for their antibacterial properties; therefore, they are utilized in wound care products, such as lotions and dressings, to manage infections by releasing silver ions under controlled conditions. Their flexibility in healthcare is shown by their ability to function as both infection control agents and specific drug delivery vehicles [89]. Nanoparticles, whether inorganic, such as silver and gold, for sophisticated therapy and imaging or polymeric for targeted drug delivery, mark a new era in precision medicine by providing solutions to a wide range of medical problems and opening the way for early disease detection, more efficient therapies, and personalized patient care (Table 3.2).

3.5 IMAGING TECHNIQUES IN DRUG DELIVERY

Imaging techniques play a crucial role in both TR and drug delivery, which bridges the gap between laboratory results and medical use. These techniques are essential for understanding the in vivo behavior of nanomaterials, providing non-invasive ways to investigate the distribution, interactions, and effectiveness of nanomaterials within biological systems. This breakthrough has transformed preclinical

TABLE 3.2
A List of Biomarkers Used for Drug Delivery Systems

Biomaterials	Structural Detail	Application	References
Polyethyleneglycol	Synthetic polymer	Conjugates enhance drug solubility and are a nanocarrier	[90]
Polylactic acid (pla)	Biodegradable polymer	Nanocarrier	[91]
Poly (lactic-co-glycolic acid) (plga)	Biodegradable polyesters formed by copolymers	Nanocarrier, implants	[92]
Liposomes	Spherical lipid vesicles that carry both hydrophilic and hydrophobic drugs	Targeted drug delivery, gene therapy	[93]
Alginate	Polysaccharide obtained from algae	Wound healing and controlled drug release	[94]
Chitosan	Cationic polysaccharides obtained from chitin	Carrier for osteomyelitis-targeting drug delivery, used in tissue engineering	[95]
Cyclodextrins	Cyclic structure composed of glucopyranose units	Enhances solubility or serves as the basis for larger drug delivery systems	[96]
Dendrimers	High-branching macromolecules having a multi-functionalized surface	Gene therapy, solubilization, and delivery of a hydrophobic drug	[97]
Polymeric micelles	Amphiphilic macromolecules with self-assemblies developed in a liquid matrix	Ideal drug delivery system for hydrophobic chemotherapeutics	[98]

research in nanomedicine and has promise for improving imaging-guided personalized medicine [99]. While CT, X-ray, and MRI are well-known for their remarkable anatomical detail, they frequently lack the sensitivity to visualize contrast compounds that are administered externally. Conversely, nuclear and optical imaging methods are particularly effective at detecting specific signals generated by labeled species, which makes them particularly useful for tracing nanomaterials that contain these contrast or signal-generating agents. Optical imaging techniques, such as fluorescence imaging, provide high spatial resolution and sensitivity, enabling the detailed observation of nanomaterial distribution and cellular interactions at the microscopic level. Similarly, positron emission tomography (PET) and single photon emission computed tomography (SPECT) are nuclear imaging methods that provide quantitative data on the biodistribution and pharmacokinetics of radiolabeled nanomaterials, although they require the administration of radiotracers. The selection of imaging technique depends on various factors such as sensitivity, cost, and spatiotemporal resolution, which impact the ability to capture multiple aspects of nanomaterial behavior [100]. With the advancement of nanomedicine research, the integration of modern data analysis and imaging technologies will further enhance personalized medicine approaches. This will ensure that therapeutic strategies are tailored to meet the specific needs of each patient and enhance overall treatment efficacy.

3.5.1 MRI (Magnetic Resonance Imaging)

MRI is a non-ionizing radiation imaging method that creates images by using the spin relaxation of nuclei, most commonly, hydrogen (^{1}H), in an external magnetic field. A key feature of MRI is the ability to provide high soft tissue contrast, which makes it helpful in tracking nanomaterials in vivo. MRI can be enhanced by incorporating paramagnetic (such as manganese or gadolinium) or superparamagnetic (like iron oxide) contrast agents into nanoparticles. This enables the visualization of nanoparticles based on their magnetic properties. However, quantifying contrast in MRIs is challenging because it depends on several variables, such as the concentration of the contrast agent, tissue characteristics, magnetic field strength, and variations in homogeneity. Furthermore, contrast quantification methods can be sequence-dependent and typically encounter nonlinearity at the endpoints of the calibration curve. As a result, they must be standardized. Despite these challenges, MRI is nonetheless a helpful technique in nanomedicine for evaluating the behavior of nanoparticles, although it might not be able to quantify tissue accumulation precisely [101].

An alternate method for tracking nanomaterials in MRI is utilizing molecules containing fluorine-19 (^{19}F), which is an MR-active nucleus with extremely low natural abundance in the [102]. Nanoparticles loaded with ^{19}F-based compounds provide an in vivo imaging technique with great sensitivity. Recent investigations have demonstrated that ^{19}F MRI efficiently visualizes nanoparticles, but the method has sensitivity limitations and requires high concentrations of ^{19}F-based compounds. The requirement for high tracer doses per nanoparticle prompts concern regarding possible human toxicity [103].

3.5.2 CT (Computed Tomography)

CT is a powerful anatomical imaging technique that creates images by using the interaction of matter with X-ray photons. CT is well-known for its speed, affordability, and user-friendliness, providing precise cross-sectional images of the body. CT imaging has benefits, but it also involves the use of ionizing radiation, which raises safety concerns. The limitation of CT imaging is its low contrast sensitivity; hence, it requires high concentrations of exogenous contrast agents for efficient nanoparticle visualization. For this reason, the utility of CT contrast agents is less well-known for tracking nanoparticles in vivo. CT imaging is very useful for obtaining high-resolution anatomical details; however, unless significant concentrations of contrast agents are utilized, it struggles to identify nanomaterials due to its low sensitivity [104].

3.5.3 Fluorescence Imaging

Near-infrared fluorescence (NIRF) imaging and fluorescence molecular tomography (FMT) are optical imaging methods that offer essential insights into the behavior of nanomaterials in vivo. The ability to multiplex or detect numerous signals simultaneously using different fluorophores makes these approaches beneficial. Using fluorophores, distinct emission channels can be obtained, which can then be demultiplexed to examine various aspects of nanoparticle dynamics. Another optical technique can be used to investigate the stability and production of nanoparticles in fluorescence resonance energy transfer (FRET) [105]. Despite these benefits, optical imaging methods have drawbacks such as quenching, bleaching of fluorophores, and restricted tissue penetration of emitted light. Because of these issues, optical imaging can only be described as semi-quantitative. Additionally, 2D methods like fluorescence reflectance imaging provide surface-weighted signals, which may complicate absolute quantification.

3.5.4 Nuclear Imaging

SPECT and positron emission tomography (PET) are nuclear imaging methods that are extremely sensitive techniques to detect gamma photons released by radiotracers. PET scanners identify positron-electron annihilation's coincidence events, whereas SPECT images identify γ-photons and X-rays linked with nuclear state transitions [106]. As SPECT and PET are sensitive, they enable the detection of low concentrations, which minimizes disturbance to biological processes. However, the absence of anatomical reference data and spatial resolution provides challenges for these methods. To overcome these challenges, SPECT and PET are frequently coupled with MRI or CT in hybrid imaging scanners. The CT co-registered data provide anatomical reference and attenuation correction maps required for accurate quantification [49].

3.5.5 Intravital Microscopy

Intravital microscopy is a method that allows visualization of nanoparticle accumulation within a cell. This technique provides comprehensive details about the interactions between specific cells and tissues and nanoparticles. However, intravital microscopy has certain drawbacks, including issues with tissue penetration and the potential for phototoxicity. Despite these challenges, this technique is effective for investigating the location and behavior of nanoparticles within organisms (Table 3.3).

3.5.6 Translation of Nanomedicine from Small to Large Preclinical Models

The advancement of nanomedicines depends significantly on preclinical research employing mouse disease models. Nevertheless, the predictive validity of these animal models for therapeutic efficacy in larger animal models and humans is often poor, which contributes to the relatively low number of clinically authorized nanotherapeutics. This is especially true for cancer nanomedicine, where the poor prediction accuracy of mouse orthotopic tumor models has been questioned, despite their wide use [112]. The in vivo performance of nanomedicines is significantly influenced by the differences between mice and humans in terms of pathophysiology, hemodynamics, metabolic activity, and immunological responses. Translational relevance can only be achieved by reproducing mouse study results in larger animal models with physiological characteristics closer to those of humans. Quantitative nuclear imaging techniques are essential to this process because they offer in vivo information on biodistribution, pharmacokinetics, delivered doses to target sites, and treatment responses. These approaches not only enhance the translational relevance of studies but also reduce the number of animals

TABLE 3.3
Imaging Techniques Commonly Used for the Development of Nanomedicine

Technique	Physical Event	Spatial Resolution	Imaging Agent	Tissue Penetration	References
MRI	Nuclear spin relaxation	<0.1 mm (for preclinical) 1–2 mm (for clinical)	Manganese, iron oxide, gadolinium	No limit	[107]
CT	X-ray absorption	1–5 mm (in vivo study)	Fluorophores	mm–cm	[108]
Optical	Infrared photon emission	<0.2 mm (for preclinical) 0.5–1 mm (for clinical)	Gold, iodine	No limit	[109]
PET	Beta annihilation-derived gamma photon emission	1–2 mm (for preclinical) 6–10 mm (for clinical)	Fluorine-18, copper-64, gallium-68, zirconium-89	No limit	[110]
SPECT	X-ray, gamma photon emission	0.5–2 mm (for preclinical) 7–15 mm (for clinical)	Indium-111 Iodine-125 Iodine-123	No limit	[111]

used, addressing ethical, logistical, and economic concerns. Additionally, the use of clinical scanners and protocols in extensive animal imaging studies facilitates the transformation from small-scale to medium/large-scale nanomedicine production, which is crucial for clinical translation [113].

3.5.6.1 Advanced Preclinical Models and Imaging Techniques in Nanomedicine

Researchers have developed advanced preclinical models that more closely resemble human pathophysiology, enabling the prediction of clinical success for nanomedicines. For instance, using dogs with spontaneous tumors has proven valuable in cancer research. Research has demonstrated that, although the effect can differ greatly among distinct solid tumors, ^{64}Cu-radiolabeled liposomes can evaluate tumor penetration in addition to enhanced retention and permeability effect in these models. Furthermore, PET imaging in dogs has been used to study the phenomenon of rapid blood clearance, which is observed after repeated infusions of PEGylated compounds. This study demonstrated the importance of control the lipid concentration of the initial dose or giving a pre-dose in order to lessen this impact [114]. These results are crucial for optimizing patient selection techniques and implementing theranostic treatments effectively. Imaging-based screening techniques are useful in the treatment of atherosclerosis for standardizing patient populations and identifying potential responders. For example, liposomal uptake in rabbit artery walls has been measured using a PET/MR imaging strategy that uses ^{89}Zr labeling for PET imaging and a comprehensive anatomical reference from MRI. Using this method, researchers accurately assessed liposomal uptake, demonstrating consistent in vivo performance across murine, rabbit, and porcine models of atherosclerosis.

3.5.6.2 Translational Imaging and Scale-Up of Nanomedicine Production

The scale-up and clinical translation of nanomedicine are critical challenges, especially in assuring consistent efficacy across different species and eventually humans. For instance, the pharmacokinetics and biodistribution of simvastatin-loaded HDL (S-HDL) have been effectively assessed by PET imaging in porcine, rabbit, and murine atherosclerosis models [115]. This study illustrated the translatability of this nanomedicine technique by demonstrating identical plaque-targeting abilities and anti-inflammatory efficacy across all three models. Furthermore, employing ^{89}Zr-labeled nano-biologics such as TRAF6i-HDL, comprehensive toxicity and biodistribution investigations have been carried out in non-human primates. Consistent organ distribution patterns between mice and non-human primates were found using longitudinal PET/MRI imaging, which adds to the findings' translatability [116]. Furthermore, research on hyaluronan-based nanotherapeutics has revealed the advantages of PET for visualizing and measuring the biodistribution of nanoparticles and tissue targeting in rabbit models. The findings above highlight the significance of sophisticated imaging techniques and larger animal models in the advancement and practical application of nanomedicines [117]. These methods help bridge the gap between experimental research and real-world, scalable treatments by offering comprehensive insights into in vivo behavior and therapeutic potential. They also present intriguing directions for future clinical use in cardiovascular disorders, oncology, and other areas.

3.6 POLYMERIC DRUG DELIVERY SYSTEM

Polymeric nanoparticles can be classified as synthetic or natural based on their origin. These nanomaterials are characterized by polymeric structure and colloidal size range (1–1,000 nm). Polymeric nanoparticles are of various types, such as nanoemulsions, dendrimers, polymeric micelles, polymersomes, biopolymer complexes, and cubosomes [118].

3.6.1 Polymer-based Nanoparticles

Polymeric nanoparticles are composed of synthetic or semisynthetic polymers, which are sub-nanosized colloidal entities with sizes typically ranging from 1 to 100 nm [119]. More drugs can be encapsulated into the core and onto the surface by adsorption because of their high surface area to volume ratio and nano size. Because of their distinct size range, these nanoparticles can circulate for longer periods of time, which prolongs drug release. Drugs encapsulated in nanoparticles (NPs) are protected from chemical degradation, enhancing their pharmacokinetic characteristics and permitting regulated release over a longer time period. Additionally, ligands can be functionalized onto NPs to boost therapeutic efficacy by site-specific drug delivery [7]. In general, biodegradable and biocompatible polymers are generally chosen for the development of polymer-based nanocarriers. Although there are numerous polymers with potential medicinal applications, only a few have received regulatory approval. Polycaprolactone (PCL), poly-DL-lactic acid (PDLLA), polyacrylate, polymethacrylate, poly-lactic-co-glycolic acid (PLGA), cellulose derivatives, polyethylene glycol [78], poly(ethylene oxide) triblock (PEO)/polypropylene oxide (PEO-PPO-PEO; poloxamers) [92], polyvinyl alcohol (PVA), and alginate are among the polymers that the United States Food and Drug Administration (USFDA) and other international authorities have approved for use in medical applications [120]. However, not all these compounds, possess the qualities required for developing useful nanocarriers. Consequently, previously approved materials are preferred for regulatory clearance, as present-day polymers and polymer conjugates are required to undergo considerable biological and toxicological evaluations before being allowed for usage in humans. For example, Yaşar et al. [121] used thiol-ene-miniemulsion photopolymerization to develop NPs based on a polymer with a particle size of 74 nm, for particular protein recognition, which demonstrated higher selectivity toward myoglobin. Similarly, Puri et al. [122] developed a pH-sensitive polymer using a dispersion polymerization approach with NPs in order to deliver bioactive compounds, where the drug release and hydrolysis were faster at pH 5 rather than at pH 7.4.

3.6.2 Polymeric Micelles

Polymeric micelles are made with amphiphilic block copolymers. These micelles are versatile spherical structures that have the potential for self-assembly and disassembly in response to concentration or temperature conditions. When exposed to an aqueous solvent, amphiphilic block copolymers self-associate to generate these nanoscale core-shell molecules. Since the micelle's cores are hydrophobic, hydrophobic drugs can be incorporated within it [123]. Compared to traditional solubilizing agents like Cremophor EL, this property makes polymeric micelles a desirable alternative for parenteral drug delivery, especially when it comes to boosting the dosage of potent but low-water-soluble drugs. However, studies of nano-bio dynamics at the molecular, cellular, and tissue levels are often overlooked [124].

An example of polymeric micelle application is found in the study by Varshosaz et al., where they used folic acid-grafted synperonic PE/F 127-cholesteryl hemisuccinate copolymer, creating polymeric micelles encapsulated with docetaxel for targeted cancer therapy. The study showed that the micelles improved the cellular uptake of docetaxel in B16F10 melanoma cells through endocytosis mediated by the folic acid receptor. This targeting technique enhances therapeutic efficacy and minimizes off-target effects. The polymeric micelles, whose concentration was evaluated to range from 0.001 to 500 μg/mL, were described as non-toxic and safe for biomedical applications. The micelles were found to have a smooth, spherical form by morphological examination [125].

3.6.3 Polymersomes

Polymersomes are emerging as a significant innovation in nanomedicine, particularly for medical diagnosis and therapy, due to their ability to control features and structure. These vesicles, which are made of amphiphilic block copolymers, have one or more bilayers with an aqueous core that can dissolve or chemically bind active substances. Compared to liposomes, polymersomes have better mechanical qualities as well as increased structural and colloidal stability in aqueous conditions. The proper design of polymersomes is significant due to their impact on cellular interactions and regulation of the immune response. Their size is also critical for efficient intracellular absorption. These polymeric vesicles are especially helpful for gene therapy and for delivering cytotoxic medications and imaging probes to specific tissues or cancerous cells [126]. For example, Zavvar et al. created gadolinium-based quantum dots in PEG-PCL nanopolymersomes for the administration of doxorubicin (DOX) and diagnostic imaging. Their work highlighted the theranostic potential of tailored polymersomes, with fewer complications, by demonstrating the regulated release of DOX, stability under physiological conditions, and increased cellular uptake and toxicity [127]. Zhou et al. produced photoresponsive-polymersomes that demonstrated efficient active targeting and regulated release of doxorubicin hydrochloride, thereby augmenting its cytotoxicity toward HeLa cells [128].

Further innovation by Li et al. involves using an MMP-cleavable peptide-linked triblock copolymer to create paclitaxel-loaded vesicles, which greatly enhanced cytotoxicity and cellular internalization against MMP-overexpressing cells. The efficacy of drug administration was enhanced by this method, which involved the morphological transformation of vesicles into smaller nanoparticles following exposure to MMP-2 [47]. Additionally, polymeric nanoreactors (NRs) containing glucose oxidase [129] were also developed by Li et al., with the intention of selectively activating in an acidic tumor environment. By efficiently integrating oxidation/chemotherapy with traditional treatments, these nanoreactors increase tumor oxidative stress and limit tumor growth [130].

A new form of vesicle (PICsome) was formulated by combining differently charged copolymer blocks in an aqueous solution. Although in the presence of serum proteins, these structures showed excellent physiological stability and therapeutic application potential. Targeted therapies are made possible by the vesicle membrane's incorporation of stimuli-sensitive linkers, which enable responsive payload release. For instance, Li et al., developed GOD-loaded nanoreactors that caused cancer cells to undergo glucose starvation and oxidative stress, which resulted in pyroptosis (an immune-stimulatory form of cell death) [131]. These developments demonstrate the adaptability and promise of polymersomes in delivering targeted drugs and diagnostic substances.

3.6.4 Dendrimers

Dendrimers are precisely designed, highly branched polymeric frameworks that have become highly effective drug nanocarriers, frequently referred to as "polymer therapeutics" [98]. These polymeric nanomaterials consist of a central core with a branching, three-dimensional macromolecular structure. They are usually smaller than 15 nm. Their distinct structure makes it possible to encapsulate and deliver drugs effectively. Folic acid-loaded dendrimer-functionalized selenium nanoparticles (FA-PAMAM-SeNPs) were created by Pillay et al., which showed a zeta potential value greater than 25 mV and a size of less than 150 nm. The cellular viability of this formulation was found to be higher (>85%) than that of placebo nanocomplexes (75%). This observation underscores the role of selenium in increasing cellular uptake and transgene expression [132]. A transtuzumab-dendrimer-fluorine DDS has been developed by Bartusik-Aebisheret al. wherein the active pharmaceutical ingredient was covalently bound to fluorinated dendrimers, demonstrating enhanced efficiency over free trastuzumab [133]. Mota et al. encapsulated folate-targeted polyurea dendrimers with I-buthionine sulfoximine. They observed an initial burst release of 60% of the drug after 1 h and reached 90% after 24 h [134]. Zhang et al. created third-generation poly(phosphorhydratzone) radical dendrimers with no reported cell toxicity for the diagnosis of brain tumors [135]. Dendrimers have potential use in gene therapy and targeted drug delivery, as demonstrated by Knauer et al.'s development of a polycationic phosphorous dendrimer-based approach for siRNA administration to target Lyn kinase [136].

3.6.5 Nanosized Hydrogels

Recent advancements in biological science have led to novel discoveries in in vitro cell culture, including cells-on-chips and cell microarrays. The materials required for these technologies should be easily manipulated, have strong cell adhesion qualities, and be biocompatible. Systemic drug distribution strategies (DDSs) built on nanomaterials are engineered to target specific areas, thereby enhancing passive targeting and ligand conjugation, which ultimately leads to improved active targeting [137]. Hydrogels, particularly those used in local DDSs, are effective in delivering drugs directly to target sites, especially when implanted near carcinoma tissues. For example, a hydrogel-based local DDS, an injectable glycol chitosan (GC) that has been cured by visible light, has shown potential in treating several solid cancers [138]. Polymer-based nanocarriers, such as nanosized hydrogels or nanogels, are produced through the physical or chemical cross-linking of polymer systems. Using lithography techniques, hydrogels can be thickened and molded, and their surfaces can be patterned. They can also be modified to acquire certain qualities using biomolecules [139]. Hydrogels, especially those derived from chitosan, are beneficial in organizing cells and tissues and serve as effective carriers of drugs, proteins, and other bioactive substances, including insulin and growth hormones. They are advantageous in various medicinal applications due to their flexible administration methods, including injectable, ocular, and nasal routes [140].

3.6.6 Polymeric Cubosomes

Recently, nanoparticles with an inverse bicontinuous cubic mesophase structure have been identified as polymeric cubosomes. Mesoporous block copolymers with unique reticulated pore networks make up these colloidal particles. Pegram developed polymer cubosomes through solution self-assembly using block copolymers made by joining hydrophobic polyisoprene polymer and hydrophilic polyethylene glycol [78] blocks together. The nonlinear structure of these block copolymers facilitates self-assembly into inverse mesophases in solution. Polymeric cubosomes have a higher capacity to encapsulate hydrophobic anticancer drugs and offer better durability because of their liquid crystalline membrane structure. Recent developments have highlighted their potential in membrane bioreactors, drug delivery, artificial cells, and biosensors, both in bulk and nanoparticle forms [141].

In a study by Zhang et al., cubosomes loaded with a mix of anticancer medications,paclitaxel, and cisplatin, were developed to effectively target cancer cells. The cubosomes were coated with poly-E-lysine that inhibited initial burst release and provided slow and sustained drug release. Transmission electron microscopy (TEM) revealed distinct spherical polyangular structures. In vitro studies demonstrated longer-lasting drug release and reduced burst discharge. Cytotoxicity studies confirmed the non-toxic nature of the cubosomes, and their therapeutic efficacy against HeLa cells was further established by impedance measurements and fluorescence imaging [142]. Boge et al. developed cubosomes for delivering the peptide LL-37 to treat skin infections. They employed three strategies: incorporating LL-37 into a gel, followed by nanoparticle dispersion; adsorbing LL-37 onto preformed cubosomes; and incorporating LL-37 during cubosome formation [143]. The pre-loaded cubosomes exhibited the highest efficacy in killing bacteria (Table 3.4).

TABLE 3.4
Polymeric Nanocarrier for Drug Delivery

Polymer	Carrier	Therapeutic Agent	Description	References
Chitosan	Nanoparticle	Retinol	This formulation enhanced the stability of retinol and decreased its toxicity and irritation.	[144]
PLGA	Nanoparticle	Antisense oligodeoxy nucleotides	It demonstrated 25%–32% drug release in a day via a diffusion-controlled method, with PLGA degradation-controlled drug release occurring after 14 days.	[145]
Gelatin	Nanoparticle	STAT6 siRNA	It has a biocompatible and stable formulation for topical siRNA delivery.	[146]
Ethyl cellulose, Eudragit®Rs	Nano capsules	Dexamethasone	*Ex vivo* studies demonstrated that the nanoparticles exhibit slow penetration and drug release when compared to standard cream. Thus, it could be an effective technique to limit the penetration and release of dexamethasone on the mucous membranes and the skin.	[147]
HPMC K15 & PEG 200	Nanoparticle	Rosuvastatin	Significantly more penetration through rat skin was seen in the ex vivo permeation assay.	[148]

(*Continued*)

TABLE 3.4 (*Continued*)
Polymeric Nanocarrier for Drug Delivery

Polymer	Carrier	Therapeutic Agent	Description	References
PCL	Nanoparticle	Risperidone	When risperidone-loaded biodegradable polymeric NPs were compared to free risperidone solution given intravenously, in vivo trials revealed a longer antipsychotic effect.	[149]
Sodium alginate	Nanoparticles	Carboplatin	Compared to the pure drug, which demonstrated drug release for up to 3 h, sodium alginate NPs showed extended drug release for up to 12 h. Sodium alginate NPs have the potential to be a valuable tool for targeted drug delivery, as evidenced by the drug's detection in the liver, lungs, and spleen of Laca mice after parenteral injection.	[150]
Methoxypoly (ethylene glycol)-b-poly(εcaprolactone-co-dioxanone)	Micelles	Curcumin	Polymeric micelles inhibited PC-3 human prostate cancer cells in a dose-dependent manner. Drug release was also shown to be sluggish.	[151]
PEG-b-PCL, PEG-b-PLA	Micelles	Sagopilone	The formulated micelles exhibited stability for up to 24 h at 37°C. The results of in vitro studies showed enhanced antiproliferative activity. In the in vivo study, no side effects were observed related to the carrier.	[152]
PLGA	Nanoparticles	Zaleplon	It showed enhanced anti-convulsant capacity when compared to free drugs.	[153]
Poly(beta-amino esters)	Nanoparticles	Genes	As for human reticuloendothelial cells, the transfection effectiveness was demonstrated to be around 85%, and for human umbilical vein endothelial cells, approximately 65%.	
PLGA	Nanoparticles	Loteprednol	An ex vivo trans-corneal penetration investigation conducted on goat corneas revealed an improved medicinal product penetration profile.	

3.7 TARGETED DRUG DELIVERY SYSTEM

The prime area of research in translational medicine is targeted drug delivery, which aims to minimize systemic exposure and adverse effects by delivering therapeutic molecules to specific disease locations.

3.7.1 Ligands for Nanoparticle Drug Delivery

3.7.1.1 Proteins and Polysaccharides

Antibodies have long been used to target specific receptors by leveraging their large size and high selectivity. Because of their specificity, antibodies are useful for both therapeutic and targeted applications. Nevertheless, their large size can limit their density on nanoparticle (NP) surfaces during modification. As an example, Roncato et al. [154] demonstrated the efficacy of using cetuximab, an anti-epidermal growth factor receptor (EGFR) antibody, in antibody-guided avidin-nucleic-acid nanoassemblies (ANANAS) for the therapy of cancer. These nanoassemblies outperformed conventional antibody-drug conjugates (ADCs) by combining biotin-conjugated molecules with poly-avidin cores, resulting in high drug-antibody ratios. Optimized with biotin-PEG-Atto488 for imaging, targeting with biotin-PEG-cetuximab, and treating with biotin-hydrazine-doxorubicin, cetuximab-guided ANANAS demonstrated better therapeutic efficacy in tumor-bearing mice and accelerated cellular internalization in EGFR-expressing cells, highlighting their potential in personalized cancer therapy.

Glycoprotein transferrin (Tf), which binds with iron, has been used because it can target tissues through Tf receptors, which are widely expressed on specific cells. In a recent study, Tf-modified NPs were utilized to break the blood-brain barrier (BBB) and deliver therapeutic substances [155]. They used an acid-cleavable linkage, DAK (2,2-bis-(aminoethoxy)-propane), to bind Tf with NP cores, enabling regulated drug release. This bond is stable at neutral pH but breaks down in an acidic environment, allowing Tf to be released from the NP core and enter the brain parenchyma. This strategy proved effective, with 200 Tf molecules per NP being the ideal ratio for brain delivery. This approach demonstrates the potential of Tf-modified NPs in targeting and treating brain diseases by ensuring specific delivery across the BBB.

The extracellular matrix's hyaluronic acid (HA) acts as a targeting moiety and hydrophilic backbone for nanoparticles (NPs). Because HA can bind to the CD44 marker, which cancer cells overexpressed, it is a great option for active targeting. The enzyme hyaluronidase 1 (Hyal-1) is substantially expressed in a number of malignancies and is capable of degrading HA, enabling targeted drug release [156]. According to the study by Choi et al., hydrophobic cholanic acids and PEG were used to generate HA-NPs, which form amphiphilic structures that resist the reticuloendothelial system (RES) and show extended blood circulation. These nanoparticles (NPs) exhibited specific cytotoxicity against CD44 overexpressing cancer cells after being loaded with the anticancer drug camptothecin (CPT). The rapid degradation of HA-NPs by Hyal-1 resulted in the quick release of CPT, thereby enhancing antitumor efficacy in animal models [157]. This approach highlights the adaptability and efficiency of HA-NPs in targeted drug delivery for cancer treatment.

3.7.1.2 Peptides

Peptides target ligands more effectively than other drug delivery methods due to their stability, can be easily conjugated to the surfaces of nanoparticles, and have inexpensive production costs. One example is the use of interleukin-4 receptor (IL-4Receptor)-binding peptide-1 (IL4RPep-1) with the sequence CRKRLDRNC, which was determined using phage-display techniques. Chi et al. showed that IL4RPep-1 binds to overexpressed IL-4R on tumor endothelium cells and lung tumor cells, and the binding is stable in serum for up to 4h. When doxorubicin-containing liposomes (IL4RPep-1-L-Dox) were labeled with this peptide, cell binding and uptake were more effective than the unlabeled liposomes (L-Dox). In vivo tests using H226 tumor-bearing mice showed that IL4RPep-1-L-Dox enhanced anticancer activity and boosted accumulation in tumor tissues, indicating its effectiveness in targeting tumor blood vessels and improving chemotherapy outcomes [158].

The arginyl-glycyl-aspartic acid (RGD) peptide, which is well-known for its capacity to bind integrins overexpressed in endothelial tumor cells, is also used to target tumors [159]. Lu et al. developed size-shrinkable NPs using RGD peptides to improve cancer therapy. Metformin (MET) and doxorubicin (DOX) were integrated into small NPs that were affixed to large gelatin nanoparticles (GNPs). RGD-functionalized tiny nanoparticles (NPs) penetrated deeply into tumors, where the large GNPs were broken down by matrix metalloproteinases-2 (MMP-2) to release MET and DOX through acid-sensitive imine bonds. With notable improvements seen in 4T1 and CT26 tumor models, this strategy not only allowed for deep tissue penetration but also produced synergistic anticancer and anti-inflammatory benefits [160].

3.7.1.3 Aptamers

Short nucleic acid sequences like DNA or RNA are called aptamers, emerging as effective active targeting ligands because of their tiny size, biodegradability, high sensitivity, and minimal immunogenicity [161]. Duo et al. [162] employed AS-1411 aptamer to target CX-5461-loaded mesoporous silica nanoparticles (MSNs) to the tumor cell's nuclei. The G-rich DNA sequence AS-1411 aptamer binds selectively to nucleolin proteins that are expressed in a few cancer cell lines and found in the cytoplasm, nucleoplasm, nucleolus, and cell surfaces. Bound nanoparticles are transported to the nucleus more easily thanks to this interaction. CX-5461 is an inhibitor of rRNA synthesis that causes tumor cells to undergo pro-death autophagy. In the study, AS-1411 aptamers were conjugated with MSNs coated with polydopamine for improved loading stability. By activating pro-death autophagy, these aptamer-modified nanoparticles significantly reduced cell proliferation by accumulating in the nucleolus of HeLa cells. The aptamer-modified nanoparticles showed noticeably increased tumor accumulation and successfully inhibited tumor growth in vivo without causing significant harm in the HeLa cell xenograft mice model (Figure 3.2).

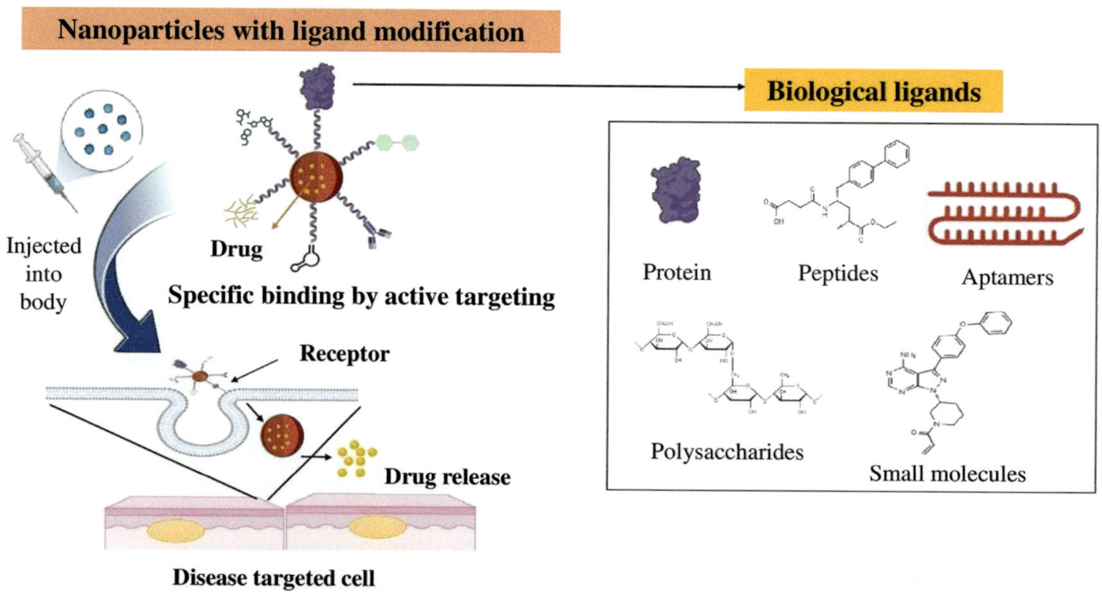

FIGURE 3.2 Illustration of biological ligands used for actively targeting nanodrug carriers.

3.7.2 Stimuli-Responsive Drug Delivery

3.7.2.1 Thermoresponsive Systems

Thermoresponsive DDSs have advanced significantly in targeted therapy, especially in oncology. To enable regulated drug release, these systems utilize materials that exhibit a nonlinear, sharp change in their characteristics in response to temperature fluctuations. These nanocarriers retain their therapeutic load at physiological temperatures (~37°C), and drug release occurs in response to localized heating (40°C–42°C), addressing the issues of rapid blood clearance and washout from the tumor. Some examples of thermoresponsive carriers are liposomes, polymer micelles, and nanoparticles. A typical polymer is poly(N-isopropyl acrylamide) (PNIPAM), which has a lower critical solution temperature. While temperature-sensitive polymeric systems often utilize polymers with well-defined transition temperatures, liposomes achieve thermoresponsiveness through phase transitions in lipid bilayers. Microwave applicators, radiofrequency oscillators, and temperature-controlled water baths are methods for applying heat in vivo, which facilitate precise drug release at the target site [163].

Thermosensitive liposomes (TSLs) are the most advanced and clinically developed thermoresponsive systems. For instance, Phase II and III trials are presently being conducted on TSLs (ThermoDox, Celsion Corporation) loaded with doxorubicin to treat hepatocellular carcinoma, colorectal liver metastases, and breast cancer. TSL formulations have recently been developed with an improved liposomal design that responds quickly to hyperthermia (~40°C–45°C) to increase drug-release efficiency. Leucine zipper peptide–liposome hybrids are another intriguing strategy that combines the advantages of TSLs with temperature-sensitive peptides that undergo shape changes upon heating. Furthermore, due to the hyperechogenic properties of carbon dioxide bubbles, permeable pores in the lipid bilayer can be produced by bubble-generating liposomal systems that generate the bubbles at a temperature of approximately 42°C. This can enhance both drug release and ultrasonic imaging [164].

Beyond liposomes, thermoresponsive polymeric drug nanocarriers utilizing PNIPAM have received significant attention in DDSs. Other polymers that have shown promise include poly(γ-2-(2-(2-methoxy ethoxy)-ethoxy) ethoxyε-caprolactone)-b-poly(γ-octyloxy-ε-caprolactone). These materials show transition temperatures that can be used for effective drug release at lower hyperthermia levels (40°C) [165]. Local hyperthermia is also employed to modulate cell-penetrating peptides (CPPs), via the formation of diblock-copolymer elastin-like polypeptides that are temperature-triggered, which improves cellular uptake. Furthermore, a decrease in temperature or cold shocks can cause nanocarriers to swell or de-swell, which increases their porosity and facilitates drug release [166]. For instance, delivering small interfering RNA (siRNA) for gene silencing has proven successful when using Pluronic F127–polyethyleneimine (PEI) nanocapsules [167]. Generally, creating materials that are both safe and sensitive to temperature changes presents a challenge in the development of thermoresponsive nanodevices. Minute temperature variations around physiological temperatures, with liposomal systems currently leading in clinical application potential.

3.7.2.2 Ultrasound-Triggered Drug Delivery

Ultrasound is a valuable tool for achieving controlled release of drugs while limiting adverse effects on healthy tissues. Ultrasound may be precisely adjusted to permeate tissues at different depths by modifying exposure time, duty cycles, and frequency. This method releases drugs from nanocarriers by using mechanical and thermal effects generated by radiation forces or cavitation phenomena. Cavitation-induced forces can cause nanocarriers to destabilize, thereby enhancing drug release and temporarily increasing vascular permeability to facilitate cell absorption. While they can be useful in crossing cavitation thresholds, low ultrasonic frequencies [168] also have some dangers, such as the possibility of metastatic dissemination [169]. To mitigate this, cavitation thresholds are lowered by the use of ultrasonic contrast agents, such as microbubbles; nevertheless, their usefulness is limited by their short lifespan and lack of extravasation.

These restrictions have been removed by developments in perfluorocarbon (PFC) nanoemulsions, which, under therapeutic ultrasonography, create microbubbles. The acoustic droplet vaporization process generates these bubbles, which improve drug delivery and cellular absorption at tumor sites, with aptamer-functionalized PFC droplets improving targeting specificity. Moreover, there is evidence of ultrasound-enhanced drug delivery through the skin, such as the notable suppression of melanocytic lesions using siRNA-loaded liposomes. By incorporating ultrasound response, echogenic liposomes, also known as bubble liposomes, enhance therapeutic efficacy [170]. For instance, in cerebral ischemia models, xenon-loaded bubble liposomes paired with ultrasound have demonstrated neuroprotective effects. Additionally, bubble liposomes have improved the efficacy of DNA transfection and gene delivery. For DNA transfections, ultrasound-triggered drug delivery is handy for avoiding degradative endocytotic pathways. Drug release with minimum temperature changes is possible when thermosensitive drug carriers are coupled with high-intensity focused ultrasound (HIFU). For example, in comparison to non-irradiated controls, doxorubicin release from TSLs produced by HIFU led to increased drug accumulation in tumor models. According to research conducted by Celsion Corporation, the combination of ThermoDox and HIFU may be effective in treating pancreatic cancer, liver metastases, and bone metastases [171].

3.7.2.3 Light-Triggered Drug Delivery

Photoresponsive systems are formed to release drugs on demand at specific visible, ultraviolet, and near-infrared light wavelengths [60]. These systems use photosensitivity-induced structural modifications of nanocarriers to release drugs as a one-time or repeatable event. For example, the reversible photoisomerization of azobenzene groups enables regulated drug release by transitioning from the trans to the cis form when exposed to UV light with a wavelength of 300–380 nm and back to the trans form when exposed to visible light. This process is used in several applications, such as azobenzene functionalization of mesoporous silica nanoparticles (MSNPs), light-controlled host-guest recognition systems, and azo-modified DNA valves. The disintegration of azo-modified surfactant micelles under UV light provides evidence that the trans-cis photoisomerization, accompanied by hydrophobic-hydrophilic transition, can also facilitate drug release [172]. An alternative strategy makes use of UV-induced spiropyran–merocyanine isomerization. This promotes deeper tissue penetration and causes spiropyran-PEGylated lipid nanoparticles to shrink reversibly [172].

For regulated drug release, researchers have also explored, the thymine photodimerization–cleavage cycle and other photocleavable groups, such as o-nitrobenzyl. For example, to accomplish targeted drug release, O-nitrobenzyl compounds have been utilized with MSNP gatekeepers and gold nanoparticles [173]. Lipid vesicles containing caged DNA have been used to illustrate the use of photolabile caging groups for light-activated protein production. This allows the synthesis of active proteins like luciferase and green fluorescent protein when exposed to UV light [174].

However, a significant obstacle to light-triggered medication distribution is the restricted permeation depth of UV-visible light (~10 mm) resulting from substantial tissue scattering, which limits its applicability to easily accessible body locations such as the skin and eyes. To overcome this, photosensitive groups that react to higher wavelengths (700–1,000 nm) have been utilized to produce NIR-responsive devices. NIR light has less scattering and can penetrate tissues more deeply. Plasmonic materials that absorb NIR light can convert photon energy into heat, which triggers drug release [175]. For instance, hollow gold nanospheres loaded with doxorubicin exhibited faster drug release when exposed to 808 nm NIR light, thereby improving anticancer efficacy while reducing systemic toxicity. Similarly, localized heating caused by NIR irradiation of gold nanorods enabled the release of doxorubicin attached to DNA helices. While gold-coated PLGA nanoparticles and gold-shell-coated lipidic nano-micelles have shown promising antitumor activity in breast cancer and glioma models, NIR-induced plasmonic nanobubbles offer a potential use similar to ultrasound-triggered drug delivery [176]. Despite these developments, the biodegradability and safety of materials utilized in light-responsive nanoparticles, like gold and silver nanoparticles, and derivatives like azobenzene and o-nitrobenzyl, remain concerns. Identifying biocompatible photosensitive materials will be essential for the clinical translation of these systems.

3.7.2.4 Ph-Sensitive Systems

Variations in pH provide a strategic means of regulating drug delivery, allowing targeted therapy in specific organs or intracellular compartments and reacting to conditions such as inflammation or cancer. There are two main approaches used: one is the use of pH-responsive polymers that alter their solubility or conformation, and the other is the construction of polymeric systems with acid-sensitive linkages that cause drug release or change the charge or ligand exposure of the polymer.

The pH-sensitive systems are utilized in anticancer therapy where pH differences (~7.4) between the acidic extracellular environment of tumors and healthy tissues (6.5–7.2) are exploited. This difference is caused by irregular angiogenesis and a shift in glycolytic metabolism of rapidly growing tumors. The tumor interstitium produces acidic metabolites due to its acidic environment. For example, at pH approx. 6.3, when amino groups are protonated, Systems based on chitosan release tumor necrosis factor alpha (TNFα) that has been encapsulated [177]. Additionally, at pH levels ranging from 6.4 to 6.8, PEG–poly(β-amino ester) micelles exhibit pH-sensitive breakdown, releasing encapsulated camptothecin. Other approaches utilize pH-sensitive changes to boost cellular uptake by exposing CPPs or to improve protein delivery into ischemic areas. Micelles based on polyhistidine and liposomes coated with Trans-Activator of Transcription (TAT) peptide showed efficient cell internalization and pH-triggered peptide exposure [178].

In the context of bacterial infections, pH-responsive DDSs have been developed to enhance antibiotic therapy, which are characterized by their low pH resulting from inflammation and anaerobic fermentation. An ionizable polyhistidine segment, for instance, can be added to the copolymer PLGA-b-polyhistidine-b-PEG to improve interaction with acidic bacterial walls and increase uptake of nanoparticles by both Gram-negative and Gram-positive bacteria [168].

Additionally, oral DDSs also utilize pH responsiveness to protect drugs from averse gastric conditions and improve intestinal absorption. Chitosan nanoparticles and copolymers based on poly (methacrylic acid) are examples of systems that disintegrate or undergo charge reversal at intestinal pH, which enhances drug release and absorption. For efficient intracellular drug delivery, pH gradients between endosomes pH 5–6 and lysosomes pH 4–5 are employed at the cellular level. The release of encapsulated drugs is facilitated by nanoparticles that are engineered to expand or disassemble in response to an acidic pH. These mechanisms may include the incorporation of protonable monomer units or the masking of hydroxyl groups [177]. Acid-degradable crosslinkers and acid-sensitive linkages, such as oximes, acetals, or hydrazones, also facilitate the breakdown of nanocarriers and drug release.

Another pH-sensitive method used to break the endosomal compartment is the proton sponge effect, which involves buffering endosomal pH to cause lysosomal enlargement and rupture. In order to achieve this, copolymers incorporating amine-containing polymers, including poly(β-amino esters) or poly-l-lysine, have been utilized [179]. Poly (β-amino ester) nanoparticles covered with lipids exhibit efficient mRNA distribution and transfection. On the other hand, endosomal disruption is enhanced by a polyhistidine region in a virus-mimetic nanogel. While chitosan's limited buffering capacity too often, necessitating complexation with polyelectrolytes to destabilize membranes and maximize its endosomolytic potential. Its ability to reverse charges has been utilized in pH-triggered drug release [180].

The pH-sensitive liposomes, formulated with lipids like 1,2-dipalmitoyl-sn-glycero-3-phosphoethanolamine or 1,2-dioleoyl-sn-glycero-3-phosphoethanolamine (DOPE), change from a lamellar to a fusogenic hexagonal phase, which helps in drug release. Through buffering and fusogenicity, conjugating low-molecular-weight PEI with DOPE improves the transport of genes and siRNA [181]. PEG-modified, positively charged liposomes also facilitate endosomal interaction. Lipid membrane disruption may occur when polymer chains that undergo phase shifts in acidic environments are anchored or caged, encouraging cargo release [182].

3.7.2.5 Enzyme-Sensitive Systems

Enzyme-mediated drug release is made possible by changing enzyme expression patterns, including those of phospholipases, proteases, and glycosidases, in conditions like cancer or inflammation, allowing more precise targeted therapy. Several methods that make use of enzyme-mediated drug delivery depend on taking advantage of the presence of these enzymes in their extracellular environment. For example, recent studies have employed short peptide sequences that matrix metalloproteinases can cleave as linkers between bioactive components. like CPP-coated dextran-coated iron oxide nanoparticles or TAT-functionalized liposomes and surface PEG chains [183]. In contrast to nanocarriers without cleavable linkers, the PEG shell breaking in the tumor microenvironment in these systems exposes the bioactive ligands, thereby improving intracellular penetration [184]. siRNA-loaded nanoparticles have produced notable findings, such as gene-silencing activity around 70% in tumor-bearing mice. Similar to this, lipopeptides or protease-sensitive polymer coatings have been developed for liposomes or porous silica nanoparticles to trigger their release, offering an effective method of targeted drug delivery [185].

Targeted drug release within intracellular compartments is another application for enzyme-mediated drug delivery. Doxorubicin is more precisely administered when lysosomal enzymes break mesoporous silica scaffolds coated with polysaccharide derivatives, which reduces the length of the polysaccharide chains and releases the drug. The lysosomal enzyme cathepsin B, highly expressed in cancerous cells, has also been utilized to facilitate cargo release through the rapid enzymatic degradation of polymersomes [186]. In other cases, medication delivery techniques based on polymers that contain a cationic peptide substrate for kinases or proteases inside cells, particularly to HIV-infected or inflammatory cells, have been utilized to produce transgenes with high cell specificity. Transcription and gene release are aided by the breakdown of polymer–DNA electrostatic interactions by enzymes. This strategy can also be used to treat bacterial infections, such as vancomycin-releasing lipase-sensitive nanogels, which have been shown to effectively limit the growth of *Staphylococcus aureus* and kill intracellular bacteria. The potential of DDSs triggered by enzymes is demonstrated by such cases [187]. However, further research is required to accurately determine target enzyme concentration at specific sites to finely uptake in the cell and establish the correlation between enzymatic activity and in vivo drug release.

3.8 REGULATORY CONSIDERATIONS IN DRUG DELIVERY SYSTEM

Despite the substantial advancements in nanomedicines in recent years, the clinical impact of nanomedicines has been hampered by the absence of standard operating procedures for preclinical development and characterization. Although efforts have been made to establish global regulatory patterns, a unified framework is still unattainable. Attempts have been made to modify current approaches from traditional medicinal products to nanomedicines, with the active pharmaceutical ingredient [173] frequently serving as the regulatory focal point. This may require compliance with regulations about biological medical products or new chemical entities (NCEs) [188]. A significant regulatory challenge is the unique nature of nanomedicines, which are extremely sensitive to changes in raw materials and production processes. Due to this sensitivity, biological properties can vary significantly, necessitating robust quality control methods to monitor physicochemical features such as size, shape, charge, and the release of the drug [189].

Biocompatibility and immunotoxicity are also important factors, as nanomedicines interact with immune cells and plasma proteins, which can impact both safety and therapeutic efficacy. Additional hurdles in the manufacturing of nanomedicines include scale-up issues and the requirement for accurate design-based quality techniques and process analytical technologies (PAT). The Conference on Harmonization (ICH) aims to address these challenges but still faces divergences, particularly between regulatory bodies such as the FDA and European Medicines Agency (EMA) [190,191]. Researchers are investigating sensitive tests and imaging methods to identify nanocarriers and distinguish them from aggregates or metabolized forms. Regulatory frameworks in Europe, the US, and Japan are gradually converging, with efforts including scientific counseling and the development of orientation documents by the EMA to guide nanomedicine development. To demonstrate the increased value of nanomedicines over current treatments, pharmacoeconomic studies —those that consider metrics such as quality-adjusted life years (QALYs) and associated healthcare costs —will be crucial [192]. Regulatory norms and business interests must continue to evolve to overcome these obstacles and guarantee the effective translation of nanomedicines into clinical practice.

3.9 CLINICAL TRIALS

During the early stages of developing new drug therapies, in vitro model systems involving cell and tissue preparations are essential for understanding the etiology and pathophysiology of diseases, which helps in creating hypotheses for prospective treatments. The purpose of these preliminary studies is to evaluate new treatment approaches that may interfere with disease mechanisms. Successful in vitro results lead to a shift toward more intricate systems, such as animal disease models. The prognostic value of these animal models depends on how closely they resemble the pathophysiology of human diseases; the more similar the model, the more accurately it can predict how humans will react to therapies [193]. Despite this, many disorders lack suitable animal models, thereby relying on patient trials for accurate evaluation of safety and efficacy. This gap emphasizes how difficult it is to extrapolate results from animal research to humans. Healthy volunteers are usually used in the first stage of human experimentation, known as Phase I testing, which assesses the pharmacology and safety of a medicine. Phase II, or therapeutic exploratory testing, is conducted in patients with the target disease if Phase I data demonstrate acceptable safety profiles. The purpose of this testing is to further evaluate potential efficacy and safety across various doses. However, due to safety concerns or lack of efficacy, a significant proportion of drugs examined in Phases I and II do not proceed to Phase III. More troubling cases arise with drugs that appear to be effective in lab settings and have been well-tolerated throughout clinical trials, yet do not show effective results in real-world clinical situations [194]. This difference frequently results from a lack of knowledge about the mechanisms underlying the disease or the effects of treatment measures, underscoring the difficulty in converting preclinical success into clinical efficacy.

Many nanomedicines (NNMs) are either commercially available or are in the clinical development stage; these NNMs mainly consist of pharmaceuticals that have already received approval and make use of DDSs like liposomes, polymeric micelles, dendrimers, and inorganic nanoparticles [195]. Liposomes, the first NNMs to be approved by the FDA, remain dominant in the market since they can transport drugs like paclitaxel and amphotericin B that are highly toxic or poorly soluble. Since their development in 1965, liposomes have been exploited as medication delivery vehicles for both small and macromolecules. This has resulted in the synthesis of prominent FDA-approved NNMs like Caelyx and Doxil [196]. Despite notable developments, the clinical benefits of liposomal NNMs have frequently been restricted to decreased toxicity rather than increased efficacy. AML with myelodysplasia-related changes (AML-MRC) or therapy-related acute myeloid leukemia (t-AML) in older individuals showed increased response and survival rates with acceptable toxicity when using Vyxeos (daunorubicin/cytarabine liposomal formulation). Biological problems, biocompatibility and safety issues, large-scale manufacturing constraints, intellectual property obstacles, regulatory affairs, and the rewards in comparison to traditional medicines are some of the complications associated with the application of NNMs in therapeutics [197]. These complexities often hinder the market entry of NNMs, regardless of their therapeutic potential.

3.9.1 Difficulties in Applying Nanomedicines to Clinical Settings

NNMs' application in therapeutic settings is more complicated and costly compared to more conventional drug formulations, such as tablets and injections. Biological problems, large-scale manufacturing difficulties, safety and biocompatibility issues, intellectual property, regulatory obstacles, and cost-effectiveness are some of the challenges. These obstacles may hinder the market entry of NNMs, regardless of the potential therapeutic benefits that NNMs offer.

In the process of translating nanomolecules clinically, biocompatibility and safety are critical factors that require thorough toxicological evaluations to ensure human safety. Regulatory bodies usually require a thorough assessment of any changes made to the drug's composition, manufacturing process, or formulation while it is in clinical development because these changes may have an impact on safety. Alterations in the synthetic pathway, impurity profiles, manufacturing processes, source materials, sterilization techniques, administration routes, drug product composition, and modifications to the drug product container closure system could all have an effect on safety [198]. Strict protocols are required to evaluate product equivalence, quality, and safety by comparing the therapeutic ingredient and product from old and modern production processes. If differences are found in the analytical data, additional qualifying or bridging studies should be conducted to support the bioavailability and safety of the new material [199].

Understanding the toxic activity of the free drugs, the way NNM delivery systems behave, how they interact with biological elements, and how drug release rates affect bioavailability, which is required for predicting probable toxicities or side effects in in vivo studies. In order to improve clinical translation potential, It is necessary to use rational design from the point of material selection to production, optimization, and purification [200]. Although research has been done on the safety of common compounds like phospholipids and biodegradable polymers, the increasing complexity of NNMs affects biodistribution, biocompatibility, and toxicity. It can be significantly impacted by synthetic compositions, coatings, and ligands. Profiles following in vivo administration. For example, an acute immunological response known as complement activation-related pseudoallergy (CARPA) can be brought on by several NNMs that cause symptoms such as anaphylaxis, facial edema, and cardiac distress. Typically, to manage symptoms, therapy is stopped, infusion rates are slowed, and allergy medications such as antihistamines, corticosteroids, and epinephrine are used [201].

To address these challenges, assays that are standardized, sensitive, and validated are required to evaluate nanotoxicology in the prior stages of clinical development. Comprehensive in vitro or ex vivo experiments are essential for identifying possible risks prior to animal model assessments [193]. Standardized protocols using several cell culture models to assess immunotoxicity, cytotoxicity, and genotoxicity are essential, especially for NNMs incorporating novel materials. The development of new approaches, such computational modeling and high-throughput screening, holds promise for the simultaneous analysis of several NNMs. Furthermore, considering the longer drug retention periods and higher circulation half-lives associated with nanoencapsulation, specialized toxicological studies in animal models are required to assess toxicity over both the long and short terms. Understanding the absorption, distribution, metabolism, and excretion (ADMET) of emerging nanomaterials is necessary to predict toxicological responses. Techniques for real-time imaging can improve understanding of NNM interactions with biological organs and tissues [202].

Regulatory supervision of NNMs encompasses several intricate concerns. Considering intellectual property (IP) is essential since NNMs are complicated, involving components that connect medical devices and medicine. IP positions need to address drug-carrier properties, encapsulated cargo, and carrier technology [203]. Complications can occur if NNM users employ commercial targeting ligands or coatings, or combine established drugs with innovative carrier technologies. Several patents and cross-licensing agreements are required, which emphasizes the necessity of efficient IP procedures and protocols to speed up commercialization [204]. Challenges include invalid patents, thickets of patents, and delays in the review process. There is a universal nano-nomenclature and improved patent search tools to prevent multiple patents on similar concepts, enabling efficient patent examination [205].

The commercialization of NNM is significantly influenced by government regulations. The current regulatory system, which is mainly based on conventional medicinal compound standards, is inadequate for the complicated structures of NNMs. Specific regulatory standards are necessary to address the distinct characteristics of NNMs, including their pharmacokinetics, pharmacodynamics, and safety profiles. It is imperative to establish international regulatory standards in partnership with academics, researchers, businesses, and key nations. This includes establishing standards for the assessment of innovative polymeric materials and intricate NNM formulations [206]. Regulatory frameworks need to take into consideration the multifunctionality and complexity of NNMs, including the integration of imaging diagnostics and treatments. It is imperative to maintain safety and quality standards without excessive regulations, thereby preventing rising approval costs and hindering the development of novel products [207].

The general pharmaceutical standards, such as stability, sterility, and purity, must be adhered to by manufacturers of NNMs. New analytical instruments and standardized protocols are necessary to evaluate key physical properties, including particle size, surface chemistry, and morphology. To ensure the effectiveness and safety of both existing and newly developed NNMs, regulatory agencies should establish standardized testing processes and protocols for toxicity studies, as well as regulatory standards. Collaboration between regulatory agencies and other stakeholders will be required for establishing these guidelines and facilitating the clinical translation of NNMs.

3.10 DEVELOPMENT OF NANOMEDICINES THROUGH TRANSLATION APPROACH

For remedial purposes, increasing targeted drug assimilation and reducing side effects is one of the prime obstacles in developing a new DDS. Despite the promising results that nanomedicine platforms (NNMs) have shown in in vitro or ex vivo investigations, it is critical to evaluate their effectiveness in vivo studies using suitable animal models of the disease. According to Hua et al. [208], these platforms frequently exhibit poor specificity, accumulation, or stability in animal models, leading to unsatisfactory results that hinder their advancement in research and development (R&D). The hurdles of drug delivery in humans are due to the variability in disease, which makes it challenging to translate the efficacy of animal models to human outcomes. Due to the complex pharmacokinetic and pharmacodynamic processes involved, the concept of a "magic bullet," which promises the precise delivery of bioactive substances to specific cells, tissues, or organs, is often unattainable [209]. Therefore, it is more prominent to develop realistic therapeutic platforms that optimize therapeutic effects, reduce dosages, and simplify dosage form design.

The intricacy of NNM formulation affects how easily it can be used in clinical settings. For large-scale manufacturing to be effective and reproducible, simplification of formulation design is essential. To defend the more expensive and intricate production method, any additional complexities in the NNM platform must provide significantly improved, consistent, and repeatable advantages in patients and animal models [210]. More study is needed to understand the benefits of ligand-based delivery systems for simple NNM platforms; to be specific, the consistency and reliability of targeted expression despite varying degrees of illness severity and patient populations should be emphasized. Formulations must be modified when transferring findings from animal models to human clinical trials to ensure they are suitable for human administration. This includes taking into account aspects like consistency and size of dosage [211]. To develop practical and effective NNMs, intricacy, effectiveness of treatment, and clinical translation must all be balanced.

To support the clinical transition of NNMs, it is important to ensure that nanosized carriers are stable in in vivo studies, scalable for production, and made of biodegradable, biocompatible, and non-toxic materials. The potential toxicity of the synthetic polymers and lipids employed in NNMs needs to be carefully assessed in all circumstances. The majority of NNMs that are now being used clinically are intended to treat cancer. They are mainly given parenterally and are based on straightforward nanomedicine platforms, such as PEGylated, surface charge-modified, and basic nanoparticles [212]. However, obstacles like cost-effectiveness, pharmacokinetic and pharmacodynamic knowledge, and manufacturing complexity have hindered the development of NNMs in non-cancer disorders. The expensive, intricate, and time-consuming development process is a significant contributor to the slow pace of clinical translation, which in turn affects investor and industry sentiment. NNMs must demonstrate clear benefits by lowering healthcare costs, increasing efficacy, enhancing quality of life, and minimizing side effects to make the investment worthwhile. Additionally, the cost-benefit analysis of NNMs must also be beneficial when compared to currently available therapies for widespread adoption [213]. To facilitate effective clinical translation, collaboration is necessary across academia and industry to develop standardized protocols and tests, along with the necessary infrastructure for NNM production and characterization, including experts in engineering, biology, pharmaceutics, medicine, and toxicology.

3.11 CLINICAL SUCCESS IN TRANSLATIONAL DRUG DELIVERY

Recent years have seen significant advancements in therapeutic areas such as oncology because of successful TR.

3.11.1 Monoclonal Antibody Treatment

Bevacizumab and trastuzumab are two FDA- and European Community-approved monoclonal antibodies, representing a significant development in TR. These monoclonal antibodies are intended for specific targeted therapy [214]. The primary target of trastuzumab is the human EGFR II (HER-II) protein; it is highly upregulated in about 25% of metastatic breast tumors. Trastuzumab has been demonstrated to reduce tumor growth and increase survival rate in patients with HER-II-positive metastatic breast cancer by targeting HER-II [215]. On the other hand, bevacizumab inhibits the growth of tumors by targeting vascular endothelial growth factor (VEGF), which is essential for angiogenesis; it prevents interaction with the tumor's blood vessels and restricts tumor growth. Bevacizumab usage in metastatic colorectal cancer has resulted in better survival rates and decreased tumor recurrence [216]. The significant advancement in this field began with the identification, sequencing, and cloning of HER2 in 1985. Advances in molecular biology, specifically proteomics, enabled the development of human anti-HER2 antibodies, which resulted in positive clinical trials and approval in 2000 [78].

3.11.2 Targeted Gene Therapy

Novel approaches to inhibiting the growth of cancer include the use of antisense technology, which targets genes related to angiogenesis, apoptosis, and cell proliferation. Gefitinib is one such medication that prevents the intracellular phosphorylation of several tyrosine kinases connected to transmembrane cell surface receptors, such as the EGFR-TK. EGFR is found on the surface of many cancer and normal cells. Patients with activating mutations in EGFR have demonstrated specific benefits from gefitinib [217]. The recent advancement in genomics, proteomics, genetics, and epigenetics is enhancing our understanding of oncogenic pathways, DNA repair processes, stem cell behavior, metabolism, and the tumor microenvironment. Additionally, advances in animal models and quantitative cancer models are fostering new approaches to cancer prevention and therapy, thereby improving the personalization of cancer treatment [218] (Table 3.5).

TABLE 3.5
Nanomedicines Currently Approved for Market [199]

Tradename	Nanoformulation	Application	Company
Doxils	Liposome-loaded doxorubicin hydrochloride	AIDS-related Kaposi's sarcoma, breast cancer, ovarian cancer, myeloma, etc.	Janssen, Centocor Ortho Biotech
Abraxanes	Paclitaxel-loaded polymeric nanocarrier	Rare metastatic breast cancer, like non-small cell lung cancer (NSCLC)	Celgene, Abraxis BioScience, AstraZeneca
Copaxones	Glatiramer acetate-loaded polymeric nanocarrier	Multiple sclerosis	Teva
Daunoxomes	Liposome-loaded anti-neoplastic medicines like daunorubicin citrate	AIDS-associated Kaposi's sarcoma	Galen, Teva, NeXstar, Gilead Sciences
Neulastas	PEGylated filgrastim	Febrile neutropenia	Amgen
Oncaspars	PEGly L asparaginase	Leukemia of lymphoid tissues	Enzon/Schering-Plow Sigma-Tau

(Continued)

TABLE 3.5 (Continued)
Nanomedicines Currently Approved for Market [199]

Tradename	Nanoformulation	Application	Company
Pegintrons	Interferons like alfa-2b which are PEGylated	Patients affected with Hepatitis-C and other liver-related diseases	Schering-Plow Merck
Rapamunes	Nanocrystal-loaded sirolimus	Immunosuppressive drugs for kidney transplants	Elan, Pfizer, Wyeth/ Alkermes
Survantas	Liposomes loaded with beractant	Respiratory problems in immature neonates.	Abbot, Abbvie
Depodurs	Liposome-loaded with morphine sulfate	Pain of a chronic nature	Pacira
Diprivans	Surfactant-derived nano formulation with propofol	Used in anesthesia	AstraZeneca, Fresenius Kabi

3.12 FUTURE PERSPECTIVES

The success of TR depends on the discovery of clinically useful biomarkers. Transferring innovative research assays to everyday practice in laboratory medicine is often a drawn-out procedure with numerous specific and sequential steps. Challenges include methodological biases in animal studies and the differences in pathophysiology between humans and animals. Sample handling and pre-analytical variability raise concerns regarding reliability and reproducibility, underscoring the need to standardize collection processes prior to the widespread adoption of novel diagnostic techniques [198]. Preclinical "proof-of-concept" investigations are hindered by factors including inconsistent data and biases in interpretation; missing data mainly affects small research. It is crucial to handle missing data and outliers effectively, and problems with statistical multiplicity need to be resolved [219]. Clinically, a one-size-fits-all approach is inadequate for molecularly targeted drugs due to inter-subject variability in drug responses. The development of personalized medicine is complex, but it has the potential to improve outcomes, benefit patients, and reduce the costs associated with healthcare [220]. TR is also influenced by non-scientific factors, including a lack of funding, regulatory hurdles, privacy concerns, fragmented infrastructure, and a shortage of qualified researchers. Better collaborations between academia and business, alternative funding sources, lower costs for clinical research, and increased public support could all aid in resolving these problems; with initiatives from private industry and government institutions like National Institute of Health (NIH) playing a crucial role.

REFERENCES

1. K. Jain, *Transdermal Drug Delivery: Technologies, Markets and Companies*. Jain Pharma Biotech Publications, Basel, 230 pp. (2008).
2. P.M. Glassman, V.R. Muzykantov, Pharmacokinetic and pharmacodynamic properties of drug delivery systems, *J. Pharmacol. Exp. Ther.* 370(3) (2019) 570–580. https://doi.org/10.1124/jpet.119.257113.
3. A.C. Anselmo, S. Mitragotri, An overview of clinical and commercial impact of drug delivery systems, *J. Control. Release* 190 (2014) 15–28. https://doi.org/10.1016/j.jconrel.2014.03.053.
4. K.K. Jain, An overview of drug delivery systems. In: Jain KK (Ed.), *Drug Delivery Systems,* 3rd edn. Humana Press, New York, pp. 1–54 (2020).
5. K. Park, Controlled drug delivery systems: past, forward and future back, *J. Control. Release* 190 (2014) 3–8.
6. M.R. Prausnitz, R. Langer, Transdermal drug delivery, *Nat. Biotechnol.* 26(11) (2008) 1261–1268.
7. M. Peracchia, E. Fattal, D. Desmaele, M. Besnard, J. Noel, J. Gomis, M. Appel, J. d'Angelo, P. Couvreur, Stealth® PEGylated polycyanoacrylate nanoparticles for intravenous administration and splenic targeting, *J. Control. Release* 60(1) (1999) 121–128. https://doi.org/10.1016/s0168-3659(99)00063-2.
8. M.E. Lobatto, V. Fuster, Z.A. Fayad, W.J. Mulder, Perspectives and opportunities for nanomedicine in the management of atherosclerosis, *Nat. Rev. Drug Discov.* 10(11) (2011) 835–852.
9. V. Mohanraj, Y. Chen, Nanoparticles-a review, *Trop. J. Pharm. Res.* 5(1) (2006) 561–573.
10. G. Amoabediny, F. Haghiralsadat, S. Naderinezhad, M.N. Helder, E. Akhoundi Kharanaghi, J. Mohammadnejad Arough, B. Zandieh-Doulabi, Overview of preparation methods of polymeric and lipid-based (niosome, solid lipid, liposome) nanoparticles: a comprehensive review, *Int. J. Polym. Mater. Polym. Biomater.* 67(6) (2018) 383–400.
11. D.R. Wilson, R. Sen, J.C. Sunshine, D.M. Pardoll, J.J. Green, Y.J. Kim, Biodegradable STING agonist nanoparticles for enhanced cancer immunotherapy, *Nanomed. Nanotechnol. Biol. Med.* 14(2) (2018) 237–246.
12. D.S. Kohane, Microparticles and nanoparticles for drug delivery, *Biotechnol. Bioeng.* 96(2) (2007) 203–209. https://doi.org/10.1002/bit.21301.
13. J. Panyam, V. Labhasetwar, Biodegradable nanoparticles for drug and gene delivery to cells and tissue, *Adv. Drug Deliv. Rev.* 55(3) (2003) 329–347. https://doi.org/10.1016/s0169-409x(02)00228-4.
14. J. Zhang, M. Saltzman, Engineering biodegradable nanoparticles for drug and gene delivery, *Chem. Eng. Prog.* 109(3) (2013) 25–30.
15. C.J. Destache, T. Belgum, M. Goede, A. Shibata, M.A. Belshan, Antiretroviral release from poly (DL-lactide-co-glycolide) nanoparticles in mice, *J. Antimicrob. Chemother.* 65(10) (2010) 2183–2187.
16. C. Mahendra, M.N. Chandra, M. Murali, M. Abhilash, S.B. Singh, S. Satish, M. Sudarshana, Phyto-fabricated ZnO nanoparticles from Canthium dicoccum (L.) for antimicrobial, anti-tuberculosis and antioxidant activity, *Process Biochem.* 89 (2020) 220–226.
17. J. Pi, L. Shen, E. Yang, H. Shen, D. Huang, R. Wang, C. Hu, H. Jin, H. Cai, J. Cai, Macrophage-targeted

isoniazid–selenium nanoparticles promote antimicrobial immunity and synergize bactericidal destruction of tuberculosis bacilli, *Angew. Chem.* 132(8) (2020) 3252–3260.
18. V. Ramalingam, S. Sundaramahalingam, R. Rajaram, Size-dependent antimycobacterial activity of titanium oxide nanoparticles against Mycobacterium tuberculosis, *J. Mater. Chem. B* 7(27) (2019) 4338–4346.
19. A. Latorre, C. Posch, Y. Garcimartín, A. Celli, M. Sanlorenzo, I. Vujic, J. Ma, M. Zekhtser, K. Rappersberger, S. Ortiz-Urda, DNA and aptamer stabilized gold nanoparticles for targeted delivery of anticancer therapeutics, *Nanoscale* 6(13) (2014) 7436–7442.
20. Q. Mu, F.M. Kievit, R.J. Kant, G. Lin, M. Jeon, M. Zhang, Anti-HER2/neu peptide-conjugated iron oxide nanoparticles for targeted delivery of paclitaxel to breast cancer cells, *Nanoscale* 7(43) (2015) 18010–18014.
21. A. Aires, S.M. Ocampo, B.M. Simões, M.J. Rodríguez, J.F. Cadenas, P. Couleaud, K. Spence, A. Latorre, R. Miranda, Á. Somoza, Multifunctionalized iron oxide nanoparticles for selective drug delivery to CD44-positive cancer cells, *Nanotechnology* 27(6) (2016) 065103.
22. V. Martinelli, G. Cellot, F.M. Toma, C.S. Long, J.H. Caldwell, L. Zentilin, M. Giacca, A. Turco, M. Prato, L. Ballerini, Carbon nanotubes promote growth and spontaneous electrical activity in cultured cardiac myocytes, *Nano Lett.* 12(4) (2012) 1831–1838.
23. A.C. Tang, M.-Y. Chang, Z.C. Tang, H.-J. Li, G.-L. Hwang, P.C. Hsieh, Treatment of acute thromboembolism in mice using heparin-conjugated carbon nanocapsules, *ACS Nano* 6(7) (2012) 6099–6107.
24. A. Chow, D.J. Stuckey, E. Kidher, M. Rocco, R.J. Jabbour, C.A. Mansfield, A. Darzi, S.E. Harding, M.M. Stevens, T. Athanasiou, Human induced pluripotent stem cell-derived cardiomyocyte encapsulating bioactive hydrogels improve rat heart function post myocardial infarction, *Stem Cell Rep.* 9(5) (2017) 1415–1422.
25. J.-Y. Min, Y. Yang, M.F. Sullivan, Q. Ke, K.L. Converso, Y. Chen, J.P. Morgan, Y.-F. Xiao, Long-term improvement of cardiac function in rats after infarction by transplantation of embryonic stem cells, *J. Thoracic Cardiovasc. Surg.* 125(2) (2003) 361–369.
26. T. Qin, R. Ma, Y. Yin, X. Miao, S. Chen, K. Fan, J. Xi, Q. Liu, Y. Gu, Y. Yin, Catalytic inactivation of influenza virus by iron oxide nanozyme, *Theranostics* 9(23) (2019) 6920.
27. X.X. Yang, C.M. Li, C.Z. Huang, Curcumin modified silver nanoparticles for highly efficient inhibition of respiratory syncytial virus infection, *Nanoscale* 8(5) (2016) 3040–3048.
28. Y. Zhang, Y. Huang, S. Li, Polymeric micelles: nanocarriers for cancer-targeted drug delivery, *AAPS Pharmscitech.* 15 (2014) 862–871.
29. C. Lu, D.J. Stewart, J.J. Lee, L. Ji, R. Ramesh, G. Jayachandran, M.I. Nunez, I.I. Wistuba, J.J. Erasmus, M.E. Hicks, Phase I clinical trial of systemically administered TUSC2 (FUS1)-nanoparticles mediating functional gene transfer in humans, *PLoS One* 7(4) (2012) e34833.
30. P. Meers, M. Neville, V. Malinin, A. Scotto, G. Sardaryan, R. Kurumunda, C. Mackinson, G. James, S. Fisher, W. Perkins, Biofilm penetration, triggered release and in vivo activity of inhaled liposomal amikacin in chronic Pseudomonas aeruginosa lung infections, *J. Antimicrob. Chemother.* 61(4) (2008) 859–868.
31. K. Pawelec, S. Best, R. Cameron, Collagen: a network for regenerative medicine, *J. Mater. Chem. B* 4(40) (2016) 6484–6496. https://doi.org/10.1039/c6tb00807k.
32. Z. Lu, T.-K. Yeh, M. Tsai, J.L.-S. Au, M.G. Wientjes, Paclitaxel-loaded gelatin nanoparticles for intravesical bladder cancer therapy, *Clin. Cancer Res.*10(22) (2004) 7677–7684.
33. Z. Noor, Nanohydroxyapatite application to osteoporosis management, *J. Osteoporos.* 2013(1) (2013) 679025.
34. R. Gillani, B. Ercan, A. Qiao, T.J. Webster, Nanofunctionalized zirconia and barium sulfate particles as bone cement additives, *Int. J. Nanomed.* (2010) 1–11.
35. A. Ricker, P. Liu-Snyder, T.J. Webster, The influence of nano MgO and BaSO4 particle size additives on properties of PMMA bone cement, *Int. J. Nanomed.* 3(1) (2008) 125–132.
36. M.D. Schofer, P.P. Roessler, J. Schaefer, C. Theisen, S. Schlimme, J.T. Heverhagen, M. Voelker, R. Dersch, S. Agarwal, S. Fuchs-Winkelmann, Electrospun PLLA nanofiber scaffolds and their use in combination with BMP-2 for reconstruction of bone defects, *PLoS One* 6(9) (2011) e25462.
37. Y. Xue, X. Xu, X.-Q. Zhang, O.C. Farokhzad, R. Langer, Preventing diet-induced obesity in mice by adipose tissue transformation and angiogenesis using targeted nanoparticles, *Proc. Nat. Acad. Sci. USA* 113(20) (2016) 5552–5557.
38. C. Jiang, M.A. Cano-Vega, F. Yue, L. Kuang, N. Narayanan, G. Uzunalli, M.P. Merkel, S. Kuang, M. Deng, Dibenzazepine-loaded nanoparticles induce local browning of white adipose tissue to counteract obesity, *Mol. Ther.* 25(7) (2017) 1718–1729.
39. W. Bao, L. Xie, X. Zeng, H. Kang, S. Wen, B. Cui, W. Li, Y. Qian, J. Wu, T. Li, A cocktail-inspired male birth control strategy with physical/chemical dual contraceptive effects and remote self-cleared properties, *ACS Nano* 13(2) (2019) 1003–1011.
40. J.H. Kim, H.J. Lee, S.H. Doo, W.J. Yang, D. Choi, J.H. Kim, J.H. Won, Y.S. Song, Use of nanoparticles to monitor human mesenchymal stem cells transplanted into penile cavernosum of rats with erectile dysfunction, *Kor. J. Urol.* 56(4) (2015) 280–287.
41. D. Scuteri, G. Sandrini, S. Tamburin, M.T. Corasaniti, P. Nicotera, P. Tonin, G. Bagetta, Bergamot rehabilitation Against agitation in dementia (BRAINAID): study protocol for a randomized, double-blind, placebo-controlled trial to assess the efficacy of furocoumarin-free bergamot loaded in a nanotechnology-based delivery system of the essential oil in the treatment of agitation in elderly affected by severe dementia, *Phytother. Res.* 35(10) (2021) 5333–5338.
42. J. Lalani, M. Rathi, M. Lalan, A. Misra, Protein functionalized tramadol-loaded PLGA nanoparticles: preparation, optimization, stability and pharmacodynamic studies, *Drug Develop. Indus. Pharm.* 39(6) (2013) 854–864.
43. L. Minati, G. Speranza, S. Torrengo, L. Toniutti, C. Migliaresi, D. Maniglio, M. Ferrari, A. Chiasera, Characterization of thiol-functionalized carbon nanotubes on gold surfaces, *Surf. Sci.* 604(17–18) (2010) 1414–1419.
44. Y. Yu, D. Chen, Y. Li, W. Yang, J. Tu, Y. Shen, Improving the topical ocular pharmacokinetics of lyophilized cyclosporine A-loaded micelles: formulation, in vitro and in vivo studies, *Drug Deliv.* 25(1) (2018) 888–899.
45. N. Venkateswaran, Y. Bian, P.K. Gupta, Practical guidance for the use of loteprednol etabonate ophthalmic suspension 0.25% in the management of dry eye disease, *Clin. Ophthalmol.* (2022) 349–355.
46. J. Zhang, L. Huang, G. Ge, K. Hu, Emerging epigenetic-based nanotechnology for cancer therapy: modulating the tumor microenvironment, *Adv. Sci.* 10(7) (2023) 2206169.

47. J. Li, S. Xiao, Y. Xu, S. Zuo, Z. Zha, W. Ke, C. He, Z. Ge, Smart asymmetric vesicles with triggered availability of inner cell-penetrating shells for specific intracellular drug delivery, *ACS Appl. Mater. Interfaces* 9(21) (2017) 17727–17735. https://doi.org/10.1021/acsami.7b02808.
48. Y. Pan, W. Tang, W. Fan, J. Zhang, X. Chen, Development of nanotechnology-mediated precision radiotherapy for anti-metastasis and radioprotection, *Chem. Soc. Rev.* 51(23) (2022) 9759–9830.
49. W.G. Kreyling, A.M. Abdelmonem, Z. Ali, F. Alves, M. Geiser, N. Haberl, R. Hartmann, S. Hirn, D.J. De Aberasturi, K. Kantner, In vivo integrity of polymer-coated gold nanoparticles, *Nat. Nanotechnol.* 10(7) (2015) 619–623. https://doi.org/10.1038/nnano.2015.111.
50. J. McCarroll, J. Teo, C. Boyer, D. Goldstein, M. Kavallaris, P.A. Phillips, Potential applications of nanotechnology for the diagnosis and treatment of pancreatic cancer, *Front. Physiol.* 5 (2014) 2.
51. F. Vahedifard, K. Chakravarthy, Nanomedicine for COVID-19: the role of nanotechnology in the treatment and diagnosis of COVID-19, *Emerg. Mater.* 4(1) (2021) 75–99.
52. L. Milane, M. Amiji, Clinical approval of nanotechnology-based SARS-CoV-2 mRNA vaccines: impact on translational nanomedicine, *Drug Deliv. Trans. Res.* 11 (2021) 1309–1315.
53. S.L. Gaytan, E. Beaven, S.S. Gadad, M. Nurunnabi, Progress and prospect of nanotechnology for cardiac fibrosis treatment, *Interdiscip. Med.* 1(4) (2023) e20230018.
54. S. La Francesca, Nano technology and stem cell therapy for CV diseases: potential applications, *Methodist DeBakey Cardiovasc. J.* 8(1) (2012) 28.
55. E.S. Smith, J.E. Porterfield, R.M. Kannan, Leveraging the interplay of nanotechnology and neuroscience: designing new avenues for treating central nervous system disorders, *Adv. Drug Deliv. Rev.* 148 (2019) 181–203.
56. Y. Gong, P. Chowdhury, P.K. Nagesh, T.J. Cory, C. Dezfuli, S. Kodidela, A. Singh, M.M. Yallapu, S. Kumar, Nanotechnology approaches for delivery of cytochrome P450 substrates in HIV treatment, *Exp. Opin. Drug Deliv.* 16(8) (2019) 869–882.
57. P. Collini, M. Noursadeghi, I. Sabroe, R. F Miller, D. H Dockrell, Monocyte and macrophage dysfunction as a cause of HIV-1 induced dysfunction of innate immunity, *Curr. Mol. Med.* 10(8) (2010) 727–740.
58. G. Coco, G. Buffon, A. Taloni, G. Giannaccare, Recent Advances in Nanotechnology for the Treatment of Dry Eye Disease, *Nanomaterials* 14(8) (2024) 669.
59. W. Liang, C. Zhou, J. Bai, H. Zhang, H. Long, B. Jiang, H. Dai, J. Wang, H. Zhang, J. Zhao, Current developments and future perspectives of nanotechnology in orthopedic implants: an updated review, *Front. Bioeng. Biotechnol.* 12 (2024) 1342340.
60. R. Shandilya, N. Pathak, N.K. Lohiya, R.S. Sharma, P.K. Mishra, Nanotechnology in reproductive medicine: opportunities for clinical translation, *Clin. Exp. Reprod. Med.* 47(4) (2020) 245.
61. Y.H. Tsou, B. Wang, W. Ho, B. Hu, P. Tang, S. Sweet, X.Q. Zhang, X. Xu, Nanotechnology-mediated drug delivery for the treatment of obesity and its related comorbidities, *Adv. Healthc. Mater.* 8(12) (2019) 1801184.
62. S. Jin, V. Labhasetwar, Nanotechnology in urology, *Urol. Clin.* 36(2) (2009) 179–188.
63. K.V. Chakravarthy, F.J. Boehm, P.J. Christo, Nanotechnology: a promising new paradigm for the control of pain, *Pain Med.* 19(2) (2018) 232–243.
64. P.M. Barbosa, L.N. Ferreira, V.T. Cruz, A. Silva, H. Szrek, Healthcare, clinical factors and rehabilitation predicting quality of life in first-time stroke patients: a 12-month longitudinal study, *J. Stroke Cerebrovasc. Dis.* 31(4) (2022) 106300. https://doi.org/10.1016/j.jstrokecerebrovasdis.2021.106300.
65. P. Trucillo, Biomaterials for drug delivery and human applications, *Materials.* 17(2) (2024) 456. https://doi.org/10.3390/ma17020456.
66. Y.T. Jadotte, D.S. Lane, Population health rounds: a novel vehicle for training in population medicine and clinical preventive medicine, *J. Public Health Manag. Pract.* 27(Supplement 3) (2021) S139–S145. https://doi.org/10.1097/PHH.0000000000001326.
67. P.C. Pires, F. Mascarenhas-Melo, K. Pedrosa, D. Lopes, J. Lopes, A. Macário-Soares, D. Peixoto, P.S. Giram, F. Veiga, A.C. Paiva-Santos, Polymer-based biomaterials for pharmaceutical and biomedical applications: a focus on topical drug administration, *Eur. Polym. J.* 187 (2023) 111868.
68. P. Dhavalikar, A. Robinson, Z. Lan, D. Jenkins, M. Chwatko, K. Salhadar, A. Jose, R. Kar, E. Shoga, A. Kannapiran, Review of integrin-targeting biomaterials in tissue engineering, *Adv. Healthcare Mater.* 9(23) (2020) 2000795.
69. S. Chatterjee, M. Saxena, D. Padmanabhan, M. Jayachandra, H.J. Pandya, Futuristic medical implants using bioresorbable materials and devices, *Biosens. Bioelectron.* 142 (2019) 111489. https://doi.org/10.1016/j.bios.2019.111489.
70. K. Zinoviev, L.G. Carrascosa, J. Sánchez del Río, B. Sepúlveda, C. Domínguez, L.M. Lechuga, Silicon photonic biosensors for lab-on-a-chip applications, *Adv. Drug Deliv. Rev.* (2008). https://doi.org/10.1155/2008/383927.
71. M.F. Zayed, H.E. Ahmed, S. Ihmaid, A.-S.M. Omar, A.S. Abdelrahim, Synthesis and screening of some new fluorinated quinazolinone–sulphonamideáhybrids as anticancer agents, *J. Taibah Univers. Med. Sci.* 10(3) (2015) 333–339.
72. E. Hutter, D. Maysinger, Gold nanoparticles and quantum dots for bioimaging, *Microsc. Res. Tech.* 74(7) (2011) 592–604. https://doi.org/10.1002/jemt.20928.
73. Z. Liang, H. Zhu, X. Wang, B. Jing, Z. Li, X. Xia, H. Sun, Y. Yang, W. Zhang, L. Shi, Adjuvants for coronavirus vaccines, *Front. Immunol.* 11 (2020) 589833. https://doi.org/10.3389/fimmu.2020.589833.
74. K. Sudesh, T. Iwata, Sustainability of biobased and biodegradable plastics, *Clean–Soil, Air, Water* 36(5-6) (2008) 433–442. https://doi.org/10.1002/clen.200700183.
75. A. Prieto, To be, or not to be biodegradable… that is the question for the bio-based plastics, *Microb. Biotechnol.* 9(5) (2016) 652–657. https://doi.org/10.1111/1751-7915.12393.
76. N.A. Oladoja, E.I. Unuabonah, O.S. Amuda, O.M. Kolawole, *Polysaccharides as a Green and Sustainable Resources for Water and Wastewater Treatment.* Springer, New York (2017).
77. K.J. Gandhi, S.V. Deshmane, K.R. Biyani, Polymers in pharmaceutical drug delivery system: a review, *Int. J. Pharm. Sci. Rev. Res.* 14(2) (2012) 57–66.
78. M.D. Pegram, A. Lipton, D.F. Hayes, B.L. Weber, J.M. Baselga, D. Tripathy, D. Baly, S.A. Baughman, T. Twaddell, J.A. Glaspy, Phase II study of receptor-enhanced chemosensitivity using recombinant humanized anti-p185HER2/neu monoclonal antibody plus cisplatin in patients with HER2/neu-overexpressing metastatic breast cancer refractory to chemotherapy treatment, *J. Clin. Oncol.* 16(8) (1998) 2659–2671. https://doi.org/10.1200/JCO.1998.16.8.2659.
79. W.B. Liechty, D.R. Kryscio, B.V. Slaughter, N.A. Peppas, Polymers for drug delivery systems, *Annu. Rev. Chem.*

Biomol. Eng. 1(1) (2010) 149–173. https://doi.org/10.1146/annurev-chembioeng-073009-100847.
80. P. Trucillo, R. Campardelli, Production of solid lipid nanoparticles with a supercritical fluid assisted process, *J. Supercrit. Fluids* 143 (2019) 16–23. https://doi.org/10.1016/j.supflu.2018.08.001.
81. Y. Malam, M. Loizidou, A.M. Seifalian, *Trends Pharmacol. Sci.* 30(11) (2009) 592–599. https://doi.org/10.1016/j.tips.2009.08.004.
82. C.-H. Tang, H.-L. Chen, J.-R. Dong, Solid lipid nanoparticles (SLNs) and nanostructured lipid carriers (NLCs) as food-grade nanovehicles for hydrophobic nutraceuticals or bioactives, *Appl. Sci.* 13(3) (2023) 1726. https://doi.org/10.3390/app13031726.
83. B. Mesure, P. Menu, J.K. Venkatesan, M. Cucchiarini, É. Velot, Biomaterials and gene therapy: a smart combination for MSC musculoskeletal engineering, *Curr. Stem Cell Res. Ther.* 14(4) (2019) 337–343. https://doi.org/10.2174/1574888X14666181205121658.
84. Y. Yu, Y. Gao, L. He, B. Fang, W. Ge, P. Yang, Y. Ju, X. Xie, L. Lei, Biomaterial-based gene therapy, *MedComm* 4(3) (2023) e259. https://doi.org/10.1002/mco2.259.
85. R.K. Fisher III, P.C. West, S.I. Mattern-Schain, M.D. Best, S.S. Kirkpatrick, R.A. Dieter III, J.D. Arnold, M.R. Buckley, M.M. McNally, M.B. Freeman, Advances in the formulation and assembly of non-cationic lipid nanoparticles for the medical application of gene therapeutics, *Nanomater.* 11(3) (2021) 825. https://doi.org/10.3390/nano11030825.
86. M. Alipour, M. Baneshi, S. Hosseinkhani, R. Mahmoudi, A. Jabari Arabzadeh, M. Akrami, J. Mehrzad, H. Bardania, Recent progress in biomedical applications of RGD-based ligand: from precise cancer theranostics to biomaterial engineering: a systematic review, *J. Biomed. Mater. Res. A* 108(4) (2020) 839–850. https://doi.org/10.1002/jbm.a.36862.
87. Y. Yang, Y. Xiao, Biomaterials regulating bone hematoma for osteogenesis, Adv. *Healthc. Mater.* 9(23) (2020) 2000726. https://doi.org/10.1002/adhm.202000726.
88. M.D. Chavanpatil, A. Khdair, Y. Patil, H. Handa, G. Mao, J. Panyam, Polymer-surfactant nanoparticles for sustained release of water-soluble drugs, *J. Pharm. Sci.* 96(12) (2007) 3379–3389. https://doi.org/10.1002/jps.20961.
89. P. Di Pietro, G. Strano, L. Zuccarello, C. Satriano, Gold and silver nanoparticles for applications in theranostics, *Curr. Top. Med. Chem.* 16(27) (2016) 3069–3102. https://doi.org/10.2174/1568026616666160715163346.
90. S. Sun, Y. Cui, B. Yuan, M. Dou, G. Wang, H. Xu, J. Wang, W. Yin, D. Wu, C. Peng, Drug delivery systems based on polyethylene glycol hydrogels for enhanced bone regeneration, *Front. Bioeng. Biotechnol.* 11 (2023) 1117647. https://doi.org/10.3389/fbioe.2023.1117647.
91. F. Rancan, D. Papakostas, S. Hadam, S. Hackbarth, T. Delair, C. Primard, B. Verrier, W. Sterry, U. Blume-Peytavi, A. Vogt, Investigation of polylactic acid (PLA) nanoparticles as drug delivery systems for local dermatotherapy, *Pharm. Res.* 26 (2009) 2027–2036. https://doi.org/10.1007/s11095-009-9919-x.
92. D.R. Perinelli, M. Cespi, G. Bonacucina, G.F. Palmieri, PEGylated polylactide (PLA) and poly (lactic-co-glycolic acid)(PLGA) copolymers for the design of drug delivery systems, *J. Pharm. Investig.* 49 (2019) 443–458.
93. P. Trucillo, R. Campardelli, E. Reverchon, Liposomes: from bangham to supercritical fluids, *Processes* 8(9) (2020) 1022. https://doi.org/10.3390/pr8091022.
94. H.H. Tønnesen, J. Karlsen, Alginate in drug delivery systems, *Drug Dev. Ind. Pharm.* 28(6) (2002) 621–630. https://doi.org/10.1081/DDC-120003853.
95. F. Tao, S. Ma, H. Tao, L. Jin, Y. Luo, J. Zheng, W. Xiang, H. Deng, Chitosan-based drug delivery systems: from synthesis strategy to osteomyelitis treatment–a review, *Carbohydr. Polym.* 251 (2021) 117063. https://doi.org/10.1016/j.carbpol.2020.117063.
96. Á. Haimhoffer, Á. Rusznyák, K. Réti-Nagy, G. Vasvári, J. Váradi, M. Vecsernyés, I. Bácskay, P. Fehér, Z. Ujhelyi, F. Fenyvesi, Cyclodextrins in drug delivery systems and their effects on biological barriers, *Sci. Pharm.* 87(4) (2019) 33. https://doi.org/10.3390/scipharm87040033.
97. J. Wang, B. Li, L. Qiu, X. Qiao, H. Yang, Dendrimer-based drug delivery systems: history, challenges, and latest developments, *J. Biol. Eng.* 16(1) (2022) 18.
98. A. Chaudhuri, K. Ramesh, D.N. Kumar, D. Dehari, S. Singh, D. Kumar, A.K. Agrawal, Polymeric micelles: a novel drug delivery system for the treatment of breast cancer, *J. Drug Deliv. Sci. Technol.* 77 (2022) 103886. https://doi.org/10.1016/j.jddst.2022.103886.
99. P. Lambin, R.T. Leijenaar, T.M. Deist, J. Peerlings, E.E. De Jong, J. Van Timmeren, S. Sanduleanu, R.T. Larue, A.J. Even, A. Jochems, Radiomics: the bridge between medical imaging and personalized medicine, *Nat. Rev. Clin. Oncol.* 14(12) (2017) 749–762. https://doi.org/10.1038/nrclinonc.2017.141.
100. C. Perez-Medina, A.J. Teunissen, E. Kluza, W.J. Mulder, R. Van der Meel, Nuclear imaging approaches facilitating nanomedicine translation, *Adv. Drug Deliv. Rev.* 154 (2020) 123–141. https://doi.org/10.1016/j.addr.2020.07.017.
101. A.A. Kislukhin, H. Xu, S.R. Adams, K.H. Narsinh, R.Y. Tsien, E.T. Ahrens, Paramagnetic fluorinated nanoemulsions for sensitive cellular fluorine-19 magnetic resonance imaging, *Nat. Mater.* 15(6) (2016) 662–668. https://doi.org/10.1038/nmat4585.
102. I. Tirotta, V. Dichiarante, C. Pigliacelli, G. Cavallo, G. Terraneo, F.B. Bombelli, P. Metrangolo, G. Resnati, 19F magnetic resonance imaging (MRI): from design of materials to clinical applications, *Chem. Rev.* 115(2) (2015) 1106–1129. https://doi.org/10.1021/cr500286d.
103. M.L. Senders, A.E. Meerwaldt, M.M. van Leent, B.L. Sanchez-Gaytan, J.C. van de Voort, Y.C. Toner, A. Maier, E.D. Klein, N.A. Sullivan, A.M. Sofias, Probing myeloid cell dynamics in ischaemic heart disease by nanotracer hot-spot imaging, *Nat. Nanotechnol.* 15(5) (2020) 398–405. https://doi.org/10.1038/s41565-020-0642-4.
104. D.P. Cormode, E. Roessl, A. Thran, T. Skajaa, R.E. Gordon, J.-P. Schlomka, V. Fuster, E.A. Fisher, W.J. Mulder, R. Proksa, Atherosclerotic plaque composition: analysis with multicolor CT and targeted gold nanoparticles, *Radiology* 256(3) (2010) 774–782. https://doi.org/10.1148/radiol.10092473.
105. A.J. Teunissen, C. Pérez-Medina, A. Meijerink, W.J. Mulder, Investigating supramolecular systems using Förster resonance energy transfer, *Chem. Soc. Rev.* 47(18) (2018) 7027–7044. https://doi.org/10.1039/C8CS00278A.
106. D.L. Bailey, K.P. Willowson, Quantitative SPECT/CT: SPECT joins PET as a quantitative imaging modality, *Eur. J. Nucl. Med. Mol. Imaging* 41 (2014) 17–25. https://doi.org/10.1007/s00259-013-2542-4.
107. Y. Xing, J. Zhao, P.S. Conti, K. Chen, Radiolabeled nanoparticles for multimodality tumor imaging, *Theranostics* 4(3) (2014) 290. https://doi.org/10.7150/thno.7341.
108. S. Kalimuthu, J.H. Jeong, J.M. Oh, B.-C. Ahn, Drug discovery by molecular imaging and monitoring therapy response in lymphoma, *Int. J. Mol. Sci.* 18(8) (2017) 1639. https://doi.org/10.3390/ijms18081639.
109. K. Heinzmann, L.M. Carter, J.S. Lewis, E.O. Aboagye, Multiplexed imaging for diagnosis and therapy, *Nat.*

Biomed. Eng. 1(9) (2017) 697–713. https://doi.org/10.1038/s41551-017-0131-8.
110. F.-M. Lu, Z. Yuan, PET/SPECT molecular imaging in clinical neuroscience: recent advances in the investigation of CNS diseases, *Quant. ImAging Med. Surg.* 5(3) (2015) 433. https://doi.org/10.3978/j.issn.2223-4292.2015.03.16.
111. J. Weber, U. Haberkorn, W. Mier, Cancer stratification by molecular imaging, *Int. J. Mol. Sci.* 16(3) (2015) 4918–4946. https://doi.org/10.3390/ijms16034918.
112. R. van der Meel, T. Lammers, W.E. Hennink, Cancer nanomedicines: oversold or underappreciated? *Expert Opin Drug Deliv.* 14(1) (2017) 1–5. https://doi.org/10.1080/17425247.2017.1262346.
113. J.M. Metselaar, T. Lammers, Challenges in nanomedicine clinical translation, *Drug Deliv. Transl. Res.* 10 (2020) 721–725. https://doi.org/10.1007/s13346-020-00740-5.
114. M.E. Lobatto, T. Binderup, P.M. Robson, L.F. Giesen, C. Calcagno, J. Witjes, F. Fay, S. Baxter, C.H. Wessel, M. Eldib, Multimodal positron emission tomography imaging to quantify uptake of 89Zr-labeled liposomes in the atherosclerotic vessel wall, *Bioconjugate Chem.* 31(2) (2019) 360–368. https://doi.org/10.1021/acs.bioconjchem.9b00256.
115. T. Binderup, R. Duivenvoorden, F. Fay, M.M. van Leent, J. Malkus, S. Baxter, S. Ishino, Y. Zhao, B. Sanchez-Gaytan, A.J. Teunissen, Imaging-assisted nanoimmunotherapy for atherosclerosis in multiple species, *Sci. Transl. Med.* 11(506) (2019) eaaw7736. https://doi.org/10.1126/scitranslmed.aaw7736.
116. M. Lameijer, T. Binderup, M.M. Van Leent, M.L. Senders, F. Fay, J. Malkus, B.L. Sanchez-Gaytan, A.J. Teunissen, N. Karakatsanis, P. Robson, Efficacy and safety assessment of a TRAF6-targeted nanoimmunotherapy in atherosclerotic mice and non-human primates, *Nat. Biomed. Eng.* 2(5) (2018) 279–292. https://doi.org/10.1038/s41551-018-0221-2.
117. T.J. Beldman, M.L. Senders, A. Alaarg, C. Pérez-Medina, J. Tang, Y. Zhao, F. Fay, J. Deichmöller, B. Born, E. Desclos, Hyaluronan nanoparticles selectively target plaque-associated macrophages and improve plaque stability in atherosclerosis, *ACS Nano* 11(6) (2017) 5785–5799. https://doi.org/10.1021/acsnano.7b01385.
118. J. Nicolas, S. Mura, D. Brambilla, N. Mackiewicz, P. Couvreur, Design, functionalization strategies and biomedical applications of targeted biodegradable/biocompatible polymer-based nanocarriers for drug delivery, *Chem. Soc. Rev.* 42(3) (2013) 1147–1235. https://doi.org/10.1039/c2cs35265f.
119. W.H. De Jong, P.J. Borm, Drug delivery and nanoparticles: applications and hazards, *Int. J. Nanomed.* 3(2) (2008) 133–149. https://doi.org/10.2147/ijn.s596.
120. R.C. Rowe, P. Sheskey, M. Quinn, *Handbook of Pharmaceutical Excipients*, Libros Digitales-Pharmaceutical Press (2009). 6th Edition, Pharmaceutical Press, 506–509.
121. M. Yaşar, F.D. Yöntem, M.V. Kahraman, N.K. Apohan, Z. Aktaş, M.O. Öncül, H. Akcakaya, Polymeric nanoparticles for selective protein recognition by using thiol-ene miniemulsion photopolymerization, *J. Biomater. Sci. Polym. Ed.* 31(16) (2020) 2044–2059. https://doi.org/10.1080/09205063.2020.1793705.
122. R. Puri, S. A Berhe, E.O Akala, pH-Sensitive polymeric nanoparticles fabricated by dispersion polymerization for the delivery of bioactive agents, *Pharm. Nanotechnol.* 5(1) (2017) 44–66. https://doi.org/10.2174/2211738505666170110102320.
123. C. Popovici, M. Popa, V. Sunel, L.I. Atanase, D.L. Ichim, Drug delivery systems based on Pluronic micelles with antimicrobial activity, *Polym.* 14(15) (2022) 3007. https://doi.org/10.3390/polym14153007.
124. S. Perumal, R. Atchudan, W. Lee, A review of polymeric micelles and their applications, *Polym.* 14(12) (2022) 2510. https://doi.org/10.3390/polym14122510.
125. J. Varshosaz, S. Taymouri, F. Hassanzadeh, S. Haghjooy Javanmard, M. Rostami, Folated synperonic-cholesteryl hemisuccinate polymeric micelles for the targeted delivery of docetaxel in melanoma, *Biomed Res. Int.* 2015(1) (2015) 746093. https://doi.org/10.1155/2015/746093.
126. Y. Men, W. Li, Y. Tu, F. Peng, G.-J.A. Janssen, R.J. Nolte, D.A. Wilson, Nonequilibrium reshaping of polymersomes via polymer addition, *ACS Nano* 13(11) (2019) 12767–12773. https://doi.org/10.1021/acsnano.9b04740.
127. T. Zavvar, M. Babaei, K. Abnous, S.M. Taghdisi, S. Nekooei, M. Ramezani, M. Alibolandi, Synthesis of multimodal polymersomes for targeted drug delivery and MR/fluorescence imaging in metastatic breast cancer model, *Int. J. Pharm.* 578 (2020) 119091. https://doi.org/10.1016/j.ijpharm.2020.119091.
128. Y. Zhou, R. Chen, H. Yang, C. Bao, J. Fan, C. Wang, Q. Lin, L. Zhu, Light-responsive polymersomes with a charge-switch for targeted drug delivery, *J. Mater. Chem. B* 8(4) (2020) 727–735. https://doi.org/10.1039/c9tb02411e.
129. D.J. Slamon, W. Godolphin, L.A. Jones, J.A. Holt, S.G. Wong, D.E. Keith, W.J. Levin, S.G. Stuart, J. Udove, A. Ullrich, Studies of the HER-2/neu proto-oncogene in human breast and ovarian cancer, *Science* 244(4905) (1989) 707–712. https://doi.org/10.1126/science.2470152.
130. J. Li, A. Dirisala, Z. Ge, Y. Wang, W. Yin, W. Ke, K. Toh, J. Xie, Y. Matsumoto, Y. Anraku, Therapeutic vesicular nanoreactors with tumor-specific activation and self-destruction for synergistic tumor ablation, *Angew. Chem.* 129(45) (2017) 14213–14218. https://doi.org/10.1002/anie.201706964.
131. J. Li, Y. Anraku, K. Kataoka, Self-boosting catalytic nanoreactors integrated with triggerable crosslinking membrane networks for initiation of immunogenic cell death by pyroptosis, *Angew. Chem.* 132(32) (2020) 13628–13632. https://doi.org/10.1002/anie.202004180.
132. N.S. Pillay, A. Daniels, M. Singh, Folate-targeted transgenic activity of dendrimer functionalized selenium nanoparticles in vitro, *Int. J. Mol. Sci.* 21(19) (2020) 7177. https://doi.org/10.3390/ijms21197177.
133. D. Bartusik-Aebisher, G. Chrzanowski, Z. Bober, D. Aebisher, An analytical study of Trastuzumab-dendrimer-fluorine drug delivery system in breast cancer therapy in vitro, *Biomed. Pharmacother.* 133 (2021) 111053. https://doi.org/10.1016/j.biopha.2020.111053.
134. P. Mota, R.F. Pires, J. Serpa, V.D. Bonifácio, l-buthionine sulfoximine detection and quantification in polyurea dendrimer nanoformulations, *Molecules* 24(17) (2019) 3111. https://doi.org/10.3390/molecules24173111.
135. S. Zhang, V. Lloveras, S. Lope-Piedrafita, P. Calero-Pérez, S. Wu, A.P. Candiota, J. Vidal-Gancedo, Metal-free radical dendrimers as MRI contrast agents for glioblastoma diagnosis: ex vivo and in vivo approaches, *Biomacromolecules* 23(7) (2022) 2767–2777. https://doi.org/10.1021/acs.biomac.2c00088.
136. N. Knauer, V. Arkhipova, G. Li, M. Hewera, E. Pashkina, P.-H. Nguyen, M. Meschaninova, V. Kozlov, W. Zhang, R.S. Croner, In vitro validation of the therapeutic potential of dendrimer-based nanoformulations against tumor stem cells, *Int. J. Mol. Sci.* 23(10) (2022) 5691. https://doi.org/10.3390/ijms23105691.
137. Y. Chabria, G.P. Duffy, A.J. Lowery, R.M. Dwyer, Hydrogels: 3D drug delivery systems for nanoparticles and

extracellular vesicles, *Biomed* 9(11) (2021) 1694. https://doi.org/10.3390/biomedicines9111694.
138. S.J. Yoon, Y.J. Moon, H.J. Chun, D.H. Yang, Doxorubicin-hydrochloride/cisplatin-loaded hydrogel/nanosized (2-hydroxypropyl)-beta-cyclodextrin local drug-delivery system for osteosarcoma treatment in vivo, *Nanomaterials* 9(12) (2019) 1652. https://doi.org/10.3390/nano9121652.
139. A. Szwajca, S. Juszczyńska, M. Jarzębski, E. Baryła-Pankiewicz, Incorporation of fluorescent fluorinated methacrylate nano-sized particles into chitosan matrix formed as a membranes or beads, *Polym.* 14(13) (2022) 2750. https://doi.org/10.3390/polym14132750.
140. L. Liu, Q. Gao, X. Lu, H. Zhou, In situ forming hydrogels based on chitosan for drug delivery and tissue regeneration, *Asian J. Pharma. Sci.* 11(6) (2016) 673–683.
141. M. Nasr, S. Saber, A.Y. Bazeed, H.A. Ramadan, A. Ebada, A.L. Ciorba, S. Cavalu, H.I. Elagamy, Advantages of cubosomal formulation for gatifloxacin delivery in the treatment of bacterial keratitis: in vitro and in vivo approach using clinical isolate of methicillin-resistant Staphylococcus aureus, *Materials* 15(9) (2022) 3374. https://doi.org/10.3390/ma15093374.
142. L. Zhang, J. Li, D. Tian, L. Sun, X. Wang, M. Tian, Theranostic combinatorial drug-loaded coated cubosomes for enhanced targeting and efficacy against cancer cells, *Cell Death Disease* 11(1) (2020) 1. https://doi.org/10.1038/s41419-019-2182-0.
143. L. Boge, K. Hallstensson, L. Ringstad, J. Johansson, T. Andersson, M. Davoudi, P.T. Larsson, M. Mahlapuu, J. Håkansson, M. Andersson, Cubosomes for topical delivery of the antimicrobial peptide LL-37. *Eur. J. Pharm. Biopharm.* 134 (2019) 60–67. doi: 10.1016/j.ejpb.2018.11.009.
144. D.-G. Kim, Y.-I. Jeong, C. Choi, S.-H. Roh, S.-K. Kang, M.-K. Jang, J.-W. Nah, Retinol-encapsulated low molecular water-soluble chitosan nanoparticles, *Int. J. Pharm.* 319(1–2) (2006) 130–138. https://doi.org/10.1016/j.ijpharm.2006.03.040.
145. Y.-S. Chen, R.G. Alany, S.A. Young, C.R. Green, I.D. Rupenthal, In vitro release characteristics and cellular uptake of poly (D, L-lactic-co-glycolic acid) nanoparticles for topical delivery of antisense oligodeoxynucleotides, *Drug Deliv.* 18(7) (2011) 493–501. https://doi.org/10.3109/10717544.2011.589088.
146. S.R. Youngren, R.K. Tekade, B. Gustilo, P.R. Hoffmann, M.B. Chougule, STAT6 siRNA matrix-loaded gelatin nanocarriers: formulation, characterization, and ex vivo proof of concept using adenocarcinoma cells, *Biomed Res. Int.* 2013(1) (2013) 858946. https://doi.org/10.1155/2013/858946.
147. B. Balzus, F.F. Sahle, S. Hönzke, C. Gerecke, F. Schumacher, S. Hedtrich, B. Kleuser, R. Bodmeier, Formulation and ex vivo evaluation of polymeric nanoparticles for controlled delivery of corticosteroids to the skin and the corneal epithelium, *Eur J Pharm Biopharm.* 115 (2017) 122–130. https://doi.org/10.1016/j.ejpb.2017.02.001.
148. M. Saadallah, O. Hamid, Formulation and evaluation of rosuvastatin calcium polymeric nanoparticles-loaded transdermal patch, *Iraqi J. Pharm.* 18 (2022) 22–38.
149. M.S. Muthu, S. Singh, Studies on biodegradable polymeric nanoparticles of risperidone: in vitro and in vivo evaluation, *Nanomed.* 3(3) (2008) 305–319.
150. K. H Singh, U. A Shinde, Development and evaluation of novel polymeric nanoparticles of brimonidine tartrate, *Curr. Drug Deliv.* 7(3) (2010) 244–251. https://doi.org/10.2174/156720110791561008.
151. L. Song, Y. Shen, J. Hou, L. Lei, S. Guo, C. Qian, Polymeric micelles for parenteral delivery of curcumin: preparation, characterization, and in vitro evaluation, *Coll. Surf. A.* 390(1–3) (2011) 25–32. https://doi.org/10.1016/j.colsurfa.2011.08.031.
152. A. Richter, C. Olbrich, M. Krause, J. Hoffmann, T. Kissel, Polymeric micelles for parenteral delivery of sagopilone: physicochemical characterization, novel formulation approaches and their toxicity assessment in vitro as well as in vivo, *Eur. J. Pharm. Biopharm.* 75(2) (2010) 80–89. https://doi.org/10.1016/j.ejpb.2010.02.010.
153. Y.A. Haggag, A.K. Abosalha, M.M. Tambuwala, E.Y. Osman, S.A. El-Gizawy, E.A. Essa, A.A. Donia, Polymeric nanoencapsulation of zaleplon into PLGA nanoparticles for enhanced pharmacokinetics and pharmacological activity, *BDDCS* 42(1) (2021) 12–23. https://doi.org/10.1002/bdd.2255.
154. F. Roncato, F. Rruga, E. Porcù, E. Casarin, R. Ronca, F. Maccarinelli, N. Realdon, G. Basso, R. Alon, G. Viola, Improvement and extension of anti-EGFR targeting in breast cancer therapy by integration with the avidin-nucleic-acid-nano-assemblies, *Nat. Commun.* 9(1) (2018) 4070. https://doi.org/10.1038/s41467-018-06602-6.
155. A.J. Clark, M.E. Davis, Increased brain uptake of targeted nanoparticles by adding an acid-cleavable linkage between transferrin and the nanoparticle core, *Proc. Nat. Acad. Sci.* 112(40) (2015) 12486–12491. https://doi.org/10.1073/pnas.1517048112
156. A. Aruffo, I. Stamenkovic, M. Melnick, C.B. Underhill, B. Seed, CD44 is the principal cell surface receptor for hyaluronate, *Cell* 61(7) (1990) 1303–1313. https://doi.org/10.1016/0092-8674(90)90694-a.
157. K.Y. Choi, H.Y. Yoon, J.-H. Kim, S.M. Bae, R.-W. Park, Y.M. Kang, I.-S. Kim, I.C. Kwon, K. Choi, S.Y. Jeong, Smart nanocarrier based on PEGylated hyaluronic acid for cancer therapy, *ACS Nano* 5(11) (2011) 8591–8599. https://doi.org/10.1021/nn202070n.
158. L. Chi, M.-H. Na, H.-K. Jung, S.M.P. Vadevoo, C.-W. Kim, G. Padmanaban, T.-I. Park, J.-Y. Park, I. Hwang, K.U. Park, Enhanced delivery of liposomes to lung tumor through targeting interleukin-4 receptor on both tumor cells and tumor endothelial cells, *J. Control. Rel.* 209 (2015) 327–336. https://doi.org/10.1016/j.jconrel.2015.05.260.
159. J. Auernheimer, C. Dahmen, U. Hersel, A. Bausch, H. Kessler, Photoswitched cell adhesion on surfaces with RGD peptides, *J. Am. Chem. Soc.* 127(46) (2005) 16107–16110. https://doi.org/10.1021/ja053648q.
160. Z. Lu, Y. Long, X. Cun, X. Wang, J. Li, L. Mei, Y. Yang, M. Li, Z. Zhang, Q. He, A size-shrinkable nanoparticle-based combined anti-tumor and anti-inflammatory strategy for enhanced cancer therapy, *Nanoscale* 10(21) (2018) 9957–9970. https://doi.org/10.1039/c8nr01184b.
161. H. Jo, C. Ban, Aptamer–nanoparticle complexes as powerful diagnostic and therapeutic tools, *Exp. Mol. Med.* 48(5) (2016) e230–e230. https://doi.org/10.1038/emm.2016.44.
162. Y. Duo, M. Yang, Z. Du, C. Feng, C. Xing, Y. Wu, Z. Xie, F. Zhang, L. Huang, X. Zeng, CX-5461-loaded nucleolus-targeting nanoplatform for cancer therapy through induction of pro-death autophagy, *Acta Biomater.* 79 (2018) 317–330. https://doi.org/10.1016/j.actbio.2018.08.035.
163. T. Tagami, W.D. Foltz, M.J. Ernsting, C.M. Lee, I.F. Tannock, J.P. May, S.-D. Li, MRI monitoring of intratumoral drug delivery and prediction of the therapeutic effect with a multifunctional thermosensitive liposome, *Biomater.* 32(27) (2011) 6570–6578. https://doi.org/10.1016/j.biomaterials.2011.05.029.
164. K.-J. Chen, H.-F. Liang, H.-L. Chen, Y. Wang, P.-Y. Cheng,

H.-L. Liu, Y. Xia, H.-W. Sung, A thermoresponsive bubble-generating liposomal system for triggering localized extracellular drug delivery, *ACS Nano* 7(1) (2013) 438–446. https://doi.org/10.1021/nn304474j.

165. J. Zhang, H. Chen, L. Xu, Y. Gu, The targeted behavior of thermally responsive nanohydrogel evaluated by NIR system in mouse model, *J. Control. Rel.* 131(1) (2008) 34–40. https://doi.org/10.1016/j.jconrel.2008.07.019.
166. S.R. MacEwan, A. Chilkoti, Digital switching of local arginine density in a genetically encoded self-assembled polypeptide nanoparticle controls cellular uptake, *Nano Lett.* 12(6) (2012) 3322–3328. https://doi.org/10.1021/nl301529p.
167. S.H. Lee, S.H. Choi, S.H. Kim, T.G. Park, Thermally sensitive cationic polymer nanocapsules for specific cytosolic delivery and efficient gene silencing of siRNA: swelling induced physical disruption of endosome by cold shock, *J. Control. Rel.* 125(1) (2008) 25–32. https://doi.org/10.1016/j.jconrel.2007.09.011.
168. A.F. Radovic-Moreno, T.K. Lu, V.A. Puscasu, C.J. Yoon, R. Langer, O.C. Farokhzad, Surface charge-switching polymeric nanoparticles for bacterial cell wall-targeted delivery of antibiotics, *ACS Nano* 6(5) (2012) 4279–4287. https://doi.org/10.1021/nn3008383.
169. C.-H. Wang, S.-T. Kang, Y.-H. Lee, Y.-L. Luo, Y.-F. Huang, C.-K. Yeh, Aptamer-conjugated and drug-loaded acoustic droplets for ultrasound theranosis, *Biomater.* 33(6) (2012) 1939–1947. https://doi.org/10.1016/j.biomaterials.2011.11.036.
170. K. Un, S. Kawakami, R. Suzuki, K. Maruyama, F. Yamashita, M. Hashida, Suppression of melanoma growth and metastasis by DNA vaccination using an ultrasound-responsive and mannose-modified gene carrier, *Mol. Pharmaceut.* 8(2) (2011) 543–554. https://doi.org/10.1021/mp100369n.
171. A. Ranjan, G.C. Jacobs, D.L. Woods, A.H. Negussie, A. Partanen, P.S. Yarmolenko, C.E. Gacchina, K.V. Sharma, V. Frenkel, B.J. Wood, Image-guided drug delivery with magnetic resonance guided high intensity focused ultrasound and temperature sensitive liposomes in a rabbit Vx2 tumor model, *J. Control. Rel.* 158(3) (2012) 487–494. https://doi.org/10.1016/j.jconrel.2011.12.011.
172. Y.-C. Liu, A.-L.M. Le Ny, J. Schmidt, Y. Talmon, B.F. Chmelka, C.T. Lee Jr, Photo-assisted gene delivery using light-responsive catanionic vesicles, *Langmuir* 25(10) (2009) 5713–5724. https://doi.org/10.1021/la803588d.
173. S.S. Agasti, A. Chompoosor, C.-C. You, P. Ghosh, C.K. Kim, V.M. Rotello, Photoregulated release of caged anticancer drugs from gold nanoparticles, *J. Am. Chem. Soc.* 131(16) (2009) 5728–5729. https://doi.org/10.1021/ja900591t.
174. J.L. Vivero-Escoto, I.I. Slowing, C.-W. Wu, V.S.-Y. Lin, Photoinduced intracellular controlled release drug delivery in human cells by gold-capped mesoporous silica nanosphere, *J. Am. Chem. Soc.* 131(10) (2009) 3462–3463. https://doi.org/10.1021/ja900025f.
175. J. Yang, J. Lee, J. Kang, S.J. Oh, H.-J. Ko, J.-H. Son, K. Lee, J.-S. Suh, Y.-M. Huh, S. Haam, Smart drug-loaded polymer gold nanoshells for systemic and localized therapy of human epithelial cancer, *Adv. Mater.* 21(43) (2009) 4339. https://doi.org/10.1002/adma.200900334.
176. J. You, R. Zhang, C. Xiong, M. Zhong, M. Melancon, S. Gupta, A.M. Nick, A.K. Sood, C. Li, Effective photothermal chemotherapy using doxorubicin-loaded gold nanospheres that target EphB4 receptors in tumors, *Cancer Res.* 72(18) (2012) 4777–4786. https://doi.org/10.1158/0008-5472.CAN-12-1003.
177. W. Qu, Y. Li, L. Hovgaard, S. Li, W. Dai, J. Wang, X. Zhang, Q. Zhang, A silica-based pH-sensitive nanomatrix system improves the oral absorption and efficacy of incretin hormone glucagon-like peptide-1, *Int. J. Nanomed.* (2012) 4983–4994. https://doi.org/10.2147/IJN.S30013.
178. G.H. Gao, M.J. Park, Y. Li, G.H. Im, J.-H. Kim, H.N. Kim, J.W. Lee, P. Jeon, O.Y. Bang, J.H. Lee, The use of pH-sensitive positively charged polymeric micelles for protein delivery, *Biomater.* 33(35) (2012) 9157–9164. https://doi.org/10.1016/j.biomaterials.2012.09.016.
179. X. Su, J. Fricke, D.G. Kavanagh, D.J. Irvine, In vitro and in vivo mRNA delivery using lipid-enveloped pH-responsive polymer nanoparticles, *Mol. Pharmaceut.* 8(3) (2011) 774–787. https://doi.org/10.1021/mp100390w.
180. V. Dehousse, N. Garbacki, A. Colige, B. Evrard, Development of pH–responsive nanocarriers using trimethylchitosans and methacrylic acid copolymer for siRNA delivery, *Biomaterials* 31(7) (2010) 1839–1849. https://doi.org/10.1016/j.biomaterials.2009.11.028.
181. R.R. Sawant, S.K. Sriraman, G. Navarro, S. Biswas, R.A. Dalvi, V.P. Torchilin, Polyethyleneimine-lipid conjugate-based pH-sensitive micellar carrier for gene delivery, *Biomater.* 33(15) (2012) 3942–3951. https://doi.org/10.1016/j.biomaterials.2011.11.088.
182. D.T. Auguste, K. Furman, A. Wong, J. Fuller, S.P. Armes, T.J. Deming, R. Langer, Triggered release of siRNA from poly (ethylene glycol)-protected, pH-dependent liposomes, *J. Control. Rel.* 130(3) (2008) 266–274.
183. L. Zhu, P. Kate, V.P. Torchilin, Matrix metalloprotease 2-responsive multifunctional liposomal nanocarrier for enhanced tumor targeting, *ACS Nano* 6(4) (2012) 3491–3498. https://doi.org/10.1021/nn300524f.
184. T.J. Harris, G. Von Maltzahn, M.E. Lord, J.H. Park, A. Agrawal, D.H. Min, M.J. Sailor, S.N. Bhatia, Protease-triggered unveiling of bioactive nanoparticles, *Small* 4(9) (2008) 1307–1312. https://doi.org/10.1002/smll.200701319.
185. H. Hatakeyama, H. Akita, E. Ito, Y. Hayashi, M. Oishi, Y. Nagasaki, R. Danev, K. Nagayama, N. Kaji, H. Kikuchi, Systemic delivery of siRNA to tumors using a lipid nanoparticle containing a tumor-specific cleavable PEG-lipid, *Biomater.* 32(18) (2011) 4306–4316. https://doi.org/10.1016/j.biomaterials.2011.02.045.
186. J.S. Lee, T. Groothuis, C. Cusan, D. Mink, J. Feijen, Lysosomally cleavable peptide-containing polymersomes modified with anti-EGFR antibody for systemic cancer chemotherapy, *Biomater.* 32(34) (2011) 9144–9153. https://doi.org/10.1016/j.biomaterials.2011.08.036.
187. M.-H. Xiong, Y. Bao, X.-Z. Yang, Y.-C. Wang, B. Sun, J. Wang, Lipase-sensitive polymeric triple-layered nanogel for "on-demand" drug delivery, *J. Am. Chem. Soc* 134(9) (2012) 4355–4362. https://doi.org/10.1021/ja211279u.
188. R.S. Gaspar, H.F. Florindo, L.C. Silva, M.A. Videira, M.L. Corvo, B.F. Martins, B. Silva-Lima, Regulatory aspects of oncologicals: nanosystems main challenges, *Nano-Oncol. New Target. Deliv. Approach.* (2014) 425–452.
189. F. Ehmann, K. Sakai-Kato, R. Duncan, D.H. Pérez de la Ossa, R. Pita, J.-M. Vidal, A. Kohli, L. Tothfalusi, A. Sanh, S. Tinton, Next-generation nanomedicines and nanosimilars: EU regulators' initiatives relating to the development and evaluation of nanomedicines, *Nanomedicine* 8(5) (2013) 849–856. https://doi.org/10.2217/nnm.13.68.
190. V. Wagner, A. Dullaart, A.-K. Bock, A. Zweck, The emerging nanomedicine landscape, *Nat. Biotechnol.* 24(10) (2006) 1211–1217.
191. R.S. Gaspar, Therapeutic products: regulating drugs and medical devices. In the *International Handbook on Regulating Nanotechnologies*, Edward Elgar Publishing (2010). In:

Maynard A, Bowman D, Hodge G, editors. International handbook on regulating nanotechnologies. Cheltenham: EE; 2010. p. 291.

192. M.A. Dobrovolskaia, S.E. McNeil, Understanding the correlation between in vitro and in vivo immunotoxicity tests for nanomedicines, *J. Control. Rel.* 172(2) (2013) 456–466. https://doi.org/10.1016/j.jconrel.2013.05.025.
193. O. Hurko, J.L. Ryan, Translational research in central nervous system drug discovery, *NeuroRx* 2(4) (2005) 671–682. https://doi.org/10.1602/neurorx.2.4.671.
194. B. Smith, L. Heywood, The international conference on harmonisation and its impact. In: Rondel RK, Varley SA, Webb CF (Eds.), *Clinical Data Management*, 2nd edn. John Wiley & Sons, Chichester, pp. 21–46 (2000).
195. J.M. Caster, A.N. Patel, T. Zhang, A. Wang, Investigational nanomedicines in 2016: a review of nanotherapeutics currently undergoing clinical trials, *WIREs Nanomed. Nanobiotechnol.* 9(1) (2017) e1416. https://doi.org/10.1002/wnan.1416.
196. T.M. Allen, P.R. Cullis, Liposomal drug delivery systems: from concept to clinical applications, *Adv. Drug Deliv. Rev.* 65(1) (2013) 36–48. https://doi.org/10.1016/j.addr.2012.09.037.
197. M. Kim, S. Williams, Daunorubicin and cytarabine liposome in newly diagnosed therapy-related Acute Myeloid Leukemia (AML) or AML with myelodysplasia-related changes, *Ann. Pharmacother.* 52(8) (2018) 792–800. https://doi.org/10.1177/1060028018764923.
198. V. Sainz, J. Conniot, A.I. Matos, C. Peres, E. Zupanőiő, L. Moura, L.C. Silva, H.F. Florindo, R.S. Gaspar, Regulatory aspects on nanomedicines, *Biochem. Biophys. Res. Commun.* 468(3) (2015) 504–510. https://doi.org/10.1016/j.bbrc.2015.08.023.
199. V.P. Torchilin, Multifunctional nanocarriers, *Adv. Drug Deliv. Rev.* 58(14) (2006) 1532–1555. https://doi.org/10.1016/j.addr.2006.09.009.
200. J.I. Hare, T. Lammers, M.B. Ashford, S. Puri, G. Storm, S.T. Barry, Challenges and strategies in anti-cancer nanomedicine development: an industry perspective, *Adv. Drug Deliv. Rev.* 108 (2017) 25–38. https://doi.org/10.1016/j.addr.2016.04.025.
201. J.A. Jackman, T. Mészáros, T. Fülöp, R. Urbanics, J. Szebeni, N.-J. Cho, Comparison of complement activation-related pseudoallergy in miniature and domestic pigs: foundation of a validatable immune toxicity model, *Nat. Biomed. Eng.* 12(4) (2016) 933–943. https://doi.org/10.1016/j.nano.2015.12.377.
202. M. Kumar Teli, S. Mutalik, G. Rajanikant, Nanotechnology and nanomedicine: going small means aiming big, *Curr. Pharm. Des.* 16(16) (2010) 1882–1892. https://doi.org/10.2174/138161210791208992.
203. J. Paradise, S.M. Wolf, J. Kuzma, A. Kuzhabekova, A.W. Tisdale, E. Kokkoli, G. Ramachandran, Developing US oversight strategies for nanobiotechnology: learning from past oversight experiences, *J. Law Med. Ethics* 37(4) (2009) 688–705. https://doi.org/10.1111/j.1748-720X.2009.00441.x.
204. J.S. Murday, R.W. Siegel, J. Stein, J.F. Wright, Translational nanomedicine: status assessment and opportunities, *Nat. Biomed. Eng.* 5(3) (2009) 251–273. https://doi.org/10.1016/j.nano.2009.06.001.
205. R. Bawa, Will the nanomedicine "patent land grab" thwart commercialization? *Nat. Biomed. Eng.* 1(4) (2005) 346–350. https://doi.org/10.1016/j.nano.2005.10.009.
206. L. Brito, M. Amiji, Nanoparticulate carriers for the treatment of coronary restenosis, *Int. J. Nanomed.* 2(2) (2007) 143–161.
207. S. Tinkle, S.E. McNeil, S. Mühlebach, R. Bawa, G. Borchard, Y. Barenholz, L. Tamarkin, N. Desai, Nanomedicines: addressing the scientific and regulatory gap, *Ann. N. Y. Acad. Sci.* 1313(1) (2014) 35–56. https://doi.org/10.1111/nyas.12403.
208. S. Hua, E. Marks, J.J. Schneider, S. Keely, Advances in oral nano-delivery systems for colon targeted drug delivery in inflammatory bowel disease: selective targeting to diseased versus healthy tissue, *Nat. Biomed. Eng.* 11(5) (2015) 1117–1132. https://doi.org/10.1016/j.nano.2015.02.018.
209. M. Barz, R. Luxenhofer, M. Schillmeier, Quo vadis nanomedicine? *Nanomed.* 10(20) (2015) 3089–3091. https://doi.org/10.2217/nnm.15.156.
210. T. Lammers, F. Kiessling, W.E. Hennink, G. Storm, Drug targeting to tumors: principles, pitfalls and (pre-) clinical progress, *J. Control. Rel.* 161(2) (2012) 175–187. https://doi.org/10.1016/j.jconrel.2011.09.063.
211. L. Sercombe, T. Veerati, F. Moheimani, S.Y. Wu, A.K. Sood, S. Hua, Advances and challenges of liposome assisted drug delivery, *Front. Pharmacol.* 6 (2015) 286. https://doi.org/10.3389/fphar.2015.00286.
212. R. Gaspar, R. Duncan, Polymeric carriers: preclinical safety and the regulatory implications for design and development of polymer therapeutics, *Adv. Drug Deliv. Rev.* 61(13) (2009) 1220–1231. https://doi.org/10.1016/j.addr.2009.06.003.
213. S. Hua, M.B. De Matos, J.M. Metselaar, G. Storm, Current trends and challenges in the clinical translation of nanoparticulate nanomedicines: pathways for translational development and commercialization, *Front. Pharmacol.* 9 (2018) 790. https://doi.org/10.3389/fphar.2018.00790.
214. D.J. Slamon, W. Godolphin, A. Ullrich, M.F. Press, Studies of the HER-2/neu proto-oncogene in human breast cancer, *Cancer Genet. Cytogenet.* 41(2) (1989) 219. https://doi.org/10.3109/07357909009017573.
215. M.A. Cobleigh, C.L. Vogel, D. Tripathy, N.J. Robert, S. Scholl, L. Fehrenbacher, J.M. Wolter, V. Paton, S. Shak, G. Lieberman, Multinational study of the efficacy and safety of humanized anti-HER2 monoclonal antibody in women who have HER2-overexpressing metastatic breast cancer that has progressed after chemotherapy for metastatic disease, *J. Clin. Oncol.* 17(9) (1999) 2639. https://doi.org/10.1200/JCO.22.02510.
216. H. Hurwitz, L. Fehrenbacher, W. Novotny, T. Cartwright, J. Hainsworth, W. Heim, J. Berlin, A. Baron, S. Griffing, E. Holmgren, Bevacizumab plus irinotecan, fluorouracil, and leucovorin for metastatic colorectal cancer, *N. Engl. J. Med.* 350(23) (2004) 2335–2342. https://doi.org/10.1056/N.Engl. J. Med.oa032691.
217. A. Borkhardt, Blocking oncogenes in malignant cells by RNA interference—new hope for a highly specific cancer treatment?, *Cancer Cell* 2(3) (2002) 167–168. https://doi.org/10.1016/s1535-6108(02)00129-0.
218. L. DeBusk, A.J. Russell, K. Imai, L.M. Matrisian, Forefront of basic and translational cancer research. In: *Seventh Joint Conference of the American Association for Cancer Research and the Japanese Cancer Association*, AACR, Nagoya, Japan (2007).
219. E.F. Unger, All is not well in the world of translational research, *J. Am. Coll. Cardiol.* 50(8) (2007) 738–740. https://doi.org/10.1016/j.jacc.2007.04.067.
220. R. Simon, Translational research in oncology: key bottlenecks and new paradigms, *Expert Rev. Mol. Med.* 12 (2010) e32. https://doi.org/10.1017/S1462399410001638.

4 Materials and Technology Involved in Neurological Mechanism Targeting

Tejas Girish Agnihotri, Shyam Sudhakar Gomte, Sandeep Kr Maharana, Naveen Singh, and Aakanchha Jain
Department of Pharmaceutics, National Institute of Pharmaceutical Education and Research (NIPER)-Ahmedabad, Palaj, Gujarat, India

4.1 INTRODUCTION

The brain, a delicate network of neurons, requires a constant supply of nutrients, oxygen, and fibers to maintain its equilibrium. One of the most promising areas of research is the development of targeted drug delivery systems for the central nervous system, specifically the brain. The brain controls the body's essential functions. Due to the presence of the blood-brain barrier (BBB), the brain presents unique challenges for drug delivery [1,2]. The BBB acts as a diffusion barrier, separating blood from the brain's extracellular fluid. In mammals, including humans, the BBB is formed by endothelial cells in the capillary walls of the brain and spinal cord. Blood vessels deliver oxygen and nutrients to tissues and organs, and the BBB meticulously regulates the transport of amino acids, glucose, nucleosides, fatty acids, ions, molecules, and cells between the bloodstream and the brain. By strictly controlling central nervous system (CNS) homeostasis, the BBB ensures proper neuronal function and protects neural tissue from toxins and pathogens. Disruptions in the BBB's integrity are crucial in the development of various neurological disorders [3]. The BBB consists of capillary endothelial cells, astrocytes, and pericytes. Neurons, microglia, and the basement membrane also support its function, especially in immune regulation. These components form the neurovascular unit (NVU), which upholds BBB integrity and CNS health. At the postcapillary venule level, the endothelial and parenchymal basement membranes form the perivascular space, which is home to bone marrow-derived cells with key immunoregulatory functions [4]. The current chapter delves into various brain-targeted strategies, including nanocarrier-based systems, receptor-mediated delivery systems, peptide-functionalized systems, and intranasal systems. Different physical BBB-disrupting techniques have also been discussed in detail in this chapter. This chapter also sheds light on some of the emerging technologies like biomimetic delivery, implants, injectable delivery, and neural tissue engineering.

4.2 BRAIN-TARGETED STRATEGIES: NOVEL APPROACHES

The brain is an incredibly complex part of the human body, protected by tightly joined cells that form a barrier to prevent the easy entry of foreign substances. This BBB is a sophisticated network that plays a crucial role in safeguarding the brain and serves as a significant obstacle to the delivery of various therapeutic agents [5]. In recent times, there have been several strategies developed for the delivery of different therapeutic agents to targeted brain sites, which will be detailed in the following section.

4.2.1 Nanotechnology-Based Drug Delivery Systems

Nanotechnology is a promising field that combines interdisciplinary branches, including physics, chemistry, biology, and engineering, to design materials within the size range of 1 to 100 nm [6]. The introduction of nanotechnology for the development of innovative drug delivery approaches leads to the fabrication of novel and suitable formulations to improve the delivery of therapeutic agents through the BBB. Several nanocarriers have been introduced for effective brain drug delivery, primarily including polymeric nanoparticles (NPs), liposomes, polymeric micelles, and hydrogel-based systems [7,8].

4.2.1.1 Polymeric Nanoparticles

Both natural and synthetic polymers offer excellent opportunities to harness nanoparticle properties. Polymeric NPs are present in different forms, such as individual polymers, polymeric blends, protein-polymer conjugates, and metal-polymer complexes, to deliver various drugs. These types of polymeric NPs can be fabricated in a way that allows them to attain varied particle sizes, different degradation and release profiles, and ease of surface functionalization [9]. Several scientists are currently engaged in the development of polymeric drug delivery systems to attain effective

DOI: 10.1201/9781003508601-4

treatment for various diseases. For instance, Garg et al. developed and optimized donepezil HCl-loaded polymeric NPs for brain drug delivery using the Quality-by-Design (QbD) approach. The *in vitro* release and *ex vivo* permeation investigation suggests the sustained release pattern of donepezil HCl through polymeric NPs. The intranasal delivery of donepezil HCl NPs showed 2.6-fold more drug concentration than pristine donepezil HCl and 10-fold more drug than oral delivery of donepezil HCl [10].

4.2.1.2 Liposomes

Liposomes are spherical-shaped vesicles with an aqueous core and a hydrophobic outer portion consisting of phospholipid bilayers [7,11]. They contain major components such as cholesterol and phospholipids. The components used for liposome preparation are non-toxic, inert, biodegradable, and non-immunogenic [12,13]. They are employed in drug delivery systems to enhance drug effects and eradicate toxic effects. Liposomes are huge molecules with lipophilic attributes, and diffusion across the BBB is difficult for them. For diffusion through the cell membrane, they utilize various transporter systems, including receptor-mediated transcytosis (RMT), adsorptive-mediated transcytosis (AMT), and carrier-mediated transcytosis (CMT). Moreover, it could be transported to the BBB using active targeting strategies [13]. Recently, Khairnar et al. developed apolipoprotein E (ApoE) surface-modified liposomes for the effective brain-targeted delivery of rivastigmine. The ApoE coating over liposomes promotes brain uptake and enhances Aβ clearance. The uptake of rivastigmine-loaded liposomes was improved by 1.3 times compared to pure rivastigmine. The biodistribution investigation suggests that the liver biodistribution of rivastigmine was highly decreased. The proposed formulation could be a promising strategy for the effective management of Alzheimer's disease (AD) [14].

4.2.1.3 Polymeric Micelles

Polymeric micelles (PMs) are a core-shell network with the basic ability to dissolve hydrophobic drugs and serve as a kinetically stable drug delivery vehicle employed for the delivery of various therapeutic agents. The PM's term is indicated for those where hydrophilic portions of the amphiphilic polymer are directed toward the outer region, and a hydrophobic portion of the amphiphilic polymer is inward to the core of the micellar system. Numerous studies have highlighted the stimuli-sensitive nature of PMs and their potential applications, emphasizing the utility of corona-like network-forming polymers and core-forming polymers. The method of fabricating PMs impacts the entrapment and percentage loading efficiency of therapeutic agents. The PMs can be fabricated using graft, di-block, tri-block polymers, and stimuli-responsive polymers. The pH, solvent, and concentration of polymers, along with their ratio and co-solvent, can effectively impact the type of micelles formed [15]. Several types of PMs are employed for the delivery of different therapeutic agents in brain drug delivery. For instance, Zhang et al. developed duo-modified lipid PMs for the delivery of anti-miRNA-21 for the oral management of glioma therapy. The lipid polymer micellar system was anchored with angiopoietin-2 (Ang-2) and transactivator of transcription (TAT), which helps to develop and stabilize anti-miR-21, thereby providing an effective treatment for the management of glioma. The cell line investigations confirmed an effective uptake of anti-miRNA-21 by glial cells, consequently resulting in enhanced apoptosis. Nude mice were used to develop an orthotopic xenograft model. The orally administered micelles led to the activation of pro-apoptotic proteins, such as p53 and caspase-3, thereby promoting apoptosis and reducing tumor growth. Overall, anti-miRNA-21 loaded micelles could be used as an effective alternative for the treatment of glioma therapy [16].

4.2.1.4 Hydrogels

Hydrogels consist of basically hydrophilic polymers with three-dimensional (3D) networks because of the availability of a large number of hydrophilic functionalities and their capacity to hold large amounts of water within a 3D network system [17]. Additionally, it possesses unique porosity, a flexible and soft structure, and the ability to respond to stimuli and exceptional resemblance to living tissues. It shows excellent attributes such as higher biocompatibility, biodegradability, and non-immunogenicity. Recently, the integration of nanosystems with hydrogel has emerged as a viable option for efficiently transporting drugs into the brain. Sayiner et al. developed Temozolomide (TMZ)-laden Poly Lactic-co-Glycolic Acid (PLGA) NPs and further loaded them into a thermoreversible hydrogel for glioblastoma (GBM) management. The PLGA NPs were loaded into a hydrogel that contained Pluronic F-127 polymer. The loading of NPs into the hydrogel resulted in delayed drug release from the gel system and exhibited sustained release behavior. The TMZ-PLGA-NPs were effectively taken up by glioma cells, and TMZ effectively reached a higher amount at the targeted site as compared to the free TMZ solution. The prepared hydrogel could be a promising approach for managing glioma after surgery [18].

4.2.2 Targeted Drug Delivery Systems

The targeted drug delivery approach enables the accurate and efficient delivery of therapeutic agents to specific sites, leveraging the benefits of nanotechnology [9,19,20]. Several targeting approaches have been introduced to achieve effective brain-targeted drug delivery, including receptor-mediated targeting, cell-penetrating peptide targeting, and antibody-based targeting.

4.2.2.1 Receptor-Mediated Targeting Systems

Receptor-mediated targeting utilizes the overexpressed receptors present on the BBB in neurodegenerative disorders or brain tumors. The process is also known as receptor-mediated transcytosis (RMT). This targeting approach promotes endocytosis using caveolae or clathrin-coated pits, similar to the process of AMT. When NPs are internalized into cells, they follow various mechanisms inside cells based on their charge, size, composition, and ligand targeting. The most common receptors used in receptor-based targeting are transferrin, lactoferrin, low-density lipoprotein, nicotinic acetylcholine receptor, and insulin receptors. The NPs could be modified in such a way that a particular ligand for receptors works as a targeting ligand, and targeted delivery of therapeutic agents could be achieved [21].

4.2.2.1.1 *Transferrin Receptor*

The transferrin receptor is highly overexpressed in different cancers, including GBM, and could be used as a specific ligand for effective targeted drug delivery. It is a membrane protein consisting of two glycosylated subunits that dimerize and attach to the circulating transferrin [22]. The binding triggers endocytosis via clathrin-coated pits, facilitating the uptake of iron into the brain. Iron acts as a co-factor for numerous enzymes and is crucial for metabolic processes, including ATP synthesis, as it is an integral part of the iron-sulfur clusters and heme groups within the mitochondrial respiratory chain complexes. Transferrin is specifically expressed in brain endothelial cells, but not in other tissues, making it a promising method for delivering medications across the BBB [23]. Ramalho et al. developed transferrin-anchored PLGA NPs for the targeted delivery of Asiatic acid in the management of glioma cells. The surface of the optimized drug-loaded NPs was anchored by transferrin for targeted delivery. The cell line studies confirmed that Asiatic acid showed enhanced anti-glioma efficacy, and transferrin-anchored NPs significantly enhanced drug cellular uptake and reduced toxicity of normal cells [24].

4.2.2.1.2 *Lactoferrin Receptor*

Lactoferrin and transferrin share similarities but differ in their receptor-binding characteristics. These two iron-binding proteins exhibit variations in their lobes and inter-lobe linker structures. Notably, lactoferrin possesses higher brain uptake than transferrin, which could be helpful in the development of innovative nanocarrier systems for brain drug delivery [25]. Recently, Qi et al. developed tripeptide Arg-Gly-Asp (RGD) and lactoferrin (Lf) modified docetaxel liposomes for the management of glioma. Developed liposomes showed excellent cellular uptake of U87MG cells and human cerebral microvascular endothelial cells (hCMEC/D3) compared to RGD-modified liposomes and Lf-modified liposomes alone. The same trend was observed on U87MG tumor spheroids. This was also reflected in orthotopic glioma models, which showed a greater brain-targeting effect and prolonged the survival time of orthotopic glioma mice [26].

4.2.2.2 Cell-Penetrating Peptide-based Targeting

The advanced mechanism for crossing the BBB by cell-penetrating peptides (CPPs) has been extensively documented in the literature [21]. CPPs are comprised of 15–25 amino acids capable of crossing cellular membranes to access the intracellular space. These peptides share common features such as positive charges, amphipathic properties, and the ability to interact with lipid membranes. The research suggests that CPP internalization does not require specific receptor binding. While the precise mechanism remains unclear, it seems that endocytosis and internalization are driven by electrostatic interactions between CPPs and lipid membranes, as well as other components such as glycoproteins and glycosaminoglycans found in the extracellular matrix [27]. CPPs strive to enhance cell internalization through their unique properties, thus acting as unique targeting ligands for effective brain drug delivery [28].

4.2.2.3 Antibody-based Targeting

The plasma B cells in the adaptive immune system produce antibodies (Abs) that detect and neutralize foreign molecules invading the body, thus protecting against threats from non-self-agent attacks. The monoclonal antibodies (mAbs) are composed of unique B cells and are monovalent binders that specifically bind to the same epitope [29]. In recent years, antibody-based targeting has gained significant attention from researchers due to its unique advantages, such as targeting ability, lower risk of side effects, and activation of the immune system, resulting in long-term immunity. For instance, El-Gamal et al. developed CPP-anchored liposomes, composed of octa arginine-modified PEGylated liposomes for targeted delivery to U87 and U251 glioma cells. These liposomes exhibited significantly enhanced cellular uptake compared to plain PEGylated liposomes. The incorporation of the CPP moiety facilitated improved cellular penetration, suggesting its potential as a promising strategy for targeted drug delivery applications [30].

4.2.3 Intranasal Drug Delivery Systems

Intranasal drug delivery to the CNS offers a promising way to bypass the highly selective BBB. This approach is non-invasive for delivering brain drugs in higher concentrations. It also makes administration easier because it doesn't require sterile dosage forms or specialized techniques, making self-administration painless and straightforward [31]. Human trials and preclinical models have demonstrated the effectiveness of the intranasal route for brain-targeted drug delivery, utilizing both solution and particle formulations. The olfactory area is the primary focus for its effectiveness, which is further enhanced by extended retention on the nasal mucosal surface, improved drug penetration through the nasal epithelia, and minimized nasal cavity drug metabolism [32]. Due to systemic drug distribution, current treatments for brain disorders are only partially successful and frequently result in major side effects. Subtherapeutic concentrations in target tissues can also result from some drugs' physicochemical characteristics, which make it more difficult for them to cross the BBB. However, technologies based on nanoparticles have demonstrated considerable promise in overcoming these obstacles. By administering the medication intranasally, they can maximize its concentration in the brain and reduce its systemic dispersion [33]. Ahmad et al. developed levodopa-loaded chitosan NPs for the treatment of Parkinson's disease (PD), and their *in vivo* studies revealed delayed T_{max} resulting from the NP's greater size relative to the free drug molecules, which takes longer to permeate over the nasal mucosa. Because of chitosan's mucoadhesive properties, the drug release is sustained since the NPs spend more time in the nasal mucosa [34].

4.2.4 Techniques Involving Physical Disruption of the BBB

4.2.4.1 Convection-Enhanced Delivery (CED) Systems

CED, which is regarded as a promising approach, employs a small-caliber catheter inserted in the brain through cranial burr holes for delivering drugs directly into the brain parenchymal tissues that later penetrate the interstitial space. This bypasses the challenges posed by the BBB to the drug administered by oral or intravenous methods (Figure 4.1a). Here, drugs are constantly administered in a diluted solution into the tumor for several days [35]. This enables intra-tumoral drug accumulation to reach a 'quasi-steady state' to kill the tumor cells. CED generally provides the ability to treat large brain volumes that include both radiographically defined tumor sites as well as their surrounding infiltrated brain tissues. Therefore, the aforementioned approach effectively delivers drugs to the brain [36]. The CED technique offers several advantages, including the minimization of systemic side effects, the effective measurement of intracranial pressure, and the maintenance of consistent and effective drug concentrations. However, certain limitations are associated with CED, including longer infusion time and unpredictable drug distribution [37].

The principle that governs CED is 'bulk flow', which involves a pressure gradient and mediates the flow of extracellular fluid. Bulk flow distribution, as compared with

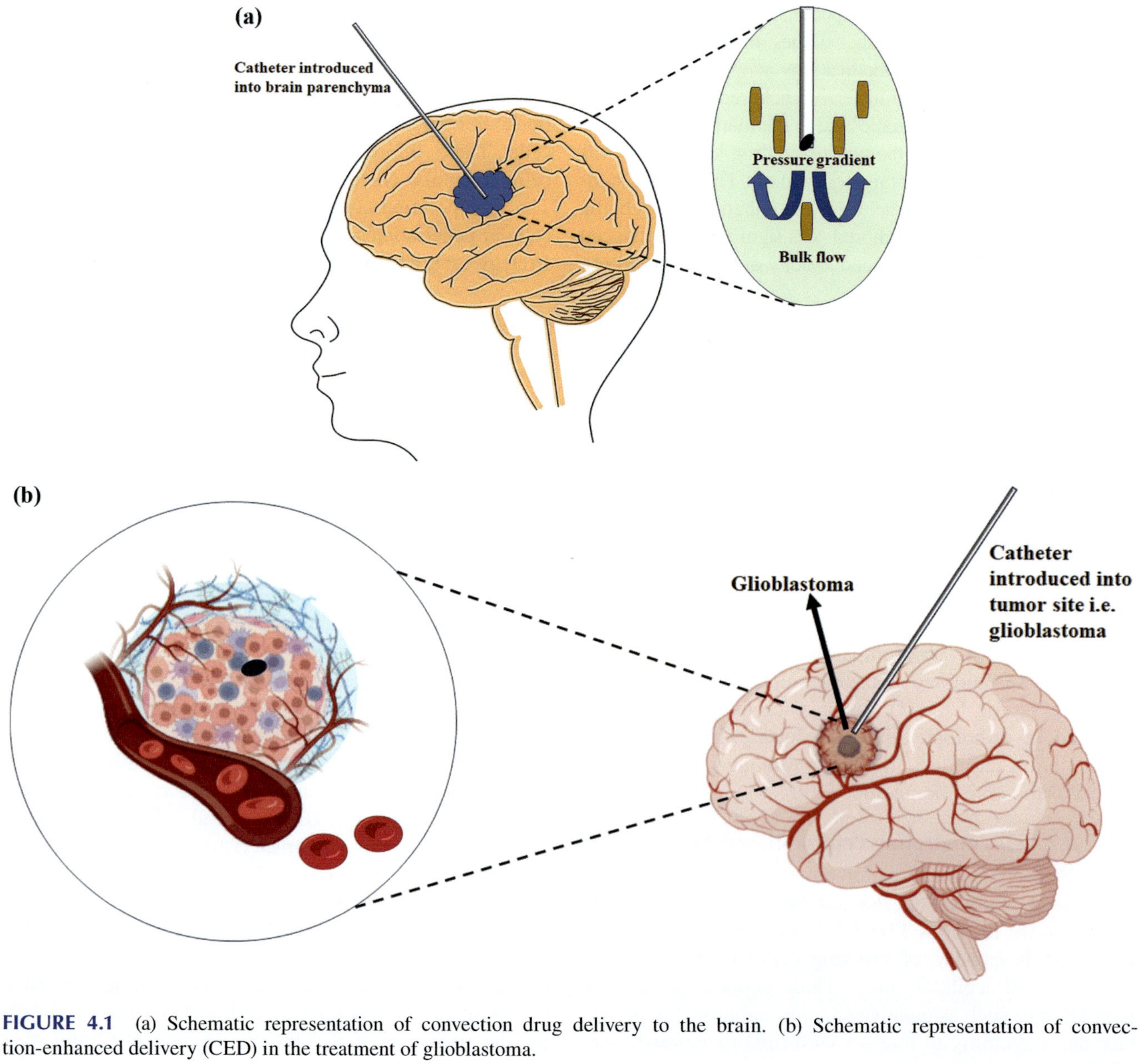

FIGURE 4.1 (a) Schematic representation of convection drug delivery to the brain. (b) Schematic representation of convection-enhanced delivery (CED) in the treatment of glioblastoma.

diffusive methods, is independent of molecular weight. Additionally, there is no requirement for higher concentrations to achieve therapeutic drug levels across a large volume of targeted tissues [38].

In CED, by employing image-guided neuronavigation, one or more catheters (stereotactic) of small diameter are either placed directly into the brain parenchyma or at the site of the tumor, i.e., GBM (Figure 4.1b). In order to prevent preferential infusate distribution into the cerebrospinal space, the catheter must be safely positioned away from the ventricles as well as the subarachnoid space. The above positioning would also ensure an adequate ratio of volume of distribution (V_d) and volume of infusion (V_i) (V_d:V_i), which is one of the important parameters governing CED [36].

Another parameter governing CED is backflow, also termed as reflux. Backflow along the catheter tract might result in infusate loss into the cerebrospinal fluid (CSF) space. This eliminates the pressure gradient and makes V_d independent of V_i [39]. Hence, various backflow-resistant catheters were developed for efficient delivery. Other parameters may include white matter vs gray matter properties, catheter-induced edema that affects the flow pattern, and V_d:V_i [40]. Therefore, further investigations are essential for clarifying V_d:V_i relationship and designing treatment protocols for different brain disorders.

Recently, Zhan and Wang investigated the suitability of six chemotherapeutic drugs, i.e., paclitaxel, cisplatin, fluorouracil, doxorubicin, methotrexate (MTX), and carmustine, from an intra-tumoral transport perspective using mathematical models. Darcy's law, coupled with convection/diffusion/reaction, was applied to a 3-dimensional (3D) realistic brain tumor model generated from magnetic resonance (MR) images. The above model shows that CED, compared to traditional methods, can increase convective intestinal fluid flow and reduce the dilution of drug concentration caused by fluid loss from the bloodstream in the tumor area around the infusion site. The physicochemical properties of the drug strongly govern its CED delivery. It was found that in brain tumors, convection played a key role in transporting paclitaxel and MTX. Results demonstrated that paclitaxel, compared to other chemotherapeutic drugs, exhibited greater penetration, accumulation, and a more effective delivery volume [41]. Some of the potential future CED innovations may include multiple-hole or hollow-fiber catheters, reflux-preventing, and ultrafine catheters.

4.2.4.2 Focused Ultrasound-Mediated Drug Delivery

Ultrasound (US) primarily consists of mechanical waves having a frequency greater than 20 kHz. The US is applied in real-time, non-invasive medical imaging. They are used in conjunction with microbubbles (MBs) for enhancing drug permeation effects, as both approaches cannot be applied independently. Under the application of the US, MBs collapse, producing cavitation effects that increase drug permeation. Research also suggests that US-induced BBB disruption evolved as an effective strategy for improving brain drug delivery to treat cancers. Also, combining intravenous MBs with low-intensity pulsed US could locally induce the opening of the BBB [42].

Among non-invasive techniques, focused ultrasound (FUS) employs low-frequency US that has superior skull penetration efficacy and causes less permanent damage to tissues. Brain targeting of drugs could also be achieved by combining FUS with MBs, which induces local heating in the focal zone [43]. FUS mainly includes the use of a frequency of 1 Hz, a US exposure pulse of 10 ms, and a time duration of 20–30 s. Cerebral blood vessel tight junctions are altered by MB-focused US to induce local reversible BBB disruption. Mechanical force on the endothelium by the MBs' interaction with low-intensity US causes the opening of tight junctions. Sound pressure exerted by FUS governs the degree of BBB disruption, which permits penetration of therapeutic agents of molecular weights up to 2,000 kDa (Figure 4.2) [44]. Several antitumor drugs, viz. 1,3-bis (2-chloroethyl)-1-nitrosourea, doxorubicin, as well as antibodies, and immune cells, could be targeted across the BBB by combining FUS with drug-encapsulated MBs as well as different NPs [45]. Additionally, MBs combined with surgically implanted pulsed US systems for BBB disruption have been used to deliver carboplatin to recurrent GBM patients without any adverse effects [46].

The determination of the area of interest and evaluation of the BBB disruption or thermal ablation area was done by combining FUS with MRI techniques. Therefore, for the induction of BBB opening, magnetic resonance-guided focused ultrasound (MRgFUS) has emerged in the recent past. When combined with injected MBs, this FUS provides a greater extent of temporal and spatial specificity. MRgFUS may detect the target site with much greater accuracy, so this technique is applied to the entire skull. In comparison to invasive surgery, MRgFUS is devoid of side effects like paresthesia and impairment in speech. Also, MRgFUS combination with drugs like Adriamycin and TMZ provides a safer and much more efficient treatment of GBM [47].

4.2.4.3 Microneedles

Microneedles (MNs) could be employed for delivering drugs to the brain. They primarily consist of short and thin needles that improve drug absorption by penetrating the skin or mucosa. MNs can be categorized into four types (Figure 4.3) [48]. Among various categories, the solid microneedles form transportation holes by piercing barriers. Dissolving MNs deliver the encapsulated drugs into the body through the erosion of the polymer matrix. Coated MNs consist of coatings of certain drugs that dissolve upon piercing. Hollow MNs deliver drugs into the body via inner hollow holes.

The above MN possesses a high barrier to humidity and moisture and displays excellent mechanical strength. Because these MNs are less dependent on environmental conditions, there may be a reduction in both piercing efficiency and mechanical stability after the coating dissolves *in vivo*. Thus, there is a requirement for process optimization,

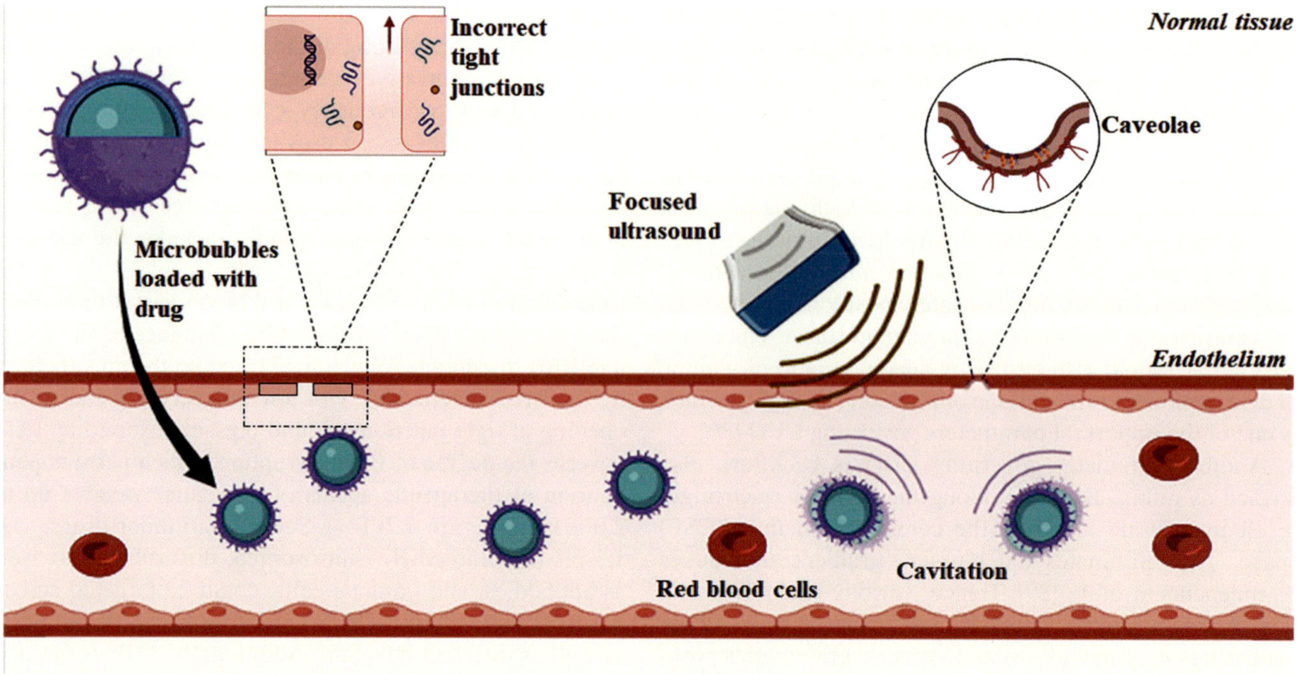

FIGURE 4.2 Focused ultrasound (FUS)-mediated disruption of the blood-brain barrier.

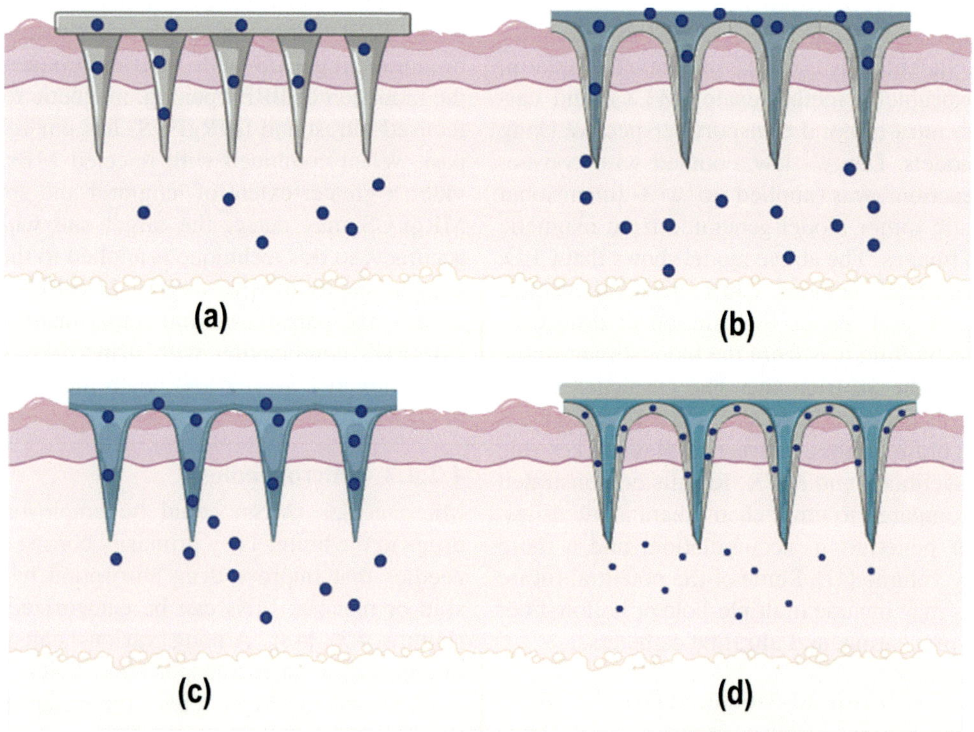

FIGURE 4.3 Schematic representation of various types of microneedles delivering drugs into the body: (a) solid microneedles, (b) hollow microneedles, (c) dissolving microneedles, and (d) coated microneedles.

like employing a heat-sensitive process for loading API onto the MNs [49]. After employing MNs to deliver NPs into the mouse brain, confocal microscopy observations demonstrated consistent particle emission sites where the MNs had penetrated. Hence, MNs had better penetration ability and were efficient in delivering drug-loaded NPs into the brain. However, it was found that dissolution rate adjustment of the fiber membrane plays a key role in controlling drug release [50]. Recently, Kim et al. fabricated an MN array (10×10) to provide a 300mm deep penetration of human dura mater to treat hydrocephalus in neurological diseases. The penetration ability was determined by porcine dura puncture tests. The *in vitro* biocompatibility test results showed that there was no blockage of 100 MNs outlets exposed to the bloodstream, indicating its potential for hydrocephalus treatment [51].

4.2.4.4 Lasers

Lasers, by stimulating the photon emission of excited atoms or molecules, generate strong beams of coherent monochromatic radiation [52]. Drug absorption is enhanced due to deeper penetration assisted by laser-assisted drug delivery. Underlying principles include applying radiation to cause reversible ablation or destruction of the barrier function. In gliomas, the BBB hinders the chemotherapeutic effects, so combining laser irradiation with ligand-mediated polymer micelles will efficiently kill tumor cells. Recently, researchers developed a disulfide bond (-S-S-) conjugated prodrug polymer comprised of camptothecin and polyethylene glycol. The above was further modified by the iRGD peptide, and the polymer could self-assemble to form micelles, which were applied to tumor cells via laser irradiation. In gliomas, the aforementioned drug delivery systems effectively permeated numerous biological barriers to reach the tumor site, as confirmed by both *in vitro* and *in vivo* studies [53].

Perfluorocarbon nanodroplets (PFCnDs) could also induce BBB opening. On applying acoustic energy or an electromagnetic field, PFCnDs absorb laser energy, causing local heating and expansion, and convert themselves from liquid to gas. US imaging, histology, etc., are used to evaluate the efficiency of laser-activated PFCnDs in opening the BBB. Thus, PFCnDs enable local and non-invasive BBB opening and could be employed for both diagnosis and treatment [54].

Considering safety concerns, the parameters of laser irradiation, such as optimal laser flux and the number of pulses, should be optimized, as they govern BBB opening. There is limited BBB opening with a certain number of pulses, whereas increasing the number of pulses and higher flux could lead to red blood cell extravasation [55]. Hence, selecting appropriate laser parameters is crucial for safe and effective BBB opening in brain drug delivery [56].

4.2.4.5 Electroporation

Electroporation utilizes an electric field to facilitate the penetration of macromolecules and ions across the cell membrane [57]. Temporary, reversible hydrophilic channels are formed in the cell membrane lipid bilayer by a high-voltage pulsed electric field, where the size and duration of the channels are governed by parameters such as pulse time, pulse number, and voltage [58]. Electroporation can be classified as either reversible or irreversible, depending on whether the cells can be rescued or killed [59]. Irreversible electroporation (IRE) primarily does not cause any thermal damage; however, for effective penetration, it might cause temporary BBB damage. IRE practically employs a single insulated intracranial needle electrode. The exposed tip and the electrode's outer surface of the intracranial needle electrode are placed on the target tissue and skin. The highest electric field occurs at the exposed tip of the intracranial electrode-tissue interface. The above then progressively weakens, producing a reversible electroporation (RE) effect, thereby destroying the BBB. Recently, electroporation emerged as a promising approach for treating intracranial solid tumors [60].

Electric fields are generated by electrical pulses acting on tissues, resulting in changes in the potential of the cell membrane, depending on pulse parameters such as pulse number, amplitude, duration, shape, and repetition frequency, as well as numerous tissue-related parameters like tissue type and cell size. Research suggests that electroporation of rats' brains causes subtle BBB disruption, thereby promoting clearance of contrast agents from accumulation, with significant BBB disruption lasting over 24h, which increased gradually with an increase in the intensity of the electric field [61].

4.2.4.6 Magnetophoresis

Magnetophoresis utilizes a magnetic field for transporting drugs across biological barriers; the stronger the magnetic field, the greater the penetration effects (Figure 4.4a) [62]. Permeation flux factor is enhanced by a fixed magnetic field, whereas penetration of drugs is increased by an oscillating magnetic field, thereby producing a massage-like effect. It has also been found that BBB opening could also be induced by thermal effects induced by magnets. Brain temperature gets raised by 1°C on exposure to a radiofrequency electromagnetic field. It then reversibly enhances the macromolecular permeability across the BBB [63]. On applying a magnetic field, magnetic nanoparticles, by combining with nanocarriers, could be guided to specific target sites for the diagnosis and treatment of diseases. However, these particles should exhibit sufficient magnetic sensitivity [64]. Covalent bonding, encapsulation, electrostatic, and hydrophobic interactions are various ways in which drug loading occurs (Figure 4.4b).

The application of direct nose-to-brain (N2B) drug delivery systems is generally restricted to the olfactory mucosa, which is directly connected to the brain, though they have been found effective for treating neurological disorders. Therefore, novel approaches are required for effective N2B delivery. Magnetophoresis has emerged as a promising aspect for simulating and improving intranasal olfactory drug delivery. Magnetic targeting mediates the *in vivo* biodistribution of magnetic particles [65].

In one experimental study led by Yeung et al, a cholesterol and phosphatidyl-choline mixture was used to develop magnetic azidothymidine 50-triphosphate (AZTTP) liposomes. Liposomes were exposed to an external magnetic field to assess their ability to cross an *in vitro* BBB, either directly or through monocyte-mediated transport. Further, as compared to free AZTTP, there was a 3-fold higher apparent permeability observed in the case of magnetic AZTTP liposomes and greater liposomal uptake by monocytes. Due to the application of an external magnetic field,

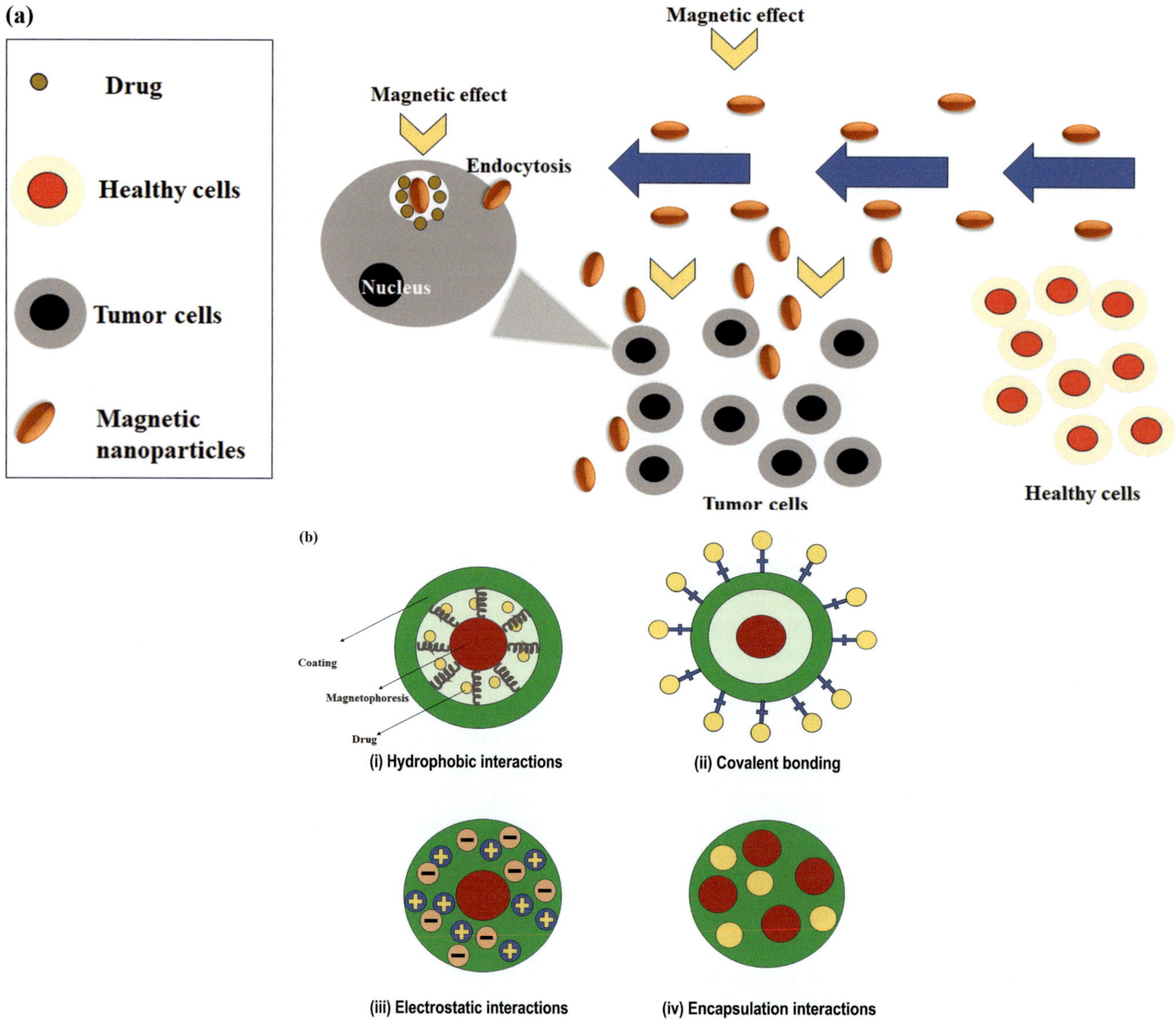

FIGURE 4.4 (a) Schematic representation of a magnetic nanoparticle-based drug delivery system targeting the tumor cells. (b) Different types of interactions during drug loading in magnetophoresis.

as compared to normal or non-magnetic monocytes, these magnetic monocytes demonstrated greater transendothelial migration. Above, magnetic nanoformulation, when applied to an external magnetic field, might be utilized for targeting active nucleotide analog reverse transcriptase inhibitors in the brain, thereby eliminating HIV reservoirs. Therefore, the above approach contributes to treating neuro-Acquired Immune Deficiency Syndrome (AIDS) [66].

4.2.5 Neural Tissue Engineering

The CNS exhibits a restricted intrinsic capacity for self-repair following injury, leading to significant functional deficits. This challenge necessitates the exploration of alternative strategies beyond traditional transplantation methods. Tissue engineering, utilizing biomaterials, offers a promising avenue for developing novel therapeutic interventions for neurological disorders. CNS injury triggers the activation and subsequent proliferation of glial cells, resulting in the formation of an inhibitory glial scar that impedes axonal regeneration. Currently, there is a paucity of clinically validated physiotherapy protocols promoting enhanced CNS repair in humans. While nerve autografts and allografts represent established techniques for peripheral nervous system (PNS) damage, clinical assessments of autograft deployment highlight the need for the development of engineered alternatives due to limitations such as donor site morbidity. Thus, significant challenges lie before neurobiologists and neurosurgeons in promoting functional restoration after PNS and CNS injuries. The key hurdles include neuronal formation, infectious diseases, immunological issues, and donor nerve scarcity. By addressing these challenges, researchers aim to achieve improved functional outcomes for patients suffering from PNS and CNS injuries [67]. There have been myriad types of polymeric materials reported for neural tissue engineering applications, including natural materials (collagen, chitin, chitosan, hyaluronic acid), and synthetic materials (poly-L-lactic acid, poly-D, L-lactic-co-glycolic acid, polyglycolic acid) [68].

Engineered polymeric scaffolds serve as prefabricated 2 or 3-D matrixes for tissue repair. This class of implantable polymers encompasses thin films, polymeric membranes, and conventional 3D constructs. Their integration with regenerative therapy offers a potential therapeutic intervention for neurodegenerative disorders [69]. As far as regenerative therapies are concerned, uniform cell distribution, sustained cell viability, and effective integration with host tissues pose significant challenges. To overcome those challenges, polymeric scaffolds and injectable hydrogels incorporating biomimetics have been formulated to mimic the extracellular matrix (ECM) environment. Notably, hyaluronic acid, a naturally occurring glycosaminoglycan, is an important constituent of the ECM. This biopolymer interacts with a variety of receptors, like CD44 and hyaluronan-mediated motility receptors, to improve cell adhesion and survival [70]. Injectable hydrogels possess mechanical properties similar to those of native CNS tissue and can be administered with minimal invasiveness. These attributes have made them a primary focus for CNS tissue regeneration and localized drug delivery research. Hyaluronan and methylcellulose-based hydrogels exhibited rapid gelation, biocompatibility, and biodegradability. Overall, these properties render them promising candidates as scaffolds for cell transplantation therapies [71].

4.3 CHALLENGES AND LIMITATIONS IN BRAIN DRUG DELIVERY

For over a decade, NPs have been enjoying their therapeutic utility because of their small size and specific physicochemical attributes. Owing to their small size, they can move across BBB either by passive diffusion or receptor-mediated endocytosis. Current research is focused on developing NPs with diagnostic or therapeutic potential for a broad spectrum of neurodegenerative disorders (AD, PD, multiple sclerosis, amyotrophic lateral sclerosis), brain cancer, and stroke. However, the enhanced diffusion properties of NPs raise concerns regarding potential cytotoxicity in living beings, particularly humans. NPs are known to interact with different subcellular components like proteins, nucleic acids, membranes, and lipids, hampering the functioning of cells and tissues. The nervous system is particularly susceptible to these interactions due to the tightly regulated and metabolically demanding nature of its cellular components, primarily neurons and glial cells [72]. NPs can also induce a variety of adverse neurological outcomes, including morphological changes, cell death, and genotoxicity. *In vitro* studies have shown the harmful effects of NPs on the nervous system [73]. For instance, polystyrene NPs have been shown to cause growth inhibition and developmental restriction in zebrafish due to disruptions in the brain-gut-microbiota axis [74]. Moreover, SiO_2 NPs have been found to penetrate the brain and accumulate in the striatum, leading to increased oxidative damage and inflammation [75]. Hence, these findings underscore the importance of further research to understand the neurotoxic mechanisms of NPs and develop strategies to mitigate the potential risks.

Similarly, polymeric NPs can have the potential for adverse effects due to chronic exposure. The high composition of drugs in polymeric nanocarriers poses threats regarding their accumulation in the CNS following repeated administration, leading to both cytotoxic and immunogenic responses. To mitigate these risks, stringent investigational protocols are imperative before clinical translation can be considered. The primary focus of research should be on determining the long-term neurotoxicity profile of NPs. This information is important for assessing the feasibility of implementing these nanocarriers in clinical settings. Furthermore, maintaining high encapsulation efficiency during the scale-up of nanoformulations from lab-scale to industrial production is paramount. The physicochemical properties and formulation strategies employed significantly impact the clinical efficacy. Thus, optimizing the formulation process for large-scale production is essential to ensure consistent encapsulation efficiency under physiologic conditions. Further, high costs associated with scaling up nanoformulations and the use of organic solvents during synthesis pose significant limitations. Developing cost-effective and environmentally friendly methods for nanomaterial production is a critical area of research [2].

4.4 EMERGING TECHNOLOGIES AND CONCLUDING REMARKS

Various new and advanced approaches have emerged recently, providing a promising aspect in brain drug delivery. As far as cell membrane-based biomimetic delivery is concerned, recently, an enormous interest has been taken in combining cell membranes and synthetic NPs that serve as initial cells and help in delivering therapeutic agents to target sites [76]. The cell membrane biomimetic vesicles can integrate the multifaceted functionalities of biological membranes into biological carriers [77]. Developing cell membrane-derived biomimetic delivery systems could serve as a promising aspect of providing targeted brain delivery. Enhanced penetration, transcytosis, and specific brain-targeting capabilities are the advantages associated with the above systems. However, stability and low yield are some issues that limit their applications [78]. Cell membrane-derived biomimetic delivery vehicles combine the cell membrane's biological functionality with the favorable physicochemical properties of synthetic NPs. There are various categories, among which erythrocyte ghosts may be considered the prototype [79]. Zhang and coworkers developed cell membrane-coated NPs by combining the membranes of erythrocytes with biodegradable polymer-based NPs (PLGA NPs). This became the most common cell membrane-derived biomimetic vehicle, which had numerous applications in drug delivery [80]. Numerous cell membranes, like cancer cell membranes and stem cell membranes, and nanoparticulate carriers like iron NPs and gold NPs, have been employed for developing biomimetic vehicles [81]. In various brain disorders, effective targeting, as well as delivery capability, are demonstrated by these biomimetic vehicles. It has also been found that stem cell membranes and immune cell membranes inherit the capability of natural inflammation from source cells, making them appropriate for developing CNS-targeting biomimetic vehicles for targeting natural lesions and BBB traversing ability [82].

Localized brain drug delivery could also be achieved with the help of injectables and the degradation or dissolution of drug-loaded implants. Intracerebral injections or implants have proven efficient for brain targeting, which employs direct injection, microencapsulated chemicals, and controlled release matrixes [83]. Fung et al. employed three chemotherapeutic drugs, viz., paclitaxel, carmustine, and 4-hydroperoxycyclophosphamide (4-HC), and developed a polymer-based implant that was implanted intracranially in cynomolgus monkeys. Polymer implants could maintain higher drug concentrations, i.e., 0.3–0.4 mM for 4-HC, 0.2–1 mM in the case of paclitaxel, and 0.5–3.5 mM for carmustine within a distance of 3 mm [84]. Benefits of this technique include localized delivery, polymer biocompatibility, low invasiveness, and predictable drug release kinetics; however, poor drug penetration ability due to implant size is one of the drawbacks associated with this technique [85]. Polysaccharide-based intracerebral implants have also demonstrated efficacy in treating brain tumors, AD, and PD. These implants consist of drugs encapsulated within a polyanhydride-based biodegradable polymer matrix. Polysaccharide implants are implanted within the cerebral portion of the brain. Drug release occurs via polymer matrix hydrolysis due to the presence of anhydride bonds on the polymer surface. In 1996, the US FDA approved Gliadel wafers, which consist of carboxy phenoxy propane and sebacic acid (20:80) and are employed for treating brain tumors [86].

PD is caused by various missense mutations like A30P, E46K, etc. or by duplication or triplication mutations in the synuclein-alpha (SNCA) gene that encodes α-synuclein [87]. Recently, Helmschrodt and coworkers developed an injection of siRNA complexed with polyethyleneimine (PEI) F25-LMW, which was administered into a human wild-type SNCA overexpressing mice. About a 65% reduction in the expression of SNCA mRNA in the striatum, as well as a 50% SNCA protein reduction, was observed after 5 days of administering 0.75 mg PEI/siRNA single injection. Thus, it was proven that PEI NPs-mediated siRNA brain delivery could serve as a therapeutic intervention for PD [88].

Brain-targeted delivery systems have emerged as a pivotal focus in the treatment of neurological disorders. Nanotechnology-based approaches, targeted drug delivery strategies, intranasal administration, and physical disruption of the BBB have shown significant promise in improving therapeutic efficacy while minimizing side effects. However, challenges such as BBB permeability, cytotoxicity, neurotoxicity, and safety concerns persist. Despite several methods developed to bypass the BBB for brain tumor targeting, their effectiveness in treating a wide range of neurological conditions remains uncertain. While nanocarriers have been extensively researched, only a few have progressed to clinical use. Emerging strategies, including cell membrane-based biomimetic delivery, neural drug discovery with iPSCs, and implantable and injectable devices, offer potential avenues for developing more effective treatments for neurological disorders.

ACKNOWLEDGMENTS

A. Jain would like to acknowledge SERB—Department of Science & Technology for Research grant (SRG/2023/000096)

REFERENCES

1. R.K. Upadhyay, Drug delivery systems, CNS protection, and the blood-brain barrier, *Biomed Res. Int.* (2014). https://doi.org/10.1155/2014/869269.
2. R. Khatoon, M.A. Alam, P.K. Sharma, Current approaches and prospective drug targeting to brain, *J. Drug Deliv. Sci. Technol.* 61 (2021) 102098. https://doi.org/10.1016/j.jddst.2020.102098.
3. G.C. Terstappen, A.H. Meyer, R.D. Bell, W. Zhang, Strategies for delivering therapeutics across the blood–brain barrier, *Nat. Rev. Drug Discov.* 20 (2021) 362–383. https://doi.org/10.1038/s41573-021-00139-y.
4. A. Alahmari, Blood-brain barrier overview: structural and functional correlation, *Neural Plast.* (2021). https://doi.org/10.1155/2021/6564585.
5. T.G. Agnihotri, G.S. Jadhav, B. Sahu, A. Jain, Recent trends of bioconjugated nanomedicines through nose-to-brain delivery for neurological disorders, *Drug Deliv. Transl. Res.* 12 (2022) 3104–3120. https://doi.org/10.1007/s13346-022-01173-y.
6. T.G. Agnihotri, A. Alexander, M. Agrawal, S.K. Dubey, A. Jain, In vitro-in vivo correlation in nanocarriers: from protein corona to therapeutic implications, *J. Control. Release* 354 (2023) 794–809. https://doi.org/10.1016/j.jconrel.2023.01.063.
7. V. Peddinti, B. Rout, T.G. Agnihotri, S.S. Gomte, A. Jain, Functionalized liposomes: an enticing nanocarrier for management of glioma, *J. Liposome Res.* (2023) 1–19. https://doi.org/10.1080/08982104.2023.2270060.
8. S.K. Dubey, G. Singhvi, K.V. Krishna, T. Agnihotri, R.N. Saha, G. Gupta, Herbal medicines in neurodegenerative disorders: an evolutionary approach through novel drug delivery system, *J. Environ. Pathol. Toxicol. Oncol.* 37 (2018) 199–208.
9. C. Janrao, S. Khopade, A. Bavaskar, S.S. Gomte, T.G. Agnihotri, A. Jain, Recent advances of polymer based nanosystems in cancer management, *J. Biomater. Sci. Polym. Ed.* 34 (2023) 1274–1335.
10. Y. Garg, M. Kumar, G. Sharma, O.P. Katare, S. Chopra, A. Bhatia, Systematic designing and optimization of polymeric nanoparticles using central composite design: a novel approach for nose-to-brain delivery of donepezil hydrochloride, *J. Clust. Sci.* 35 (2024) 1007–1019.

11. V.S. Chaudhari, U.S. Murty, S. Banerjee, Lipidic nanomaterials to deliver natural compounds against cancer: a review, *Environ. Chem. Lett.* 18 (2020) 1803–1812.
12. A. Akbarzadeh, R. Rezaei-Sadabady, S. Davaran, S.W. Joo, N. Zarghami, Y. Hanifehpour, M. Samiei, M. Kouhi, K. Nejati-Koshki, Liposome: classification, preparation, and applications, *Nanoscale Res. Lett.* (2013). https://doi.org/10.1186/1556-276X-8-102.
13. F. Juhairiyah, E.C.M. de Lange, Understanding drug delivery to the brain using liposome-based strategies: studies that provide mechanistic insights are essential, *AAPS J.* 23 (2021) 1–16.
14. P. Khairnar, A. Singh, K. Ahirwar, R. Shukla, ApoE3 anchored liposomal delivery of rivastigmine for brain delivery: formulation, characterization, and in vivo pharmacokinetic evaluation, *AAPS PharmSciTech* 24 (2023) 223.
15. S. Kotta, H.M. Aldawsari, S.M. Badr-Eldin, A.B. Nair, K. Yt, Progress in polymeric micelles for drug delivery applications, *Pharmaceutics* 14 (2022) 1636.
16. C. Zhang, X. Zhao, J. Zhang, Q. Zhang, C. Jin, D. Li, X. Chen, X. Hai, X. Chen, X. Tang, Dual-modified brain-targeting lipid polymer micelle delivery of anti-miRNA-21 for achieving oral treatment of glioma, *J. Drug Deliv. Sci. Technol.* 96 (2024) 105721.
17. A. Raeisi, F. Farjadian, Commercial hydrogel product for drug delivery based on route of administration, *Front. Chem.* 12 (2024) 1–14. https://doi.org/10.3389/fchem.2024.1336717.
18. O. Sayiner, S. Arisoy, T. Comoglu, F.G. Ozbay, G. Esendagli, Development and in vitro evaluation of temozolomide-loaded PLGA nanoparticles in a thermoreversible hydrogel system for local administration in glioblastoma multiforme, *J. Drug Deliv. Sci. Technol.* 57 (2020) 101627.
19. S.S. Gomte, T.G. Agnihotri, S. Khopade, A. Jain, Exploring the potential of pH-sensitive polymers in targeted drug delivery, *J. Biomater. Sci. Polym. Ed.* 35 (2024) 228–268.
20. T.G. Agnihotri, S.S. Gomte, A. Jain, Emerging theranostics to combat cancer: a perspective on metal-based nanomaterials, *Drug Dev. Ind. Pharm.* 48 (2022) 585–601. https://doi.org/10.1080/03639045.2022.2153862.
21. R.G.R. Pinheiro, A.J. Coutinho, M. Pinheiro, A.R. Neves, Nanoparticles for targeted brain drug delivery: what do we know? *Int. J. Mol. Sci.* 22 (2021) 11654.
22. T. Moos, E.H. Morgan, Transferrin and transferrin receptor function in brain barrier systems, *Cell. Mol. Neurobiol.* 20 (2000) 77–95.
23. T.A. Rouault, Iron metabolism in the CNS: implications for neurodegenerative diseases, *Nat. Rev. Neurosci.* 14 (2013) 551–564.
24. M.J. Ramalho, M. Bravo, J.A. Loureiro, J. Lima, M.C. Pereira, Transferrin-modified nanoparticles for targeted delivery of Asiatic acid to glioblastoma cells, *Life Sci.* 296 (2022) 120435.
25. J. Wally, S.K. Buchanan, A structural comparison of human serum transferrin and human lactoferrin, *Biometals.* 20 (2007) 249–262.
26. N. Qi, S. Zhang, X. Zhou, W. Duan, D. Gao, J. Feng, A. Li, Combined integrin αvβ3 and lactoferrin receptor targeted docetaxel liposomes enhance the brain-targeting effect and anti-glioma effect, *J. Nanobiotechnology.* 19 (2021) 1–17. https://doi.org/10.1186/s12951-021-01180-0.
27. S. Console, C. Marty, C. García-Echeverría, R. Schwendener, K. Ballmer-Hofer, Antennapedia and HIV transactivator of transcription (TAT)"protein transduction domains" promote endocytosis of high molecular weight cargo upon binding to cell surface glycosaminoglycans, *J. Biol. Chem.* 278 (2003) 35109–35114.
28. S.M. Ghorai, A. Deep, D. Magoo, C. Gupta, N. Gupta, Cell-Penetrating and Targeted Peptides Delivery Systems as Potential Pharmaceutical Carriers for Enhanced Delivery across the Blood–Brain Barrier (BBB), *Pharmaceutics* 15 (2023) 1999.
29. L. Qian, X. Lin, X. Gao, R.U. Khan, J.-Y. Liao, S. Du, J. Ge, S. Zeng, S.Q. Yao, The dawn of a new era: targeting the "undruggables" with antibody-based therapeutics, *Chem. Rev.* 123 (2023) 7782–7853.
30. F.R. El-Gamal, M.A. Akl, H.A. Mowafy, H. Mukai, S. Kawakami, M.I. Afouna, Synthesis and evaluation of high functionality and quality cell-penetrating peptide conjugated lipid for octaarginine modified PEGylated liposomes in U251 and U87 glioma cells, *J. Pharm. Sci.* 111 (2022) 1719–1727.
31. Q. Huang, X. Chen, S. Yu, G. Gong, H. Shu, Research progress in brain-targeted nasal drug delivery, *Front. Aging Neurosci.* 15 (2023) 1–12. https://doi.org/10.3389/fnagi.2023.1341295.
32. S. Ul Islam, A. Shehzad, M. Bilal Ahmed, Y.S. Lee, Intranasal delivery of nanoformulations: a potential way of treatment for neurological disorders, *Molecules* 25 (2020) 1–27. https://doi.org/10.3390/molecules25081929.
33. M.L. Formica, D.A. Real, M.L. Picchio, E. Catlin, R.F. Donnelly, A.J. Paredes, On a highway to the brain: a review on nose-to-brain drug delivery using nanoparticles, *Appl. Mater. Today* 29 (2022) 101631. https://doi.org/10.1016/j.apmt.2022.101631.
34. M.Z. Ahmad, A.H. Bin Sabri, Q.K. Anjani, J. Domínguez-Robles, N. Abdul Latip, K.A. Hamid, Design and development of levodopa-loaded polymeric nanoparticles for intranasal delivery, *Pharmaceuticals* 15 (2022) 370.
35. Y. Mardor, Y. Roth, A. Orenstein, J.S. Cohen, Z. Lidar, M. Hadani, Z. Ram, T. Jonas, M. Faibel, R. Pfeffer, D. Nass, S.E. Maier, Monitoring response to convection-enhanced taxol delivery in brain tumor patients using diffusion-weighted magnetic resonance imaging, *Cancer Res.* 61(13) (2001) 4971–4973.
36. R.S. D'Amico, M.K. Aghi, M.A. Vogelbaum, J.N. Bruce, Convection-enhanced drug delivery for glioblastoma: a review, *J. Neurooncol.* (2021). https://doi.org/10.1007/s11060-020-03408-9.
37. R. Hunt Bobo, D.W. Laske, A. Akbasak, P.F. Morrison, R.L. Dedrick, E.H. Oldfield, Convection-enhanced delivery of macromolecules in the brain, *Proc. Natl. Acad. Sci. USA* (1994). https://doi.org/10.1073/pnas.91.6.2076.
38. R.R. Lonser, M. Sarntinoranont, P.F. Morrison, E.H. Oldfield, Convection-enhanced delivery to the central nervous system, *J. Neurosurg.* (2015). https://doi.org/10.3171/2014.10.jns14229..
39. P.Y. Chen, T. Ozawa, D.C. Drummond, A. Kalra, J.B. Fitzgerald, D.B. Kirpotin, K.C. Wei, N. Butowski, M.D. Prados, M.S. Berger, J.R. Forsayeth, K. Bankiewicz, C.D. James, Comparing routes of delivery for nanoliposomal irinotecan shows superior anti-tumor activity of local administration in treating intracranial glioblastoma xenografts, *Neuro. Oncol.* (2013). https://doi.org/10.1093/neuonc/nos305.
40. A. Healy, M. Vogelbaum, Convection-enhanced drug delivery for gliomas, *Surg. Neurol. Int.* (2015). https://doi.org/10.4103/2152-7806.151337.
41. W. Zhan, C.H. Wang, Convection enhanced delivery of chemotherapeutic drugs into brain tumor, *J. Control. Release.* (2018). https://doi.org/10.1016/j.jconrel.2017.12.020.
42. K. Matsuo, H. Okamoto, Y. Kawai, Y.S. Quan, F. Kamiyama, S. Hirobe, N. Okada, S. Nakagawa, Vaccine

efficacy of transcutaneous immunization with amyloid β using a dissolving microneedle array in a mouse model of Alzheimer's disease, *J. Neuroimmunol.* (2014). https://doi.org/10.1016/j.jneuroim.2013.11.002.
43. D. McMahon, C. Poon, K. Hynynen, Evaluating the safety profile of focused ultrasound and microbubble-mediated treatments to increase blood-brain barrier permeability, *Expert Opin. Drug Deliv.* (2019). https://doi.org/10.1080/17425247.2019.1567490.
44. Y. Sun, L. Du, M. Yang, Q. Li, X. Jia, Q. Li, L. Zhu, Y. Zhang, Y. Liu, S. Liu, Brain-targeted drug delivery assisted by physical techniques and its potential applications in traditional Chinese medicine, *J. Tradit. Chinese Med. Sci.* (2021). https://doi.org/10.1016/j.jtcms.2021.07.003.
45. S.A. Nour, N.S. Abdelmalak, M.J. Naguib, H.M. Rashed, A.B. Ibrahim, Intranasal brain-targeted clonazepam polymeric micelles for immediate control of status epilepticus: in vitro optimization, ex vivo determination of cytotoxicity, in vivo biodistribution and pharmacodynamics studies, *Drug Deliv.* 23 (2016) 3681–3695. https://doi.org/10.1080/10717544.2016.1223216.
46. J. Oh, K. Liu, T. Medina, F. Kralick, H. Noh, A novel microneedle array for the treatment of hydrocephalus, *Microsyst. Technol.* (2014). https://doi.org/10.1007/s00542-013-1988-4.
47. B. Patel, P.H. Yang, A.H. Kim, The effect of thermal therapy on the blood-brain barrier and blood-tumor barrier, *Int. J. Hyperth.* (2020). https://doi.org/10.1080/02656736.2020.1783461.
48. M. Izco, J. Blesa, M. Schleef, M. Schmeer, R. Porcari, R. Al-Shawi, S. Ellmerich, M. de Toro, C. Gardiner, Y. Seow, A. Reinares-Sebastian, R. Forcen, J.P. Simons, V. Bellotti, J.M. Cooper, L. Alvarez-Erviti, Systemic exosomal delivery of shRNA minicircles prevents Parkinsonian pathology, *Mol. Ther.* (2019). https://doi.org/10.1016/j.ymthe.2019.08.010.
49. J. Jampilek, K. Zaruba, M. Oravec, M. Kunes, P. Babula, P. Ulbrich, I. Brezaniova, R. Opatrilova, J. Triska, P. Suchy, Preparation of silica nanoparticles loaded with nootropics and their in vivo permeation through blood-brain barrier, *Biomed Res. Int.* (2015). https://doi.org/10.1155/2015/812673.
50. H.R. Jeong, H. Jun, H.R. Cha, J.M. Lee, J.H. Park, Safe coated microneedles with reduced puncture occurrence after administration, *Micromachines* (2020). https://doi.org/10.3390/MI11080710.
51. E. Kim, G. Erdos, S. Huang, T.W. Kenniston, S.C. Balmert, C.D. Carey, V.S. Raj, M.W. Epperly, W.B. Klimstra, B.L. Haagmans, E. Korkmaz, L.D. Falo, A. Gambotto, Microneedle array delivered recombinant coronavirus vaccines: immunogenicity and rapid translational development, EBioMedicine. (2020). https://doi.org/10.1016/j.ebiom.2020.102743.
52. N. Lee, D. Yoo, D. Ling, M.H. Cho, T. Hyeon, J. Cheon, Iron oxide based nanoparticles for multimodal imaging and magnetoresponsive therapy, *Chem. Rev.* (2015). https://doi.org/10.1021/acs.chemrev.5b00112.
53. S. Lehrer, P.H. Rheinstein, Transspinal delivery of drugs by transdermal patch back-of-neck for Alzheimer's disease: a new route of administration. *Discov. Med.* (2019).
54. E.C. Leuthardt, C. Duan, M.J. Kim, J.L. Campian, A.H. Kim, M.M. Miller-Thomas, J.S. Shimony, D.D. Tran, Hyperthermic laser ablation of recurrent glioblastoma leads to temporary disruption of the peritumoral blood brain barrier, *PLoS One* (2016). https://doi.org/10.1371/journal.pone.0148613.
55. N. Lipsman, Y. Meng, A.J. Bethune, Y. Huang, B. Lam, M. Masellis, N. Herrmann, C. Heyn, I. Aubert, A. Boutet, G.S. Smith, K. Hynynen, S.E. Black, Blood–brain barrier opening in Alzheimer's disease using MR-guided focused ultrasound, *Nat. Commun.* (2018). https://doi.org/10.1038/s41467-018-04529-6.
56. M.F. Lorenzo, S.C. Thomas, Y. Kani, J. Hinckley, M. Lee, J. Adler, S.S. Verbridge, F.C. Hsu, J.L. Robertson, R. V. Davalos, J.H. Rossmeisl, Temporal characterization of blood–brain barrier disruption with high-frequency electroporation, *Cancers (Basel)* (2019). https://doi.org/10.3390/cancers11121850.
57. M. Qu, Q. Lin, L. Huang, Y. Fu, L. Wang, S. He, Y. Fu, S. Yang, Z. Zhang, L. Zhang, X. Sun, Dopamine-loaded blood exosomes targeted to brain for better treatment of Parkinson's disease, *J. Control. Release* (2018). https://doi.org/10.1016/j.jconrel.2018.08.035.
58. X. Zhuang, X. Xiang, W. Grizzle, D. Sun, S. Zhang, R.C. Axtell, S. Ju, J. Mu, L. Zhang, L. Steinman, D. Miller, H.G. Zhang, Treatment of brain inflammatory diseases by delivering exosome encapsulated anti-inflammatory drugs from the nasal region to the brain, *Mol. Ther.* (2011). https://doi.org/10.1038/mt.2011.164.
59. O. Rascol, T. Zesiewicz, K.R. Chaudhuri, M. Asgharnejad, E. Surmann, E. Dohin, S. Nilius, L. Bauer, A randomized controlled exploratory pilot study to evaluate the effect of rotigotine transdermal patch on parkinson's disease–associated chronic pain, *J. Clin. Pharmacol.* (2016). https://doi.org/10.1002/jcph.678.
60. A. Rodriguez, S.B. Tatter, W. Debinski, Neurosurgical techniques for disruption of the blood–brain barrier for glioblastoma treatment, *Pharmaceutics* (2015). https://doi.org/10.3390/pharmaceutics7030175.
61. S. Sharabi, D. Last, D. Daniels, S. Liraz Zaltsman, Y. Mardor, The effects of point-source electroporation on the blood-brain barrier and brain vasculature in rats: an MRI and histology study, *Bioelectrochemistry* (2020). https://doi.org/10.1016/j.bioelechem.2020.107521.
62. K.F. Timbie, B.P. Mead, R.J. Price, Drug and gene delivery across the blood-brain barrier with focused ultrasound, *J. Control. Release* (2015). https://doi.org/10.1016/j.jconrel.2015.08.059.
63. W. Wang, N.I. Marin-Ramos, H. He, S. Zeng, H.Y. Cho, S.D. Swenson, L. Zheng, A.L. Epstein, A.H. Schonthal, F.M. Hofman, L. Chen, T.C. Chen, NEO100 enables brain delivery of blood-brain barrier impermeable therapeutics, *Neuro. Oncol.* (2021). https://doi.org/10.1093/neuonc/noaa206.
64. J. Xi, Z. Zhang, X.A. Si, J. Yang, W. Deng, Optimization of magnetophoretic-guided drug delivery to the olfactory region in a human nose model, *Biomech. Model. Mechanobiol.* (2016). https://doi.org/10.1007/s10237-015-0730-9.
65. M.L. Yarmush, A. Golberg, G. Serša, T. Kotnik, D. Miklavčič, Electroporation-based technologies for medicine: principles, applications, and challenges, *Annu. Rev. Biomed. Eng.* (2014). https://doi.org/10.1146/annurev-bioeng-071813-104622.
66. C. Yeung, S. Chen, B. King, H. Lin, K. King, F. Akhtar, G. Diaz, B. Wang, J. Zhu, W. Sun, A. Khademhosseini, S. Emaminejad, A 3D-printed microfluidic-enabled hollow microneedle architecture for transdermal drug delivery, *Biomicrofluidics* (2019). https://doi.org/10.1063/1.5127778.
67. A. Akhtar, A. Andleeb, T.S. Waris, M. Bazzar, A.R. Moradi, N.R. Awan, M. Yar, Neurodegenerative diseases and effective drug delivery: a review of challenges and novel therapeutics, *J. Control. Release.* 330 (2021) 1152–1167. https://doi.org/10.1016/j.jconrel.2020.11.021.

68. Y. Kim, Z. Zharkinbekov, K. Raziyeva, L. Tabyldiyeva, K. Berikova, D. Zhumagul, K. Temirkhanova, A. Saparov, Chitosan-Based Biomaterials for Tissue Regeneration, (2023).
69. T. Patel, J. Zhou, J.M. Piepmeier, W.M. Saltzman, Polymeric nanoparticles for drug delivery to the central nervous system, *Adv. Drug Deliv. Rev.* 64 (2012) 701–705. https://doi.org/10.1016/j.addr.2011.12.006.
70. R.Y. Tam, T. Fuehrmann, N. Mitrousis, M.S. Shoichet, Regenerative therapies for central nervous system diseases: a biomaterials approach, *Neuropsychopharmacology* 39 (2014) 169–188. https://doi.org/10.1038/npp.2013.237.
71. M.J. Caicco, T. Zahir, A.J. Mothe, B.G. Ballios, A.J. Kihm, C.H. Tator, M.S. Shoichet, Characterization of hyaluronan-methylcellulose hydrogels for cell delivery to the injured spinal cord, *J. Biomed. Mater. Res. A.* 101 (2013) 1472–1477. https://doi.org/10.1002/jbm.a.34454.
72. K.M. Asha Spandana, M. Bhaskaran, V.V.S.N.R. Karri, J. Natarajan, A comprehensive review of nano drug delivery system in the treatment of CNS disorders, *J. Drug Deliv. Sci. Technol.* 57 (2020) 101628. https://doi.org/10.1016/j.jddst.2020.101628.
73. L. Xuan, Z. Ju, M. Skonieczna, P.K. Zhou, R. Huang, Nanoparticles-induced potential toxicity on human health: applications, toxicity mechanisms, and evaluation models, *MedComm.* 4 (2023) 1–39. https://doi.org/10.1002/mco2.327.
74. M. Teng, X. Zhao, C. Wang, C. Wang, J.C. White, W. Zhao, L. Zhou, M. Duan, F. Wu, Polystyrene nanoplastics toxicity to Zebrafish: dysregulation of the brain-intestine-microbiota axis, *ACS Nano* 16 (2022) 8190–8204. https://doi.org/10.1021/acsnano.2c01872.
75. J. Wu, C. Wang, J. Sun, Y. Xue, Neurotoxicity of silica nanoparticles: Brain localization and dopaminergic neurons damage pathways, *ACS Nano* 5 (2011) 4476–4489. https://doi.org/10.1021/nn103530b.
76. B.T. Luk, L. Zhang, Cell membrane-camouflaged nanoparticles for drug delivery, *J. Control. Release* (2015). https://doi.org/10.1016/j.jconrel.2015.07.019.
77. B. Rout, T.G. Agnihotri, A. Jain, Advancement in triple-negative breast cancer therapeutics: a comprehensive review on the potential of cell membrane-coated nanoparticles, *J. Drug Deliv. Sci. Technol.* 98 (2024) 105935. https://doi.org/10.1016/j.jddst.2024.105935.
78. V. Bourganis, O. Kammona, A. Alexopoulos, C. Kiparissides, Recent advances in carrier- mediated nose-to-brain delivery of pharmaceutics, *Eur. J. Pharm. Biopharm.* 128 (2018) 337–362. https://doi.org/10.1016/j.ejpb.2018.05.009.
79. M.L. Steer, C. Baldwin, A. Levitzki, Preparation and characterization of hormone-sensitive, resealed erythrocyte ghosts, *J. Biol. Chem.* (1976). https://doi.org/10.1016/s0021-9258(17)33204-0.
80. S. Aryal, C.M.J. Hu, R.H. Fang, D. Dehaini, C. Carpenter, D.E. Zhang, L. Zhang, Erythrocyte membrane-cloaked polymeric nanoparticles for controlled drug loading and release, *Nanomedicine.* (2013). https://doi.org/10.2217/nnm.12.153.
81. J. Gao, D. Chu, Z. Wang, Cell membrane-formed nanovesicles for disease-targeted delivery, *J. Control. Release.* (2016). https://doi.org/10.1016/j.jconrel.2016.01.024.
82. H. Wu, X. Jiang, Y. Li, Y. Zhou, T. Zhang, P. Zhi, J. Gao, Engineering stem cell-derived biomimetic vesicles for versatility and effective targeted delivery, *Adv. Funct. Mater.* 30 (2020) 1–15. https://doi.org/10.1002/adfm.202006169.
83. A. Bjorklund, U. Stenevi, Intracerebral neural implants: neuronal replacement and reconstruction of damaged circuitries, *Annu. Rev. Neurosci.* (1984). https://doi.org/10.1146/annurev.neuro.7.1.279.
84. L.K. Fung, M.G. Ewend, A. Sills, E.P. Sipos, R. Thompson, M. Watts, O.M. Colvin, H. Brem, W.M. Saltzman, Pharmacokinetics of interstitial delivery of carmustine, 4-hydroperoxycyclophosphamide, and paclitaxel from a biodegradable polymer implant in the monkey brain, *Cancer Res.* (1998).
85. M. Westphal, D.C. Hilt, E. Bortey, P. Delavault, R. Olivares, P.C. Warnke, I.R. Whittle, J. Jääskeläinen, Z. Ram, A phase 3 trial of local chemotherapy with biodegradable carmustine (BCNU) wafers (Gliadel wafers) in patients with primary malignant glioma, *Neuro. Oncol.* (2003). https://doi.org/10.1093/neuonc/5.2.79.
86. J.P. Jain, S. Modi, A.J. Domb, N. Kumar, Role of polyanhydrides as localized drug carriers, *J. Control. Release* (2005). https://doi.org/10.1016/j.jconrel.2004.12.021.
87. A. Singh, S.K. Maharana, R. Shukla, P. Kesharwani, Nanotherapeutics approaches for targeting alpha synuclien protein in the management of Parkinson's disease, *Process Biochem.* 110 (2021) 181–194. https://doi.org/10.1016/j.procbio.2021.08.008.
88. C. Helmschrodt, S. Höbel, S. Schöniger, A. Bauer, J. Bonicelli, M. Gringmuth, S.A. Fietz, A. Aigner, A. Richter, F. Richter, Polyethylenimine nanoparticle-mediated siRNA delivery to reduce α-synuclein expression in a model of Parkinson's disease, *Mol. Ther. Nucleic Acids* 9 (2017) 57–68. https://doi.org/10.1016/j.omtn.2017.08.013.

5 Specific Applications and Systems

Protein/Peptide Drug Delivery Systems

Virender Kumar[1], *Ashwani Kumar*[2], *Shweta Saloni Kakkar*[3], *Vandana Garg*[3], *Shiv Kumar Yadav*[4], *and* Davinder Kumar[4]

[1]College of Pharmacy, Pandit Bhagwat Dayal Sharma University of Health Sciences, Rohtak, India

[2]Department of Pharmaceutical Sciences, Gurukul Kangri Vishwavidyalaya, Haridwar, India

[3]Department of Pharmaceutical Sciences, M.D. University, Rohtak, Haryana, India

[4]B.S. Anangpuria Institute of Pharmacy, Faridabad, Haryana, India

5.1 INTRODUCTION

The evolution of drug delivery systems has transformed therapeutic approaches, facilitating targeted and efficient treatment for various diseases [1]. This compilation presents a variety of drug delivery modalities, showcasing innovative strategies that improve the bioavailability and effectiveness of therapeutic agents [2]. Protein/Peptide Drug Delivery Systems employing nanoparticles offer a state-of-the-art solution for enhancing the stability of biologics and targeted delivery, addressing issues related to their natural instability and rapid clearance from circulation. Similarly, nucleic acid-based therapeutic delivery systems are making strides in gene therapy and RNA-based treatments, enabling precise alterations in cellular function and promising significant advancements in the treatment of genetic disorders and cancers [3]. Beyond these foundational systems. Colonic, nasopulmonary, ocular, and brain drug delivery systems underscore the importance of route-specific strategies, which focus on the unique physiological barriers and therapeutic needs of each system. Additionally, we examine the intricacies of transdermal drug therapy and the challenges posed by viral infections and suitable delivery systems [4]. Investigating effective delivery systems for viral treatments emphasizes the urgent need for innovative solutions considering emerging viral threats [5]. This chapter offers a comprehensive perspective on the current landscape of drug delivery systems [6].

1. **Protein/Peptide Drug Delivery Systems by Nanoparticle**: A range of nanoparticle NPDDS (nanoparticulate drug delivery system) has been developed, both of which have unique physicochemical and structural properties that improve the transport of proteins and peptides [7]. Micelles, dendrimers, liposomes, mesoporous silica, and gold nanoparticles are a few examples of them. Several lipids that are unique to receptors have been utilized for nanoparticle NDDS (nanoparticulate drug delivery system) platforms that target users' active features regarding ovarian cancer [8–12]. Bioactive small molecules, hyaluronic acid, peptides and proteins (such as antibodies), and steroids are some examples of these targeted agents, as shown in Figure 5.1.

2. **Nucleic Acid-Based Therapeutic Delivery System:** Chemicals based on nucleic acids, including DNA and RNA, and their analogues, have emerged as groundbreaking treatments for several illnesses, such as infectious diseases, malignancies, and genetic abnormalities [13]. The ability to manipulate genetic material enables targeted therapeutic interventions [14]. However, delivering nucleic acids effectively to the intended cells poses significant challenges, highlighting the need for a specialized delivery system. Nucleic acids are naturally unstable and can be quickly degraded by nucleases in the bloodstream [15]. Their large size and negative charge hinder their ability to cross cell membranes. Successful delivery necessitates targeting specific cell types to ensure therapeutic effects while sparing healthy cells. Additionally, delivery systems may trigger immune responses, which can reduce efficacy or cause adverse effects. Furthermore [16]. Several delivery methods have been developed to overcome these obstacles [17]. Modified viruses, like adenoviruses and lentiviruses, can effectively deliver nucleic acids into host cells [18]. Although they offer high transfection efficiency, concerns exist regarding their safety and potential immune responses. RNA-based therapies, such as siRNA and mRNA, can be utilized to silence oncogenes or to deliver tumor-associated antigens, thereby stimulating immune responses [19]. Nucleic acid vaccines, such as mRNA vaccines, have shown significant potential in eliciting robust immune responses against infectious diseases, including COVID-19 [20]. Nucleic acid-based approaches can also inhibit viral replication and modulate the immune response during viral infections. Additionally, nucleic acid delivery

DOI: 10.1201/9781003508601-5

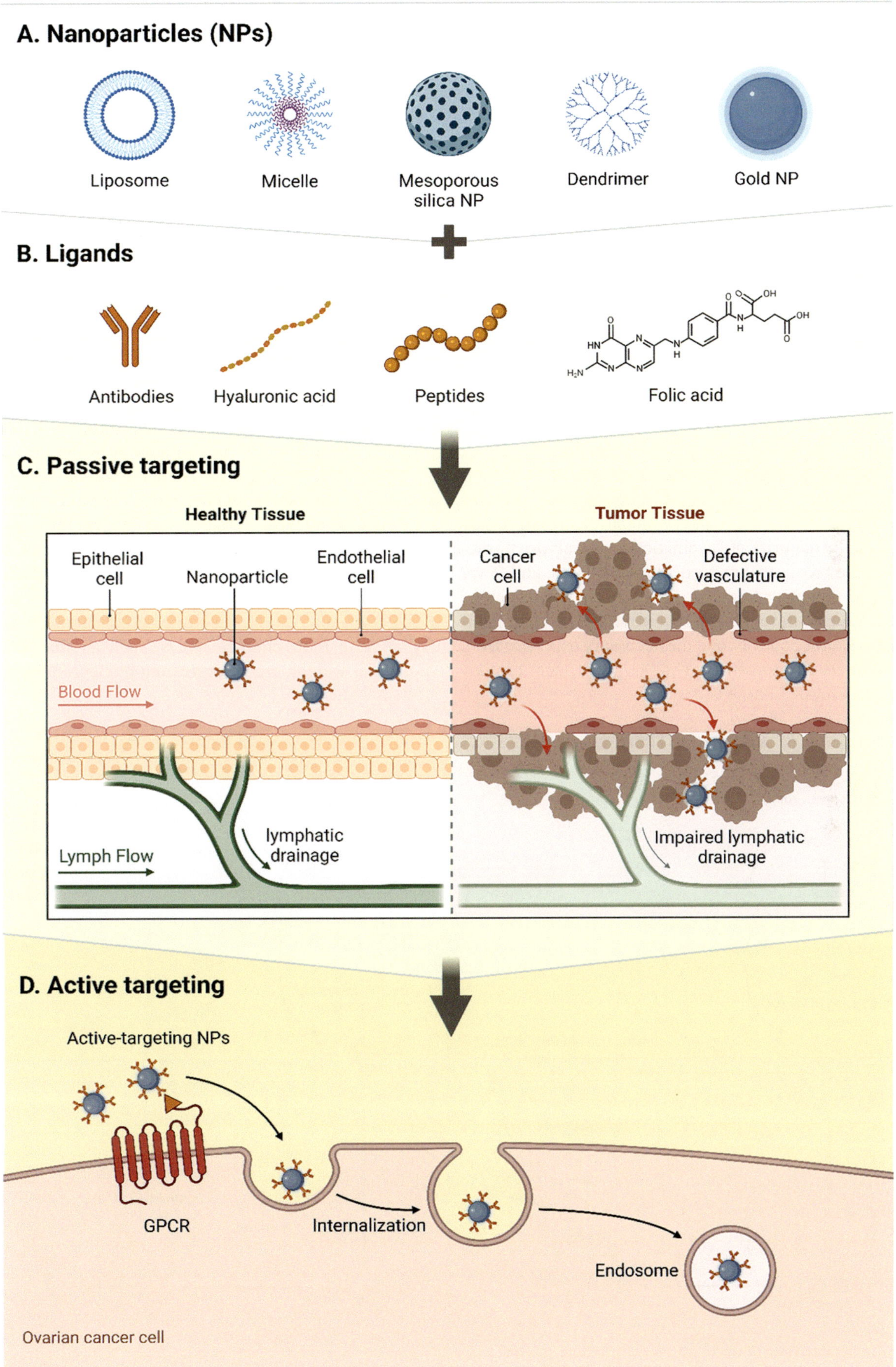

FIGURE 5.1 Protein/peptide drug delivery systems by nanoparticle.

systems can facilitate tissue repair and regeneration by delivering genes that promote cell survival and proliferation [21].

Gene therapy involves transferring foreign genetic material into host cells to achieve therapeutic effects. Initially aimed at treating specific genetic disorders, the scope of gene therapy has broadened to include various inherited and acquired diseases [22–28].

In nonviral gene therapy, carriers made from lipids, like solid lipid nanoparticles or lipid carriers with nanostructure, are utilized to transport genetic material into cells. After the endosomes disintegrate, the nanoparticles expel microRNA (miRNA) into the cytoplasm. After that, this miRNA is integrated into the RNA-induced silencing complex. (RISC), which then targets and decomposes the chosen messenger RNA (mRNA). This action effectively halts the production of specific proteins within the cell [29–31]. The viral and non-gene therapy is shown in Figure 5.2.

Vaccine development: Vaccine development is a challenging, intricate, and expensive endeavor that involves clinical, process, and assay development. It is also highly risky, since the majority of vaccination proposals are unsuccessful in the preclinical or early clinical phases, and fewer than 1 in 15 that enter Phase II clinical trials are ultimately licensed. Vaccine development requires robust project management systems, controls, and the necessary expertise from both engineers and scientists. The "target product profile" is an essential strategic instrument in this approach (TPP). The TPP outlines the desired characteristics and features of the product being developed, and its main advantages over competitors, and provides a high-level plan for the clinical and nonclinical research required to evaluate the product's safety and effectiveness in the intended use group. A clearly defined TPP ensures that all parties involved, including process development and research, manufacturing, clinical teams, regulatory bodies, and senior management, have a clear understanding of the expected outcomes of the product development program [34].

Lipid Nanoparticles: Liposome nanoparticles (LNPs) have emerged as one of the most effective nano-delivery systems for the administration of chemotherapeutic drugs, antibiotics, and nucleic acid therapeutics. Pfizer/BioNTech and Moderna's LNP-based mRNA vaccines were remarkably successful during the COVID-19 pandemic, with global sales of $37 billion and $17.7 billion, respectively, in 2021. There is growing momentum in the development of different LNP-based RNA therapies, building on this success, given their potential applications in vaccines and treatments for genetic diseases and cancers [35].

Polymeric Nanoparticles: Because they may condense and form complexes with anionic RNA molecules, cationic polymers are the most commonly used polymers for RNA transport, which facilitates endosomal escape [36]. By altering their characteristics, such as charge density, molecular weight, architecture, and biodegradability, these

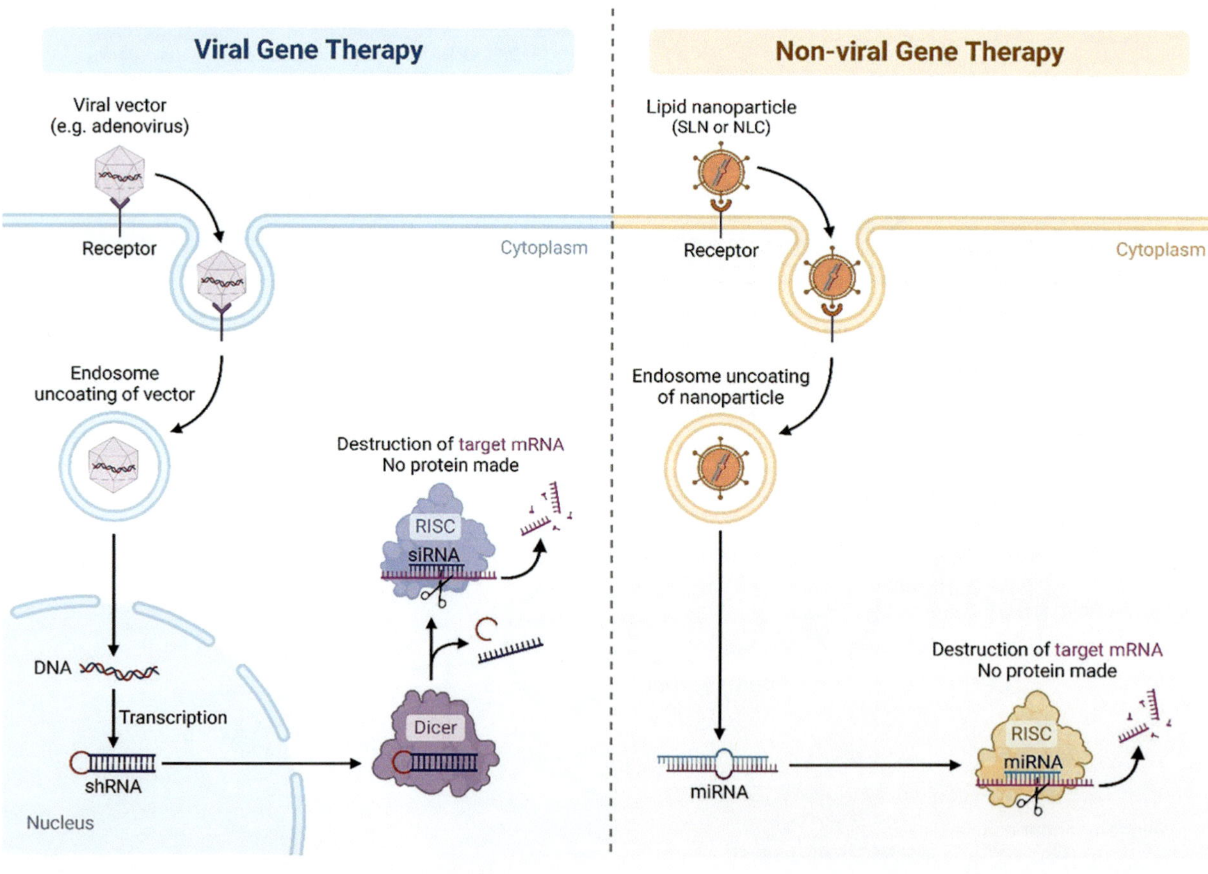

FIGURE 5.2 Viral and nonviral gene therapy [32,33].

cationic polymers can produce different degrees of cytotoxicity [37], in addition to the introduction of copolymers. Even though non-cationic polymers are more biocompatible, they usually need to covalently conjugate with RNA molecules or co-formulate with positively charged species in order to improve RNA loading. More accurate and regulated RNA distribution is made possible by stimuli-responsive polymers, which can modify their physicochemical characteristics in response to internal or external stimuli. There are several methods for producing RNA-loaded polymeric nanoparticles, including polymeric micelles, polymersomes, and solid polymeric nanoparticles. Numerous therapeutic uses, such as the therapy of cancer, heart disease, autoimmune disorders, and infectious diseases, have been studied for RNA-loaded polymeric nanoparticles [38–40].

Cationic Lipids: Numerous clinical disorders associated with abnormalities in gene expression may be prevented or cured via gene therapy. Chemical, biological, and mechanical DNA transfection techniques are the primary methods for delivering genes to target cells [41]. Most research has been done on viral vectors; however, their limitations and related negative effects—including a limited host range, a significant risk of immunogenicity and mutagenicity, a low manufacturing yield, and a limited gene-loading capability—sparked curiosity about creating nonviral vectors [42]. Cationic lipids show promise as nucleic acid delivery vehicles in this regard. Low immunogenicity, high loading capacity, a wide variety of host cells, affordability, and ease of production are some of the benefits they have over viral vectors. Application possibilities for cationic lipids in therapeutic delivery based on nucleic acids [43].

Gold Nanoparticles: There are several reasons why gold nanoparticles (AuNPs) have become the most widely used choice in numerous studies. Key factors include their chemical stability, resistance to enzymatic degradation, low cytotoxicity, and favorable physicochemical properties. AuNPs have been demonstrated to assist in mapping lymph nodes and assessing vascular blood flow [44]. They are also utilized as tools for cancer imaging and as core elements for specific mRNA, forming dual probes to detect melanoma, prostate, or carcinoma cells, as well as circulating breast cancer cells and brain tumors [45,46].

Exosomes: Different types of cells secrete extracellular vesicles, or exosomes. Typically having a diameter of 30–150 nm, these vesicles contain a wide variety of bioactive substances containing different signaling molecules, proteins, lipids, messenger RNA (mRNA), microRNA (miRNA), small interfering ribonucleic acid (siRNA), deoxyribonucleic acid (DNA), and others [47].

5.2 COLONIC DRUG DELIVERY SYSTEMS

The process of injecting drugs in a method that allows the formulation to pass through the upper gastrointestinal system without altering the drug's ability to dissolve and absorb in the colon is known as colon-targeted drug delivery, or CTDD. Some gastrointestinal conditions, including irritable bowel syndrome (IBS), ulcerative colitis (UC), and Crohn's disease (CD), require a medication with a local effect. These conditions are treated with metronidazole, prednisolone, sulfasalazine, hydrocortisone, and dexamethasone [48,49]. Because the medicine disintegrates and absorbs in the colon, colon-targeting drug delivery systems prevent drug loss and provide a higher concentration of the dosage to the targeted region, resulting in fewer systemic side effects. Drug absorption is improved by the colonic mucosa, particularly for peptides and proteins. Because the contents of the colon can be retained for up to five days, the colon is a perfect organ for absorbing medications [50]. The discovery of numerous pharmaceutical advantages and possibilities in recent years has led to a resurgence of colonic medication administration. Medications that are targeted to the colon can better cure local illnesses, make targets available for local therapy, minimize systemic medication exposure and the harm it causes, and perhaps make drugs more bioavailable. Colic medication distribution has typically concentrated on treating local conditions, including colorectal cancer and inflammatory bowel disease (IBD). Colonic medication delivery maximizes drug concentration at the intended location while reducing exposure across the body, which can enhance the treatment of local illnesses [51–59]. As the colonic and rectal environments are better characterized, novel regions, such as the lymphatic system, the gut's immunological system, and microbiota, are becoming recognized [60,61]. These new targets may make it easier to treat systemic diseases and disorders affecting distant organs in addition to colonic illnesses. Neurological conditions like Alzheimer's and Parkinson's illnesses, as well as traumatic spinal cord injuries, dyslipidemia, and urinary tract infections, have all been linked to intestinal microbiota modulation [62]. Additionally, targeting the mesenteric lymphatic system may help cure insulin resistance and improve brain accessibility. Probiotics, postbiotics, vaccines, oligonucleotides, and [63] biologics are just a few of the novel therapy modalities that have generated increased interest in colonic administration due to their development as possible therapeutic targets and advancements in pharmaceutical sciences [64].

Sulfasalazine, a prodrug initially used in 1941 to treat IBD, was the first instance of colonic drug administration [65]. For colonic administration, the prodrug approach is still used. For example, a recent study demonstrated that an azo prodrug of tofacitinib could successfully treat a mouse model of IBD. Similarly, by blocking systemic absorption, gut restriction of chemicals, especially peptides, may facilitate local colonic activity [66] Prodrug and gut-restricted treatment design, however, is drug-specific and may take a while to receive regulatory approval [67] Rather because a single formulation can aid in the colonic release of numerous different types of medications, this study will concentrate on formulation-based colonic drug delivery. Furthermore, the modified-release formulations of licensed medications can obtain market authorization more quickly, as they don't need to redo studies proving their inherent pharmacological properties, such as toxicity or pharmacology [68].

5.2.1 Drug Absorption Mechanism in Colon

Colon targeted drug delivery system (CTDDS) it is intended to deliver the medication at the appropriate and efficient concentration in the colon while preserving the small intestine's composition [69]. The majority of medications either follow paracellular or transcellular routes. The large intestine is not optimal for drug absorption through conventional formulations when these villi are absent, which is why drug administration methods that target the colon are required to increase drug absorption. The villi's presence improves drug absorption in the small intestine [70]. The colonic mucosa has a tighter epithelium than the small intestine, which contributes to its lower paracellular permeability and higher electrical resistance compared to the small intestine, even though drug absorption occurs primarily in the small intestine [71]. Because the colon has a shorter transit time, drugs can remain in the colon for longer because they can remain in contact with the colonic mucosa for longer. Although medications can theoretically be absorbed throughout the whole gastrointestinal tract, they are typically absorbed in the small intestine's duodenum and proximal jejunum [72]. The purpose of delayed-release dose forms is to provide a drug's effective release into the colon. This formulation, which targets the colon, demonstrates a "burst release," a sustained or protracted release, or a focused release. Single-unit CDDS and multi-particulate dosage form systems are the two primary types of colon-targeted medication formulations. The disadvantage of single-unit CTDDS deems the formulation unsatisfactory. Unintentional disintegration, which can result from improper manufacturing processes due to the unique GI physiology, is one of these drawbacks [73]. Other non-disease-dependent factors include age (pediatrics and geriatrics), ethnicity, genetics, obesity, pregnancy, and gastrointestinal tract (GIT) diseases, all of which affect the drug's bioavailability in the colon [74–76].

5.2.2 Colonic Transit Time

In a few instances, such as UC, the typical transitory time of the colon was impacted, with the transient period being 24 h as opposed to 52 h for a normal person. The colonic transit time was altered whether fasting or fed, and it was delayed during sleep. Therefore, a critical factor affecting the bioavailability of medications in the colon is the colonic transitory time [77].

5.2.3 Colonic Fluid Time

The colon absorbs about 90% of the water that enters it. Consequently, the intestinal volume decreases, which makes it harder for the medication to dissolve from the dose form. Undigested food, including proteins, carbohydrates, and lipids, enters the colon and serves as a substrate for microbial flora, or microbial enzymes, which in turn influence colon medication absorption and metabolism [78].

5.2.4 Colonic Luminal Viscosity

Compared to the upper gastrointestinal system, the colon's high water absorption capacity causes the high viscosity of colonic material in the lumen. An increase in lumen contents viscosity results in a decrease in medication absorption and dissolution. The viscosity modifies how well medicine molecules penetrate the site of absorption or the bacteria in the colon that cause sickness [79].

5.2.5 Advanced CDDs

A multi-stimulus method for oral colonic medication administration. The targeting performance of formulation techniques that depend on a single physiological stimulus (pH, transit duration, or microbial enzymes) for colonic drug delivery might vary significantly between and within patients [80]. Therefore, it has been demonstrated in recent years that combining drug release mechanisms based on many physiological stimuli provides a more reliable method [81]. Sequential or parallel trigger systems are two possible classifications for these combination or multifaceted systems, depending on their architecture. Multiple matrixes are divided into distinct layers from the basis of a sequential trigger system, in which the drug release mechanism develops layer by layer, beginning at the outermost layer and working its way toward the core, and in this approach, drug liberation is contingent on each layer being sequentially activated by its physiological signal. A parallel system, in juxtaposition, involves several drug release mechanisms operating concurrently. Independent trigger mechanisms are usually activated simultaneously in parallel systems by combining them into a single layer. This means that drug liberation should still happen via the other mechanism(s) even if one process malfunctions due to physiological variability (for instance, not being digested by intestinal flora or failing to reach a pH-mediated dissolving threshold). As such, parallel systems are the most dependable approach to developing novel colon-targeted medications. In terms of colonic drug delivery, the greatest translational success to date has been attained by multi-stimuli coatings for tablets, capsules, and pellets [82].

pH and microbiota-triggered combination system:

1. **Phloral**: The initial dual-triggered colonic drug delivery technology in order to achieve solid was the Phloral system. It is a coating method with a single layer composed of resistant starch (amylose and amylopectin) and a homogeneous mixture of Eudragit S because each trigger mechanism works independently, the two parts complement one another and can make up for one another's activity if one doesn't activate with this technique, Eudragit® S makes sure the tablet's integrity is

preserved throughout its passage through the small intestine and stomach. It also controls the swelling of the starch by acting as a structural agent. In contrast, resistant starch serves as a substrate for the colon's flora, offering a backup plan for initiating medication release in the event that Eudragit S's essential pH threshold is not reached. In fact, regardless of the patients' feeding status, the Phloral technology has demonstrated positive results in the treatment of IBD. Phloral has also been effectively used to treat infections caused by Clostridium difficile and obesity [83].

2. **Opticore**: Abbreviated as optimized colonic release, this combination method was created recently with the goal of delivering medications quickly to the ileocolonic area, which has higher fluid volumes than the mid-to distal colon the foundation of the system is made from with of two coating layers: the base layer consist of a buffering salt and an enteric polymer that is neutral called Eudragit® S. This layer is further encircled by an external Phloral layer because the base layer is alkaline, acidic medication formulation might need an extra HPMC layer to separate the acidic medication from the alkalinity of the basal layer by speeding up the Phloral coating's disintegration, the OPTICORETM system produces rapid ileocolonic medication release. Through holes in the Phloral coating, the formulation dissolves the enteric base layer and permits luminal fluid to enter the distal gastrointestinal system. This dissolution causes the inner surface of the remaining Phloral layer to increase in pH, ionic strength, and buffer capacity, which accelerates ionization and dissolves the Eudragit® S component of Phlora [84].
3. **Nasopulmonary Drug Delivery System:** The most convenient and favored method of medicine delivery is oral since it is simple to produce and administer. As a result, other medication delivery techniques emerged. Due to insufficient absorption in the GIT, alternative medication delivery technique has emerged. The nasal route was utilized in the ancient Indian Ayurvedic medical system to administer a variety of medications; this technique is called "Nasya." These days, oral and parenteral drug administration approaches are viewed as less efficient and less reliable than medication administered intranasally. Without a doubt, intranasal drug delivery has been widely used over a considerable amount of time for the reduction of symptoms, avoidance, or management of various nasal conditions that are topical [85]. To alleviate the issues associated with the oral and parenteral routes, alternative routes of drug delivery, such as transdermal, rectal, buccal, and nasal, are being researched. All of these pathways provide an alternate pathway for the systemic distribution of medicines, circumventing the hepatic first-pass metabolism. Currently, nasal administration devices are used to treat osteoporosis, migraine headache, nocturnal enuresis, smoking cessation, and vitamin B12 deficiency. Treatments for cancer, epilepsy, antiemetics, rheumatoid arthritis, and diabetes that require insulin are extra areas of therapy that are being [86].

5.2.6 Nasal Anatomy Physiology

The nasal channel that runs between the nasopharynx and nasal vestibule is around 12–14 cm deep. In this channel, the nasal cellular machinery is in intimate contact with the mucus that protects the mucosa from inspired air. Nestled within the nasal cavities are three unique functional zones: the vestibular, respiratory, and olfactory [87].

The volume and total surface area of the human nasal cavity are roughly 16–19 mL and 180 cm^2, respectively. It is divided by the septum into two nasal chambers. Some of the regions' descriptions are provided below:

1. **The Respiratory Region**: With the maximum level of vascularity, the respiratory area is the largest and primarily responsible for systemic medication absorption.
2. **The Vestibular Region**: It is in charge of filtering out airborne pollutants and is situated at the nasal passage's opening. In terms of medication absorption, it is regarded as the least significant of the three zones.
3. **The Olfactory Region**: Its surface area is roughly 10 cm^2, and it is essential to the movement of medications to the brain and CSF (cerebrospinal fluid). The area of the human nose consists of a thick lamina propria of connective tissue, which supports the olfactory epithelium. Bowen's bundle, axons, and blood are all found in the lamina propria. Vessels, despite the fact that the epithelium is composed of three separate cells, such as supporting cells, basal cells, neurons, and the olfactory receptor cells, are scattered throughout the supporting cells. Bipolar neurons that have a single dendrite that extends from the free apical surface to the cell body and terminates in a knob for smell with nonmotile cilia that stretch above the epithelium are known as olfactory receptor cells [88]. A layer of mucus covers the nasal passage's epithelium, trapping particles. The cilia that clean the nasal cavity of the mucus layer do it every ten to fifteen minutes. Adult mucosal secretions have a pH between 5.5 and 6.5. The nasal cavity contains a variety of enzymes, including Glutathione S-transferase, Carboxylesterase, and Cytochrome P-450 gap. [89,90].

4. **NASAL Drug Delivery: Formulation Development Research**: The majority of OTC (over the counter) nose medications are designed as solutions to address typical cold and allergic rhinitis symptoms because they facilitate improved dispersion across a greater surface area. A simple medicine solution is effective in this regard. Such formulations have a short nasal residence time (3–20 min) and exhibit significant inter-individual variability. Fast circulation peak levels can be achieved using this route. In experimental animal models, a large range of medications has been assessed for systemic bioavailability following transnasal injection. For improved residence and bioavailability, transnasal delivery of medications in various dose forms, including sprays, powders, and microspheres, has been explored. In order to control postoperative pain, nasal delivery is being studied; compared to intravenous administration, mucosal delivery requires a dosage of fentanyl that is only 1.1–1.5 times higher. When it comes to efficacy, nasal vaccine delivery is a highly desirable method of administration [91].

5.2.7 Noval Intranasal Drug Delivery to Target CNS

a. **Microemulsion**: The application of the microemulsion technology for intranasal administration shows promise. A microemulsion is a clear, stable, isotropic mixture of water, oil, and surfactant that is often mixed with a cosurfactant. Pharmaceutical scientists are currently interested in these systems due to their significant potential to function as drug delivery vehicles by integrating a variety of pharmacological compounds. Thermodynamic stability, ease of manufacturing and scaling up, spontaneous production, and improved drug absorption and solubility are their advantages. Setting up a requirement for a dose form that is acceptable to pharmaceuticals requires a thorough comprehension of the microemulsion structure, phase behavior, and contributing elements of its thermodynamic stability, variables impacting medication liberation from the formulation, optimum conditions, Microemulsion excipients: possible applications, and the microemulsion system's limitations [92].
b. **Nanoparticles**: polymer, lipid, or both together. Nanosystems utilized in the creation of a nano medication delivery method for central nervous system (CNS) therapy disorders include solid lipid, carbon nanotubes, nanofibers, nanorobots, polymeric nanoparticles, nanospheres, nanosuspension, nanoemulsions, nanogels, nanomicelles, and nanoliposomes, nanoparticles (SLN), lipid carriers with a nanostructure (NLC) and conjugates of lipid drugs (LDC). The precise method by which nanoparticles open barriers is unknown. However, the delivered nanoparticles pass beyond the blood-brain barrier (BBB) to enter the blood-brain barrier by a diversity of endocytotic processes. The albumin- or polymeric nanoparticle-based polymers. It is said that poly (butylcyanoacrylate) enters the brain by endocytosis mediated by its tiny size. When the medication is released into the brain milieu by these intact nanoparticles, it eventually undergoes biodegradation as a result of endocytotic absorption due of BBB's extremely small size [93].
c. **Microsphere**: One of the specialized techniques that is gaining popularity for creating nasal goods is microsphere technology, which offers longer contact with the nasal mucosa, which improves both bioavailability and absorption. When there is a microsphere, the dehydration of the nasal mucosa is caused by the moisture that the microspheres absorb. This led to reversible cell shrinkage, offering a momentary physical division of the tight connections between cells that raise the medication absorption. Nasal microsphere medication delivery systems are soluble in water yet absorb it into the matrix, causing enlargement of the spheres to create a gel. The ingredients utilized to create the microspheres are hyaluronic acid, albumin, dextran, and starch. Repeated administration of dextran microspheres with starch. By forming microspheres, the bioavailability of proteins and peptides has been enhanced in various animal species. A few low-molecular-weight medications have also been effectively administered in the form of microspheres. Depending on the type of bioadhesive used in the formulation, microspheres may remain in the nasal cavity for three to five hours. Since smaller particles are more likely to enter the lungs, the optimal microsphere particle size requirement for nasal distribution should be between 10 and 50 μm [94].
d. **Nasal In Situ Gel**: Drug delivery systems known as "in situ gel formulations" are those that are in solution form prior to being administered to the body, but afterward to create gel, they go through gelation. This might be accomplished by utilizing different polymers, including PVA (polyvinyl alcohol), poloxamers, carbopol, and chitosan [95,96].
e. **Ocular Drug Delivery:** The eye is a complicated organ in humans with unique anatomical and physiological features that make drug delivery to targeted ocular areas challenging. For many years, scientists have been interested in the effective topical administration of medications. Their difficult task is to extend the duration of drug presence in the eye and ensure proper ocular penetration. Various ocular barriers hamper effective

medication distribution, such as the barriers of the cornea, blood-corneal, and precorneal. Ocular medication delivery can be accomplished via topical, intravitreal, intraocular, juxta-scleral, subconjunctival, intracameral, and retrobulbar methods. More than 95% of available products are in liquid form [97]. However, other formulations may be semi-solid (such as ointments and gels), solid (such as powders, inserts, and lenses), or mixed (such as in situ gels, which have the capacity to encapsulate). Recently, there has been a growing interest in nanotechnology-based carriers for both lipophilic and hydrophilic drugs, as they improve drug stability, increase ocular permeability, and extend residence time [98], thereby enhancingbioavailability. Several important aspects of ocular medication administration are shown in Figure 5.3

Nanotechnology has given rise to a variety of novel nanocarriers for treating eye disorders by modifying and formulating existing drugs. It has prompted more commercial nano-based medications for ocular drug delivery to be developed. Nanostructured carriers have developed into minimally invasive drug delivery systems that maintain therapeutic drug concentrations in the eye for extended periods, reducing both the frequency of administration and potential side effects. To date, a range of nanocarriers, including polymeric nanoparticles, lipid nanoparticles, liposomes, niosomes, micelles, dendrimers, and nanofibers, have been introduced as innovative technologies to surpass ocular barriers and enhance the delivery of therapeutic drugs to target areas with improved ocular bioavailability [99]. The nanotechnology-based ocular delivery systems are shown in the in Figure 5.4.

Every year, biomaterials are inserted into millions of people globally. These materials, whether naturally sourced or man-made, provoke a response in the body that typically results in the formation of fibrous tissue around the material, which can decrease its effectiveness over time. In the field of eye care, devices known as glaucoma drainage devices (GDIs) are used to lower the pressure inside the eye, helping to halt the progression of glaucoma and preserve sight. However, even with recent advancements in reducing their size and altering their surface properties, the GDIs currently used in clinical practice often become enveloped in fibrous tissue, leading to a significant risk of surgical failure. Aditya et al. carried out a study where the performance of glaucoma drainage implants (GDIs) with either nanofiber coatings or smooth surfaces was assessed to determine how surface

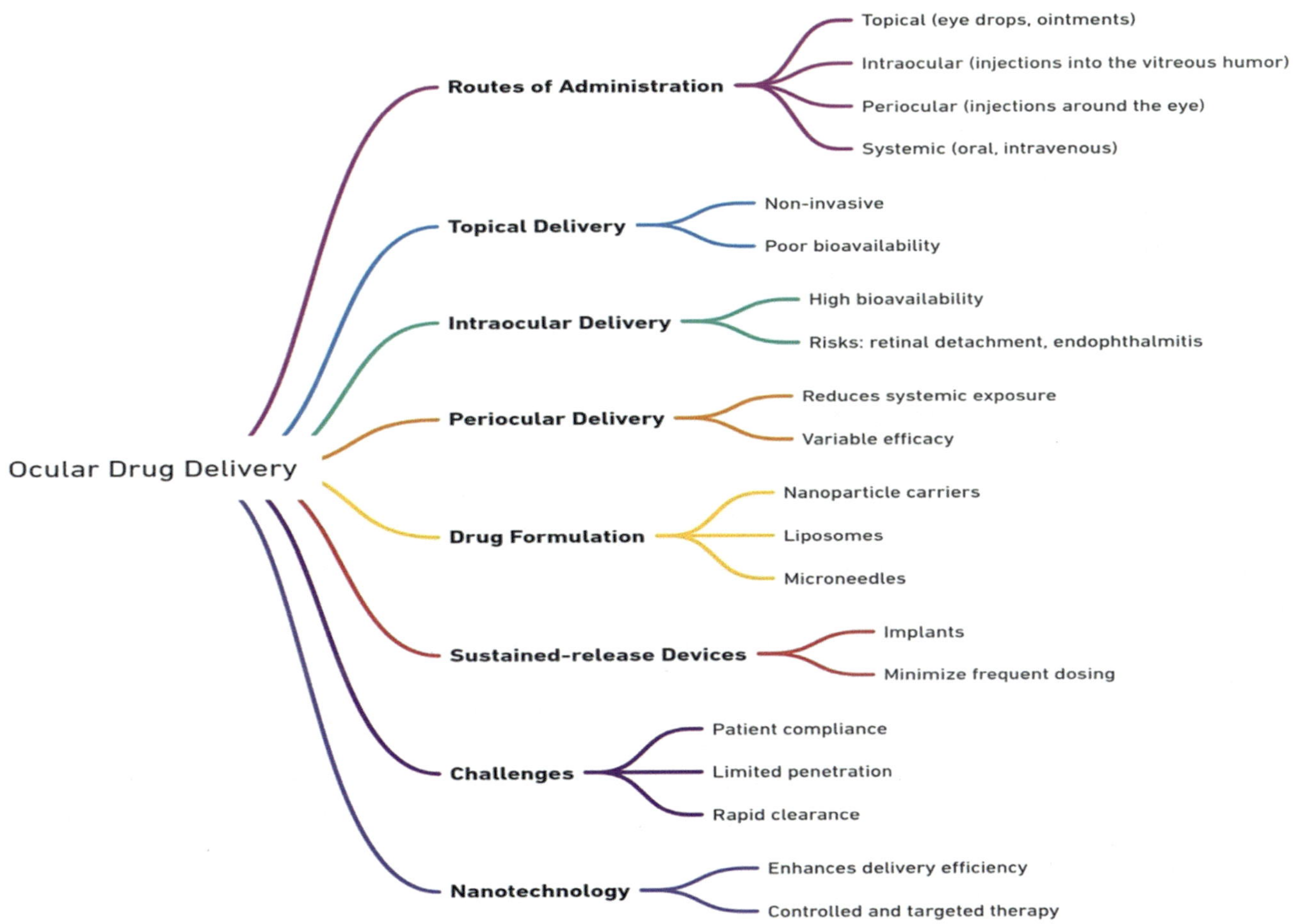

FIGURE 5.3 Key points of ocular drug delivery.

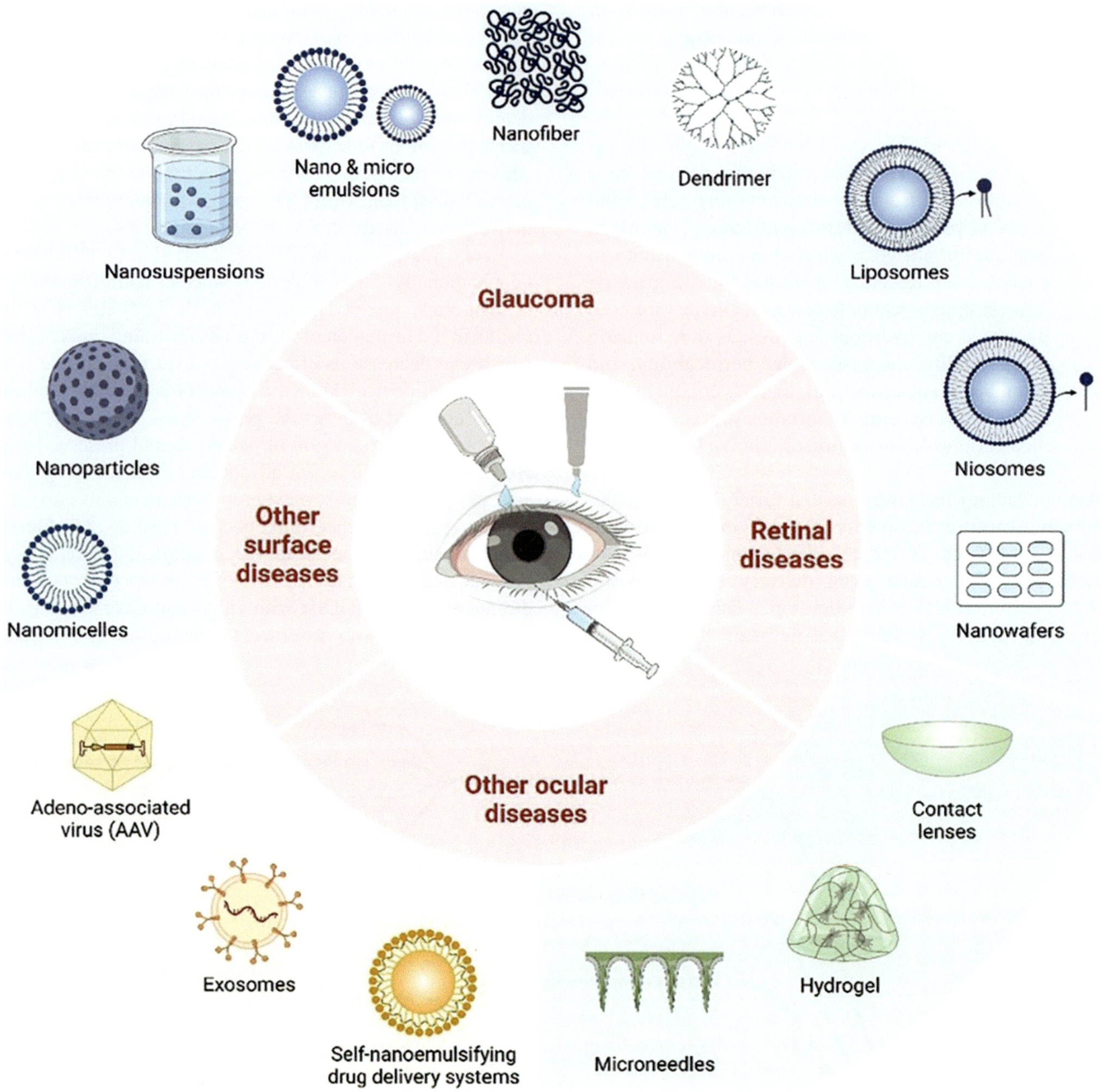

FIGURE 5.4 Ocular drug delivery systems utilizing nanotechnology.

topography influences implant functionality. In vitro tests demonstrated that nanofiber surfaces facilitated fibroblast integration and maintained them in a non-proliferative state, even when exposed to pro-fibrotic stimuli, unlike smooth surfaces [100]. In experiments with rabbits, the GDIs designed with a nanofiber structure were found to be compatible with biological tissues, averted abnormally low intraocular pressure (hypotony), and facilitated an outflow of aqueous humor on par with that of GDIs available on the market. Moreover, these nanofiber GDIs exhibited a substantial decrease in both the development of fibrous tissue around the implant and the activity of critical fibrotic indicators within adjacent tissues [101]. A comparison of conventional GDIs and nanofiber-based GDIs is shown in Figure 5.5.

1. **Brain Drug Delivery:**
   The BBB presents a major obstacle to the delivery of medications to the brain. Within a sound brain, the BBB serves as a diffusion barrier that protects, preventing the majority of chemicals from entering the brain from the blood, hence promoting appropriate brain function, permitting only tiny molecules to pass through. However, in some medical conditions—like multiple sclerosis, diabetes, stroke, seizures, Parkinson's disease, and Alzheimer's disease—the BBB can become compromised [102]. To create novel medications to treat brain disorders, whether from biologics or tiny molecules that are soluble in water (such as

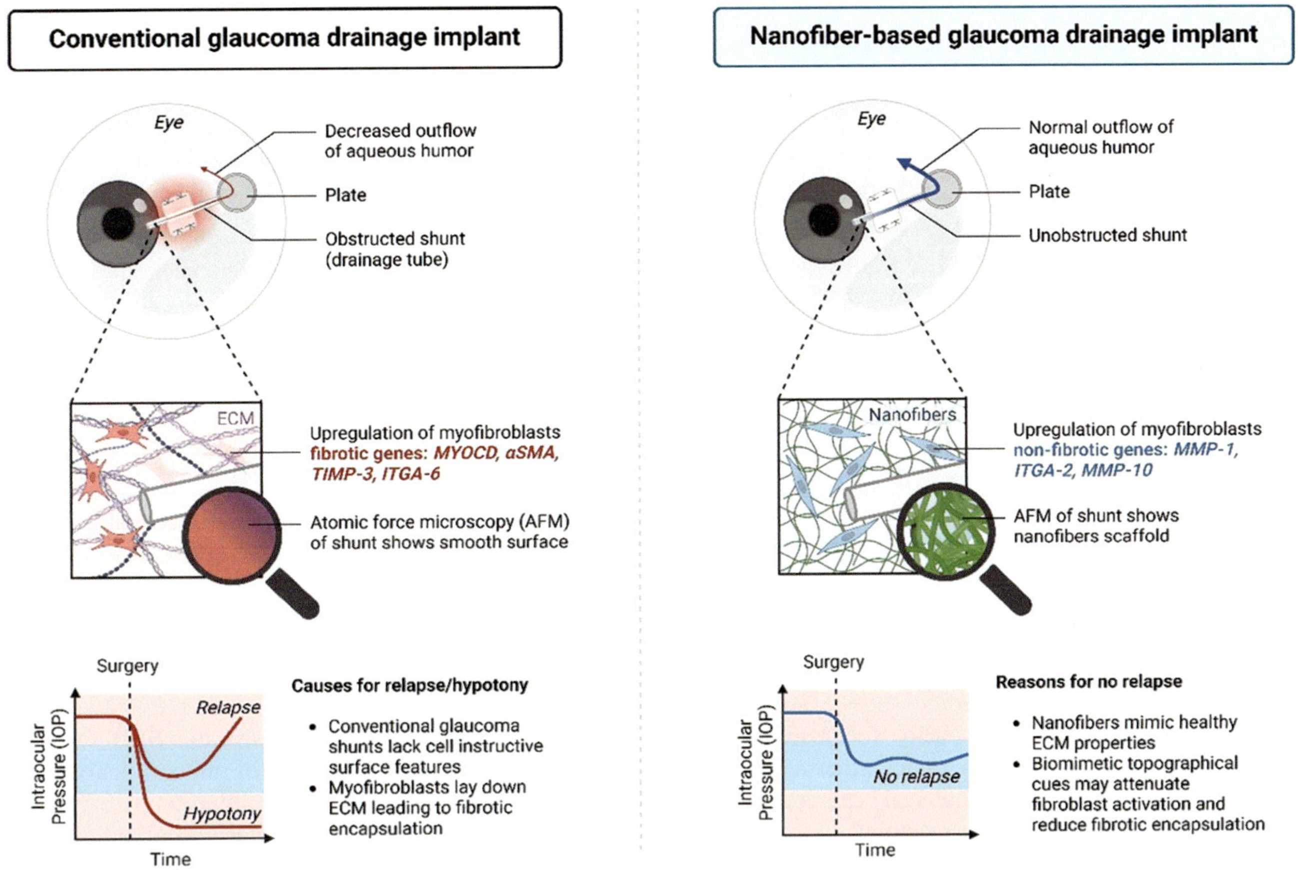

FIGURE 5.5 A comparison of conventional glaucoma drainage implants and nanofiber-based glaucoma drainage implants.

medications containing nucleic acids or recombinant proteins), a variety over the past 40 years, a variety of brain medication delivery systems have been developed. These technological advancements can be roughly classified as:

a. Brain-invasive medicine delivery: IBDS (brain invasive drug delivery system) involves bypassing the BBB through methods such as injecting drugs directly into the fluid around the brain (CSF) via administering medication intrathecal, transnasally, or directly injecting medicines into the brain via the use of intracranial implants or convection-enhanced diffusion (CED) [103].
b. BB interferes with the administration of brain drugs: When it comes to the administration of brain drugs, disruption of the BBB occurs when there is damage to the brain capillary endothelial cells' tight interconnections. This can be accomplished by either injecting harmful substances intra-arterially or by sonicating the brain after administering microbubbles intravenously [104].
c. Transvascular medication delivery to the brain: Small molecule transporters (CMTs) or biologics (RMTs) are examples of carrier-mediated transporters. This approach also encompasses the development of CNS-targeted drugs. Alongside the passive diffusion lipid-soluble small molecules across the BBB, active efflux transporter- such as P-glycoprotein (P-gp).

Presently, the BBB a significant difficulty to delivering drugs into the CNS microenvironment. Tight junctions (TJs) and specialized cells in the barrier provide high electrical resistance, which prevents therapeutic chemicals from passing paracellularly [105] because they can readily pass through the BBB by passive diffusion. Tiny molecules with high lipophilicity are ideal for delivering medications to the brain. However, high lipophilicity can be a double-edged sword, resulting in issues such as limited solubility, poor absorption, and quick drug degradation [106]. Small lipophilic drugs under 500 kDa often show limited therapeutic effects because they typically target a single molecular pathway. In contrast, many macromolecular drugs and numerous signaling pathways, including therapeutic genes, have limited efficacy due to their high molecular weight and poor lipophilicity [107]. Consequently, the ability of materials with a mean size of 0.1–100 nm to pass through the BBB has attracted a lot of attention. This interest stems from their uniqueness; their

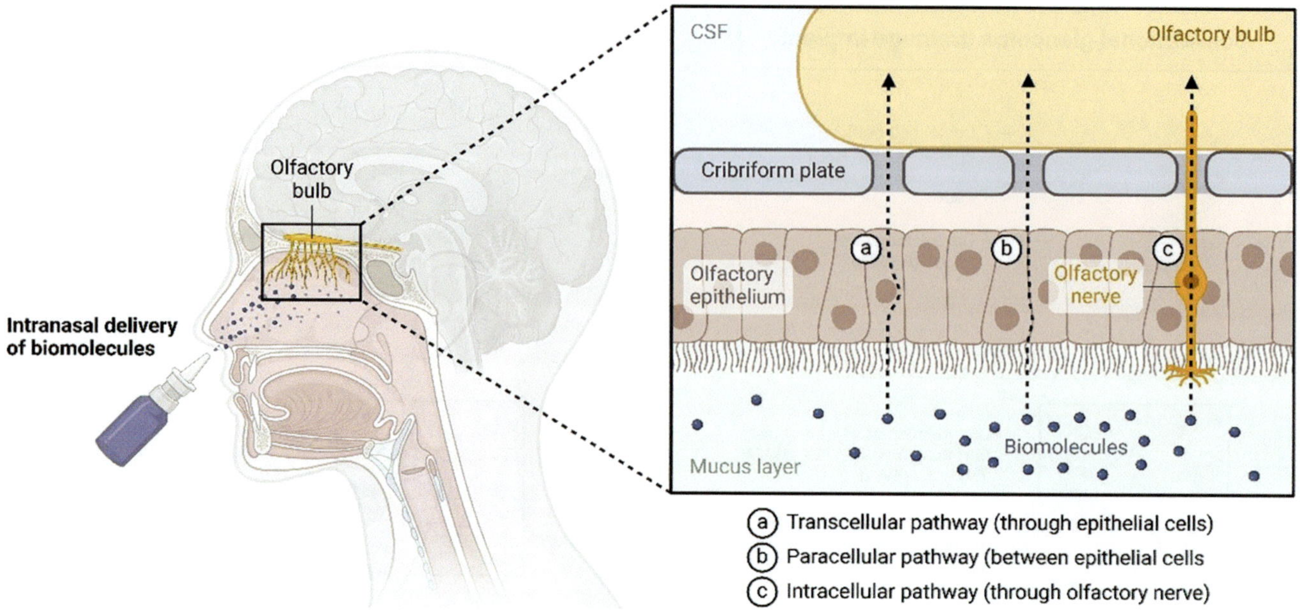

FIGURE 5.6 Drug delivery to the brain.

small size, huge specific surface area, and strong adsorption capacity for macromolecules are among their optical, thermal, magnetic, and physicochemical characteristics, and high chemical reactivity. Leveraging these advantageous properties, nanomedicines are anticipated to make the passage of drugs easier, transcending the BBB, thereby enhancing their bioavailability [108]. To date, a variety of nanomedicines have been developed using nanomaterials, carbon-based, polymeric, inorganic, and organic materials, together with metal nanostructures, red blood cell membranes, liposomes, and extracellular vesicles. Delivering drugs to the brain is crucial for effectively preventing and treating brain-related diseases. However, traditional administration routes like oral and intravenous methods face challenges due to the BBB and significant metabolism on the first pass [109]. As a result, the nasal-brain drug delivery route is gaining interest as a viable alternative for effective brain delivery, as shown in Figure 5.6.

Recent research has investigated the application of poly(lactide-co-glycolic) acid increasing therapeutic agents-encapsulated nanoparticles to treat Alzheimer's illness [110] and brain cancer [111]. In vitro studies demonstrated that these nanoparticles of polymers improved drug delivery to the brain, leading to less plaque, inflammation, and oxidative stress buildup when delivering curcumin for Alzheimer's treatment. Additionally, they facilitated the effective internalization of doxorubicin into human glioma cells, which causes cancer cells to be cytotoxic. Dendrimers have been utilized in strategies based on nanotechnology to bypass the BBB for treating circulatory arrest, stroke, brain tumors, neurodegenerative illnesses, and neuroinflammation. Researchers synthesized ethylene glycol polymer and glioma-homing peptide-dendrimers of conjugated polyamidoamines specifically for glioblastomas, with in vivo findings showing enhanced tumor penetration [112].

5.3 VIRUS INFECTION AND SUITABLE DELIVERY SYSTEM

Viruses are obligate intracellular pathogens characterized by a basic structure made up of proteins that encase their nucleic acid genome, which may or may not be surrounded by a lipid envelope. Although rapid nucleic acid amplification tests for laboratory diagnosis are becoming more accessible, along with effective antiviral treatments and safe vaccines, managing most viral infections still relies heavily on established surveillance and infection control practices. Additionally, many viruses can be easily eliminated with standard disinfectants [113–115].

5.3.1 Viral Gene Delivery System

Many viruses possess effective mechanisms for transferring genetic material into target cells, leading current gene technologies to focus on viral vectors that offer increased transduction efficiency and enhanced gene expression. The type of virus being used determines the best design for these vectors. To ensure the safe use of in vivo gene therapy, certain challenges must first be addressed. Only the appropriate cell types should express therapeutic factors, and healthy cells shouldn't experience any negative side effects [116].

5.3.1.1 Adenoviral Systems

First identified in 1953 from human adenoid tissue cultures, adenoviruses are frequently employed as gene carriers. Among the adenoviruses, types Ad2 and Ad5 from group C are the most extensively studied. The virus's capsid determines its tropism, with the highly expressed Coxsackievirus B-adenovirus receptor being bound by groups A and C-F, enabling high infectivity across various tissues. On the other hand, group B attaches itself to the complement-regulatory protein CD46. The nucleus of the infected cell is included in the virions that are produced by adenoviruses during replication. Replication is possible with these vectors, and purification usually entails quicker and easier procedures [117]. Known to be linear dsDNA viruses, adenoviruses are non-integrating, somewhat large (~26–40 kb) particles that are icosahedral and non-enveloped, with a diameter of roughly 950 Å and a molecular weight of roughly 150 MDa (not including elongated fiber proteins). Although they can swiftly infect cells, they do not reproduce there. Because they don't have any lipids or membranes, adenoviral particles are stable in solvents like ethanol or ether. The development of adenoviruses as one of the earliest vector systems for gene expression has resulted in notable advancements [118]. Because of their well-established molecular biology, their potential in order to transfer DNA into various cell types, and their ability to transfer foreign DNA chunks up to 36 kb, adenoviruses are crucial for gene transfer. Within the cell nucleus, a viral genome is found in a compressed form, and the proteins V, VII, and X are associated with it. The initiating primer for viral DNA replication is a terminal protein (TP), It is joined to the adenoviral DNA's 5′ end via a covalent bond. One late transcriptional unit makes up the adenoviral genome. Five early transcription units (E1A, E1B, E2, E3, and E4] and four intermediate units (IVa2, IX, VAI, and VAII) are included. The first viral transcriptional unit to be produced upon cell entry, E1A, pushes the host cell into the S phase, activates other early transcription units, and creates the ideal conditions for viral replication. By blocking p53-dependent apoptosis, together with p53, Bak, and Bax proteins, E1B proteins enable the infected cell to survive. DNA polymerase and single-stranded DNA-binding proteins, which are essential for viral genome replication, are encoded by the E2A unit. The E3 proteins aid the virus in avoiding the host's immunological response, which guarantees the survival of the infected cell. On the other hand, the E4 unit encodes proteins that influence transformation and cell cycle regulation [119]. A symmetrical icosahedral capsid enclosing the seven proteins connects non-covalently to form the nucleus (II, III, IIIa, IV, VI, VIII, and IX). The nucleoprotein nucleus is encapsulated by these adenoviral structural proteins, which also stabilize the DNA. Seven peptides make up the capsid: a penton base (III), Trimeric hexons that bind to three minor capsid proteins (VI, VIII, and IX) for stability (II), penton-bound proteins that bind to the hexon base (III), and a receptor-binding fiber (IV). A globular knob at the C-terminus, a rod-like shaft, and a tail at the N-terminus make up the fiber's three domains. Through receptor-mediated endocytosis, adenoviruses infiltrate their target cells. Effective transduction is made possible by adenoviruses existing as episomes within the infected cell rather than integrating their DNA with the host's genome [10]. Yet this also restricts the amount of time transgenes may be expressed, which causes protein expression to stop abruptly [120]. One promising strategy for successful in vivo gene therapy is the use of adenoviruses. Hepatocytes have demonstrated notable transduction, for example, following systemic vector delivery. These vectors have several significant disadvantages, even though they are effective in delivering genes to various organs. High dosages are required to produce lethal effects because certain target cells have low amounts of the proper adenoviral receptors. Additionally, the virus's non-specific tropism may cause non-target cells to be transduced. The ability of adenovirus vectors to elicit potent immunological and inflammatory reactions at high dosages is a major concern [121].

5.3.1.2 Retroviral Systems

Members of the Retroviridae family, Retroviruses are circularly enclosed, diploid, single-stranded RNA viruses with genomic sizes between 7 and 11 kb and dimensions between 80 and 120 nm. There are two subfamilies within the Retroviridae family: Orthoretrovirinae and Spumaretrovirinae. Despite their potential to cause diseases including cancer, leukemia, and AIDS, retroviruses have greatly improved treatment thanks to their employment as vectors for gene therapy. In the early 1980s, it was discovered that gene delivery vectors based on retroviruses could carry foreign genetic material. Retroviruses incorporate themselves into the host genome to produce viral proteins (gag, pol, and env) that are utilized for gene delivery. As the most widely utilized retrovirus, the Moloney murine leukemia virus can transfer exogenous genetic material up to approximately 9 kb in size. Retroviral vectors can transfer DNA up to 8 kb in size. Enzymes such as reverse transcriptase and integrase, which are encoded by pol, are found within a retroviral virion. This virion contains round or triangular gag-encoded capsid proteins covered with gag-encoded nucleocapsid proteins. One of the distinctive characteristics of retrovirus genomes is that they are genuinely diploid and generated by the cellular transcriptional machinery. Furthermore, for replication, these viruses need cellular tRNA. The structure of their genomes distinguishes simple and sophisticated retroviruses. Complex retroviruses contain extra genes, whereas simple retroviruses only encode the genes for protease, pol (polymerase), and gag (group-specific antigen). Since they were the first altered viruses to be utilized for gene delivery, retroviruses have been used in most clinical gene therapy studies. Based on the well-characterized and researched Moloney murine leukemia virus (MMLV), the majority of retroviral vectors are utilized in therapeutic settings. Its genome encodes gag, pol, and env, three polyproteins necessary for replication and packaging [121]. By interacting with cellular receptors and viral proteins embedded in the membrane, the lipid envelope of retroviruses facilitates their entry into host cells. Retroviruses can attach themselves to receptor-positive target cells by combining chimeric envelope proteins with viral particles. Some macrophages and T cells have a cell-surface protein called CD4, which is bound by the surface protein SUCell-surface G-proteins that function as chemokine receptors. These cell-surface G-proteins are known as chemokine receptors (CKR). Further binding to CKR is facilitated by conformational changes in the SU protein domain brought on by this interaction. When the viral and cellular membranes unite, the viral nucleus begins to internalize. The glycoprotein in the retroviral envelope determines the virus's tropism (Env). The initial phases of infection are characterized by the fusion of the viral and cellular membranes following Env binding to the cellular receptor. Targeting retroviruses using Env cell-binding ligands or single-stranded antibodies has been unsuccessful because these large ligands can prevent Env from attaching to the virion. After the Env protein becomes particle-coated, the viral nucleus is discharged into the cytoplasm, where it produces double-stranded DNA through reverse transcription. In their own replication cycle, retroviruses transform single-stranded RNA into double-stranded DNA inside the infected cell [122]. Receptors on the cell surface interact with the envelope protein of a retrovirus to infect target cells. The virus transforms its single-stranded RNA into double-stranded DNA once it gets inside. Following its transportation, this DNA merges with the host cell's genome in the nucleus. For many retroviruses, the delivery of their DNA into the nucleus depends on the nuclear envelope and the mitotic spindle breaking down. Integrating reverse-transcribed genomes into the host cell's chromosomes is a crucial feature of retroviruses. Retroviral vectors accomplish transfection by incorporating transgenes into the DNA of the target cell, as opposed to adenoviral vectors. Nevertheless, transgenic expression frequently declines in a matter of days or weeks. Retroviral vectors may also be ineffective in a variety of in vivo settings, in part because the human complement system swiftly deactivates them. Although this silencing process is not well understood, it is believed to be related to promoter or DNA methylation, or the interaction of chromosome condensation with splice regions [123].

5.4 CONCLUSION

In summary, the advancement of drug delivery systems marks a crucial development in medicine, facilitating more precise and effective therapeutic strategies tailored to individual patients. The innovative methods explored, including protein/peptide drug delivery systems with nanoparticles and advanced nucleic acid therapies, demonstrate significant enhancements in bioavailability and therapeutic effectiveness. Additionally, the examination of route-specific delivery systems underscores the importance of overcoming unique physiological barriers, which in turn improves treatment outcomes for a variety of medical conditions. As we confront new challenges, such as viral infections, the demand for innovative and effective delivery solutions has become increasingly urgent. This compilation not only highlights current advancements but also envisions a future filled with potential, where ongoing innovations in drug delivery can greatly enhance patient care and therapeutic success. The insights derived from these discussions highlight the crucial role of drug delivery systems in transforming modern medicine, ultimately leading to enhanced health outcomes for patients worldwide.

REFERENCES

1. Jeong, Woo Yeup et al. "Recent advances in transdermal drug delivery systems: a review." *Biomaterials Research* 25(1) (2021) 24.
2. Yeo, Yoon, ed. *Nanoparticulate Drug Delivery Systems: Strategies, Technologies, and Applications*. Wiley, 2013.
3. Agrahari, Vibhuti, Vivek Agrahari, and Ashim K. Mitra. "Nanocarrier fabrication and macromolecule drug delivery: challenges and opportunities." *Therapeutic Delivery* 7(4) (2016) 257–278.
4. Limenh, Liknaw Workie. "Advances in the transdermal delivery of antiretroviral drugs." *SAGE Open Medicine* 12 (2024). doi:10.1177/20503121231223600.
5. Zhang, Wentao, et al. "Transdermal gene delivery." *Journal of Controlled Release* 371 (2024) 516–529.
6. Pradhan, Madhulika, and Krishna Yadav, eds. *3D Printing and Microfluidics in Dermatology: Innovations in Drug Delivery*. CRC Press, Boca Raton, FL (2024).
7. Gharat, Sankalp Ashok, Munira Mumtaz Momin, and Tabassum Khan. "Pharmacokinetic, pharmacodynamic, preclinical and clinical models for evaluation of nanoparticles." In: Sankalp Ashok Gharat, Munira Mumtaz Momin, and Tabassum Khan (Eds.), *Pharmacokinetics and Pharmacodynamics of Novel Drug Delivery Systems: From Basic Concepts to Applications: A Machine-Generated Literature Overview*. Springer Nature (2024), pp. 81–178.
8. Georgeous, Joel et al. "Review of gold nanoparticles: synthesis, properties, shapes, cellular uptake, targeting, release mechanisms and applications in drug delivery and therapy." *Pharmaceutics* 16(10) (2024) 1332.
9. Lam, Brandon et al. "In situ vaccination via tissue-targeted cDC1 expansion enhances the immunogenicity of chemoradiation and immunotherapy." *Journal of Clinical Investigation* 134(1) (2024) e171621.
10. Ahanger, Ishfaq A. et al. "Comprehensive perspective towards the management of proteinopathies by elucidating protein misfolding and aggregation." *CNS & Neurological Disorders-Drug Targets* 23(2) (2024) 153–180.
11. Zhang, Zhiyuan et al. "Alginate-based electrospun nanofibers and the enabled drug controlled release profiles: a review." *Biomolecules* 14(7) (2024) 789.
12. Deshmukh, Rohitas et al. "Recent review on biological barriers and host–material interfaces in precision drug delivery: advancement in biomaterial engineering for better treatment therapies." *Pharmaceutics* 16(8) (2024) 1076.
13. Musa, Dickson Achimugu, et al. "Promising RNA-based therapies for viral infections, genetic disorders and cancer." *Academia Molecular Biology and Genomics* 1(1) (2024).
14. Liu, Beilei et al. "Exploring treatment options in cancer: tumor treatment strategies." *Signal Transduction and Targeted Therapy* 9(1) (2024) 175.
15. Shi, Yesi et al. "Chemically modified platforms for better RNA therapeutics." *Chemical Reviews* 124(3) (2024) 929–1033.
16. Elumalai, Karthikeyan, Sivaneswari Srinivasan, and Anandakumar Shanmugam. "Review of the efficacy of

nanoparticle-based drug delivery systems for cancer treatment." *Biomedical Technology* 5 (2024) 109–122.
17. Judijanto, Loso et al. "Application of integrated logistics networks in improving the efficiency of distribution and delivery of goods in Indonesia a literature review." *Sciences du Nord Economics and Business* 1(1) (2024) 01–10.
18. Zhou, Xun et al. "Adeno-associated virus engineering and load strategy for tropism modification, immune evasion, and enhanced transgene expression." *International Journal of Nanomedicine* 19 (2024) 7691–7708.
19. Zhou, Xuefei et al. "RNA-based modulation of macrophage-mediated efferocytosis potentiates antitumor immunity in colorectal cancer." *Journal of Controlled Release* 366 (2024) 128–141.
20. Pardi, Norbert, and Florian Krammer. "mRNA vaccines for infectious diseases—advances, challenges, and opportunities." *Nature Reviews Drug Discovery* 23 (2024) 1–24.
21. Li, Jiaxuan et al. "Advancing mRNA therapeutics: the role and future of nanoparticle delivery systems." *Molecular Pharmaceutics* 21(8) (2024) 3743–3763.
22. Taghdiri, Maryam, and Claudio Mussolino. "Viral and non-viral systems to deliver gene therapeutics to clinical targets." *International Journal of Molecular Sciences* 25(13) (2024) 7333.
23. Verma, Abhishek, and Ankit Awasthi. "Developing non-viral or viral vectors for efficient and targeted delivery of genetic material, such as DNA or RNA, for gene therapy applications." *Pharmaspire* 15 (2024) 243–256.
24. Wang, Jiang-Hui et al. "Adeno-associated virus as a delivery vector for gene therapy of human diseases." *Signal Transduction and Targeted Therapy* 9(1) (2024) 78.
25. Lu, Bowen et al. "The next-generation DNA vaccine platforms and delivery systems: advances, challenges, and prospects." *Frontiers in immunology* 15 (2024) 1332939.
26. Guo, Mengke et al. "Review on nucleic acid nanodelivery systems for inflammatory diseases." *Journal of Drug Delivery Science and Technology* 91 (2024) 105987.
27. Wang, Tianjiao et al. "Nucleic acid drug and delivery techniques for disease therapy: present situation and future prospect." *Interdisciplinary Medicine* 2(1) (2024) e20230041.
28. Ambros, Victor. "MicroRNA-mediated gene regulation and the resilience of multicellular animals." *Postępy Biochemii* 70(1) (2024) 62–70.
29. Silvestri, Alessandro, Chandni Bansal, and Ignacio Rubio-Somoza. "After silencing suppression: miRNA targets strike back." *Trends in Plant Science* (2024).
30. Li, Quan et al. "The biosynthesis process of small RNA and its pivotal roles in plant development." *International Journal of Molecular Sciences* 25(14) (2024) 7680.
31. Kim, Jeonghwan, et al. "Strategies for non-viral vectors targeting organs beyond the liver." *Nature Nanotechnology* 19(4) (2024) 428–447.
32. S. Ahmed et al., "Vaccine Development against COVID-19: Study from Pre-Clinical Phases to Clinical Trials and Global Use." *Vaccines* 9(8) (2021). doi: 10.3390/VACCINES9080836.
33. Anderson, Allan, Chad McAllister, and Ernie Harris. *Product Development and Management Body of Knowledge: A Guidebook for Product Innovation Training and Certification*. Wiley (2024).
34. Croitoru, George-Alexandru, et al. "Nanostructured drug delivery systems in immunotherapy: an updated overview of nanotechnology-based therapeutic innovations." *Applied Sciences* 14(19) (2024) 8948.
35. Pawar, Smita et al. "Unlocking the potential of nanocarrier-mediated mRNA delivery across diverse biomedical frontiers: a comprehensive review." *International Journal of Biological Macromolecules* 234 (2024) 131139.
36. Mehta, Mohit J. et al. "Recent progress in the endosomal escape mechanism and chemical structures of polycations for nucleic acid delivery." *Macromolecular Bioscience* 24(4) (2024) 2300366.
37. Porello, Ilaria et al. "Advancing nucleic acid delivery through cationic polymer design: non-cationic building blocks from the toolbox." *Polymer Chemistry* 15 (2024) 2800–2826.
38. Beach, Maximilian A. et al. "Polymeric nanoparticles for drug delivery." *Chemical Reviews* 124(9) (2024) 5505–5616.
39. Pourmadadi, Mehrab et al. "Polymeric nanoparticles as delivery vehicles for targeted delivery of chemotherapy drug fludarabine to treat hematological cancers." *Inorganic Chemistry Communications* 167 (2024) 112819.
40. Haidar, Laura Libnan, Marcela Bilek, and Behnam Akhavan. "Surface bio-engineered polymeric nanoparticles." *Small* 20(21) (2024) 2310876.
41. Berezianko, Ivan Alexandrovich, and Sergei Vladimirovich Kostjuk. "Ionic liquids in cationic polymerization: a review." *Journal of Molecular Liquids* 397 (2024) 124037.
42. Xiong, Jiaofeng et al. "Low-hysteresis and high-toughness hydrogels regulated by porous cationic polymers: the effect of counteranions." *Angewandte Chemie International Edition* 63(1) (2024) e202316375.
43. Ghobadlou, Mohammad Zarei, Naser Asadzadeh, and Majid Ahmadlouydarab. "Impact of cationic surfactant-coated hydrophilic nanoparticles on polymeric solution performance in chemical enhanced oil recovery (CEOR) from a two-dimensional porous medium." *Chemical Engineering Research and Design* 204 (2024) 427–437.
44. Ghobashy, Mohamed Mohamady, et al. "Gold nanoparticles in microelectronics advancements and biomedical applications." *Materials Science and Engineering* B 301 (2024) 117191.
45. Karnwal, Arun, et al. "Gold nanoparticles in nanobiotechnology: from synthesis to biosensing applications." *ACS Omega* 9(28) (2024) 29966–29982.
46. Hang, Yingjie, Anyang Wang, and Nianqiang Wu. "Plasmonic silver and gold nanoparticles: shape-and structure-modulated plasmonic functionality for point-of-caring sensing, bio-imaging, and medical therapy." *Chemical Society Reviews* 53 (2024) 2932–2971.
47. Zeng Wei, Zhiyong Wen, Hui Chen, and Yaping Duan. Exosomes as carriers for drug delivery in cancer therapy. *Pharmaceutical Research* 40 (2023) 873–887. doi:10.1007/s11095-022-03224-y.
48. Jha, Daya Kishore et al. "Antibiotics for inflammatory bowel disease: current status." *Indian Journal of Gastroenterology* 43(1) (2024) 145–159.
49. Friedman, Sonia. General principles of medical therapy of ulcerative colitis." In: Gary Richard Lichtenstein (Ed.), *Ulcerative Colitis: The Complete Guide to Medical Management*. CRC Press, 2024, pp. 121–134.
50. Singh, Manvi et al. "Introduction to colonic disorders." In: Harish Dureja, Raimar Loebenberg, Sachin Kumar Singh, Muralikrishnan Dhanasekaran, and Kamal Dua (Eds.), *Advanced Drug Delivery Systems for Colonic Disorders*. Academic Press, 2024, pp. 1–22.
51. Arif Muhammed, Rozhan et al. "A review on development of colon targeted drug delivery system." *International Journal of Applied Pharmaceutics* 16(2) (2024) 12–27.
52. Manna, Sreejan et al. "Recent updates on guar gum derivatives in colon-specific drug delivery." *Carbohydrate Polymers* 334 (2024) 122009.

53. Corrie, Leander, et al. "Polysaccharide-fecal microbiota-based colon-targeted self-nanoemulsifying drug delivery system of curcumin for treating polycystic ovarian syndrome." *Naunyn-Schmiedeberg's Archives of Pharmacology* (2024) 1–23.
54. Behera, Anindita. "Colon-responsive oral drug delivery for combating colonic disorders." *Advanced Drug Delivery Systems for Colonic Disorders*. Academic Press, 2024. 153–179.
55. A. K. Philip and B. Philip, "Colon Targeted Drug Delivery Systems: A Review on Primary and Novel Approaches." *Oman Med J* 25(2) (2010) 79. doi: 10.5001/OMJ.2010.24.
56. Kumar, Ankaj, et al. "Nanomedicine for colon-targeted drug delivery: strategies focusing on inflammatory bowel disease and colon cancer." *Nanomedicine* (2024) 1–22.
57. Tarannum, H. et al. "Colon targeted nano drug delivery systems." *Nanoparticle Toxicity and Compatibility* 161 (2024) 122–152.
58. Moghimipour, Eskandar, and Somayeh Handali. "Functionalized liposomes as a potential drug delivery system for colon cancer treatment: a systematic review." *International Journal of Biological Macromolecules* (2024) 132023.
59. Agrawal, Nimesh et al. "Colon-specific and targeted drug delivery system (CDDS) A Review." *Scholar Academic Journal of Pharmacy* 6 (2024) 208–214.
60. Ding, Guoao, et al. "Gut microbiota regulates gut homeostasis, mucosal immunity, and influences immune-related diseases." *Molecular and Cellular Biochemistry* (2024) 1–13.
61. Shakhpazyan, Nikolay K. et al. "Exploring the role of the gut microbiota in modulating colorectal cancer immunity." *Cells* 13(17) (2024) 1437.
62. Long, Jing et al. "Intratumoral microbiota in colorectal cancer: focus on specific distribution and potential mechanisms." *Cell Communication and Signaling* 22(1) (2024) 455.
63. Y. B. Miao et al., "Cracking the intestinal lymphatic system window utilizing oral delivery vehicles for precise therapy." *J Nanobiotechnology* 21(1) (2023) 263. doi: 10.1186/S12951-023-01991-3.
64. S. Kumar, N. K. Garg, A. Jain, P. Pandey, A. Khopade, and K. K. Sawant, "Recent progress in macromolecules: From current therapeutic strategies to theranostic applications." *J Drug Deliv Sci Technol* 96 (2024) 105664, doi: 10.1016/J.JDDST.2024.105664.
65. R. Sagar Kumar, T. S. Kumar, and R. Siva Kumar, "Targeting the Colon: Innovative Drug Delivery Systems for Local and Systemic Therapy." *International Journal of Scientific Research and Technology* 2 (2025). doi: 10.5281/ZENODO.17328167.
66. Jagtap, Ruchiket Kiran, Manisha Savle, and Gajanan Sanap. "A review on colon targeted drug delivery system." *International Journal of Pharmaceutical Sciences* 2 (2024) 142–154.
67. McCoubrey, Laura Elizabeth et al. "The colon targeting efficacies of mesalazine medications and their impacts on the gut microbiome." *Journal of Controlled Release* 369 (2024) 630–641.
68. S. Amidon, J. E. Brown, and V. S. Dave, "Colon-Targeted Oral Drug Delivery Systems: Design Trends and Approaches." *AAPS PharmSciTech* 16(4) (2015) 731. doi: 10.1208/S12249-015-0350-9.
69. Steigert, Sebastian et al. "Characterization of luminal contents from the fasted human proximal colon." *European Journal of Pharmaceutical Sciences* 200 (2024) 106821.
70. Z. Wang et al., "Impact of Food Physical Properties on Oral Drug Absorption: A Comprehensive Review." *Drug Des Devel Ther* 19 (2025) 267. doi: 10.2147/DDDT.S497515.
71. G. Lemmens, A. Van Camp, S. Kourula, T. Vanuytsel, and P. Augustijns, "Drug Disposition in the Lower Gastrointestinal Tract: Targeting and Monitoring." *Pharmaceutics* 13(2) (2021) 161. doi: 10.3390/PHARMACEUTICS13020161.
72. Sardou, Hossein Shahdadi et al. "Colon-targeted delivery systems of budesonide as second-line therapy in inflammatory bowel disease." *Journal of Drug Delivery Science and Technology* (2024) 105472.
73. Chen, Wen et al. "Drug delivery systems for colorectal cancer chemotherapy." *Chinese Chemical Letters* 35(2) (2024) 109159.
74. Widiani Batubara, Risa, Fadlina Chany Saputri, and Raditya Iswandana. "Current update of clinical therapeutic strategies for colon-targeted delivery systems." *Pharmaceutical Sciences and Research* 11(1) (2024) 1.
75. Wang, Nuoya et al. "Current advances of nanomaterial-based oral drug delivery for colorectal cancer treatment." *Nanomaterials* 14(7) (2024) 557.
76. E. Fernandez, R. Perez, A. Hernandez, P. Tejada, M. Arteta, and J. T. Ramos, "Factors and mechanisms for pharmacokinetic differences between pediatrie population and adults." *Pharmaceutics* 3(1) (2011) 53–72. doi: 10.3390/PHARMACEUTICS3010053.
77. Lupianez-Merly, Camille et al. "Additional criteria on scintigraphic testing for diagnosis of rapid colonic transit in patients with chronic diarrhea." *Neurogastroenterology & Motility* (2024) e14917.
78. Abdi, Fatma et al. "pH-dependent pressure-sensitive colonic capsules for the delivery of aqueous bacterial suspensions." *Journal of Controlled Release* 365 (2024) 688–702.
79. M. E. V. Johansson, J. M. Holmén Larsson, and G. C. Hansson, "The two mucus layers of colon are organized by the MUC2 mucin, whereas the outer layer is a legislator of host–microbial interactions." *Proc Natl Acad Sci U S A* 108(1) (2010) 4659. doi: 10.1073/PNAS.1006451107.
80. Liu, Ye et al. "Advancements in hydrogel-based drug delivery systems for the treatment of inflammatory bowel disease: a review." *Biomaterials Science* 12 (2024) 837–862.
81. Shi, Yulong et al. "Tailored drug delivery platforms: stimulus-responsive core–shell structured nanocarriers." *Advanced Healthcare Materials* 13(1) (2024) 2301726.
82. Waqar, Muhammad Ahsan et al. "Advanced polymers and recent advancements on gastroretentive drug delivery system; a comprehensive review." *Journal of Drug Targeting* (2024) 1–17.
83. Labelle, Marc-André et al. "Carboxymethyl starch films as enteric coatings: processing and mechanistic insights." *Journal of Pharmaceutical Sciences* 113(3) (2024) 725–734.
84. N. R. Anderson, N. G. Minutolo, S. Gill, and M. Klichinsky, "Macrophage-Based Approaches for Cancer Immunotherapy." *Cancer Res* 81(5) (2021) 1201–1208. doi: 10.1158/0008-5472.CAN-20-2990.
85. Dhaker, Gayatri et al. "Review on Nasya Karma and its pharmacodynamics: a conceptual study." *World Journal of Pharmaceutical Research* 13 (2024) 1126–1135.
86. Katare, Pratik et al. "Nasal drug delivery system and devices: an overview on health effects." *ACS Chemical Health & Safety* 31(2) (2024) 127–143.

87. Tariyal, Kanchan. "Review on nasal drug delivery system and its application." *Journal for Research in Applied Sciences and Biotechnology* 3(1) (2024) 281–285.
88. Jadhav, Avinash et al. "A novel approach for nasal drug delivery system." *Asian Journal of Pharmaceutical Research and Development* 12(2) (2024) 96–106.
89. Subbaiah, Murugaiah AM, Jarkko Rautio, and Nicholas A. Meanwell. "Prodrugs as empowering tools in drug discovery and development: recent strategic applications of drug delivery solutions to mitigate challenges associated with lead compounds and drug candidates." *Chemical Society Reviews* 53 (2024) 2099–2210.
90. Chen, Yaoxing et al. "Intranasal drug delivery: the interaction between nanoparticles and the nose-to-brain pathway." *Advanced Drug Delivery Reviews* 207 (2024) 115196.
91. Bava, Bharat, Kruti Sharma, and Vikas Yadav. "Intranasal drug delivery system: a review." *Research Journal of Science and Technology* 16(1) (2024) 51–58.
92. Huang, Qianqian et al. "Nanotechnology for enhanced nose-to-brain drug delivery in treating neurological diseases." *Journal of Controlled Release* 366 (2024) 519–534.
93. Khope, Rashmi G. et al. "Intranasal drug delivery: a non-invasive method for improving neurotherapeutic delivery." *Asian Journal of Pharmaceutical Research* 14(1) (2024) 53–61.
94. M. Butola and N. Nainwal, "Non-Invasive Techniques of Nose to Brain Delivery Using Nanoparticulate Carriers: Hopes and Hurdles." *AAPS PharmSciTech* 25(8) (2024). doi: 10.1208/S12249-024-02946-Z.
95. Xu, Dong et al. "Advances and future perspectives of intranasal drug delivery: a scientometric review." *Journal of Controlled Release* 367 (2024) 366–384.
96. A. Pires, A. Fortuna, G. Alves, and A. Falcão, "Intranasal drug delivery: how, why and what for?." *J Pharm Pharm Sci* 12(3) (2009) 288–311. doi: 10.18433/J3NC79.
97. Bisen, Amol C. et al. "A review on polymers in ocular drug delivery systems." *MedComm–Biomaterials and Applications* 3(2) (2024) e77.
98. Assiri, Ahmad A. et al. "Block copolymer micelles as ocular drug delivery systems." *Drug Discovery Today* 29 (2024) 104098.
99. Giri, Bhupendra Raj et al. "Advancements in ocular therapy: a review of emerging drug delivery approaches and pharmaceutical technologies." *Pharmaceutics* 16(10) (2024) 1325.
100. Mandal, Suraj, Prabhakar Vishvakarma, and Km Bhumika. "Developments in emerging topical drug delivery systems for ocular disorders." *Current Drug Research Reviews Formerly: Current Drug Abuse Reviews* 16(3) (2024) 251–267.
101. Weber, Constance et al. "Ocular drug delivery systems: glaucoma patient perceptions from a German university hospital eye clinic." *Graefe's Archive for Clinical and Experimental Ophthalmology* 262(2) (2024) 545–556.
102. Taha, Esraa et al. "Versatile nanoparticulate systems as a prosperous platform for targeted nose–brain drug delivery." *Molecular Pharmaceutics* 21(3) (2024) 999–1014.
103. T. G. Schreiner, M. Menéndez-González, O. D. Schreiner, and R. C. Ciobanu, "Intrathecal Therapies for Neurodegenerative Diseases: A Review of Current Approaches and the Urgent Need for Advanced Delivery Systems. *Biomedicines* 13(9) (2025) 2167. doi: 10.3390/BIOMEDICINES13092167.
104. R. Daneman and A. Prat, "The Blood–Brain Barrier." *Cold Spring Harb Perspect Biol* 7(1) (2015) 020412. doi: 10.1101/CSHPERSPECT.A020412.
105. Ruan, Shuyao et al. "Microneedle-mediated nose-to-brain drug delivery for improved Alzheimer's disease treatment." *Journal of Controlled Release* 366 (2024) 712–731.
106. Okmen Altas, Burcu et al. "Tadpole-like anisotropic polymer/lipid Janus nanoparticles for nose-to-brain drug delivery: importance of geometry, elasticity on mucus-penetration ability." *Molecular Pharmaceutics* 21(2) (2024) 633–650.
107. Gao, Jingjing et al. "Precision drug delivery to the central nervous system using engineered nanoparticles." *Nature Reviews Materials* 9(8) (2024) 567–588.
108. Zheng, Xiaoxiao et al. "Current non-invasive strategies for brain drug delivery: overcoming blood–brain barrier transport." *Molecular Biology Reports* 51(1) (2024) 25.
109. Pinkiewicz, Miłosz et al. "Breaking barriers in neuro-oncology: a scoping literature review on invasive and non-invasive techniques for blood–brain barrier disruption." *Cancers* 16(1) (2024) 236.
110. Tian, Mingzhen, Zhichao Ma, and Guang-Zhong Yang. "Micro/nanosystems for controllable drug delivery to the brain." *The Innovation* 5(1) (2024) 100548.
111. Tashima, Toshihiko, and Nicolas Tournier. "Non-invasive device-mediated drug delivery to the brain across the blood–brain barrier." *Pharmaceutics* 16(3) (2024) 361.
112. Wilson, Matthew G. et al. "Remotely controlled drug release in deep brain regions of non-human primates." *Journal of Controlled Release* 369 (2024) 775–785.
113. Ibañez-Escribano, Alexandra et al. "Microbial matryoshka: addressing the relationship between pathogenic flagellated protozoans and their RNA viral endosymbionts (family totiviridae)." *Veterinary Sciences* 11(7) (2024) 321.
114. Wongchitrat, Prapimpun, Theerawut Chanmee, and Piyarat Govitrapong. "Molecular mechanisms associated with neurodegeneration of neurotropic viral infection." *Molecular Neurobiology* 61(5) (2024) 2881–2903.
115. Li, Hongyi R. et al. "Engineering viral vectors for acoustically targeted gene delivery." *Nature Communications* 15(1) (2024) 4924.
116. Taghdiri, Maryam, and Claudio Mussolino. "Viral and non-viral systems to deliver gene therapeutics to clinical targets." *International Journal of Molecular Sciences* 25(13) (2024) 7333.
117. Scarsella, Luca et al. "Advances of recombinant adenoviral vectors in preclinical and clinical applications." *Viruses* 16(3) (2024) 377.
118. Lu, Shao-Chia et al. "FastAd: a versatile toolkit for rapid generation of single adenoviruses or diverse adenoviral vector libraries." *Molecular Therapy Methods & Clinical Development* 32 (2024) 101356.
119. Fischer, Julian et al. "Expanding the scope of adenoviral vectors by utilizing novel tools for recombination and vector rescue." *Viruses* 16(5) (2024) 658.
120. Lee, Hyeon Dong et al. "Comparative biodistribution study of baculoviral and adenoviral vector vaccines against SARS-CoV-2." *Journal of Microbiology and Biotechnology* 34(1) (2024) 185.
121. Obeagu, Emmanuel Ikechukwu, Chukwudi Nnamdi Anyanwu, Obeagu, and Godwin Uche. "Challenges and considerations in managing blood transfusion for individuals with HIV." *Elite Journal of HIV* 2(2) (2024) 1–7.
122. J. Mayer et al., "2024 taxonomy update for the family Retroviridae." *Arch Virol* 170(8) (2025) 164. doi: 10.1007/S00705-025-06353-Y.
123. Rojekar, Satish et al. "Current status of mannose receptor-targeted drug delivery for improved anti-HIV therapy." *Journal of Controlled Release* 372 (2024) 494–521.

6 Role of Polymers Used in Nanoparticles for Controlled Drug Delivery

Sumeyra Sena Dag[1], *Cem Varan*[2], *and Erem Bilensoy*[1]

[1]Department of Pharmaceutical Technology, Faculty of Pharmacy, Hacettepe University, Sıhhiye Ankara, Turkey

[2]Department of Nanotechnology and Nanomedicine, Graduate School of Science and Engineering Beytepe, Hacettepe University, Ankara, Turkey

6.1 INTRODUCTION

Polymers have become indispensable raw materials for many industries nowadays. Compared to other raw materials, the prominent properties of polymeric raw materials, such as versatility, lightness, durability, low cost, and recyclability, are included in all areas of daily life, especially textiles, construction, transportation, electronics, and packaging. The use of polymers in the biomedical field is also quite widespread. Nowadays, polymers are widely used in the preparation of prostheses, implants, stents, and wound dressings, as well as tissue engineering scaffolds, biosensors for diagnosis, and controlled drug delivery systems, which is also the focus of this episode. In this section, information about the classification, properties, use in drug delivery systems, superiority, and future perspective of polymers is compiled.

6.2 POLYMERS

Polymers are made up of smaller units called monomers, just as living organisms are composed of cells. In other words, each unit present in a polymer chain is considered a monomer. Monomers can form polymer chains by bonding to each other in various configurations and numbers through chemical bonds (Figure 6.1). The process by which the functional groups in the structure of monomers interact with one another to form macromolecules is called polymerization (Odian, 2004).

Molecules formed by the polymerization of a small number of monomers (such as 2–10) are called oligomers, while polymerizations of many monomers (over 10) form polymers, which are macromolecular structures with high molecular weight (usually over 700 Da) (Callister, 2020).

Polymers can be of natural or synthetic origin, and have a wide range of applications in various industrial areas such as food, cosmetics, textiles, aerospace, automotive, and the pharmaceutical industry. These versatile uses of polymers are due to their chemical, physical, and mechanical properties, making them indispensable in modern industry (Saldívar-Guerra, 2013).

FIGURE 6.1 Schematic representation of monomer and polymer chains.

6.2.1 Classification of Polymers

There are different approaches to the classification of polymers based on the source or the bonding geometries of the monomers, and the structures of the polymer chains.

6.2.1.1 According to the Structure of Polymer Chains

Polymers are classified into four groups based on their chain structure: linear polymers, branched polymers, crosslinked polymers, and network polymers (Figure 6.2). This structural diversity is critical in the physicochemical properties of the polymer.

a. **Linear Polymer:** They are chain structures in which monomers are linked together consecutively in a long and straight manner. They are long, flat, and flexible. They often exhibit crystalline and strong structures, such as high-density polyethylene (HDPE).
b. **Branched Polymer:** They are polymeric structures to which short side chains are attached in a linear main chain. This branching pattern can reduce the packaging efficiency of the polymer and impact its density and melting point. This

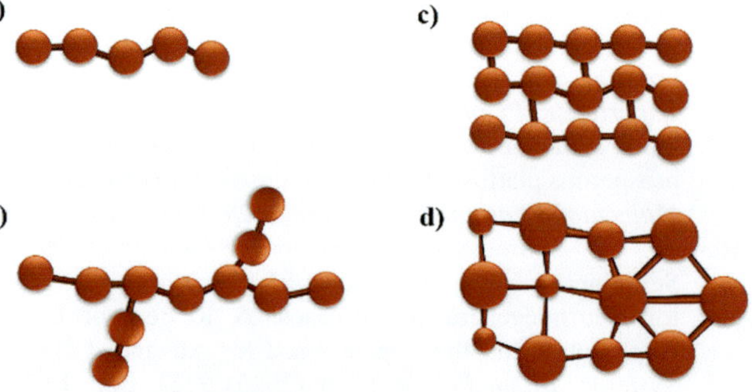

FIGURE 6.2 Schematic representation of (a) linear, (b) branched, (c) crosslinked, and (d) network polymers.

DOI: 10.1201/9781003508601-6

structure is seen in examples such as low-density polyethylene (LDPE).

c. **Crosslinked Polymer**: They are structures in which linear polymer chains are linked together by covalent bonds. The crosslinking process is a chemical reaction and irreversible. Cross-links provide flexibility to the polymer and determine many properties of the formulation, especially the mechanical properties. It also provides the polymer with greater durability and resistance to heat. Rubber is an example of crosslinked polymers.

d. **Network Polymer**: They are three-dimensional polymer chains formed by increasing cross-link density. These structures often exhibit high rigidity and heat resistance. Epoxy and phenolic resins are examples of network polymers. In addition, hydrogels have the ability to swell, thanks to the cross-links in their structure.

This structural diversity enables polymers to be tailored to various applications, offering a wide range of uses in the biomedical industry and technology.

6.2.1.2 According to the Arrangement of Monomers

Polymers are classified into two main groups, homopolymers and copolymers, according to their monomer content (Figure 6.3).

a. **Homopolymer:** The polymer chain is made up of a single type of monomer.

b. **Copolymer:** The polymer chain is made up of two or more multiple-type monomers. They have a more complex molecular structure, which allows them to acquire unique properties. There are different subclasses according to the position and structure of the monomer in the chain.

Random Copolymer: Monomers along the chain are randomly arranged.

Block Copolymer: Monomer is arranged in combined blocks.

Alternating Copolymer: Monomers are arranged sequentially in an order.

Gradient Copolymer: Monomers are arranged to form a gradient.

Grafted Copolymer: The chain consisting of a single type of monomer is grafted with other types of monomers.

Periodic Copolymer: Monomers are arranged in a particular order of repetition.

Aperiodic Copolymer: Although they are not in a simple repetitive pattern as in alternative and/or block copolymers, they are polymers that are not random and have a well-defined pattern within the structure.

6.2.1.3 According to their Thermal Behavior

Polymers are divided into three groups as thermoplastic, thermoset, and elastomer according to their thermodynamic behavior (Figure 6.4).

a. **Thermoplastic Polymers:** They are polymers that can be softened and physically change shape by heating to the glassy transition temperature (Tg). When the effect of temperature and pressure is eliminated, they can regain their former physical properties. They are suitable for use in many industrial applications thanks to their easy formability. Polystyrene (PS), polyethylene (PE), polycarbonate (PC), polypropylene (PP), polymethylmethacrylate (PMMA), polylactic acid (PLA), polytetrafluoroethylene (PTFE), and polycaprolactone (PCL) are among these kinds of

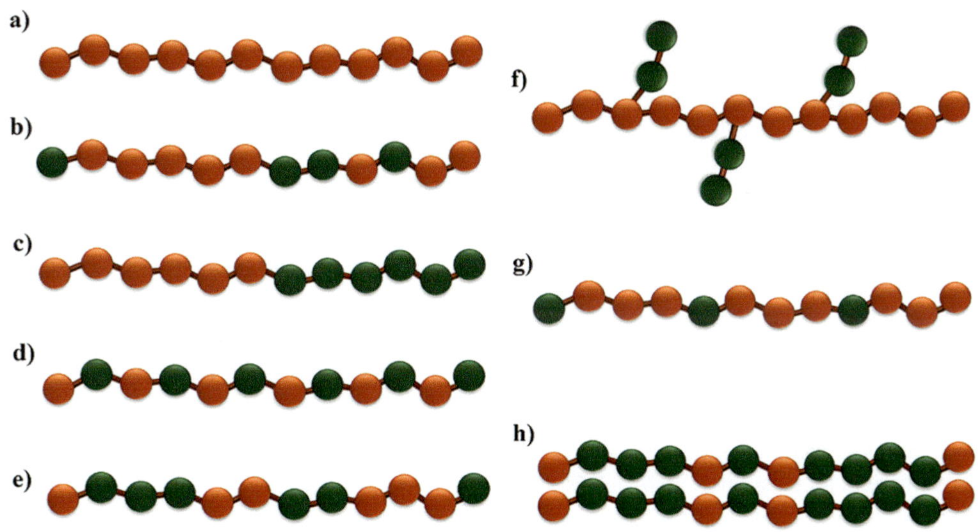

FIGURE 6.3 Schematic representation of (a) homopolymer, (b) random copolymer, (c) block copolymer, (d) alternating copolymer, (e) gradient copolymer, (f) grafted copolymer, (g) periodic copolymer, and (h) aperiodic copolymer.

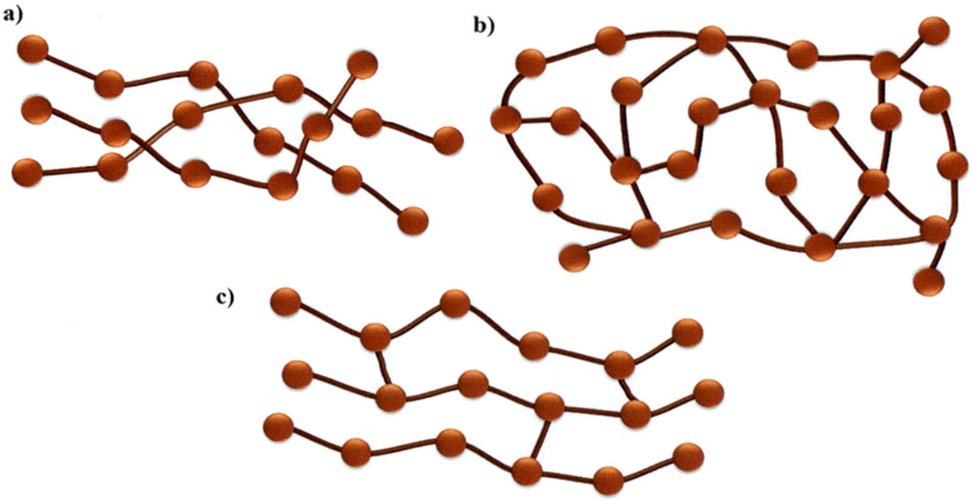

FIGURE 6.4 Schematic representation of (a) thermoplastic, (b) thermoset, and (c) elastomer polymer.

polymers. Thermoplastic polymers do not contain cross-links and thus have noticeably shorter processing times (Shenoy, 1996; MacDermott, 1997). They can be modified according to their intended use with chemical modifications. Biocompatible and inert thermoplastic polymers are used in the production of various medical devices, including microfluidic chips, pacemakers, artificial hearts, and hemodialysis membranes in the fields of medicine and pharmacy (Biron, 2013).

b. **Thermoset Polymers:** They are polymers that do not change shape when temperature and pressure are applied and harden by forming permanent covalent bonds through chemical crosslinking reactions. Unlike thermoplastic polymers, they cannot return to their former state when the application of temperature and pressure is eliminated. This is because the chain structures of thermoset polymers contain cross-links. The mobility is also restricted by the bonds established between the leading chains of the thermoset polymer. Thermoset polymers decompose at elevated temperatures. Polyurethane, polyesters, epoxy, and silicone are examples of thermoset polymers (Biron, 2013).
c. **Elastomers:** They are polymers that can change shape when pressure is applied and regain their former shape when pressure is eliminated, thanks to their high flexibility and elasticity properties (**Yoda, 1998**). They retain their elastic properties at low temperatures (from −40°C to 100°C). They are widely used in many industries thanks to their reversible deformation properties. Rubber is a typical example of an elastomer; also, polyisoprene and polybutadiene can be given as examples (Ducrot, 2014).

6.2.1.4 According to Monomer Source

Polymers are divided into two groups: natural sources and chemically synthesized.

a. **Natural Polymers:** They are polymers obtained by extraction from various natural sources such as plants, animals, and microorganisms. Their most important advantage is that they are biocompatible and biodegradable. In addition, the fact that renewable resources can be used in production and are low toxicity is another advantage. In this way, degradation and sustainability are ensured without causing environmental toxicity. However, the content of the extract obtained from natural sources varies, causingstandardization to become difficult and differences between series. In addition, these polymers are difficult to process, production costs are high, and their resistance to mechanical and environmental conditions is low. Natural polymers also have an immunogenicity risk. Nucleotides (especially DNA and RNA), peptides, proteins (keratin, collagen, albumin, and gelatin), polysaccharides (such as cellulose, cyclodextrins, chitosan, alginate, hyaluronic acid, carrageenan), and natural rubbers can be given as examples of natural polymers (Satchanska, 2024).
b. **Synthetic Polymers:** They are petroleum-derived polymers produced by chemical synthesis. Their most significant advantage is that it can be modified easily for the purpose to be used. For this reason, their industrial use is quite wide. Polyethylene (PE), polypropylene (PP), polylactic acid (PLA), polyglycolic acid (PGA), polyvinyl chloride (PVC), polytetrafluoroethylene (PTFE), and polystyrene are commonly used synthetic polymers. It is particularly important that synthetic polymers, used in

systems applied to patients as drug delivery systems or medical devices, are biocompatible, biodegradable, non-toxic, non-immunogenic, and approved by authorities such as the FDA. In this context, PLA, PGA, Polylactic-Co-Glycolic Acid (PLGA), PCL, and Poly-L-lactide (PLLA) are synthetic polymers widely used in pharmaceutical technology and the biomedical industry, as their medical applications have been approved by the FDA (FDA, 2024).

6.3 POLYMERIC NANOPARTICLES

Polymeric nanoparticles are colloidal systems with sizes between 1 and 1,000 nm and containing active pharmaceutical ingredients (API) in adsorbed, encapsulated, or dissolved form. Polymeric nanoparticles can be prepared using natural or synthetic polymers. It is important that the polymer be biocompatible, biodegradable, and show low toxicity to prepare a polymeric nanoparticulate system suitable for use in treatment.

The physicochemical properties of polymeric nanoparticles, such as particle size, surface charge, and surface modifications, are especially important in terms of the behavior of the nanosystem and its effect on the body. Particle size and surface charge directly affect many features, such as the passage of nanoparticles through biological membranes, biodistribution, and interaction with the mononuclear phagocytic system (MPS). Modifications on particle size and surface properties may allow the nanoparticle to passively target tissue or organ and remain in the systemic circulation for longer periods without being recognized by MPS. Polymeric nanoparticles can encapsulate the API and ensure that they reach the targeted area stably. Thus, it can be assured that the API is protected from enzymatic degradation and its stability is increased. In addition, drug release from polymeric nanoparticles can be modified to ensure that the API is released in the targeted area for the desired period and amount (Elumalai, 2024).

The surface of polymeric nanoparticles can be decorated with ligands, proteins, and antibodies. This decoration process allows nanoparticles to reach target organs, tissues, cells, and even organelles through active targeting. Thus, by preventing the passage of APIs to areas other than the target, side effects can be minimized, and a more effective drug treatment can be provided. Thanks to all these approaches, cost-effective production and an optimum treatment regimen at the most effective dose can be ensured.

The surfaces of the nanoparticles can also be decorated so as not to be recognized by MPS. For example, a nanoparticle whose surface is decorated like a brush with polyethylene glycol (PEG) chains cannot be recognized and opsonized by MPS because it cannot be labeled by plasma proteins. This process, in which the surfaces of nanoparticles are coated with PEG chains, is called PEGylation and is often used to obtain "stealth" drug delivery systems that cannot be recognized by MPS. Since PEG is a biocompatible, non-toxic, and FDA-approved polymer, it is the most preferred polymer in the preparation of these systems (Abu Lila, 2018). For example, Genexol-PM®, a nanoparticle formulation loaded with paclitaxel, is an approved drug prepared from a PEG-based block copolymer. However, polymers such as poloxamer-type block copolymers, polyvinylpyrrolidone (PVP), polyglycerols, and polyacrylamides can also be used (Hadjesfandiari, 2018).

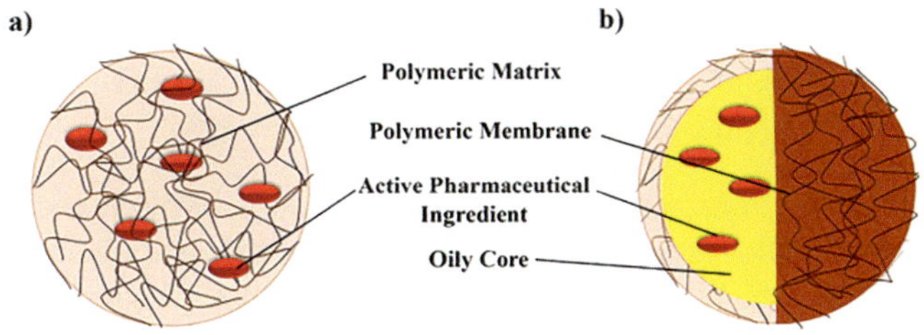

FIGURE 6.5 Schematic representation of polymeric (a) nanosphere and (b) nanocapsule.

6.3.1 Classification of Polymeric Nanoparticles

Polymeric nanoparticles are divided into two groups, nanospheres and nanocapsules (Figure 6.5), depending on their structural properties and interactions with API.

6.3.1.1 Nanosphere

Nanospheres are matrix-type systems in which API is homogeneously distributed in the polymeric matrix or is adsorbed or conjugated to the surface. Their particle size is smaller when compared to that of a nanocapsule. With their porous structure, they release API faster than nanocapsules. The polymer concentration affects and allows modification of the release rate. Drug release from nanospheres may be by different mechanisms. If the API is adsorbed or conjugated to the surface, it is released by the desorption of the polymer. If the API is dispersed or dissolved in the polymeric matrix, drug release occurs through the erosion of the polymeric matrix, diffusion mechanisms within the matrix, or a combination of both mechanisms (Dristant, 2023).

6.3.1.2 Nanocapsule

Nanocapsules are core-shell-type drug delivery systems consisting of a lipophilic inner core that encapsulates API and a polymeric shell surrounding this core. Particle sizes are larger than nanospheres. Due to their polymeric membrane, the drug release rate is slower. The release rate can be controlled by changing the membrane porosity. Thanks to their physical structure, they are more stable systems than nanospheres.

Similar preparation techniques, such as polymer dispersion and monomer polymerization, are used in the preparation of nanospheres and nanocapsules. The primary difference in the preparation of nanocapsules lies in the addition of oil to the organic phase during the preparation process. Thus, the polymer forms a membrane by coating around this oil, and the nanocapsule with an oily core is formed (Mora-Huertas, 2010).

6.4 POLYMERS AND POLYMERIC NANOPARTICLES FOR CONTROLLED DRUG DELIVERY

Polymers have a wide range of applications in biomedical fields, including controlled drug delivery, medical devices, and tissue engineering. It is important that the polymers to be used in these applications have certain properties and fulfill various functions. Polymers to be used in controlled drug delivery systems should be biocompatible, biodegradable, and non-toxic, and approved by authorities such as the FDA for biomedical usage. Properties of polymers, such as solubility, crystallinity, molecular weight, and glassy transition temperature (Tg), are the factors that determine the release properties of the drug delivery system and its use in the biomedical field.

Biocompatible polymers are preferred in drug delivery systems. The body can tolerate these polymers and have low side effects due to their low immunogenic properties. Similarly, it is essential that the polymer to be used is biodegradable in terms of toxicity. Biodegradable polymers degrade in the body over time, allowing implants or carrier systems applied to the body to be excreted without the need for surgical intervention and without any potential toxic build-up. A crucial point to be considered in the selection of biodegradable polymers is that the polymer is non-toxic, and the decomposition products are non-toxic and do not cause inflammatory responses. For this reason, it is also important to choose biodegradable polymers that are approved for biomedical use by authorities such as the FDA.

The solubility of API with low water solubility can be increased by using hydrophilic polymers. In this way, solubility and absorption properties are improved, which directly affects bioavailability. In addition, the controlled-release properties of nanoparticles can be modified by leveraging the degradation rates of polymers. Thus, it is possible to release the API at the desired speed for a desired period. This is particularly important to keep the API concentration within the therapeutic range and to minimize side effects.

In polymeric drug delivery systems, the stability of the API can be improved by increasing the resistance of the API to biological conditions and enzymatic degradation. Different system designs, such as controlled drug delivery systems, implants, and microneedles, can be made by selecting suitable polymers for this purpose. In addition, specially designed polymers called "tailor substances", which have appropriate properties for targeted diseases, are also being studied today.

As mentioned, polymer selection is important in preparing a successful controlled drug delivery system. Nowadays, natural polymers such as chitosan, alginate, albumin, cyclodextrin, and synthetic polymers such as PLA, PLGA, PGA, and PCL are frequently studied in the literature because they meet these properties.

6.4.1 Natural Polymers in Controlled Drug Delivery Systems

6.4.1.1 Polysaccharides

Polysaccharides are natural polymers that are widely used in the pharmaceutical, food, textile, and biomaterials industries. Their source can be plants (such as starch and cellulose) or animals (including glycogen and chitin). They are long-chain polymeric structures formed by the bonding of many monosaccharides through glycosidic bonds (Gandini, 2021).

6.4.1.1.1 Cyclodextrin

Cyclodextrin is one of the frequently used polysaccharides in the preparation of controlled drug delivery systems. They are cyclic oligosaccharides and are obtained by enzymatic degradation of starch. Cyclodextrins have been widely used for many years in the pharmaceutical, food, and cosmetic industries. They have low toxicity, low immunogenicity, and are cost-effective. In addition, cyclodextrins have been defined as substances with "Generally Recognized as Safe" status and have been used as drug excipients in pharmaceutical technology. The fact that they are biocompatible and biodegradable increases their use in the pharmaceutical field. Cyclodextrins have a hydrophobic inner cavity and a hydrophilic outer surface. They have three derivatives, α, β-, and γ-, which are formed by the bonding of 6, 7, and 8 glucopyranose units by α-(1,4) glycosidic bonds, respectively. Thanks to their truncated cone-shaped geometric structure, APIs with low water solubility can be loaded into their hydrophobic inner cavities, thereby improving water solubility. This feature is also used to protect from enzymatic degradation, maintain stability, and mask bad taste or odor. Drug delivery systems prepared from cyclodextrins are also used to cross biological barriers such as the blood-brain barrier (BBB) and to deliver the drug to the target area (Vecsernyés, 2014). Cyclodextrins are used to improve the solubility and stability of drugs, to increase drug absorption, and to provide controlled release (Varan et al., 2017)

In a study by Beňová et al., a drug delivery system consisting of cyclodextrin and silica loaded with 5-fluorouracil (5-FU) was prepared. The pores of this mesoporous silica drug delivery system were clogged with β-cyclodextrin, thereby entrapping 5-FU. When the drug delivery system is in an acidic pH, drug release occurs by the release of β-cyclodextrins from these pores. Therefore, a controlled drug delivery system has been developed in which 5-FU is encapsulated at neutral pH (7.4), but 80% of the drug is released within 24 h at the tumoral acidic pH (5). In addition, the anticancer efficacy of this system has been demonstrated in human glioma U87 MG cells (Beňová et al., 2020).

In another study with cyclodextrins, Ünal et al. developed cyclodextrin-based cationic nanoparticles loaded with camptothecin (CPT) for the treatment of colorectal cancer. They successfully improved the optimal formulations containing CPT, with a diameter of 135 nm and a zeta potential of +40 mV. When nanoparticles were administered orally, 52% of the CPT, which has an oral bioavailability problem, successfully reached the colon. They also reported that CPT-loaded cyclodextrin nanoparticles show high cytotoxicity on HT-29 colon cancer cells and high mucosal penetration in Caco-2 cells (Ünal et al, 2020).

6.4.1.1.2 Alginate

Alginate is another polysaccharide obtained from brown seaweed (such as *Laminaria hyperborea*, *Laminaria digitata*, *Laminaria japonica*, *Ascophyllum nodosum*, and *Macrocystis pyrifera*), chemically composed of β-D-mannuronic acid and α-L-guluronic acid monomers. Thanks to advantages such as high biocompatibility, biodegradability, and low toxicity, it has a prominent place, especially in controlled drug delivery systems. Alginate is also an FDA-approved polymer and is used as a tissue scaffold in tissue engineering as well as drug delivery systems. In addition to drug delivery systems and tissue scaffolds, it has been used as an excipient in the pharmaceutical technology industry for many years due to its stabilizer properties.

The solubility of alginate in neutral and basic media was high, and it can gel in aqueous environments or in the presence of +2 valent cations such as Ca^{2+}. Due to this gel-forming property, it is frequently used in hydrogel studies. (Hariyadi, 2020). This property of alginate also causes a significant advantage in the development of drug delivery systems. The release of the drug can be controlled by the swelling of the polymer depending on the pH. Alginate nanoparticles are widely used in oral, nasal, and dermal drug delivery systems and have become an indispensable polymer.

For example, Zhang et al. prepared pH-sensitive Astaxanthin (AST) encapsulated alginate-based drug delivery systems using biodegradable calcium alginate for the treatment of hepatocellular carcinoma. It has been shown that these drug delivery systems with an encapsulation efficiency of 87% can release pH-sensitive AST. Cell culture studies for hepatocellular carcinoma have shown that AST-loaded drug delivery systems inhibit HepG2 cell proliferation, too (Zhang, 2020)

Boi et al. prepared a drug delivery system in which they combined doxorubicin (DOX) $CaCO_3$ with ALG microbeads. In this study, controlled drug release was achieved by using the crosslinked structure of alginate, and the burst release effect was prevented. The anticancer effect of the drug delivery system has been demonstrated in the MCF-7 cell line of human breast adenocarcinoma. Compared to free DOX-loaded alginate beads, it has been observed that a continuous and slower release is obtained with combined DOX $CaCO_3$ with ALG microbeads (Boi et al., 2020).

6.4.1.1.3 Chitosan

Chitosan (poly-(β-1/4)-2-amino-2-deoxy-D-glucopyranose) is a natural polymer obtained by deacetylation of chitin, which is found in the exoskeletons of shellfish such as crabs, shrimps, and lobsters, or in the cell walls of algae and fungi. Chitosan, which has cationic properties, is also a biocompatible and biodegradable polymer. It has high solubility in an acidic environment but limited solubility at neutral pH (e.g., pH 7.4 physiological environment). The physicochemical properties of the polymer, such as solubility and viscosity, may vary depending on the molecular weight and degree of acetylation of the polymer. For example, as the degree of deacetylation increases, more amino groups become protonable, and the solubility of chitosan increases accordingly. Chitosan with low molecular weight dissolves more easily than high molecular weight, while its viscosity values decrease accordingly. Depending on this feature of chitosan, the properties of the drug delivery system can be modified.

The antibacterial, antioxidant, and anti-inflammatory properties of chitosan, along with its biocompatibility and biodegradability, have led to its use in nanoparticulate drug delivery systems (Aranaz et al., 2021).

Additionally, it is a polymer with high cellular uptake due to its cationic properties and surface amine groups. For this reason, chitosan is also used to modify the surfaces of polymeric nanoparticles and to increase cellular uptake. Chitosan is also used in the preparation of nanoparticulate systems that provide effective transport and long-term release of drugs with its mucoadhesive properties. The mucoadhesive properties of chitosan offer a significant advantage in drug applications for the mouth, nose, eyes, and skin (George, 2019).

Alqahtani et al. prepared chitosan nanoparticles for the treatment of gonorrhea caused by antibiotic-resistant *Neisseria gonorrhea* strains. Chitosan nanoparticles were prepared by the ionic gelation method. According to the results, it was observed that chitosan nanoparticles exhibited antibacterial activity against all strains tested, including those resistant to multiple antibiotics. This study employed an alternative antibiotic approach using chitosan nanoparticles (Alqahtani et al., 2020).

6.4.1.2 Proteins

6.4.1.2.1 Albumin

Albumin is a protein-based polymer derived from the blood plasma of humans and animals. It is a biodegradable, non-toxic, and non-immunogenic polymer with high biocompatibility. Human serum albumin (HSA) and bovine serum albumin are frequently used in medical applications and controlled drug delivery systems. Albumin has a high solubility in water and increases the solubility of hydrophobic API. It has a high API binding capacity and provides targeted drug transport thanks to its surface modifications. Albumin is used in various delivery systems such as polymeric nanoparticles, liposomes, and polymer-drug conjugates, and provides controlled release of the drug, thanks to all these advantages.

Albumin-based nanoparticles in the literature are biocompatible and non-immunogenic, as well as systems with a high tendency to accumulate in tumor tissues (Elsadek, 2012)

At the same time, studies have shown that albumin nanoparticles can increase the solubility of lipophilic substances in the blood and deliver APIs to the target tissues. Albumin-based nanoparticles increase the absorption and bioavailability of APIs. Cell culture studies prove that HSA is non-cytotoxic and safe. (Watcharin, 2014). The surface of albumin-based nanoparticle drug delivery systems can be modified with various functional groups (thiol, amino, and carboxylic groups, etc.) with high binding capacity. In addition, the covalent bonding of various drug molecules to albumin enables the preparation of stable drug delivery systems. Thus, it is possible to prepare nanoparticulate drug delivery systems that only release drugs in targeted tumor tissues and/or cancer cells. In addition to increasing the effectiveness of treatment, these nanoparticles also prevent healthy cells from being affected by active drug substances (Anhorn et al., 2008).

Significant advances are being made in cancer treatment with the preparation of albumin-based nanoparticles, which contain an anticancer drug with known in vitro efficacy but low solubility and permeability. For example, paclitaxel's efficacy in the treatment of metastatic breast cancer is known, but it has a solubility problem. The solubility problem of paclitaxel is being overcome with Cremophor EL, which is also found in Taxol®. However, Cremophor EL is a highly toxic substance. To eliminate this toxicity, Abraxane®-containing albumin-conjugated paclitaxel nanoparticles are used. Abraxane® is the first polymeric nanoparticle containing HSA and was approved by the FDA in 2005 in the treatment of breast cancer (Hassanin et al., 2020). Studies have shown that the maximum tolerated dose (MTD) for Abraxane® is 70%–80% higher than for Taxol® (Sparreboom, 2005). This success with Abraxane® later led to the study of many different albumin-based nanoparticles. Albumin-based nanoparticles containing docetaxel, rapamycin, and thiocolchicine can be examples of these studies (George, 2019).

Drug delivery systems for active targeting can be prepared by modifying the surfaces of albumin nanoparticles. Sun et al. developed cabazitaxel (CTX) loaded nanoparticles by conjugating folic acid (FA) to the surface of albumin. Due to the presence of highly expressed folic acid receptors (FR) in tumor cells, FA-modified albumin nanoparticles were actively targeted to prostate cancer treatment. In the results of the cytotoxicity study, it has been shown that FA-modified albumin nanoparticles highly inhibit FR overexpressing cancer cells. In addition, cellular uptake studies show that FA modification enhanced FR-mediated endocytosis in HeLa cells (Sun, 2019).

6.4.1.2.2 Collagen

Collagen is the most abundant protein in the human body and forms the building block of tissues such as bone, cartilage, joints, tendons, fibers, muscles, and skin. It provides moisturizing and elasticity to the skin, as well as the durability of muscles and bones. It consists of three alpha chains with a helical structure, and the central molecule is tropocollagen. Although there are more than 30 types of collagens, the most important ones are type I, type II, type III, type V, and type X collagen. Type I and type III collagen are mostly from cattle or fish; type II collagen is obtained from chicken, while type V and type X collagen are obtained from eggshell membrane. Collagen is used in pharmaceutical technologies, controlled drug delivery systems, tissue engineering, and nutraceuticals. It is a biodegradable, biocompatible, non-toxic polymer with low antigenicity properties (Arun, 2021). In the literature, collagen is used in the preparation of controlled drug delivery systems such as polymeric nanoparticles, hydrogels, liposomes, niosomes, microspheres, and microneedles in combination with various neutral polymers such as chitosan, keratin, elastin, and silk fibroin (Sionkowska, 2021).

Malathi et al. prepared ZnO-loaded niosome nanocomposites using type I collagen. The wound healing and antimicrobial effects of nanocomposites were examined, and it was seen that they facilitated wound healing and skin regeneration. In cytotoxicity studies with a 3T3 healthy fibroblast cell line, collagen nanocomposites have been shown to be biocompatible. In addition, these nanocomposites have proven to be highly effective against *Bacillus subtilis*, *Staphylococcus aureus*, *Pseudomonas aeruginosa*, and *Escherichia coli* (Malathi, 2021).

Siddiqui et al. prepared resveratrol-loaded collagen nanoparticles using type I collagen. Collagen nanoparticles are designed for the treatment of epilepsy and are coated with glutathione to facilitate their passage across the BBB. In animal studies, nano-resveratrol was applied to the hippocampus of mice. It has been shown that the effectiveness of resveratrol treatment with collagen nanoparticles has significantly increased. The bioavailability of resveratrol increased with the use of collagen nanoparticles. As a result, resveratrol-loaded collagen nanoparticles were more effective than free resveratrol in preventing cognitive impairment and controlling epileptic seizures in mice (Siddiqui, 2022).

6.4.2 Synthetic Polymers in Controlled Drug Delivery Systems

6.4.2.1 Polylactic Acid (PLA)

PLA, an aliphatic polyester, is a lipophilic synthetic polymer. It is synthesized from lactic acid, obtained by fermentation from natural sources such as sugar cane or corn starch, by polymerization reactions. PLA is a thermoplastic and high-strength polymer. The degradation time is long. PLA has two different optical forms (D-lactide and L-lactide), and degradation properties can be regulated by racemization of D- and L-isomers. Poly-L-lactide has higher tensile strength than poly (DL-lactide). PLA is also a biocompatible, biodegradable, non-toxic polymer and suitable for preparing controlled drug delivery systems. The degradation products of PLA are known to be safe (Das, 2023). It is possible to prepare controlled drug delivery systems by modifying the degradation time of PLA. There are studies on the potential of release profile adjustable systems for cancer treatment to increase therapeutic efficacy. Due to its low immunogenicity, PLA also has a wide range of uses as medical devices (surgical sutures, stents, fillers, screws) and drug delivery systems (such as liposomes, polymeric nanoparticles, dendrimers, and micelles) in medical fields such as pharmaceutical technology, tissue engineering, dentistry, and orthopedics. It is also approved by the FDA for use as sutures, bone implants, and screws, as well as vaccine antigens (proteins, peptides, and DNA), and as drug delivery systems (Tyler, 2016).

Numerous studies in the literature have examined the use of PLA in drug delivery systems. Oliveira Silva et al. prepared donezepyl (DNZ) loaded nanoparticles and extracellular vesicles for the treatment of Alzheimer's disease using PLA. With these systems, they aimed to enhance the ability of DNZ to cross the BBB, target the brain, and increase the effectiveness of treatment. In this context, a series of *in vitro* and *in vivo* studies were carried out. With the prepared nanoparticle systems, continuous drug release of DNZ was ensured. It was also shown that both formulations were non-toxic and had high cellular uptake by endothelial bEnd.3 cells (Oliveira Silva, 2024).

Shapira et al. provided a stereo-complexation of peptides and proteins using poly(D-lactide)-polyethylene glycol polymer. In this way, it has prepared nanoparticles (≈400 nm) to increase insulin stability and ensure its controlled release. According to the study's results, it was demonstrated that continuous insulin release from the stereo complex was sustained for 14 weeks (approximately 3 months). The therapeutic efficacy of PLA-insulin stereo complex nanoparticles was evaluated on mice, and it has been shown to be an effective and safe release system (Shapira-Furman, 2023).

Malik et al. prepared antitumor drugs Palbociclib (PAL) and Regorafenib (REG) encapsulated in d-α-Tocopheryl. Polyethylene glycol succinate (TPGS) PLA nano micelles were prepared by the thin-film hydration method. The particle size of nanomicelles for the treatment of lung cancer was found to be in the range of 120–190 nm, polydispersity index (PDI) was 0.2–0.3 and Zeta Potential was −18 to −30 mV. *in vitro* characterization and release studies were carried out in comparison with pure drug dispersions. It was seen that continuous drug release could be achieved with the prepared nano micelles. It has also been demonstrated that the antitumoral efficacy of drug-loaded nanosystems is enhanced through cell culture studies (Malik, 2024).

6.4.2.2 Polyglycolic Acid (PGA)

PGA is a biodegradable polymer obtained by the polymerization of glycolic acid. PGA is a hydrophilic polymer that is biocompatible, non-toxic, and non-immunogenic. PGA, which is an FDA-approved polymer like PLA, is used in the preparation of controlled drug delivery systems. Degradation products are non-toxic and are harmlessly excreted from the body. PGA, thanks to its crystal structure, slows down drug release and provides a long-release effect. When PGA nanoparticles are used as carriers for chemotherapy drugs, they ensure that the drugs reach the targeted areas directly, thereby reducing systemic side effects. To adjust the release profile, nanoparticle systems are prepared by using PGA in combination with other polymers, especially PLA. PGA is one of the ideal polymers for tissue engineering, wound healing, and various pharmaceutical applications.

In a study by Zhang et al., the effect of PGA degradation on cytokine release was investigated. For this purpose, the degradation kinetics of PGA are regulated by modifying the system components and adding sodium tripolyphosphate to the PGA scaffold. The effect of PGA on the expression of macrophages in acidic environments has been investigated, revealing that the degradation of PGA inhibits the secretion of IL-1β and IL-6 in both PGA and PGA-TPP groups, regardless of pH, thereby further reducing leukocyte infiltration in PGA-TPP scaffolds. In summary, this study concluded that the inclusion of Sodium Tripolyphosphate (TPP) in PGA implants can alleviate the inflammatory response and improve the clinical outcomes of implants (Zhang, 2020).

In a study, Shuai et al. prepared a 3D-printed hydroxyapatite/PLA bone scaffold that exhibits bone-like bioactivity and osteoconductivity. PGA was added to the formulation to accelerate the degradation of the 3D-Printed HAP/PLA scaffold. As a result of this addition, the scaffold's degradation rate increased. And so, the 3D-printed scaffold was made into a viable system with abundant new bone tissue and blood vessel tissue formation (Shuai, 2021).

6.4.2.3 Polylactic-Co-Glycolic Acid (PLGA)

PLGA is a polymer obtained by the copolymerization of PLA and PGA at different rates. In addition to its biocompatibility and biodegradable properties, the fact that it is an FDA-approved copolymer allows the preparation of drug delivery systems (Pawar, 2023). By changing the molar ratios of PLA and PGA, the degradation rate of PLGA and thus the API release profile can be adjusted. In addition, properties critical to the drug delivery system, such as amorphous crystallinity, molecular weight, degradation rate, flexibility, durability, glass transition temperature, and degree of crystallinity, can be altered by manipulating the lactic/glycolide ratio (Tyler, 2016). As the PLA ratio increases in PLGA, degradation rate and solubility decrease, so the release rate decreases. As the PGA ratio increases, the degradation rate and solubility increase, and the release rate increases accordingly

By changing the ratios of PLA and PGA, PLGA polymers with amorphous or crystalline properties can also be obtained. Since amorphous PLGA polymer is more flexible, it is suitable for use in nanoparticulate drug delivery systems. However, crystalline PLGA is used in the biomedical field as surgical sutures, screws, threads, and plates due to its more complex structure (Mittal, 2007).

Sunazuka et al. prepared curcumin-loaded nanoparticles using PLGA and PLA. In this study, the effect of temperature-dependent PLGA/PLA NPs on curcumin release was examined. By adjusting the lactide-glycolide polymer (L: G) ratio and molecular weight, systems with appropriate glass transition temperatures were obtained. With the optimized PLGA/PLA NPs, it was observed that CUR exhibited a first burst release (within the first 30 minutes) followed by diffusion-controlled release (over 24h) with increasing temperature (Sunazuka, 2024).

Yao et al. used PLGA polymers to prepare isoliensin-loaded PLGA-PEG polymeric nanoparticles for the treatment of hypertension and protection of vascular lesions. The stability, *in vitro* drug release profiles, and *in vivo* pharmacokinetics of isoliensin-loaded nanoparticles were examined. According to results, PEG-PLGA-isoliensine has been shown to significantly increase the absorption of isoliensine and provide targeted *in vivo* distribution (Yao, 2024).

6.4.2.4 Polycaprolactone (PCL)

PCL is an aliphatic polyester-based synthetic polymer. Thanks to biocompatible, biodegradable, and non-toxic properties, it is suitable for biomedical applications. PCL is a hydrophobic, semi-crystalline polymer. It has high solubility and is easy to process. PCL is a polymer approved by the FDA and European Medicine Agency (EMA) and used in the preparation of drug delivery systems. In addition to drug delivery systems, they are used in tissue engineering and in the biomedical field for purposes like wound dressing and sutures. Nowadays, Capronor®, a PCL-based implant, provides long-term release of levonorgestrel, a type of progestin, and is used as a contraceptive by subdermal application. Trusopt® (dorzolamide hydrochloride), which is used in the treatment of glaucoma and ocular hypertension to lower intraocular pressure, is also a PCL-based ophthalmic solution (Pawar, 2023).

In a study by Ahmady et al., capsaicin-loaded alginate nanoparticles embedded in polycaprolactone-chitosan nanofiber were prepared. The release of capsaicin from nanofibers and its anticancer activity were investigated. In capsaicin-loaded alginate nanoparticles embedded in polycaprolactone-chitosan nanofibers, capsaicin release was extended from 120h (about 5 days) to 500h (about 3 weeks). Based on these results, the nanocomposite drug delivery system was found to be promising for the long-term and controlled release of capsaicin in cancer treatment (Ahmady, 2023).

Huo et al. prepared artemisinin (ART) loaded nanofibers using PCL and collagen polymers. The release properties of ART-loaded PCL/Col nanofibers, prepared for the treatment of malaria, were investigated. In the results obtained, it was observed that the nanofibers exhibited continuous release behavior for up to 48h (about 2 days), with the increase of the hydrophobic PCL component, the nanofiber diameter increased, pilling decreased, and ART from PCL/Col nanofibers provided a slower release (Huo, 2021).

6.4.2.4.1 Advantages and Disadvantages of Polymers in Controlled Drug Delivery Systems

The superior properties of polymers provide a series of advantages in their use in the biomedical field, especially in pharmaceutical technology. In addition, there are also some disadvantages of polymeric drug delivery systems. The key factor causing the advantages and disadvantages of polymeric drug delivery systems is the properties of the polymers, such as biocompatibility, biodegradability, solubility, immunogenicity, and stability. In addition, properties of nanoparticles such as particle size, surface charge, and surface modifications stand out as parameters that determine their advantages and disadvantages in drug delivery systems.

6.4.3 Biocompatibility

Biocompatibility refers to the ability of a substance applied to the body to function in harmony with cells, tissues, and organs, without causing any harmful effects. Simply, it can be summarized as the compatibility of the material with the biological systems. Biocompatible polymers are tolerated by the body and are important in the preparation of drug delivery systems with their low immunogenic properties. For this reason, biocompatible polymers emerge as key components in the preparation of controlled drug delivery systems. The fact that the prepared drug delivery system is also biocompatible is considered a significant advantage in terms of safe use.

6.4.4 Biodegradation

Another important property is that polymers are biodegradable. Biodegradation is the process that involves the breakdown of substances into simpler compounds by the action of enzymes, microorganisms, and other biological mechanisms in the environment.

Biodegradable systems are crucial for the development of drug delivery systems. Even if a drug delivery system is taken into the body in case of repeated use as a drug, it is biocompatible, its accumulation in the body over time can cause various side effects. This problem emerges as a fundamental problem, especially in metallic nanoparticles. The breakdown of biodegradable polymeric nanoparticles within biological cycles and their non-accumulation in the body is seen as one of the most important advantages. In addition, the drug release profile can be controlled by modifying the degradation rate of polymers. Thus, the drug can be released at the desired period. Controlled and/or sustained-release drug delivery systems prepared with this approach not only increase treatment effectiveness but also reduce the frequency of drug dosing and improve patient compliance. With these systems, drug concentration fluctuations in the blood can be reduced, and a more effective drug treatment can be provided with less API.

A critical point to be considered in biodegradable systems is that the degradation units are also non-toxic and biocompatible. FDA-approved natural or synthetic biodegradable polymers used in drug delivery systems meet this requirement and are therefore preferred in the preparation of drug delivery systems (George, 2019).

Biocompatible but non-biodegradable polymers such as polyethylene (PE), polypropylene (PP), polyvinyl chloride (PVC), polystyrene (PS), Teflon (PTFE), polymethylmethacrylate (PMMA), and polycarbonate (PC) are frequently preferred, especially in areas where it is desired to remain stable in the body for a long time. These polymers are often used in biomedical applications such as implants. In such applications, biodegradation emerges as a disadvantage. However, biodegradation is a significant advantage in the preparation of systems that are desired to leave the body after treatment, such as drug delivery systems. In controlled drug delivery systems, biocompatible and biodegradable polymers such as PLA, PGA, PLGA, PCL, polyhydroxy butyrate, polypropylene fumigate, polydioxanone, cyclodextrin, alginate, chitosan, collagen, albumin, and other polysaccharides and proteins are frequently used (Das, 2023; George, 2019). These polymers degrade in the body over time, so it is a particularly important advantage that the implants or carrier systems applied to the body are removed from the body without the need for surgical intervention. In addition, the fact that the degradation products do not cause toxicity or inflammatory responses is among the advantages of these polymers. The fact that the polymers used in the preparation of drug delivery systems are FDA-approved is also an important feature in terms of the safety of these systems.

6.4.5 Solubility

Many APIs had low solubility and/or low permeability. This situation adversely affects the therapeutic efficacy of the disease. Polymeric drug delivery systems are advantageous in enhancing the solubility and permeability properties of APIs, thereby paving the way for their effective treatment. Nanoparticles are small systems compared to conventional drug delivery systems. Particle size increases the surface area of the nanoparticle system. With the rise in the surface area, the solubility of drugs with solubility problems is increased. The absorption and bioavailability of drugs also increase. Increasing the solubility is important in providing a more effective treatment with a lower amount of API and preventing side effects. For example, paclitaxel is a highly effective anticancer drug against cancer cells; however, it has a solubility issue. Paclitaxel molecules form micro-sized crystals in aqueous media. These crystals may cause occlusion and/or damage to the vessels in the application area. Excipients such as Cremophor EL, used to provide paclitaxel solubility, also have serious toxicity problems. Nowadays, the solubility of paclitaxel and the elimination of these side effects by using polymeric nanoparticles have been ensured.

6.4.6 Immunogenicity

Whether the immune system recognizes it and causes an immune response is also important for drug delivery systems. Non-immunogenic polymers and nanoparticles are not recognized as foreign by the body and therefore do not trigger an immune response. With these properties, they also do not cause undesirable immune reactions in repeated doses and/or long-term treatments. To ensure non-immunogenicity, biocompatible polymers such as PEG can also be modified on the surface of the nanoparticles. This process, known as PEGylation, prevents the designed nanoparticles from being recognized by the immune system, prolonging their circulation time and increasing bioavailability (Abu Lila, 2018).

Synthetic polymers are seen as more advantageous systems in terms of immunogenicity compared to natural polymers. Natural polymers are extracted from natural sources such as plants, fungi, algae, shellfish, and other animals. Although they are biocompatible and abundant in nature, they are obtained from various sources, and the purification steps are long and complex; they always have the risk of an immunogenic response.

6.4.7 Stability

Stability is the ability of a substance to maintain its physical, chemical, microbiological, therapeutic, and toxicological properties for a specific period. In pharmaceutical technology, stability means maintaining the efficacy, safety, and quality characteristics of the active drug or the product obtained and is of critical importance. After the drug formulation is produced, it must have the characteristics of the first day it was made during storage, transportation, use, and storage in hospitals and pharmacies.

A physically stable drug means that it retains its properties such as appearance, hardness, taste, and smell. Factors such as temperature, pH, phase separation, viscosity, abrasion, and dispersion all affect physical stability. Chemically stable means the drug retains its potency, does not undergo chemical reactions like hydrolysis or oxidation, and does not form degradation products. Chemical degradation of drugs can reduce the potency of the drug or form toxic degradation products.

Due to the stability issues of APIs and drug formulations, situations such as insufficient dissolution, dispersion, and absorption can occur with changes in physicochemical properties. This directly affects the drug's therapeutic effectiveness in the biological environment (Williams, R. L, 2006).

Nanoparticulate drug delivery systems have a high surface area-to-volume ratio due to their nanoscale size. For this reason, they can often present stability problems. The stability of the nanoparticles directly affects the physical and chemical properties of the formulation, the efficacy, shelf life, and release profile of the drug. In addition to storage stability, nanoparticle systems need to maintain their biological stability within the body. The ability of nanoparticles to remain in circulation for an extended period, effectively reach targeted tissues, and maintain their effectiveness at the cellular level depends on their biological stability. Opsonization, immune system responses, protein adsorption, or enzymatic degradation of nanoparticles are among the factors that affect their biological stability. To reduce the immune system's response and inhibit opsonization, the surface of the nanoparticles can be decorated with polymers, such as PEG, thereby preventing recognition by the MPS. In addition, biodegradable polymers allow the preparation of drug delivery systems resistant to enzymatic degradation. (Abu Lila, 2018). These systems also improved the resistance of API to body conditions and enzymatic degradation. Thus, APIs maintain their effectiveness until they reach the target area, and a more effective drug treatment is provided with a lower API.

6.4.8 Particle Size

Particle size comes first among the physicochemical advantages of polymeric nanoparticle drug delivery systems. and the most important critical quality parameter of a nano system. It affects many properties, including its behavior, stability, duration of stay in circulation, and the target site it will reach in the body. In addition to particle size, the PDI, which indicates the particle size distribution, is also a crucial parameter. The PDI value ranges from 0 to 1 and reflects the system's polydispersity properties. As the PDI value approaches 0, the monodispersity increases, while as the PDI approaches 1, the polydispersity increases. Polymeric nanoparticles are colloidal systems and are affected by Brownian motion due to their size. With the decrease in particle size, Brownian motion increases and allows them to disperse in aqueous environments (Singh, 2009). This allows the resulting nanoparticle system to disperse in aqueous media without precipitation, even when prepared with a hydrophobic polymer and API.

The particle size also plays a significant role in targeting the polymeric drug delivery system. For example, nanoparticulate systems must be smaller than 400 nm in size to benefit from the enhanced permeability retention effect, pass through the pores in the blood vessel, reach the cancerous tissue, and accumulate. If it is below 100 nm, it can escape from MPS and remain in the blood for longer periods of time. However, systems with dimensions less than 50 nm can accumulate in reticuloendothelial system organs such as the liver, spleen, and kidney, causing toxicity (Pathak, 2009).

The approach of accumulating nanoparticles in target tissues using their size is called passive targeting. The most well-known passive targeting approach is to utilize the enhanced permeability and retention (EPR) effect. Due to diseases such as cancer, natural physiological processes are disrupted. During this process of deterioration, the tumor tissue becomes abnormally vascularized. In healthy and impermeable vessels with tight junctions, there is an increase in vascular permeability due to vascular mediators expressed with the disease. When nanoparticle systems are applied to the body, they cannot leak from healthy vessels to tissues due to tight junctions. However, when they reach the tumor tissue with an impaired vascular structure, they can leak out of the vessels and reach the tumor area. In addition, due to the insufficient functioning of the lymphatic system in the tumor tissue, the fluid pressure between the cells is higher. Therefore, the leaked nanoparticles cannot be excreted from the tumor area by the lymphatic system and begin to accumulate. This cancer-induced condition is called the EPR effect. The small size of the nanoparticles provides an advantage in terms of benefiting from the EPR effect, also known as passive targeting. Thus, the effectiveness of the treatment increases, and the side effects on healthy tissues are reduced (Kang, 2020; Lahooti, 2023).

6.4.9 Surface Charge

The surface charge of the nanoparticulate system is expressed by the zeta potential. Zeta potential affects many properties of the nanoparticle system, such as storage and physiological stability, drug distribution in the body, interaction with the vascular endothelium, and intracellular uptake.

Zeta potential is highly decisive in terms of the shelf-life stability and aggregate potential of the nanoparticle. The fact that the zeta potential is in the high negative or positive value ranges makes the inter-particle repulsive forces strong. In this way, polymers are prevented from agglomerating. In colloidal systems, it is sufficient for them to be above ± 30 mV of zeta potential. But at low zeta potential below ± 10 mV, the repulsive forces are weak, and the nanoparticles tend to be attracted to each other. Therefore, stability problems such as agglomeration and aggregation are encountered (Singh, 2009).

Nanoparticles with a negative surface charge are systems that are perceived as foreign substances by the body in the bloodstream, opsonized, and removed from the blood quickly. Nanoparticles with a positive surface charge have a high affinity for the vascular endothelium. Due to the negative charge of the cell membrane, cationic nanoparticles that interact highly with the vascular endothelium may not be able to reach the desired area by adhering to the vessel. This may lead to a decrease in the effectiveness of treatment. It is especially important to adjust the zeta potential in a deliberate manner so that the nanoparticle system can remain in the blood for a longer period and reach the site of action without being recognized by macrophages. (Singh & Mishra, 2009).

6.4.10 Surface Modification

Polymeric nanoparticle surface properties can be modified easily due to their physicochemical properties. By using polymers in the preparation of drug delivery systems, "tailored drug delivery systems" with properties suitable for the patient or disease can be prepared.

The surface morphology (shape, surface area, structure) of the nanoparticulate system directly determines its properties. Features such as the uptake mechanisms, the release rate, and the amount of drug reaching the target area are related to the nanoparticle surface properties. With surface modifications, active targeting of the nanoparticle system can be achieved. Thus, a more effective and safe treatment can be provided by delivering API to only the target area. Active targeting is based on the decoration of the nanoparticle surface with molecules such as ligands, antibodies, or proteins with high affinity for receptors in the target region. In this way, the nanoparticulate drug delivery system is directed to a specific organ, tissue, or cell (Li, 2022). With this strategy, it is ensured that the drug accumulates more effectively in the targeted diseased area than in other areas. Since the most effective treatment is provided with the drug dose applied, the side effects caused by high-dose applications are reduced. Thanks to systems decorated only with tumor-specific ligands, healthy tissues are prevented from the toxic effects caused by chemotherapy, and patient compliance is also increased. With active targeting approaches, drug passage can be achieved through biological barriers that have low drug permeability and are difficult overcome. In addition, drug release can be carried out in the target area, at the desired time, and in the desired amount for the desired period. Despite all these advantages, the development processes of these systems are quite complex, complicated, and expensive. The mechanisms by which nanoparticle systems pass through cell membranes have not yet been fully elucidated.

Currently, active targeted delivery systems are designed to target overexpressed receptors in cancer cells. For example, nanoparticles whose surface is decorated with folate receptor-specific ligands can specifically bind to cancer cells that express excess folate receptor. In addition to the specific binding mechanisms of nanoparticles to receptors, antigens, and proteins, active targeting mechanisms are also supported by physical (magnetic field, heat) and chemical factors (pH, reactive oxygen species) (Chen, 2023).

6.4.11 Advantages

In summary, polymeric nanoparticle drug delivery systems that provide controlled release are an important innovation in modern medicine. These systems enhance therapeutic efficacy by increasing the absorption and bioavailability of drugs, reducing their side effects, and adjusting the amount and frequency of administered doses. Polymeric nanoparticles offer controlled and/or continuous drug release, reducing the dose and frequency of medication required, thereby increasing patient compliance. Thanks to this superiority, it provides a great convenience and an increase in the effectiveness of treatment, especially in the treatment of diseases that involve a heavy treatment regimen, such as cancer or chronic diseases that require years of drug use (Verma & Garg, 2010). Biocompatible and biodegradable polymers approved by the FDA, used in drug delivery systems, provide controlled drug release without causing toxicity in the body, are broken down by natural processes, and do not release toxic breakdown products. This is a significant advantage that ensures long-term safety and effectiveness (Singh, 2009). Polymers enhance treatment efficacy by enabling the delivery of drugs specifically targeted to tissues or cells, while minimizing potential side effects that can occur in healthy cells. By using polymers, a nanoparticulate drug delivery system can be prepared that encapsulates a variety of drugs, including both hydrophobic and hydrophilic ones. This offers great flexibility in terms of the effective use of various therapeutic agents (Sánchez & Villaverde, 2020). By making polymeric surface modifications, it can be ensured that API is protected from biological and environmental factors (such as pH and enzymes) and increases its stability and effectiveness. This is especially important for harsh conditions such as enzymatic and acidic environments in the gastrointestinal tract (Xu, 2020).

6.4.12 Disadvantages

The production of polymeric nanoparticles can be a complex and costly process. Although low-scale production is possible, industrial production, scale-up, and repeatability problems may be experienced. This can be an obstacle for clinical studies. Another risk factor is that some polymers or their degradation products cause toxicity. Nanoparticles obtained from natural polymers carry the risk of inducing various immunogenic reactions due to the insufficient purification of the polymer. Therefore, it is necessary to carefully evaluate the safety profiles of polymers and ensure that they are biocompatible (Sánchez, 2020). Polymeric nanosystems may have limited capacity to overcome some biological barriers (such as the BBB) that are difficult to overcome in the body. This can make it difficult for drugs to reach the target tissue. The opposite is also possible, and it is also a risk factor for polymeric nanoparticles to unpredictably cross biological barriers and reach unwanted tissues due to their size. Precisely controlling the release kinetics of polymer nanoparticles is difficult or may not always be possible. It can lead to early release of the drug from the drug delivery nanosystem at an undesirable rate and time, or the polymeric nanosystem not releasing the drug. In this case, an undesirable release profile occurs, and the appropriate therapeutic effect cannot be achieved. Polymeric drug delivery systems can also be difficult to remove or degrade from the body, and accumulation can be observed, especially in the liver and kidneys. To overcome this situation, the structural properties of the polymer should be carefully investigated, and the system should be designed with appropriate features (Xu, 2020).

6.5 FUTURE OF POLYMER AND POLYMERIC NANOPARTICLES

Polymers used in controlled-release nanoparticle drug delivery systems promise significant advances in the future. With the increase in studies on smart polymers, the number of systems that provide targeted and controlled drug release by showing sensitivity to environmental conditions (such as pH, temperature, light, and magnetic field) will also increase (Alsehli, 2020). The use of polymeric delivery systems specially designed for the patient's genetic structure and disease status will provide a significant advance in personalized medicine. These developments are expected to increase patient compliance and treatment efficacy (Han, 2019). In addition, polymeric systems designed for combined therapies can create synergistic effects by carrying more than one therapeutic agent simultaneously. This is especially promising in the treatment of multifactorial diseases such as cancer. Innovations in polymer chemistry will enable the discovery of new biocompatible and biodegradable polymers, allowing for the development of safer and more effective carrier systems (Dristant, 2023). Hybrid systems involving nano- and micro-scale particles will enable the implementation of more complex and precise treatment strategies, ensuring that drugs are delivered effectively to the targeted tissue. Such systems will also allow for more complex treatment strategies in the future. These innovations can revolutionize treatment processes by enhancing the efficiency and reliability of controlled-release systems.

REFERENCES

Abu Lila, A. S., Shimizu, T., & Ishida, T. (2018). "PEGylation and anti-PEG antibodies." In *Engineering of Biomaterials for Drug Delivery Systems*, Woodhead Publishing, 51–68.

Ahmady, A. R., Solouk, A., Saber-Samandari, S., Akbari, S., Ghanbari, H., & Brycki, B. E. (2023). "Capsaicin-loaded alginate nanoparticles embedded polycaprolactone-chitosan nanofiber as a controlled drug delivery nanoplatform for anticancer activity." *Journal of Colloid and Interface Science*, 638, 616–628.

Alqahtani, F., Aleanizy, F., El Tahir, E., Alhabib, H., Alsaif, R., Shazly, G., ... & Mahdavi, J. (2020). "Antibacterial activity of chitosan nanoparticles against pathogenic *N. gonorrhoea*." *International Journal of Nanomedicine*, 15, 7877–7887.

Alsehli, M. (2020). "Polymeric nanocarriers as stimuli-responsive systems for targeted tumor (cancer) therapy: Recent advances in drug delivery." *Saudi Pharmaceutical Journal*, 28(3), 255–265.

Anhorn, M. G., Wagner, S., Kreuter, J., Langer, K., & von Briesen, H. (2008). "Specific targeting of HER2 overexpressing breast cancer cells with doxorubicin-loaded trastuzumab-modified human serum albumin nanoparticles." *Bioconjugate Chemistry*, 19(12), 2321–2331.

Aranaz, I., Alcántara, A. R., Civera, M. C., Arias, C., Elorza, B., Heras Caballero, A., & Acosta, N. (2021). "Chitosan: An overview of its properties and applications." *Polymers*, 13(19), 3256.

Arun, A., Malrautu, P., Laha, A., Luo, H., & Ramakrishna, S. (2021). "Collagen nanoparticles in drug delivery systems and tissue engineering." *Applied Sciences*, 11(23), 11369.

Beňová, E., Bergé-Lefranc, D., Zeleňák, V., Almáš^, M., Huntošová, V., & Hornebecq, V. (2020). "Adsorption properties, the pH-sensitive release of 5-fluorouracil and cytotoxicity studies of mesoporous silica drug delivery matrix." *Applied Surface Science*, 504, 144028.

Biron, M. (2013). "Outline of the actual situation of plastics compared to conventional materials." *Thermoplastics and thermoplastic composites*, 1–29.

Boi, S., Rouatbi, N., Dellacasa, E., Di Lisa, D., Bianchini, P., Monticelli, O., & Pastorino, L. (2020). "Alginate microbeads with internal microvoids for the sustained release of drugs." *International Journal of Biological Macromolecules*, 156, 454–461.

Callister Jr, W. D., & Rethwisch, D. G. (2020). *Materials science and engineering: an introduction*. John wiley & sons.

Chen, Z., Kankala, R. K., Long, L., Xie, S., Chen, A., & Zou, L. (2023). "Current understanding of passive and active targeting nanomedicines to enhance tumor accumulation." *Coordination Chemistry Reviews*, 481, 215051.

Das, A., Ringu, T., Ghosh, S., & Pramanik, N. (2023). "A comprehensive review on recent advances in preparation, physicochemical characterization, and bioengineering applications of biopolymers." *Polymer Bulletin*, 80(7), 7247–7312.

Dristant, U., Mukherjee, K., Saha, S., & Maity, D. (2023). "An overview of polymeric nanoparticles-based drug delivery system in cancer treatment." *Technology in cancer research & treatment*, 22, 15330338231152083.

Ducrot, E., Chen, Y., Bulters, M., Sijbesma, R. P., & Creton, C. (2014). "Toughening elastomers with sacrificial bonds and watching them break." *Science*, 344(6180), 186–189.

Elsadek, B., & Kratz, F. (2012). "Impact of albumin on drug delivery—New applications on the horizon." *Journal of Controlled Release*, 157(1), 4–28.

Elumalai, K., Srinivasan, S., & Shanmugam, A. (2024). "Review of the efficacy of nanoparticle-based drug delivery systems for cancer treatment." *Biomedical Technology*, 5, 109–122.

Gandini, A., & M. Lacerda, T. (2021). "Monomers and macromolecular materials from renewable resources: state of the art and perspectives." *Molecules*, 27(1), 159.

George, A., Shah, P. A., & Shrivastava, P. S. (2019). "Natural biodegradable polymers based nano-formulations for drug delivery: A review." *International Journal of Pharmaceutics*, 561, 244–264.

Hadjesfandiari, N., & Parambath, A. (2018). "Stealth coatings for nanoparticles: Polyethylene glycol alternatives." In *Engineering of biomaterials for drug delivery systems* Woodhead Publishing, 345–361.

Han, H. J., Ekweremadu, C., & Patel, N. (2019). "Advanced drug delivery system with nanomaterials for personalised medicine to treat breast cancer." *Journal of Drug Delivery Science and Technology*, 52, 1051–1060.

Hariyadi, D. M., & Islam, N. (2020). "Current status of alginate in drug delivery." *Advances in pharmacological and pharmaceutical sciences*, 2020(1), 8886095.

Hassanin, I., & Elzoghby, A. (2020). "Albumin-based nanoparticles: A promising strategy to overcome cancer drug resistance." *Cancer drug resistance*, 3(4), 930.

Huo, P., Han, X., Zhang, W., Zhang, J., Kumar, P., & Liu, B. (2021). "Electrospun nanofibers of polycaprolactone/collagen as a sustained-release drug delivery system for artemisinin." *Pharmaceutics*, 13(8), 1228.

Kang, H., Rho, S., Stiles, W. R., Hu, S., Baek, Y., Hwang, D. W., ... & Choi, H. S. (2020). "Size-dependent EPR effect of polymeric nanoparticles on tumor targeting." *Advanced healthcare materials*, 9(1), 1901223.

Lahooti, B., Akwii, R. G., Zahra, F. T., Sajib, M. S., Lamprou, M., Alobaida, A., ... & Mikeleis, C. M. (2023). "Targeting endothelial permeability in the EPR effect." *Journal of Controlled Release*, 361, 212–235.

Li, W., Zhang, Y., Kankala, R. K., Zou, L., & Chen, Z. (2022). "Antibody and cellular-based therapies for pediatric acute lymphoblastic leukemia: mechanisms and prospects." *Pharmacology*, 107(7-8), 368–375.

MacDermott, C. P. (1997). *Selecting Thermoplastics for Engineering Applications* (2nd ed.). CRC Press. (Kitap)

Malathi, S., Balashanmugam, P., Devasena, T., & Kalkura, S. N. (2021). "Enhanced antibacterial activity and wound healing by a novel collagen blended ZnO nanoparticles embedded niosome nanocomposites." *Journal of Drug Delivery Science and Technology*, 63, 102498.

Malik, M., Ali, Z., Khan, S., Zeb, A., ud Din, F., Alamri, A. H., & Lahiq, A. A. (2024). "TPGS-PLA nanomicelles for targeting lung cancer; synthesis, characterization, and in vitro antitumor efficacy." *Journal of Drug Delivery Science and Technology*, 91, 105238.

Mittal, G., Sahana, D. K., Bhardwaj, V., & Kumar, M. R. (2007). "Estradiol loaded PLGA nanoparticles for oral administration: effect of polymer molecular weight and copolymer composition on release behavior in vitro and in vivo." *Journal of controlled release*, 119(1), 77–85.

Mora-Huertas, C. E., Fessi, H., & Elaissari, A. (2010). "Polymer-based nanocapsules for drug delivery." *International Journal of Pharmaceutics*, 385(1-2), 113–142.

Odian, G. (2004). *Principles of polymerization*. John Wiley & Sons.

Oliveira Silva, R., Counil, H., Rabanel, J. M., Haddad, M., Zaouter, C., Ben Khedher, M. R., ... & Ramassamy, C. (2024). "Donepezil-Loaded Nanocarriers for the Treatment of Alzheimer's Disease: Superior Efficacy of Extracellular Vesicles Over Polymeric Nanoparticles." *International Journal of Nanomedicine*, 19, 1077–1096.

Pathak, Y., & Thassu, D. (2009). "Drug delivery nanoparticles formulation and characterization." *Drugs and Pharmaceutical Science Series*. Informa healthcare USA, 67,1-30

Pawar, R., Pathan, A., Nagaraj, S., Kapare, H., Giram, P., & Wavhale, R. (2023). "Polycaprolactone and its derivatives for drug delivery." *Polymers for Advanced Technologies*, 34(10), 3296–3316.

Williams, R. L., & Shao, Z. (2006). Bioavailability and bioequivalency testing. In J. P. Remington (Ed.), Remington: The Science and Practice of Pharmacy (21st ed., pp. 1037–1045). Lippincott Williams & Wilkins

Saldívar-Guerra, E., & Vivaldo-Lima, E. (2013). "Introduction to polymers and polymer types." *Handbook of polymer synthesis, characterization, and processing*, 1–14.

Sánchez, A., Mejía, S. P., & Orozco, J. (2020). "Recent advances in polymeric nanoparticle-encapsulated drugs against intracellular infections." *Molecules*, 25(16), 3760.

Satchanska, G., Davidova, S., & Petrov, P. D. (2024). "Natural and Synthetic Polymers for Biomedical and Environmental Applications." *Polymers*, 16(8), 1159.

Shapira-Furman, T., & Domb, A. J. (2024). "Insulin Extended Release from PLA-PEG Stereocomplex Nanoparticles." *Macromolecular Bioscience*, 24(5), 2300497.

Shenoy, A. (1997). *Thermoplastic Melt Rheology and Processing* (1st ed.). CRC Press.

Shuai, C., Yang, W., Feng, P., Peng, S., & Pan, H. (2021). "Accelerated degradation of HAP/PLLA bone scaffold by PGA blending facilitates bioactivity and osteoconductivity." *Bioactive Materials*, 6(2), 490–502.

Siddiqui, M. A., Asad, M., Akhter, J., Hoda, U., Rastogi, S., Arora, I., ... & Samim, M. (2022). "Resveratrol-loaded glutathione-coated collagen nanoparticles attenuate acute seizures by inhibiting HMGB1 and TLR-4 in the hippocampus of mice." *ACS Chemical Neuroscience*, 13(8), 1342–1354.

Singh, R., & Lillard Jr, J. W. (2009). "Nanoparticle-based targeted drug delivery." *Experimental and molecular pathology*, 86(3), 215–223.

Sionkowska, A. (2021). "Collagen blended with natural polymers: Recent advances and trends." *Progress in Polymer Science*, 122, 101452.

Sparreboom, A., Scripture, C. D., Treu, V., Williams, P. J., De, T., Yang, A., ... & Desai, N. (2005). "Comparative preclinical and clinical pharmacokinetics of a cremophor-free, nanoparticle albumin-bound paclitaxel (ABI-007) and paclitaxel formulated in Cremophor (Taxol)." *Clinical cancer research*, 11(11), 4136–4143.

Sun, Y., Zhao, Y., Teng, S., Hao, F., Zhang, H., Meng, F., … Teng, L. (2018). "Folic acid receptor-targeted human serum albumin nanoparticle formulation of cabazitaxel for tumor therapy." *International Journal of Nanomedicine*, 14, 135–148.

Sunazuka, Y., Ueda, K., Higashi, K., Wada, K., & Moribe, K. (2024). "Mechanistic Analysis of Temperature-Dependent Curcumin Release from Poly (lactic-co-glycolic acid) / Poly (lactic acid) Polymer Nanoparticles." *Molecular Pharmaceutics*, 21(3), 1424-1435.

Tyler, B., Gullotti, D., Mangraviti, A., Utsuki, T., & Brem, H. (2016). "Polylactic acid (PLA) controlled delivery carriers for biomedical applications." *Advanced drug delivery reviews*, 107, 163–175.

U.S. Food and Drug Administration. (n.d.). "Considering whether an FDA-regulated product involves the application of nanotechnology." Retrieved July 1, 2024, from https://www.fda.gov/regulatory-information/search-fda-guidance-documents/considering-whether-fda-regulated-product-involves-application-nanotechnology.

Ünal, S., Aktaş, Y., Benito, J. M., & Bilensoy, E. (2020). «Cyclodextrin nanoparticle bound oral camptothecin for colorectal cancer: Formulation development and optimization.» *International Journal of Pharmaceutics*, 584, 119468.

Varan, G., Varan, C., Erdoğar, N., Hıncal, A. A., & Bilensoy, E. (2017). "Amphiphilic cyclodextrin nanoparticles." *International Journal of Pharmaceutics*, 531(2), 457–469.

Vecsernyés, M., Fenyvesi, F., Bácskay, I., Deli, M. A., Szente, L., & Fenyvesi, É. (2014). "Cyclodextrins, blood–brain barrier, and treatment of neurological diseases." *Archives of medical research*, 45(8), 711–729.

Verma, A., & Stellacci, F. (2010). "Effect of surface properties on nanoparticle–cell interactions." *small*, 6(1), 12–21.

Watcharin, W., Schmithals, C., Pleil, T., Köberle, V., Korkusuz, H., Huebner, F., ... & Kreuter, J. (2014). "Biodegradable human serum albumin nanoparticles as contrast agents for the detection of hepatocellular carcinoma by magnetic resonance imaging." *European Journal of Pharmaceutics and Biopharmaceutics*, 87(1), 132–141.

Xu, Y., Shrestha, N., Préat, V., & Beloqui, A. (2020). "Overcoming the intestinal barrier: a look into targeting approaches for improved oral drug delivery systems." *Journal of controlled release*, 322, 486–508.

Yao, M., Wu, M., Yuan, M., Wu, M., Shen, A., Chen, Y., ... & Peng, J. (2024). "Enhancing the therapeutic potential of isoliensinine for hypertension through PEG-PLGA nanoparticle delivery: A comprehensive in vivo and in vitro study." *Biomedicine & Pharmacotherapy*, 174, 116541.

Yoda, R. (1998). "Elastomers for biomedical applications." *Journal of Biomaterials Science, Polymer Edition*, 9(6), 561–626.

Zhang, J., Song, C., Han, Y., Xi, Z., Zhao, L., Cen, L., & Yang, Y. (2020). "Regulation of inflammatory response to polyglycolic acid scaffolds through incorporation of sodium tripolyphosphate." *European Polymer Journal*, 122, 109349.

Zhang, X., Li, W., Dou, X., Nan, D., & He, G. (2020). "Astaxanthin encapsulated in biodegradable calcium alginate microspheres for the treatment of hepatocellular carcinoma in vitro." *Applied biochemistry and biotechnology*, 191, 511–527.

7 Colonic Drug Delivery Systems as Multiunit Potential

Therapeutic Strategies and Opportunities

Mazen Al-Mohaya[1,2] *and Burcu Demiralp*[2]

[1]Institute of Health Sciences, Istanbul University, Istanbul, Türkiye

[2]Pharmaceutical Technology Department, Faculty of Pharmacy, Istanbul University, Istanbul, Türkiye

7.1 INTRODUCTION

The field of colonic drug delivery systems is currently undergoing a significant resurgence, presenting promising prospects for targeted and effective treatments. This resurgence is propelled by a large number of recently identified pharmaceutical benefits and opportunities [1]. Direct delivery of medications to the colon provides numerous advantages, such as improved treatment for local diseases, targeted therapeutic options, reduced systemic drug exposure and associated toxicity, and enhanced bioavailability [2–4]. Initially, this approach was mainly used for treating local conditions, including inflammatory bowel disease (IBD) and colorectal cancer [5,6]. These systems are designed to concentrate drugs at the target site and minimize their systemic presence, offering significant enhancements in the management of local diseases [7,8]. In addition, with an enhanced comprehension of the colon and rectal medium, novel local targets, including the microbiome, enteric immune system, and lymphatic system, have been identified [9,10]. The identification of these promising targets not only contributes to the advancement of colonic disorder management but also provides opportunities to address systemic conditions and diseases that impact distant organs [11–13].

As global living standards improve, access to a wide variety of foods has increased. Nonetheless, a considerable portion of the population lacks an understanding of balanced nutrition and tends to overconsume processed foods. The increase in the digestive system relative to health problems has resulted in more than 1.93 million new instances of colorectal cancer on a global scale in 2020, according to data from the International Agency for Research on Cancer (IARC). 9.7% of all recently diagnosed cancer cases can be attributed to this type [14]. The spectrum of diseases known as IBD, which involves Crohn's disease (CD) and ulcerative colitis (UC), represents a prevalent inflammatory disorder of the digestive tract. While initially more frequently reported in Europe and North America, there is an increasing trend of IBD cases being identified in areas such as South America, Eastern Europe, Africa, and Asia. Globally, an estimated 3.9 million women and 3 million men are impacted by IBD, presenting significant challenges to public health frameworks and the worldwide economy [15–17].

The exact triggers of gastrointestinal diseases remain unclear, likely involving a complex interplay of factors, including environmental influences, genetic variations, and enzymatic factors. However, dietary imbalance is widely recognized as a risk factor contributing to damage in the gastrointestinal microenvironment and impairment of the gastrointestinal barrier. The initial example of colonic drug delivery can be traced back to the introduction of sulfasalazine in 1941, a prodrug formulated for the management of IBD [18]. This compound comprises 5-aminosalicylic acid (5-ASA, commonly referred to as mesalamine or mesalazine), linked to sulfapyridine via an azo linkage [19]. In the colon, bacterial enzymes break down this bond, releasing 5-ASA to locally address the inflammation. Despite its longstanding use in treating IBD, nearly half of the patients experience allergic reactions or other adverse side effects from sulfasalazine [20]. These undesirable reactions have been linked to its sulfapyridine moiety [21], leading to the creation of new 5-ASA-based prodrugs such as olsalazine and balsalazide [22,23]. The prodrug concept continues to be a cornerstone in colonic medication delivery; for instance, an azo-linked prodrug of tofacitinib has displayed efficacy in a mouse IBD model [24]. Moreover, designing drugs that remain within the gastrointestinal tract, particularly peptides, could ensure localized action in the colon while minimizing systemic exposure [25]. Nonetheless, crafting such prodrugs and gut-specific treatments requires precise tailoring for each medication and might involve extensive regulatory scrutiny [26].

Oral dosage forms designed for colonic drug delivery include those that specifically release medication in the colon and rectally administered treatments. In 1982, a pioneering formulation targeted to the ileum, as well as the proximal colon via oral administration, was unveiled. This formulation showcased a pH-responsive methacrylate polymer known as Eudragit® S [27]. This synthesized copolymer, composed of poly(methacrylic acid) and methyl methacrylate in a 1:2 ratio, dissolves at intestinal

DOI: 10.1201/9781003508601-7

pH levels above 7.0 [27], making it suitable for use as an enteric coating that releases drugs in the terminal ileum. Its application is common in numerous commercial drug products, especially those delivering mesalazine for IBD management, including Asacol® and Lialda®/Mezavant®. Nevertheless, there have been concerns regarding the reliability of pH-responsive polymers for drug release in the colon [1,27,28]. Clinical studies have indicated inconsistent dissolution patterns with enteric coatings, with some dissolving prematurely or not at all in certain patients [29,30]. Such variability is due to fluctuations in intestinal pH and can result in some patients not receiving their full medication dose. Additionally, the altered physiology in IBD may further complicate the effectiveness of enteric coating, as these patients often have lower intestinal pH levels compared to healthy individuals [30].

In this chapter, we turn our focus to the development of formulation-driven methods for delivering drugs to the colon. It is important to note that a versatile formulation could be key in enabling the colonic administration of a diverse range of pharmaceuticals. Additionally, creating modified-release versions of drugs that have already received approval could streamline the market entry process, as it may eliminate the need for repetitive studies on established drug characteristics, such as toxicity or pharmacodynamics [26]. This discussion will cover the latest progress in various formulation strategies aimed at colonic drug delivery and examine their use in pharmaceutical practices.

7.2 ROUTE OF ADMINISTRATION FOR COLONIC DRUG DELIVERY SYSTEMS

The treatment of colonic ailments often involves established methods such as intravenous injections (for example, administering conventional medications like aminosalicylic acid and glucocorticoids), subcutaneous injections (for instance, delivering biologics like vedolizumab and ustekinumab), and rectal approaches (utilizing treatments such as mesalamine, hydrocortisone suppositories, and glucocorticoid enemas) [31–34]. Despite their widespread use, these traditional therapies have notable drawbacks. Intravenous drug delivery can lead to systemic distribution, which diminishes the precision of targeting specific areas. Treatments via subcutaneous injections can quickly reach the affected region but may cause severe skin irritation and other issues, such as the alteration of subcutaneous fat and the development of nodules with repeated use [34]. Rectal methods are often considered inconvenient and typically require assistance from another person. Therefore, oral delivery has become a favored approach due to its ease and speed of administration. Nonetheless, oral colonic drug delivery systems (OCDS) face several hurdles in ensuring targeted drug release within the colon amidst the complexities of the digestive system.

However, rectal drug delivery presents benefits over colonic delivery via the oral route, as it bypasses the upper GI tract, resulting in more predictable drug release and

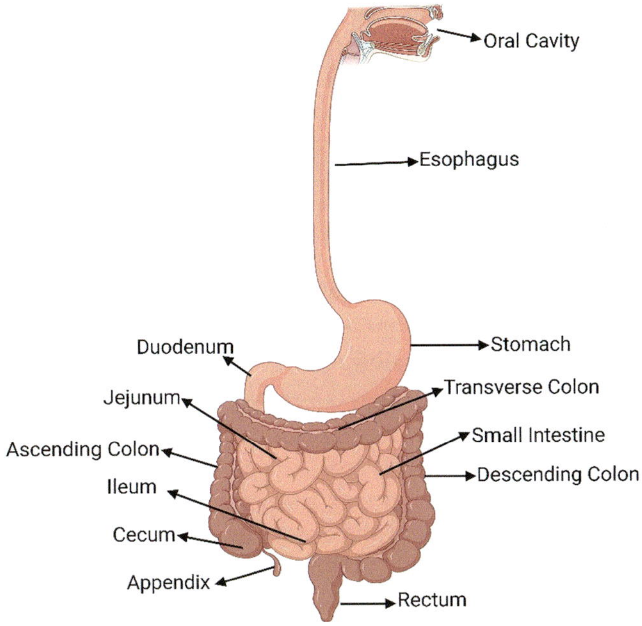

FIGURE 7.1 Gastrointestinal anatomy. (According to [42].)

absorption. The stable rectal environment, especially in a fasting state, enhances the predictability of drug delivery and absorption [35]. The rectal route can provide benefits for drugs with unfavorable taste profiles, which may cause gastrointestinal irritation, instability in the upper gastrointestinal tract, or significant hepatic first-pass metabolism [36]. This route may also benefit individuals with compromised safe swallowing, such as those with dysphagia, unconsciousness, or advanced age. Rectal formulations are often met with reduced acceptance compared to oral formulations, primarily due to cultural perceptions and challenges associated with administration. Their primary suitability lies in the treatment of pathologies in the distal colon and rectum, although there is ongoing exploration of their potential for systemic applications such as vaccines [37–39].

The oral administration of OCDS entails traversing through multiple stages, encompassing the oral cavity and the gastrointestinal tract (GIT), and ultimately arriving at the colon, as presented in Figure 7.1. However, prior to delivering the colon, OCDS confronts various factors that do not directly influence the targeted release in situ [40,41].

7.3 FACTORS INFLUENCING ORAL COLONIC DRUG DELIVERY SYSTEM

7.3.1 pH

The digestive system comprises a complex network of organs and tissues subjected to various pH levels. The stomach is located within an environment characterized by high acidity, with a pH (1.5–3.5) [43]. In the acidic environment of the stomach, the activation of gastric amylase, monosaccharides, and disaccharidase triggers the secretion of pepsin and stimulates the release of digestive juices (pancreatic juice, bile juice, and small intestinal juice). The duodenum is positioned at the superior part of the small intestine and is connected to the stomach. It functions as the smallest and widest canal within the small intestine. The organ principally receives gastric juice as well as pancreatic and bile juices, thereby sustaining a moderately acidic environment with a pH of 5.4–6.1. During transit through the lower part of the small intestine, the pH increases steadily to approximately 6.5, while the regional pH within the colon ranges from 6.4 to 7.4 [43,44].

7.3.2 Gastric Emptying and Emptying Time

The mechanism by which food transitions from the stomach to the duodenum is known as gastric emptying. This process typically commences around 5 min post ingestion. Gastric emptying occurs when the stomach's internal pressure exceeds that of the duodenum, propelling chyme into it and consequently slowing gastric motility by activating certain receptors in the intestinal lining. The entry of acidic content into the duodenum and its subsequent neutralization by chyme gradually lessens its suppressive impact on stomach activity, allowing for an increase in gastric motions until additional chyme is introduced into the duodenum [45]. The duration of this process, referred to as gastric emptying time, varies with different materials but is generally between 2 and 6 h. Solids tend to remain in the stomach longer than liquids, and larger chunks take more time to empty than smaller granules. Once in the duodenum, it takes about 25 cm and roughly 3–4 h for chyme to advance to the jejunum and ileum. The chyme that follows from the initial emptying then proceeds to the colon [46,47]. Therefore, an effective oral colonic drug delivery system should account for a transit time of 3–8 h post-gastric emptying to reach the colon. Ideally, such a system would be designed to release drugs 4 h after stomach entry, showcasing a time-sensitive release mechanism. However, since foods of varying volumes and physicochemical properties have different gastric emptying rates, relying solely on time-sensitive carriers may lead to premature or delayed breakdown, which does not align with the requirements of OCDS. Thus, time-sensitive OCDS are often designed with a blend of time-controlled-release mechanisms and other carrier types to achieve accurate colon targeting [48].

7.3.3 Gastrointestinal Mucosal Barrier

The gastrointestinal mucosal barrier is composed of a tightly bound membrane of epithelial cells, serving as a critical boundary between the internal and external environments of the gastrointestinal tract. This barrier is enveloped in a significant layer of mucus, secreted by various sources, including epithelial cells and glands found in the oxyntic, gastric cardia, and pyloric regions [49]. The mucus's strong adhesive quality forms a protective film, approximately 0.5 mm thick, over the chyme surface, providing lubrication and safeguarding the epithelial lining from mechanical damage. Scientists have utilized the commonly negative charge of the intestinal mucosa surface to create positively charged nanocarriers that preferentially bind to the colon [50,51]. Notably, anionic nanoparticles, while unable to directly bind with the negative charge of the intestinal mucosa, demonstrate the capability to permeate through the mucus owing to diminished electrostatic repulsion [52].

7.3.4 Colonic Microbiota

The colonic microbial ecosystem is the most extensive and intricate within the human body, containing around 400–500 different types of microorganisms, collectively known as the gut microbiota [53]. A variety of natural bioactive compounds are known to interact with the gut microbiota or their secreted enzymes, through processes like anaerobic fermentation, enzymatic breakdown, and metabolic reactions. These interactions are vital for preserving colonic health and preventing the onset of colonic diseases [31,54]. Such compounds can be directly absorbed through the intestinal lining or may be transformed and metabolized by the gut microbiota before absorption [55,56].

7.4 ASSESSMENT OF COLONIC

7.4.1 Basic Anatomy

Understanding the distinct physiological characteristics of the colon within the gastrointestinal tract (GIT) is crucial for effectively targeting drug delivery to this region. The colon spans approximately 1.6 m on average in the human GIT, with slight variations between males (1.66±0.36 m) and females (1.55±0.29 m) [57,58]. The colon is divided into proximal and distal segments: the proximal colon includes the cecum, ascending colon, and transverse colon, while the distal colon consists of the descending colon, sigmoid colon, rectum, and anus. The hepatic flexure connects the ascending and transverse colon, while the splenic flexure connects the transverse and descending colon. Compared to the small intestine, the colon has a wider diameter, ranging from 1.87 cm at the terminal ileum to approximately 4.8 cm at the widest point in the caecum [57], narrowing toward the sigmoid colon, and widening again at the rectum [59]. The cecum, transverse colon, and sigmoid colon are mobile and suspended by mesentery in the peritoneal cavity, while the ascending colon, descending colon, and rectum are immobile and positioned behind the peritoneum [57]. The colon is responsible for several essential physiological functions such as the absorption of water, minerals, and vitamins, compaction of feces, digestion of polysaccharides, and regulation of intestinal immunity [10,60]. The appendix is typically connected to the caecum near the ileocaecal junction, approximately 1–2 cm below it [59]. Recent studies have challenged the traditional view of the appendix as a vestigial structure, highlighting its role as a reservoir for colonic microbiota and potentially serving as a niche for the enteric immune system [61,62]. This evolving understanding underscores the importance of considering the appendix in strategies targeting the colon for therapeutic interventions.

7.4.2 Transit Time

The time required for drugs to reach the colon depends on their journey across the superior GIT. This process is influenced by several elements, involving the properties of the drug form, the stage of peristalsis, and simultaneous food consumption. Without food, liquid medications typically leave the stomach in 10–20 min, but this time can increase significantly for larger solid forms [58]. The ingestion of food, especially meals rich in fat, can notably postpone the stomach's release of drugs [63]. Stomach transit times vary among individuals, with pre-menopausal women, particularly during their menstrual cycle's luteal phase, showing considerably more extended periods than post-menopausal women and men [64]. Illnesses can extend these times further; for example, individuals with type 2 diabetes may have gastric transit times up to three times longer than healthy persons [65]. These differences are crucial when creating formulations aimed at the colon since the anticipated stomach residence time of the dosage is often determined by the characteristics of the patient group being targeted. On average, the transit time for medications to traverse the small intestine is approximately 3–4 h, though there is significant variation among and within individuals [66]. The Migrating Motor Complex (MMC) governs small intestine movement in a fasting state with rhythmic contractions. When fed, more regular contractions occur to mix intestinal contents for digestion and nutrient absorption, leading to quicker movement of drugs through the small intestine [66,67].

The caecum marks the colon's beginning and links to the terminal ileum via the ileocaecal valve. This valve controls the flow of ileal content into the colon and blocks colonic bacteria from entering the ileum [68]. The ileocaecal valve relaxes when the terminal ileum expands, permitting digested substances and drug forms to move into the caecum. Conversely, when the caecum expands, the valve tightens to stop contents from moving backward [69]. Drug movement through the colon is orchestrated by smooth muscle contractions and relaxations. Rare but potent neurogenic contractions occur 6–20 times daily in healthy people, propelling contents along extensive sections of the colon [63,66,70]. These contractions are thought to be triggered by colonic distension that activates mechanosensitive nerves in smooth muscles. Additionally, pacemaker cells within the deep circular muscle of the colonic wall produce slower but more frequent myogenic contractions (2–6 times/minute), primarily in the lower colon. The coordinated contractions are facilitated through the collaboration of these contractions with propulsive forces, resulting in the formation of clusters of synchronized contractions [71].

The journey of a drug form through the colon is impacted by a multitude of pharmaceutical and physiological factors, including the patient's gender (including menstrual cycle phases), age, overall health, the composition of their microbiome, and the characteristics of the drug form itself [64,72]. Liquids tend to navigate through the colon faster than solids because they can trigger more robust neurogenic contractions [71]. In contrast, larger solid forms may pass through more swiftly than smaller ones, as the latter might get caught in the mucosal creases [63]. Electromagnetic tracking in a study of 111 healthy adults revealed that an $8.3 \times 21.5\ mm^2$ capsule had an average colonic transit time of 21 h, although individual times varied widely. This research confirmed previous findings that women generally experience longer transit times in the transverse and descending sections of the colon. Additionally, it was found that aging correlates with increased total colonic and entire gut transit times. Such data are crucial for the design of colon-specific therapeutic agents and could lead to personalized drug formulations for distinct patient demographics [73].

7.4.3 Vascular and Lymphatic System

The colon is mainly supplied with blood loaded with oxygen from the superior and inferior mesenteric blood vessels, along with their branches [59]. Deoxygenated blood originating from the colon is transported via the superior and inferior mesenteric blood vessels, which conjoin to form the splenic blood vessel, and thereafter, connect with the hepatic portal vein. The venous pathway is of the highest importance in the improvement of colon-targeted therapies. It is essential to consider that drugs absorbed into the bloodstream via the colon undergo hepatic metabolism before reaching other organs [69]. Therefore, drugs intended for systemic effects through colon targeting must either resist hepatic metabolism or undergo metabolic activation to maintain their bioavailability. However, when pharmaceuticals are administered directly to the lower rectum, they bypass hepatic metabolism due to the venous drainage of the distal rectum merging with the iliac vein, which in turn connects to the inferior vena cava, facilitating direct transport of blood to the right atrium of the heart [35].

Beyond liver metabolism, drugs administered to the colon can also enter the bloodstream through the mesenteric lymph nodes (MLNs), a pathway that is particularly relevant for lipophilic medications in lipid-rich formulations [74]. The lymphatic system that surrounds the colon runs parallel to its blood vessels and is responsible for removing chyle, antigens, interstitial fluid, and plasma proteins from the colonic tissues [59]. Lymph from the initial part of the colon travels from the MLNs to the intestinal lymphatic trunk and subsequently to the bloodstream via the thoracic duct, internal jugular, and left subclavian veins. The MLNs are essential for regulating the immune functions of the colon. They facilitate tolerogenic and inflammatory reactions through the adaptive immune system, which operates distinctly compared to the small intestine [75,76]. The immune systems in the small intestine and colon are different. This enables targeted adjustments to immune responses in these specific areas of the gut.

7.4.4 The Colonic Epithelium

The colonic epithelium exhibits distinct differences from that of the small intestine. Unlike the small intestine, which has villi, the colon does not; however, it does possess microvilli on the apical surface of its epithelial cells. The mucosa of the colon is irregularly folded, and together with colonic crypts, it expands the surface area by 10–15 folds more than if it were a flat tube [1,58]. Nevertheless, drug permeability in the colon is typically lower than in the small intestine. Drugs classified as Biopharmaceutics Classification System (BCS) class I generally show higher absorption rates than those in BCS classes III and IV due to the reduced surface area [77,78]. The columnar epithelial cells of the colon are arranged in a single layer and joined by tight junctions. The permeability of these junctions is influenced by the expression of claudins, with notably higher permeability observed in the descending colon compared to the ascending colon [79,80].

Colonocytes, neuroendocrine cells, goblet cells, and immunoregulatory cells make up the colonic epithelium and are regenerated every 4–5 days by intestinal stem cells located in the crypts [81,82]. Immune cells such as microfold (M) cells and tuft cells play roles in immune tolerance and open avenues for innovative vaccines and treatments for immune-related conditions like IBD [83,84]. In contrast to the small intestine, the colon has fewer drug transporters and metabolic enzymes. There's also a variation in transporter expression between the ascending and descending parts of the colon [79].

A double layer of mucus coats the colonic epithelium, acting as a barrier between the microbiota and epithelial cells while facilitating chyme movement but hindering drug absorption [57]. This mucus layer is predominantly made up of water and MUC2 mucin. It's continuously regenerated every 24–48 h and its thickness varies across different colonic regions [57]. Drug absorption through the colonic epithelium predominantly occurs through transcellular diffusion, passive paracellular transport, and active transcellular transport [35]. The absorption route and degree are impacted by factors, for instance, drug pKa, lipophilicity, and the pH level of the gastrointestinal tract [80].

7.4.5 Mucosal Fluid and Gas

The composition of the luminal fluid within the GIT is pivotal for oral drug administration, offering both advantages and obstacles [1]. Gastric fluid in the stomach varies in pH from 0.4 to 4.5, which can promote the dissolution of alkaline drugs but might deactivate specific treatments such as peptides and probiotics [64,85]. Moving into the duodenum, the pH ascends to 5.0–7.0 due to bicarbonate secretion [58] and persists to increase in the pH of the jejunum to 6.63 and the pH of the ileum to 7.49. There's notable variability in luminal pH between individuals, which affects drug solubility and allows for targeted gastrointestinal release [86]. The fluid within the ascending colon typically demonstrates a lower pH in comparison to the ileocecal region due to the presence of lactate and short-chain fatty acids (SCFAs) produced by colonic bacteria. Additionally, this fluid exhibits greater acidity than that found in the distal colon, where the production of lactate and SCFAs is considerably lower [1,29,87]. Colonic fluid is a complex blend of water, chyme, microbiota, electrolytes, lipids, proteins, bile acids, and SCFAs, among other metabolites [88], with its composition influenced by dietary absorption and the specific colonic region sampled. While direct in situ pH measurement is deemed more accurate than ex vivo approaches, inconsistencies in measurement methods pose a challenge and underscore the necessity for additional research to deepen our comprehension of the colonic milieu for drug delivery [89]. The volume of colonic fluid significantly influences drug bioavailability, averaging approximately 560 mL in fasted healthy adults but fluctuating post-meal and throughout the colon due to water reabsorption [90].

The majority of colonic fluid is generally bound with microbiota or chyme, which renders it inaccessible for drug interaction [1]. The amount of free colonic fluid is minimal yet varies greatly among individuals, influencing drug dissolution and uptake [90]. Clinical investigations indicate that drugs with low solubility may experience suboptimal absorption in the colon due to scarce free fluid availability [78].

Gas pockets within the colon (100–300 mL), shaped by factors like microbiome composition, diet, and transit time, can also affect drug delivery. The colonic lumen is an anaerobic space with oxygen levels diminishing along its length. Predominant gases contain CO_2, H, NH_3, N, CH_4, and S gases [91]. The ascending colon is characterized by significantly low oxygen levels, measuring approximately 11 mmHg (about 2%). In contrast, the sigmoid colon exhibits 3 mmHg (about 0.4%), thereby creating a gradient that diminishes as it progresses toward the rectum [92]. The colonic epithelial cells and bacteria have acclimated to the hypoxic conditions, thus accentuating the distinctive environment confronted by pharmaceutical agents within the colon. A thorough understanding of colonic fluid composition and behavior is crucial for developing effective drug delivery systems tailored to specific areas of the gastrointestinal tract [93].

7.4.6 The Microbiota

The colon is home to a vast and varied microbial ecosystem, including bacteria, fungi, viruses, and archaea, along with numerous enzymes and metabolites. It's estimated that the bacterial population in a person's colon can vary from 10^{10} to 10^{12} cells per mL of colonic fluid, with a genetic encoding capacity that exceeds the human genome by 150 times, encompassing potentially over 1,000 species [94,95]. The makeup of this microbiome differs throughout the colon, influenced by variables such as oxygen partial pressure, pH levels, bile acid concentrations, nutrient presence, immune responses, and the rate of passage [96]. The predominant bacterial phyla identified in the colon are Firmicutes, Bacteroidetes, Actinobacteria, and Proteobacteria [10]. The diversity of microbes escalates from the proximal to the distal colon but then declines near the rectum, with particular species exerting more dominance in specific areas [35]. Bacteria that ferment dietary fibers into lactic acid and SCFAs are predominantly located in the proximal colon. While our understanding of the distribution of viruses and fungi is limited, certain bacteriophages and *Candida albicans* are acknowledged as familiar residents [97,98]. The microbiome's metabolic activities can notably influence drug pharmacokinetics (PK), thereby affecting their therapeutic effectiveness [94,99,100]. The gut microbiota can deplete drug concentrations directly or chemically modify them, which may impact both drug efficacy and safety. Developers of colon-targeted drugs need to account for potential microbiome-mediated drug depletion effects [100,101]. Although microbial degradation of drugs in the rectum is typically considered minimal compared to upstream colon regions, the rectum still presents a potential opportunity for localized drug administration [35]. Moreover, drugs can alter the composition and function of the colonic microbiota. Antibiotics particularly have profound effects on the gut microbiome that persist well beyond the treatment period [102,103]. Understanding the intricate interactions between drugs and the microbiome is vital for developing effective treatments targeting the colon [104]. Medications that adversely affect the microbiome might be less appropriate for colon delivery; conversely, those that exert beneficial impacts could open new avenues for therapy [104].

7.5 COLONIC DRUG DELIVERY APPROACHES

7.5.1 Innovative Multi-Trigger Systems Developed for Efficient Delivery of Drugs to the Colon via the Oral Route

Strategies for the formulation of colonic drug delivery have undergone evolution to address the myriad challenges presented by the diversity of gastrointestinal (GI) conditions exhibited among patients. Instead of relying on a single physiological stimulus, such as pH, transit time, or microbial enzymes [1], recent approaches focus on combining multiple release mechanisms based on different stimuli [105]. These combination systems can be classified as either sequential or parallel trigger systems. Moreover, sequential trigger systems comprise multiple layers of matrixes. In this configuration, drug release occurs sequentially from the external layer to the core, with each layer being activated by a specific physiological stimulus [106]. In contrast, parallel systems incorporate multiple drug release mechanisms into a single layer, allowing for the simultaneous activation of independent triggers [105]. This parallel approach ensures that if one mechanism fails as a consequence of physiological change, drug release can still occur via other mechanisms, making it a more reliable strategy [105,107]. Tablets, capsules, and pellets with multi-stimuli coatings have demonstrated efficacy in delivering drugs to the colon. There is an ability for the development of multi-stimuli colon-targeted nanomedicines [108,109], although they are currently infrequent and have not exhibited significant efficacy in human studies. Safety considerations are of paramount importance in the development of new formulations. Various materials, such as Eudragit® polymers (pH-triggered release), hydroxypropyl methylcellulose (HPMC, time-triggered release), and polysaccharides (microbiota-triggered release), have been classified as generally recognized as safe (GRAS) by regulatory agencies like the Food and Drug Administration (FDA) for colonic delivery [110]. In order to achieve efficient systemic delivery of biologics, it may be necessary to integrate permeation enhancers into formulations to facilitate passing through the colonic epithelium [111–113]. There are currently approved products available for the oral delivery of peptides to produce systemic effects, exemplified by Rybelsus® and Mycapssa® [25]. Briefly, advanced parallel systems incorporate multiple independent release mechanisms, thereby enhancing the probability of effective colon targeting even amidst variations in gastrointestinal conditions due to patient- or disease-specific factors. It is crucial to consider the specific requirements of each colonic drug delivery project whenever developing a multi-trigger system.

7.5.2 pH-Triggered with Microbiota-Triggered Systems

7.5.2.1 Phloral®

The Phloral® framework represents a milestone in colonic drug delivery, being the pioneering dual-triggered technology to hit the market. The single-layer coating system is meticulously crafted by blending Eudragit® S with starch that resists digestion, which includes amylopectin and amylose [114,115]. Thanks to their distinct triggering mechanisms, these ingredients synergize, offering a backup in case one mechanism fails [105]. Eudragit® S plays a dual role: preserving tablet integrity during its journey via the stomach and small intestine while also acting as a structural support for the starch, regulating its swelling. Conversely, resistant starch acts as a carrier for colonic microbiota, stepping in to trigger drug release should Eudragit® S fail to reach its critical pH threshold. Undoubtedly, Phloral® has demonstrated noteworthy efficacy in treating IBD, regardless of patients' feeding conditions [105,116]. Its success also extends to treating Clostridioides difficile disease [4] as well as fighting obesity [3].

7.5.2.2 OPTICORE™

The OPTICORE™ system, derived from OPTImised COlonic RElease, represents an advanced dual-layered drug delivery system formed to efficiently release medications in the ileocolonic area, where the amount of fluid in this area is significantly higher than in the middle to lower part of the colon [117,118]. The innovative system encompasses two coating layers. This includes an inner layer containing an impartial enteric polymer (Eudragit® S) blended with a salt for buffering purposes, enclosed by a surface Phloral® coat [118]. Because of the alkaline nature of the base layer, a supplemental HPMC layer may be needed to isolate acidic drugs from the alkaline base layer. The OPTICORE™ system facilitates quick drug release in the ileocolonic region by enabling the quick dissolution of the Phloral® coating. As the formulation progresses through the lower GIT, pores develop in the coating of Phloral®, allowing luminal fluid to enter and leading to the subsequent dissolution of the enteric layer. The dissolution of the Phloral® layer results in an elevation of ionic strength, pH, and buffer capacity on the inner surface, thereby facilitating the rapid ionization and dissolution of the Eudragit® S component [119,120]. Significantly, the OPTICORE™ system has demonstrated efficacy in treating IBD by delivering 5-ASA selectively to sites of colonic inflammation [107]. It serves as the colon-targeted delivery mechanism in Asacol™ 1,600 mg, an IBD medication approved in Europe after successful clinical trials (Phase III) [121]. Particularly, these multi-stimuli systems enable colonic delivery of 5-ASA with 1,600 mg, the highest dose approved and intended for oral administration. Given the acidic nature of 5-ASA, the buffering action of the OPTICORE™ foundational level speeds up its release via the Phloral® layer, ensuring rapid colonic delivery. This capability to administer significant doses of 5-ASA decreases the medication burden on patients. Additionally, OPTICORE™ was studied for addressing Clostridioides difficile infection. It achieves this by delivering metronidazole benzoate to the ileocolonic region, which provides the optimal acidity for the drug's effectiveness [122].

7.5.3 pH-Triggered and Time-Triggered Systems

The pH- and drug delivery systems that depend on time for their combination, though less common, have been explored in several studies. For instance, one study coated 5-ASA pellets with a blend of two Eudragit® S and L (enteric polymers) alongside an Eudragit® RS(time-dependent polymer), with or without inulin [123]. Subsequent in vivo X-ray imaging verified that the tablets maintained their structural integrity throughout the GIT [124]. Researchers have implemented a dual-coating approach involving the application of an alkaline aqueous solution and an organic solution of Eudragit® S. The inner layer is composed of an alkaline aqueous solution containing buffering agents, while the outer layer is coated with an organic solution [125]. This method enhanced drug dissolution at pH levels above 7 and demonstrated more invariant tablet disintegration in the lower intestinal tract during human trials [121]. Microspheres with pH- and time-dependent release systems were produced for delivering prednisolone to the colon. A combination of Eudragit® S and ethyl cellulose was used to improve colonic drug delivery and prevent early release in the superior intestine [126]. Another example, Eudracol®, utilizes a multiunit technology with coatings of Eudragit® RL/RS and Eudragit® FS 30D to achieve delayed and uniform drug release specific to the colon in a pH- and time-dependent release system [127].

7.5.4 Microbiota-Triggered and Time-Triggered Systems

Recent research inquiry has extensively investigated individual or subsequent microbiota- and time-dependent mechanisms for delivering drugs to the colon. However, it is uncommon to observe concurrent combinations of these mechanisms [6]. Materials commonly employed as stimuli for microbiota-sensitive responses comprise chitosan, guar gum, resistant starch, and pectin. Time-dependent mechanisms encompass reservoir, osmotic designs, or capsular. Reservoir systems, primarily employing erodible or diffusive polymers, namely HPMC, are very extensively researched [1]. In the context of tablet coating, a parallel system uses a blend of pectin and HPMC in an 80:20 ratio. This formulation leverages the microbial degradation of pectin and HPMC-induced gastrointestinal swelling [128]. The preliminary investigation involving six healthy male subjects demonstrated the effective delivery of the coating to the ascending or transverse colon. An alternative method involved the development of capsule shells formed from a composite of HPMC and high-amylose starch. The subsequent liberation of paracetamol from these shells was evaluated using in vitro and ex vivo models [129].

7.5.5 pH-Triggered, Microbiota-Triggered, and Time-Triggered Systems

Recently, Moutaharrik et al. introduced a sophisticated drug delivery system designed to target the colon, which incorporates three distinct triggering mechanisms. This development represents a significant advancement in the field of drug delivery. This system incorporates dual-layer coating. The inner layer consists of a swellable, time-sensitive cellulose derivative like HPMC or hydroxypropyl cellulose. The outer layer consists of a mixture of a pH-responsive polymer (Eudragit® S) and a polysaccharide that is activated by microbiota. In addition, this novel approach is the first of its kind to utilize a triple mechanism for colon targeting, showing effective colon-specific delivery in both in vitro and ex vivo models. The promising initial results suggest that further development and testing in in vivo models could pave the way for clinical use [130].

7.5.6 Polymer Nanoparticle and Microparticle Systems

Numerous studies have shown the effectiveness of pH-responsive polymeric nanoparticles as systems for delivering drugs to the colon [131,132]. Mutalik et al. employed a novel pH-sensitive hydrolyzed polyacrylamide-grafted-xanthan gum (PAAm-g-XG) to deliver curcumin nanoparticles to the colon. The PAAm-g-XG-developed nanoparticles exhibited minimal drug release within acidic environments (pH 1.2 and 4.5), whereas, at pH 7.2, a more rapid and substantial drug release was observed. This demonstrated the efficacy of the nanoparticles in mitigating colonic inflammation and ameliorating weight loss in IBD rat models [133]. Furthermore, the drug release rate can be controlled by combining two different pH-sensitive polymers. Moreover, Sahu and Pandey have formulated HBsAg-loaded nanoparticles using a blend of Eudragit® L100 and S100 for efficient colonic immunization. This innovative approach has led to improved distribution of nanoparticles in the colon and a heightened immune response [134]. In efforts to enhance site-specific targeting of the colon. A study performed by Naeem et al. developed budesonide-loaded nanoparticles specifically designed to respond to changes in pH and time for the treatment of IBD. Present nanoparticles were produced using Eudragit® FS30D (a pH-dependent polymer) and RS100 (a time-dependent polymer) via the emulsion solvent evaporation technique (oil-in-water). Eudragit® FS30D exhibits dissolution behavior in environments with a pH exceeding 7.0, whereas Eudragit® RS100 functions as a controlled-release polymer, characterized by low permeability. Joining these polymers successfully reduces the drug release in the superior GIT and carries out sustained drug release all over the colon. Additionally, the pH- and time-dependent nanoparticles efficaciously facilitate drug delivery to inflamed colonic sites in colitis mouse models, suggesting promise for targeted colonic therapy [49].

7.5.7 Lipid-Based Delivery System

Liposomes serve as a highly effective drug delivery system with double-layered phospholipids [135]. They offer biodegradability, biocompatibility, and the capability to encapsulate drugs (hydrophilic and lipophilic) [136]. To bolster their stability in acid environments and refine site-specificity, the exteriors of liposomes can be enveloped with pH-responsive polymers or ligands. Particularly, Zhao et al. formulated colon-targeted anionic liposomes for sorafenib via coating liposomes with glycol chitosan and Eudragit® S100. This formulation demonstrated exceptional stability across varying pH levels, thereby reducing drug leakage and ultimately enhancing systemic sorafenib delivery in rats [137]. Solid lipid nanoparticles offer exceptional drug protection and high entrapment efficiency with the ability to release drugs at specific target sites [135]. Their lipid matrix degrades gradually, allowing for sustained drug release [135]. SMEDDS (self-microemulsifying drug delivery systems) showed promise in enhancing the oral bioavailability of hydrophobic drugs, making them suitable candidates for colon targeting [138]. Zhang et al. formulated folate-modified self-microemulsifying drug delivery systems (FSMEDDS) incorporating curcumin, followed by encapsulation within soft capsules coated with a pH-dependent polymer (Eudragit® S 100). These curcumin-loaded FSMEDDS capsules demonstrated efficient targeting of folate receptors on colon tumor cells, showcasing their promise for the precise delivery of drugs to the colon [139].

7.5.8 Ligand-Receptor-Mediated Drug Delivery System

In order to optimize the treatment of colonic diseases while minimizing associated toxicities, researchers have investigated the use of ligand-receptor-mediated drug delivery systems. These systems are designed to enhance target specificity by utilizing the interaction between ligands on the carrier surface and specific receptors present in the diseased area [140]. In relation to the demonstration profiles of definite receptors or proteins in the targeted cells or organs, a variety of ligands is made available for application. The incorporation of these ligands into pH-sensitive systems enhances gastrointestinal stability and facilitates precise targeting as required. The following are several ligands commonly used for delivering drugs to the colon.

7.5.8.1 Antibodies

Harel et al. formulated antibody-conjugated liposomes that had enhanced cellular internalization in comparison with unconjugated liposomes. These liposomes exhibited a preference for distribution to inflamed mucosa, resulting in significant accumulation at the site of inflammation, with levels more than four times higher than in normal mucosa [141]. In another study, Xiao et al. developed nanoparticles featuring CD98 antibodies on their surface, referred to as scCD98-functionalized nanoparticles, for the treatment of IBD. In mouse models of colitis, research findings have shown that CD98 is overexpressed in intestinal macrophages as well as colonic epithelial cells. The scCD98-functionalized nanoparticles demonstrated a notable affinity for CD98-overexpressing cells. When loaded with CD98 siRNA (siCD98), these nanoparticles effectively attenuated the expression levels of CD98 and alleviated the symptoms of colitis in mouse models [142].

7.5.8.2 Folic Acid

Folic acid (FA) serves as a ligand for selective tumor targeting, owing to the prevalent overexpression of the folate receptor across various cancer types [143]. Numerous research studies have indicated that the use of folic acid-decorated nanoparticles can significantly improve the targeting of drugs to tumors. As an example, Xiong et al. demonstrated that FA-conjugated liposomes can enhance the antitumor efficacy of daunorubicin by promoting folate receptor-mediated drug uptake [144]. Similarly, Handali et al. involved the improvement of FA-conjugated liposomes loaded with 5-fluorouracil (5-FU). These FA-liposomes loaded with 5-FU demonstrated enhanced cytotoxicity and a significant decrease in tumor volume compared to conventional free drug administration. These findings underscore the potential of folic acid-targeted liposomes as effective drug carriers, facilitating the selective delivery of drugs to cancer cells [145].

7.5.8.3 Hyaluronic Acid

Hyaluronic acid (HA), a natural polysaccharide, has garnered significant attention due to its pronounced affinity for a specific receptor CD44, which is frequently overexpressed in various malignancies. Consequently, HA-conjugated drug delivery systems are under scrutiny for their potential in targeted drug delivery [146]. Research has shown that mesoporous silica nanoparticles modified with HA effectively target cancer cells that overexpress CD44 [146]. In a similar vein, Vafaei et al. prepared self-assembled HA nanoparticles to serve as vehicles for budesonide, with the specific aim of targeting the inflamed intestinal mucosa in cases of IBD. Their findings revealed that these budesonide-loaded HA nanoparticles displayed an increased absorption by inflamed cells that overexpressed CD44 receptors, ultimately resulting in the reduction of IL-8 and TNF-α secretion in an inflamed cell model [147–151].

7.5.8.4 Peptides

Peptides have been the focus of considerable attention as potential ligands for targeted drug delivery owing to their numerous advantages, including biocompatibility, cost efficiency, chemical variety, and responsiveness to stimuli [150]. Peptides have a greater affinity and specificity for binding compared to small molecule ligands because they have a larger binding affinity with receptors [152]. Peptides are readily accessible for high-throughput testing and can be manufactured using automated solid-phase peptide synthesis devices. Modifying peptide sequences has the potential to mitigate metabolic instability caused by proteases, thereby enhancing their applicability in targeted drug delivery systems. This underscores the potential of the Tachykinini (TK) peptide as a targeting ligand for colon-specific therapy [153]. Guo et al. have prepared nanoparticles that are specifically targeted to the colon and have been co-modified with amphipathic chitosan derivatives (ACS) and cell-penetrating peptides (CPP) to enhance the oral bioavailability of insulin. The ACS alteration plays a critical role in protecting the CPPs from impairment in the upper GIT and facilitating the colonic drug delivery system. Upon reaching the colon, the gradual degradation of the ACSs present on the nanoparticle surface exposes the CPPs, which in turn facilitate drug penetration throughout the colonic epithelium. Both in vitro and in vivo assessments have unequivocally demonstrated the capacity of these CS-CPP nanoparticles to enhance the oral absorption of proteins and peptides significantly, thus highlighting their ability as a highly effective colonic drug delivery system [154].

7.6 RECTAL DRUG DELIVERY

Rectal dosage forms, encompassing memory, foams, enemas, gels, plus suppositories, are extensively employed in the handling of various conditions, including IBD followed with aminosalicylates and corticosteroids, and also for addressing constipation, hemorrhoids, discomfort, and feeling sick/throwing up [155]. Recently, there has been notable advancement in the development of rectal preparations. This progress entails a shift from traditional formulations toward more sophisticated systems designed to improve local drug maintenance or systemic absorption, addressing special medical requirements [40].

Epithelial tissue actively transports sodium ions into the colon lumen to facilitate the absorption of water, a natural mechanism that can be harnessed to direct drugs into the epithelium osmotically. By lowering the tonicity of rectal medications, an osmotic difference is established, which enables hydrophilic drugs and mucus-penetrating nanoparticles to move across the epithelial surface without causing harm [156,157].

In a comparative study of three kinds of tenofovir-loaded enemas, it was observed that hypotonic sodium-based enemas resulted in a noticeable increase in drug concentration in the colorectal tissues of mice when compared to isotonic and hypertonic preparations. This heightened concentration consequently facilitated rapid systemic drug absorption. Rectal formulations can result in local irritation, annoyance owing to leakage, or an increased urge to defecate, thereby diminishing acceptability among certain patients [158]. Effectively addressing these challenges is paramount to enhancing patient acceptance.

An effective strategy for addressing rectal leakage involves the utilization of thermosensitive polymers such as Pluronic or Poloxamers. These polymers have a transformation from a liquid to a gel state at body temperature [159]. This thermosensitive gel can decrease leakage, improve drug bioavailability over mucoadhesion, and alleviate the unpleasantness associated with the insertion of solid dosage forms. Despite promising results from multiple animal analyses and also the human analyzsis in 1998, a thermosensitive rectal formulation for gelation has not been produced until now in the market, likely owing to challenges in attaining economical production on a large scale [159–161].

7.7 IN VITRO FOR STUDYING

Evaluating the breakdown, solubility, and consistency of systems aimed at the colon is essential for creating new pharmaceutical products. Nonetheless, aligning laboratory-based findings with anticipated biological effects poses difficulties due to the colon's distinct physiological traits. Industry standards often employ United States Pharmacopeia (USP) dissolution testing to examine the performance of fast-acting drug formulations, especially those with readily dissolvable medications, as indicated by the presence of biowaiver initiatives [162]. Yet, standard USP tools like the multi-container USP 3 device might not be entirely appropriate for assessing the biorelevance of drugs designed for the colon. This is due to several issues: nonrepresentative fluid amounts and agitation speeds, unsuitability for disintegrating drug forms, and the presumption that dissolving is the primary factor in drug uptake [163]. In contrast, drug permeation through the colonic lining tends to be slower than through the small intestine, suggesting that for many drugs targeting the colon, permeability may be the limiting factor in absorption [77].

These challenges have spurred the creation of specialized methods tailored for the colon, which integrate sophisticated systems for drug stability and dissolution, including bicarbonate buffers, simulated gastrointestinal fluids, animal-derived fluids and tissues, mixtures of human fecal matter, and biorelevant fluid dynamics through customized permeability tests [164]. The use of standardized USP buffers, including hydrochloric acid, acetate, citrate, and phosphate, is crucial in evaluating the release of drugs from pharmaceutical compounds. These solutions span a wide pH spectrum and possess specified ionic concentrations and buffering capabilities [165]. Yet, despite their widespread application, these solutions typically do not reflect the biological conditions of the human gut.

In 2005, proposals were made to employ bicarbonate-based solutions for more biologically accurate dissolution tests for drugs targeting the colon. This suggestion stemmed from the understanding that bicarbonate solutions better replicate the buffering power of gut fluids than the traditionally used phosphate solutions [166]. Fadda and colleagues evaluated the predictive capacity of bicarbonate environments for in vivo actions of extended and prolonged-release mesalazine formulas [167]. Their research indicated that the release patterns in Krebs bicarbonate solution were more aligned with human disintegration and pharmacokinetic profiles than those in standard phosphate solution. This was corroborated by an analysis on Lialda® (Mezavant® XL, 5-ASA), where in vitro release data displayed a significant link with human results obtained through gamma-scintigraphy [168].

7.8 IN VIVO ANIMAL MODELS FOR STUDYING

For the innovation of colon-specific drugs, choosing animal models that mirror the human digestive system is crucial. Yet, factors like cost, accessibility, and manageability often dictate the selection of these models over physiological congruence [169]. Notable variations in digestive structure and function across species result in different drug absorption rates and bioavailability. Since no animal model is an exact match for the human digestive system, a variety of animals are employed in colonic drug delivery research, ranging from small creatures like rats, mice, rabbits, and guinea pigs to bigger animals, for instance, pigs, dogs, and non-human primates [170].

7.9 IN SILICO FOR STUDYING

The process of drug development is notorious for its high expenses and inherent uncertainties, making any technique that can expedite the formulation process and predict biological responses while conserving resources extremely beneficial [171]. A range of predictive methods is available to assist in the creation of new drugs aimed at the colon, such as design of experiments (DoE), molecular dynamics (MD), mechanistic modeling (MM), machine learning (ML), and more specialized approaches like limited element analysis, as well as computational fluid dynamics [172].

DoE is an extensively adopted approach in drug research for modeling how various procedural factors influence specific results. It was applied in several studies to fine-tune the makeup of enteric polymer coatings for colon-targeted drugs to ensure their effectiveness in biological systems [173,174].

Molecular modeling is the virtual representation of molecular interactions, utilizing principles from both classical and quantum physics [175]. Depending on the specific experimental requirements and available resources, molecular systems can be depicted with varying degrees of complexity, from detailed atomistic to more generalized models [176]. MD stands out for its accuracy in simulating dynamic atomic-level interactions and plays a crucial role as an in silico technique in drug exploration, particularly in examining how drugs bind to their targets [177,178]. Within the realm of colonic drug discovery, MD is instrumental for pinpointing active compounds against novel therapeutic targets, including those within the microbiome [179].

MM and ML excel at processing a wide array of data sizes, from atomic dimensions to macroscopic levels. MM involves devising mathematical models that establish a mechanistic connection between independent and dependent variables [180]. These models are streamlined interpretations, based on hypotheses derived from the present trial evidence, and intentionally omit certain factors that influence interactions. MM encompasses physiologically based pharmacokinetic (PBPK) models, which simulate the influence of absorption, distribution, metabolism, and excretion (ADME) on drug levels in blood plasma and bodily tissues [181]. PBPK models can adjust critical ADME parameters to precisely forecast how physiological differences affect pharmacokinetics across various patient demographics [182].

ML has recently been acknowledged for its capacity to reduce costs and enhance success rates across all stages of drug development, from initial discovery through clinical testing [183,184]. ML comprises a suite of technologies that employ mathematical algorithms to discern patterns within data sets, enabling predictions about experimental results for new data sets [172].

7.10 CONCLUSIONS AND FUTURE VIEWPOINT

Colonic drug delivery is undergoing a transformative period, capitalizing on modern understandings of the colon's special physiological attributes, effective formulation techniques, and sophisticated tools spanning in vitro, in vivo, and in silico platforms to develop novel and potent drugs. This analysis underscores the crucial need to synchronize every aspect of colon-targeted drug development with gastrointestinal environments, from identifying colon-specific targets to creating formulations that ensure accurate drug delivery at the desired site of action, to thorough evaluation of formulations in preclinical stages. The emergence of dual-trigger formulation technologies has greatly enhanced the dependability of oral colonic drug delivery, paving the way for customized engagement with novel systemic targets or improved management of localized conditions. Future research could further explore these formulation methods to design drugs specific to particular diseases, with release mechanisms tailored to the respective pathophysiology or personalized for distinct patient groups. There's also a vast potential for innovation in rectal drug delivery, with possibilities to apply cutting-edge techniques like precision formulation and computational design to overcome current obstacles in this delivery pathway. Additionally, enriching the collection of physiologically pertinent tools for analyzing rectal formulation pharmacokinetics and harmonizing in vitro and in vivo models for evaluating orally administered colon-targeted drugs are crucial directions for progress. Further studies validating the effectiveness of sophisticated computational tools in optimizing the development pathway for colon-targeted drugs could act as an impetus for their wider acceptance in the research field. Recent advancements in crafting colon-targeted drugs hold promise for introducing safer and more effective treatments into clinical settings.

REFERENCES

1. Awad A, Madla CM, McCoubrey LE, Ferraro F, Gavins FKH, Buanz A et al. Clinical translation of advanced colonic drug delivery technologies. *Adv Drug Deliv Rev.* 2022;181:114076. doi:10.1016/j.addr.2021.114076.
2. Campbell C, McKenney PT, Konstantinovsky D, Isaeva OI, Schizas M, Verter J et al. Bacterial metabolism of bile acids promotes generation of peripheral regulatory T cells. *Nature.* 2020;581(7809):475–9.
3. Peiris M, Aktar R, Reed D, Cibert-Goton V, Zdanaviciene A, Halder W et al. Decoy bypass for appetite suppression in obese adults: role of synergistic nutrient sensing receptors GPR84 and FFAR4 on colonic endocrine cells. *Gut.* 2022;71(5):928–37.
4. Allegretti JR, Fischer M, Sagi S V, Bohm ME, Fadda HM, Ranmal SR et al. Fecal microbiota transplantation capsules with targeted colonic versus gastric delivery in recurrent Clostridium difficile infection: a comparative cohort analysis of high and low doses. *Dig Dis Sci.* 2019;64:1672–8.
5. Ciftci K, Groves MJ. Delivery of antitumor compounds to the rat colon: in vitro and in vivo evaluation. *Int J Pharm.* 1996;145(1–2):157–64.
6. Foppoli A, Maroni A, Moutaharrik S, Melocchi A, Zema L, Palugan L et al. In vitro and human pharmacoscintigraphic evaluation of an oral 5-ASA delivery system for colonic release. *Int J Pharm.* 2019;572:118723.
7. Yadav V, House A, Matiz S, McCoubrey LE, Bettano KA, Bhave L et al. Ileocolonic-targeted JAK inhibitor: a safer and more effective treatment for inflammatory bowel disease. *Pharmaceutics.* 2022;14(11):2385.
8. Verstockt B, Vermeire S. Tofacitinib tissue exposure correlates with endoscopic outcome. *J Crohn's Colitis.* 2022;16:123–30.
9. Vuik FER, Dicksved J, Lam SY, Fuhler GM, van der Laan LJW, van de Winkel A,et al. Composition of the mucosa-associated microbiota along the entire gastrointestinal tract of human individuals. *United Eur Gastroenterol J.* 2019;7(7):897–907.
10. James KR, Gomes T, Elmentaite R, Kumar N, Gulliver EL, King HW et al. Distinct microbial and immune niches of the human colon. *Nat Immunol.* 2020;21(3):343–53.
11. Winkler ES, Shrihari S, Hykes BL, Handley SA, Andhey PS, Huang Y-JS et al. The intestinal microbiome restricts alphavirus infection and dissemination through a bile acid-type I IFN signaling axis. *Cell.* 2020;182(4):901–18.
12. Ferrocino I, Ponzo V, Pellegrini M, Goitre I, Papurello M, Franciosa I et al. Mycobiota composition and changes across pregnancy in patients with gestational diabetes mellitus (GDM). *Sci Rep.* 2022;12(1):9192.
13. van Tilburg Bernardes E, Pettersen VK, Gutierrez MW, Laforest-Lapointe I, Jendzjowsky NG, Cavin J-B et al. Intestinal fungi are causally implicated in microbiome assembly and immune development in mice. *Nat Commun.* 2020;11(1):2577.
14. WHO. Latest global cancer data: cancer burden rises to 19.3 million new cases and 10.0 million cancer deaths in 2020. *Int Agency Res Cancer.* 2020;13–5. https://www.iarc.who.int/fr/news-events/latest-global-cancer-data-cancer-burden-rises-to-19-3-million-new-cases-and-10-0-million-cancer-deaths-in-2020/.
15. Wang R, Li Z, Liu S, Zhang D. Global, regional, and national burden of inflammatory bowel disease in 204 countries and territories from 1990 to 2019: a systematic analysis based on the Global Burden of Disease Study 2019. *BMJ Open.* 2023;13(3):e065186.
16. Kaplan GG, Windsor JW. The four epidemiological stages in the global evolution of inflammatory bowel disease. *Nat Rev Gastroenterol Hepatol.* 2021;18(1):56–66.
17. King D, Reulen RC, Thomas T, Chandan JS, Thayakaran R, Subramanian A, et al. Changing patterns in the epidemiology and outcomes of inflammatory bowel disease in the United Kingdom: 2000-2018. *Aliment Pharmacol Ther.* 2020;51(10):922–34.
18. Svartz N. Behandlung der ulzerösen Kolitis mit Salazopyrin. *Gastroenterologia.* 1941;66(5–6):312–22.
19. Khan AKA, Piris J, Truelove SC. An experiment to determine the active therapeutic moiety of sulphasalazine. *Lancet.* 1977;310(8044):892–5.
20. Riley SA. What dose of 5-aminosalicylic acid (mesalazine) in ulcerative colitis? *Gut.* 1998;42(6):761–3.
21. Klotz U. Clinical pharmacokinetics of sulphasalazine, its metabolites and other prodrugs of 5-aminosalicylic acid. *Clin Pharmacokinet.* 1985;10(4):285–302.
22. Wadworth AN, Fitton A. Olsalazine: a review of its pharmacodynamic and pharmacokinetic properties, and therapeutic potential in inflammatory bowel disease. *Drugs.* 1991;41:647–64.
23. Chan RP, Pope DJ, Gilbert AP, Sacra PJ, Baron JH, Lennard-Jones JE. Studies of two novel sulfasalazine analogs, ipsalazide and balsalazide. *Dig Dis Sci.* 1983;28:609–15.
24. Zhao J, Zhang B, Mao Q, Ping K, Zhang P, Lin F et al. Discovery of a colon-targeted azo prodrug of tofacitinib through the establishment of colon-specific delivery systems constructed by 5-ASA–PABA–MAC and 5-ASA–PABA–diamine for the treatment of ulcerative colitis. *J Med Chem.* 2022;65(6):4926–48.
25. Brayden DJ. Localised delivery of macromolecules to the large intestine: translation to clinical trials. *BioDrugs.* 2022;36(6):687–700.
26. European Medicines Agency. *Guideline on the Pharmacokinetic and Clinical Evaluation of Modified Release Dosage Forms*, EMA/CHMP/EWP/280/96 Rev 1, 2014. European Medicines Agency, London, UK, 2014.
27. Thakral S, Thakral NK, Majumdar DK. Eudragit®: a technology evaluation. *Expert Opin Drug Deliv.* 2013;10(1):131–49.

28. Ibekwe VC, Fadda HM, Parsons GE, Basit AW. A comparative in vitro assessment of the drug release performance of pH-responsive polymers for ileo-colonic delivery. *Int J Pharm.* 2006;308(1–2):52–60.
29. Ibekwe VC, Fadda HM, McConnell EL, Khela MK, Evans DF, Basit AW. Interplay between intestinal pH, transit time and feed status on the in vivo performance of pH responsive ileo-colonic release systems. *Pharm Res.* 2008;25:1828–35.
30. Hatton GB, Madla CM, Rabbie SC, Basit AW. All disease begins in the gut: influence of gastrointestinal disorders and surgery on oral drug performance. *Int J Pharm.* 2018;548(1):408–22.
31. Esmeeta A, Adhikary S, Dharshnaa V, Swarnamughi P, Maqsummiya ZU, Banerjee A et al. Plant-derived bioactive compounds in colon cancer treatment: an updated review. *Biomed Pharmacother.* 2022;153:113384.
32. Liu EY, Loomes DE. Ustekinumab and vedolizumab dual biologic therapy in the treatment of Crohn's disease. *Case Rep Med.* 2017;2017.
33. Rahimi R, Nikfar S, Rezaie A, Abdollahi M. Pregnancy outcome in women with inflammatory bowel disease following exposure to 5-aminosalicylic acid drugs: a meta-analysis. *Reprod Toxicol.* 2008;25(2):271–5.
34. Pan Z, Zhou Z, Zhang H, Zhao H, Song P, Wang D et al. CD90 serves as differential modulator of subcutaneous and visceral adipose-derived stem cells by regulating AKT activation that influences adipose tissue and metabolic homeostasis. *Stem Cell Res Ther.* 2019;10:1–18.
35. Hua S. Physiological and pharmaceutical considerations for rectal drug formulations. *Front Pharmacol.* 2019;10:1196.
36. Purohit TJ, Hanning SM, Wu Z. Advances in rectal drug delivery systems. *Pharm Dev Technol.* 2018;23(10):942–52.
37. Hua S. Distinct microbial and immune niches of the human colon. *Nat Immunol.* 2020;21:343–53.
38. McConnell EL, Basit AW, Murdan S. Colonic antigen administration induces significantly higher humoral levels of colonic and vaginal IgA, and serum IgG compared to oral administration. *Vaccine.* 2008;26(5):639–46.
39. Cho C-S, Hwang S-K, Gu M-J, Kim C-G, Kim S-K, Ju D-B et al. Mucosal vaccine delivery using mucoadhesive polymer particulate systems. *Tissue Eng Regen Med.* 2021;18:693–712.
40. Kurakula M, Gorityala S, Moharir K. Recent trends in design and evaluation of chitosan-based colon targeted drug delivery systems: update 2020. *J Drug Deliv Sci Technol.* 2021;64:102579.
41. Amidon S, Brown JE, Dave VS. Colon-targeted oral drug delivery systems: design trends and approaches. *Aaps Pharmscitech.* 2015;16:731–41.
42. Kornum DS, Terkelsen AJ, Bertoli D, Klinge MW, Høyer KL, Kufaishi HHA et al. Assessment of gastrointestinal autonomic dysfunction: present and future perspectives. Vol. 10, *J Clin Med.* 2021.
43. Vandamme TF, Lenourry A, Charrueau C, Chaumeil JC. The use of polysaccharides to target drugs to the colon. *Carbohydr Polym.* 2002;48(3):219–31.
44. Cheng H, Huang S, Huang G. Design and application of oral colon administration system. *J Enzyme Inhib Med Chem.* 2019;34(1):1590–6.
45. Liu W, Kong Y, Ye A, Shen P, Dong L, Xu X et al. Preparation, formation mechanism and in vitro dynamic digestion behavior of quercetin-loaded liposomes in hydrogels. *Food Hydrocoll.* 2020;104:105743.
46. Feng K, Wei Y, Hu T, Linhardt RJ, Zong M, Wu H. Colon-targeted delivery systems for nutraceuticals: a review of current vehicles, evaluation methods and future prospects. *Trends Food Sci Technol.* 2020;102:203–22.
47. Naeem M, Awan UA, Subhan F, Cao J, Hlaing SP, Lee J et al. Advances in colon-targeted nano-drug delivery systems: challenges and solutions. *Arch Pharm Res.* 2020;43(1):153–69.
48. Naeem M, Choi M, Cao J, Lee Y, Ikram M, Yoon S et al. Colon-targeted delivery of budesonide using dual pH- and time-dependent polymeric nanoparticles for colitis therapy. *Drug Des Devel Ther.* 2015;9:3789–99.
49. Turner K, Genta RM. The nonneoplastic stomach. In: C Fenoglio-Preiser (Ed.), *Gastrointestinal Pathology*, 4th edn. Springer, New York, 2017, pp. 136–223.
50. Yurtdaş-Kırımlıoğlu G, Görgülü Ş. Surface modification of PLGA nanoparticles with chitosan or Eudragit® RS 100: characterization, prolonged release, cytotoxicity, and enhanced antimicrobial activity. *J Drug Deliv Sci Technol.* 2021;61:102145.
51. Murgia X, Loretz B, Hartwig O, Hittinger M, Lehr C-M. The role of mucus on drug transport and its potential to affect therapeutic outcomes. *Adv Drug Deliv Rev.* 2018;124:82–97.
52. Bandi SP, Bhatnagar S, Venuganti VVK. Advanced materials for drug delivery across mucosal barriers. *Acta Biomater.* 2021;119:13–29.
53. Rachid R, Chatila TA. The role of the gut microbiota in food allergy. *Curr Opin Pediatr.* 2016;28(6):748–53.
54. Gupta J, Ahuja A, Gupta R. Green approaches for cancers management: an effective tool for health care. *Anti-Cancer Agents Med Chem* (Formerly *Curr Med Chem Agents*). 2022;22(1):101–14.
55. Fabbrini M, D'Amico F, Barone M, Conti G, Mengoli M, Brigidi P et al. Polyphenol and tannin nutraceuticals and their metabolites: how the human gut microbiota influences their properties. *Biomolecules.* 2022;12(7):875.
56. Marín L, Miguélez EM, Villar CJ, Lombó F. Bioavailability of dietary polyphenols and gut microbiota metabolism: antimicrobial properties. *Biomed Res Int.* 2015;2015.
57. Barducci L, Norton JC, Sarker S, Mohammed S, Jones R, Valdastri P et al. Fundamentals of the gut for capsule engineers. *Prog Biomed Eng.* 2020;2(4):42002.
58. Aulton ME, Taylor K. *Aulton's Pharmaceutics: The Design and Manufacture of Medicines.* Elsevier Health Sciences, Amsterdam, 2013.
59. Kalra A, Wehrle CJ, Tuma F. *Anatomy, Abdomen and Pelvis, Peritoneum.* StatPearls Publishing, Island, FL. 2023.
60. Kahai P, Mandiga P, Wehrle CJ, Lobo S. *Anatomy, Abdomen and Pelvis: Large Intestine.* StatPearls Publishing, Island, FL. 2017.
61. Cai S, Fan Y, Zhang B, Lin J, Yang X, Liu Y et al. Appendectomy is associated with alteration of human gut bacterial and fungal communities. *Front Microbiol.* 2021;12:724980.
62. Vitetta L, Chen J, Clarke S. The vermiform appendix: an immunological organ sustaining a microbiome inoculum. *Clin Sci.* 2019;133(1):1–8.
63. Weitschies W, Blume H, Mönnikes H. Magnetic marker monitoring: high resolution real-time tracking of oral solid dosage forms in the gastrointestinal tract. *Eur J Pharm Biopharm.* 2010;74(1):93–101.
64. Madla CM, Gavins FKH, Merchant HA, Orlu M, Murdan S, Basit AW. Let's talk about sex: differences in drug therapy in males and females. *Adv Drug Deliv Rev.* 2021;175:113804.
65. Hatton GB, Madla CM, Rabbie SC, Basit AW. Gut reaction: impact of systemic diseases on gastrointestinal

physiology and drug absorption. *Drug Discov Today.* 2019;24(2):417–27.
66. McConnell EL, Fadda HM, Basit AW. Gut instincts: explorations in intestinal physiology and drug delivery. *Int J Pharm.* 2008;364(2):213–26.
67. Müller M, Canfora EE, Blaak EE. Gastrointestinal transit time, glucose homeostasis and metabolic health: modulation by dietary fibers. *Nutrients.* 2018;10(3):275.
68. Feldman M, Friedman LS, Brandt LJ. Sleisenger and Fordtran's gastrointestinal and liver disease E-book: pathophysiology, diagnosis, management. *Health Sci.* 2015.
69. Campbell J, Berry J, Liang Y. Anatomy and physiology of the small intestine. In: MJ Zinner, SW Ashley, and JE Fischer (Eds.), *Shackelford's Surgery of the Alimentary Tract*, 2 Volume Set. Elsevier, Amsterdam, 2019, pp. 817–41.
70. Corsetti M, Costa M, Bassotti G, Bharucha AE, Borrelli O, Dinning P et al. First translational consensus on terminology and definitions of colonic motility in animals and humans studied by manometric and other techniques. *Nat Rev Gastroenterol Hepatol.* 2019;16(9):559–79.
71. Spencer NJ, Dinning PG, Brookes SJ, Costa M. Insights into the mechanisms underlying colonic motor patterns. *J Physiol.* 2016;594(15):4099–116.
72. Tian H, Chen Q, Yang B, Qin H, Li N. Analysis of gut microbiome and metabolite characteristics in patients with slow transit constipation. *Dig Dis Sci.* 2021;66:3026–35.
73. Nandhra GK, Mark EB, Di Tanna GL, Haase A, Poulsen J, Christodoulides S et al. Normative values for region-specific colonic and gastrointestinal transit times in 111 healthy volunteers using the 3D-Transit electromagnet tracking system: influence of age, gender, and body mass index. *Neurogastroenterol Motil.* 2020;32(2):e13734.
74. Jewell A, Brookes A, Feng W, Ashford M, Gellert P, Butler J et al. Distribution of a highly lipophilic drug cannabidiol into different lymph nodes following oral administration in lipidic vehicle. *Eur J Pharm Biopharm.* 2022;174:29–34.
75. Houston SA, Cerovic V, Thomson C, Brewer J, Mowat AM, Milling S. The lymph nodes draining the small intestine and colon are anatomically separate and immunologically distinct. *Mucosal Immunol.* 2016;9(2):468–78.
76. Esterházy D, Canesso MCC, Mesin L, Muller PA, de Castro TBR, Lockhart A et al. Compartmentalized gut lymph node drainage dictates adaptive immune responses. *Nature.* 2019;569(7754):126–30.
77. Yang Y, Zhao Y, Yu A, Sun D, Yu LX. Oral drug absorption: evaluation and prediction. In: Y Qiu, Y Chen, GZ Zhang, L Yu, and RV Mantri (Eds.), *Developing Solid Oral Dosage Forms.* Elsevier, Amsterdam, 2017, pp. 331–54.
78. Tannergren C, Bergendal A, Lennernas H, Abrahamsson B. Toward an increased understanding of the barriers to colonic drug absorption in humans: implications for early controlled release candidate assessment. *Mol Pharm.* 2009;6(1):60–73.
79. Martini E, Krug SM, Siegmund B, Neurath MF, Becker C. Mend your fences: the epithelial barrier and its relationship with mucosal immunity in inflammatory bowel disease. *Cell Mol Gastroenterol Hepatol.* 2017;4(1):33–46.
80. Lemmens G, Van Camp A, Kourula S, Vanuytsel T, Augustijns P. Drug disposition in the lower gastrointestinal tract: targeting and monitoring. *Pharmaceutics.* 2021;13(2):161.
81. Santos AJM, Lo Y-H, Mah AT, Kuo CJ. The intestinal stem cell niche: homeostasis and adaptations. *Trends Cell Biol.* 2018;28(12):1062–78.
82. Kong S, Zhang YH, Zhang W. Regulation of intestinal epithelial cells properties and functions by amino acids. *Biomed Res Int.* 2018;2018.
83. Mörbe UM, Jørgensen PB, Fenton TM, von Burg N, Riis LB, Spencer J et al. Human gut-associated lymphoid tissues (GALT); diversity, structure, and function. *Mucosal Immunol.* 2021;14(4):793–802.
84. Hendel SK, Kellermann L, Hausmann A, Bindslev N, Jensen KB, Nielsen OH. Tuft cells and their role in intestinal diseases. *Front Immunol.* 2022;13:822867.
85. Wang J, Yadav V, Smart AL, Tajiri S, Basit AW. Toward oral delivery of biopharmaceuticals: an assessment of the gastrointestinal stability of 17 peptide drugs. *Mol Pharm.* 2015;12(3):966–73.
86. Evans DF, Pye G, Bramley R, Clark AG, Dyson TJ, Hardcastle JD. Measurement of gastrointestinal pH profiles in normal ambulant human subjects. *Gut.* 1988;29(8):1035–41.
87. Fadda HM, Sousa T, Carlsson AS, Abrahamsson B, Williams JG, Kumar D et al. Drug solubility in luminal fluids from different regions of the small and large intestine of humans. *Mol Pharm.* 2010;7(5):1527–32.
88. Pyper K, Brouwers J, Augustijns P, Khadra I, Dunn C, Wilson CG et al. Multidimensional analysis of human intestinal fluid composition. *Eur J Pharm Biopharm.* 2020;153:226–40.
89. Diakidou A, Vertzoni M, Goumas K, Söderlind E, Abrahamsson B, Dressman J et al. Characterization of the contents of ascending colon to which drugs are exposed after oral administration to healthy adults. *Pharm Res.* 2009;26:2141–51.
90. Pritchard SE, Marciani L, Garsed KC, Hoad CL, Thongborisute W, Roberts E et al. Fasting and postprandial volumes of the undisturbed colon: normal values and changes in diarrhea-predominant irritable bowel syndrome measured using serial MRI. *Neurogastroenterol Motil.* 2014;26(1):124–30.
91. Barber C, Mego M, Sabater C, Vallejo F, Bendezu RA, Masihy M et al. Differential effects of western and mediterranean-type diets on gut microbiota: a metagenomics and metabolomics approach. *Nutrients.* 2021;13(8):2638.
92. Singhal R, Shah YM. Oxygen battle in the gut: hypoxia and hypoxia-inducible factors in metabolic and inflammatory responses in the intestine. *J Biol Chem.* 2020;295(30):10493–505.
93. Jalili-Firoozinezhad S, Gazzaniga FS, Calamari EL, Camacho DM, Fadel CW, Bein A et al. A complex human gut microbiome cultured in an anaerobic intestine-on-a-chip. *Nat Biomed Eng.* 2019;3(7):520–31.
94. McCoubrey LE, Basit AW. Addressing drug–microbiome interactions: the role of healthcare professionals. *Pharm J.* 2022;308(7958):1–27.
95. Martinez-Guryn K, Leone V, Chang EB. Regional diversity of the gastrointestinal microbiome. *Cell Host Microbe.* 2019;26(3):314–24.
96. Tian Y, Gui W, Koo I, Smith PB, Allman EL, Nichols RG et al. The microbiome modulating activity of bile acids. *Gut Microbes.* 2020;11(4):979–96.
97. Liang G, Bushman FD. The human virome: assembly, composition and host interactions. *Nat Rev Microbiol.* 2021;19(8):514–27.
98. Strati F, Di Paola M, Stefanini I, Albanese D, Rizzetto L, Lionetti P et al. Age and gender affect the composition of fungal population of the human gastrointestinal tract. *Front Microbiol.* 2016;7:1227.

99. McCoubrey LE, Gaisford S, Orlu M, Basit AW. Predicting drug-microbiome interactions with machine learning. *Biotechnol Adv.* 2022;54:107797.
100. Vinarov Z, Abdallah M, Agundez JAG, Allegaert K, Basit AW, Braeckmans M et al. Impact of gastrointestinal tract variability on oral drug absorption and pharmacokinetics: an UNGAP review. *Eur J Pharm Sci.* 2021;162:105812.
101. McCoubrey LE, Elbadawi M, Orlu M, Gaisford S, Basit AW. Machine learning uncovers adverse drug effects on intestinal bacteria. *Pharmaceutics.* 2021;13:1026.
102. Maier L, Goemans C V, Wirbel J, Kuhn M, Eberl C, Pruteanu M et al. Unravelling the collateral damage of antibiotics on gut bacteria. *Nature.* 2021;599(7883):120–4.
103. Mulder M, Radjabzadeh D, Kiefte-de Jong JC, Uitterlinden AG, Kraaij R, Stricker BH et al. Long-term effects of antimicrobial drugs on the composition of the human gut microbiota. *Gut Microbes.* 2020;12(1):1791677.
104. Nayak RR, Alexander M, Deshpande I, Stapleton-Gray K, Rimal B, Patterson AD et al. Methotrexate impacts conserved pathways in diverse human gut bacteria leading to decreased host immune activation. *Cell Host Microbe.* 2021;29(3):362–77.
105. Varum F, Freire AC, Fadda HM, Bravo R, Basit AW. A dual pH and microbiota-triggered coating (Phloral™) for fail-safe colonic drug release. *Int J Pharm.* 2020;583:119379.
106. Kamm MA, Sandborn WJ, Gassull M, Schreiber S, Jackowski L, Butler T et al. Once-daily, high-concentration MMX mesalamine in active ulcerative colitis. *Gastroenterology.* 2007;132(1):66–75. https://www.sciencedirect.com/science/article/pii/S0016508506022347.
107. Varum F, Freire AC, Bravo R, Basit AW. OPTICORE™, an innovative and accurate colonic targeting technology. *Int J Pharm.* 2020;583:119372.
108. Brar B, Ranjan K, Palria A, Kumar R, Ghosh M, Sihag S et al. Nanotechnology in colorectal cancer for precision diagnosis and therapy. *Front Nanotechnol.* 2021;3. doi:10.3389/fnano.2021.699266.
109. Chaubey P, Momin M, Sawarkar S. Significance of ligand-anchored polymers for drug targeting in the treatment of colonic disorders. *Front Pharmacol.* 2020;10:480659.
110. U.S. Food and Drug Administration. *Generally Recognized as Safe Notices (GRN).* U.S. Food and Drug Administration, 2016.
111. Gregory AL, Pensinger DA, Hryckowian AJ. A short-chain fatty acid–centric view of Clostridioides difficile pathogenesis. *PLoS Pathog.* 2021;17(10):e1009959.
112. Kneiszl R, Hossain S, Larsson P. In silico-based experiments on the mechanistic interactions between several intestinal permeation enhancers with a lipid Bilayer model. *Mol Pharm.* 2021;19(1):124–37.
113. Twarog C, Fattah S, Heade J, Maher S, Fattal E, Brayden DJ. Intestinal permeation enhancers for oral delivery of macromolecules: a comparison between salcaprozate sodium (SNAC) and sodium caprate (C10). *Pharmaceutics.* 2019;11(2):78.
114. Dodoo CC, Wang J, Basit AW, Stapleton P, Gaisford S. Targeted delivery of probiotics to enhance gastrointestinal stability and intestinal colonisation. *Int J Pharm.* 2017;530(1–2):224–9.
115. Ibekwe VC, Khela MK, Evans DF, Basit AW. A new concept in colonic drug targeting: a combined pH-responsive and bacterially-triggered drug delivery technology. *Aliment Pharmacol Ther.* 2008;28(7):911–6.
116. Yadav V, House A, Matiz S, McCoubrey LE, Bettano KA, Bhave L et al. Ileocolonic-targeted JAK inhibitor: a safer and more effective treatment for inflammatory bowel disease. *Pharmaceutics.* 2022;14(11):2385.
117. Major G, Murray K, Singh G, Nowak A, Hoad CL, Marciani L, et al. Demonstration of differences in colonic volumes, transit, chyme consistency, and response to psyllium between healthy and constipated subjects using magnetic resonance imaging. *Neurogastroenterol Motil.* 2018;30(9):e13400.
118. Varum F, Bravo R, Basit A. *OPTICORE™: A First-in-Class Colonic Targetin g Technology.* ONdrugDelivery Mag Frederick Furness Publ Ltd, Lewes, UK, 2020, pp. 40–44.
119. Varum FJO, Hatton GB, Freire AC, Basit AW. A novel coating concept for ileo-colonic drug targeting: proof of concept in humans using scintigraphy. *Eur J Pharm Biopharm.* 2013;84(3):573–7.
120. Varum FJO, Veiga F, Sousa JS, Basit AW. Mucoadhesive platforms for targeted delivery to the colon. *Int J Pharm.* 2011;420(1):11–9.
121. D'haens GR, Sandborn WJ, Zou G, Stitt LW, Rutgeerts PJ, Gilgen D et al. Randomised non-inferiority trial: 1600 mg versus 400 mg tablets of mesalazine for the treatment of mild-to-moderate ulcerative colitis. *Aliment Pharmacol Ther.* 2017;46(3):292–302.
122. Preisig D, Varum F, Bravo R, Hartig C, Spleiss J, Abbes S et al. Colonic delivery of metronidazole-loaded capsules for local treatment of bacterial infections: a clinical pharmacoscintigraphy study. *Eur J Pharm Biopharm.* 2021;165:22–30.
123. Sardou HS, Akhgari A, Mohammadpour AH, Namdar AB, Kamali H, Jafarian AH et al. Optimization study of combined enteric and time-dependent polymethacrylates as a coating for colon-targeted delivery of 5-ASA pellets in rats with ulcerative colitis. *Eur J Pharm Sci.* 2022;168:106072.
124. Vemula SK, Veerareddy PR, Devadasu VR. Pharmacokinetics of colon-specific pH and time-dependent flurbiprofen tablets. *Eur J Drug Metab Pharmacokinet.* 2015;40:301–11.
125. Liu F, Moreno P, Basit AW. A novel double-coating approach for improved pH-triggered delivery to the ileo-colonic region of the gastrointestinal tract. *Eur J Pharm Biopharm.* 2010;74(2):311–5.
126. Hashem FM, Shaker DS, Nasr M, Ragaey R. In vitro and in vivo evaluation of combined time and pH-dependent oral colonic targeted prednisolone microspheres. *Br J Pharm Res.* 2013;3(3):420.
127. Patel MM. Cutting-edge technologies in colon-targeted drug delivery systems. *Expert Opin Drug Deliv.* 2011;8(10):1247–58.
128. Hodges LA, Connolly SM, Band J, O'Mahony B, Ugurlu T, Turkoglu M, et al. Scintigraphic evaluation of colon targeting pectin–HPMC tablets in healthy volunteers. *Int J Pharm.* 2009;370(1–2):144–50.
129. Casati F, Melocchi A, Moutaharrik S, Uboldi M, Foppoli A, Maroni A, et al. Injection molded capsules for colon delivery combining time-controlled and enzyme-triggered approaches. *Int J Mol Sci.* 2020;21(6):1917.
130. Moutaharrik S, Maroni A, Melocchi A, Zema L, Foppoli A, Cerea M et al. Oral colon delivery platform based on a novel combination approach: design concept and preliminary evaluation. *J Drug Deliv Sci Technol.* 2021;66:102919.
131. Zeeshan M, Ali H, Khan S, Khan SA, Weigmann B. Advances in orally-delivered pH-sensitive nanocarrier systems; an optimistic approach for the treatment of inflammatory bowel disease. *Int J Pharm.* 2019;558:201–14.
132. Ma X, Williams III RO. Polymeric nanomedicines for poorly soluble drugs in oral delivery systems: an update. *J Pharm Investig.* 2018;48(1):61–75.
133. Mutalik S, Suthar NA, Managuli RS, Shetty PK, Avadhani K, Kalthur G et al. Development and performance evaluation

of novel nanoparticles of a grafted copolymer loaded with curcumin. *Int J Biol Macromol*. 2016;86:709–20.
134. Sahu KK, Pandey RS. Development and characterization of HBsAg-loaded Eudragit nanoparticles for effective colonic immunization. *Pharm Dev Technol*. 2019;24(2):166–75.
135. Guo Y, Zong S, Pu Y, Xu B, Zhang T, Wang B. Advances in pharmaceutical strategies enhancing the efficiencies of oral colon-targeted delivery systems in inflammatory bowel disease. *Molecules*. 2018;23(7):1622.
136. Yoon S-W, Shin DH, Kim J-S. Liposomal itraconazole formulation for the treatment of glioblastoma using an inclusion complex with HP-β-CD. *J Pharm Investig*. 2019;49:477–83.
137. Zhao M, Lee SH, Song JG, Kim HY, Han H-K. Enhanced oral absorption of sorafenib via the layer-by-layer deposition of a pH-sensitive polymer and glycol chitosan on the liposome. *Int J Pharm*. 2018;544(1):14–20.
138. Madhav KV, Kishan V. Self microemulsifying particles of loratadine for improved oral bioavailability: preparation, characterization, and in vivo evaluation. *J Pharm Investig*. 2018;48:497–508.
139. Zhang L, Zhu W, Yang C, Guo H, Yu A, Ji J et al. A novel folate-modified self-microemulsifying drug delivery system of curcumin for colon targeting. *Int J Nanomed*. 2012 Jan 9;7:151–62. doi:10.2147/IJN.S27639.
140. Si X-Y, Merlin D, Xiao B. Recent advances in orally administered cell-specific nanotherapeutics for inflammatory bowel disease. *World J Gastroenterol*. 2016;22(34):7718.
141. Harel E, Rubinstein A, Nissan A, Khazanov E, Nadler Milbauer M, Barenholz Y et al. Enhanced transferrin receptor expression by proinflammatory cytokines in enterocytes as a means for local delivery of drugs to inflamed gut mucosa. *PLoS One*. 2011;6(9):e24202.
142. Xiao B, Laroui H, Viennois E, Ayyadurai S, Charania MA, Zhang Y et al. Nanoparticles with surface antibody against CD98 and carrying CD98 small interfering RNA reduce colitis in mice. *Gastroenterology*. 2014;146(5):1289–300.
143. Shia J, Klimstra DS, Nitzkorski JR, Low PS, Gonen M, Landmann R et al. Immunohistochemical expression of folate receptor α in colorectal carcinoma: patterns and biological significance. *Hum Pathol*. 2008;39(4):498–505.
144. Xiong S, Yu B, Wu J, Li H, Lee RJ. Preparation, therapeutic efficacy and intratumoral localization of targeted daunorubicin liposomes conjugating folate-PEG-CHEMS. *Biomed Pharmacother*. 2011;65(1):2–8.
145. Handali S, Moghimipour E, Rezaei M, Ramezani Z, Kouchak M, Amini M et al. A novel 5-fluorouracil targeted delivery to colon cancer using folic acid conjugated liposomes. *Biomed Pharmacother*. 2018;108:1259–73.
146. Yu M, Jambhrunkar S, Thorn P, Chen J, Gu W, Yu C. Hyaluronic acid modified mesoporous silica nanoparticles for targeted drug delivery to CD44-overexpressing cancer cells. *Nanoscale*. 2013;5(1):178–83.
147. Vafaei SY, Esmaeili M, Amini M, Atyabi F, Ostad SN, Dinarvand R. Self assembled hyaluronic acid nanoparticles as a potential carrier for targeting the inflamed intestinal mucosa. *Carbohydr Polym*. 2016;144:371–81.
148. Xiao B, Han MK, Viennois E, Wang L, Zhang M, Si X et al. Hyaluronic acid-functionalized polymeric nanoparticles for colon cancer-targeted combination chemotherapy. *Nanoscale*. 2015;7(42):17745–55.
149. Xiao B, Zhang Z, Viennois E, Kang Y, Zhang M, Han MK et al. Combination therapy for ulcerative colitis: orally targeted nanoparticles prevent mucosal damage and relieve inflammation. *Theranostics*. 2016;6(12):2250.
150. Tesauro D, Accardo A, Diaferia C, Milano V, Guillon J, Ronga L et al. Peptide-based drug-delivery systems in biotechnological applications: recent advances and perspectives. *Molecules*. 2019;24(2):351.
151. Prajapati SK, Jain A, Shrivastava C, Jain AK. Hyaluronic acid conjugated multi-walled carbon nanotubes for colon cancer targeting. *Int J Biol Macromol*. 2019;123:691–703.
152. Jiang Z, Guan J, Qian J, Zhan C. Peptide ligand-mediated targeted drug delivery of nanomedicines. *Biomater Sci*. 2019;7(2):461–71. doi:10.1039/C8BM01340C.
153. Ren Y, Mu Y, Song Y, Xie J, Yu H, Gao S et al. A new peptide ligand for colon cancer targeted delivery of micelles. *Drug Deliv*. 2016;23(5):1763–72.
154. Guo F, Ouyang T, Peng T, Zhang X, Xie B, Yang X et al. Enhanced oral absorption of insulin using colon-specific nanoparticles co-modified with amphiphilic chitosan derivatives and cell-penetrating peptides. *Biomater Sci*. 2019;7(4):1493–506.
155. Christophi GP, Rengarajan A, Ciorba MA. Rectal budesonide and mesalamine formulations in active ulcerative proctosigmoiditis: efficacy, tolerance, and treatment approach. *Clin Exp Gastroenterol*. 2016;125–30.
156. Hoang T, Date AA, Ortiz JO, Young T-W, Bensouda S, Xiao P et al. Development of rectal enema as microbicide (DREAM): preclinical progressive selection of a tenofovir prodrug enema. *Eur J Pharm Biopharm*. 2019;138:23–9.
157. Xiao P, Gumber S, Marzinke MA, Date AA, Hoang T, Hanes J et al. Hypo-osmolar formulation of tenofovir (TFV) enema promotes uptake and metabolism of TFV in tissues, leading to prevention of SHIV/SIV infection. *Antimicrob Agents Chemother*. 2017;62(1):21. doi:10.1128/aac.01644-17.
158. Maisel K, Chattopadhyay S, Moench T, Hendrix C, Cone R, Ensign LM et al. Enema ion compositions for enhancing colorectal drug delivery. *J Control Release*. 2015;209:280–7.
159. Bialik M, Kuras M, Sobczak M, Oledzka E. Achievements in thermosensitive gelling systems for rectal administration. *Int J Mol Sci*. 2021;22(11):5500.
160. Abd El Razek AM, Hasan AA, Sabry SA, Mahdy MA, Hamed EE. Metoclopramide hydrochloride thermally sensitive rectal in situ gelling system, a novel out-patient treatment for vomiting in pediatric age. *J Drug Deliv Sci Technol*. 2019;50:9–17. https://www.sciencedirect.com/science/article/pii/S1773224718312036.
161. El-Sonbaty MM, Ismail HR, Kassem AA, Samy AM, Akl MA. Mucoadhesive thermoreversible formulation of metoclopramide for rectal administration: a promising strategy for potential management of chemotherapy-induced nausea and vomiting. *Pharm Dev Technol*. 2020;25(5):535–46.
162. Kumpusiri MS, Maneesatid P. The dissolution procedure: Development and validation. *Pharmacopeial Forum*. 2005;31(5):26–31.
163. Batchelor H, Butler J. Dissolution. In: H Batchelor (Ed.), *Biopharmaceutics: From Fundamentals to Industrial Practice*. Wiley, Chichester,, 2022, pp. 73–98.
164. Stamatopoulos K, Batchelor HK, Simmons MJH. Dissolution profile of theophylline modified release tablets, using a biorelevant dynamic colon model (DCM). *Eur J Pharm Biopharm*. 2016;108:9–17.
165. Wahlgren M, Axenstrand M, Håkansson Å, Marefati A, Lomstein Pedersen B. In vitro methods to study colon release: state of the art and an outlook on new strategies for better in-vitro biorelevant release media. *Pharmaceutics*. 2019;11(2):95.
166. Fadda HM, Basit AW. Dissolution of pH responsive formulations in media resembling intestinal fluids: bicarbonate

versus phosphate buffers. *J Drug Deliv Sci Technol.* 2005;15(4):273–9.
167. Fadda HM, Merchant HA, Arafat BT, Basit AW. Physiological bicarbonate buffers: stabilisation and use as dissolution media for modified release systems. *Int J Pharm.* 2009;382(1–2):56–60.
168. Wray H, Joseph R, Palmen M, Pierce D. P-0030: combined pharmacokinetic and scintigraphic analyses for the comparison of 5-ASA release profiles from MMXTM mesalamine and another delayed-release mesalamine formulation. *Inflamm Bowel Dis.* 2008;14(suppl_3):S19–20.
169. Graham H, Åman P. The pig as a model in dietary fibre digestion studies. *Scand J Gastroenterol.* 1987;22(sup129):55–61.
170. Hatton GB, Yadav V, Basit AW, Merchant HA. Animal farm: considerations in animal gastrointestinal physiology and relevance to drug delivery in humans. *J Pharm Sci.* 2015;104(9):2747–76.
171. Wong CH, Siah KW, Lo AW. Estimation of clinical trial success rates and related parameters. *Biostatistics.* 2019;20(2):273–86.
172. Elbadawi M, McCoubrey LE, Gavins FKH, Ong JJ, Goyanes A, Gaisford S et al. Harnessing artificial intelligence for the next generation of 3D printed medicines. *Adv Drug Deliv Rev.* 2021;175:113805.
173. Hales D, Tefas LR, Tomuță I, Moldovan C, Gulei D, Munteanu R et al. Development of a curcumin-loaded polymeric microparticulate oral drug delivery system for colon targeting by quality-by-design approach. *Pharmaceutics.* 2020;12(11):1027.
174. Al-Mohaya M, Mesut B. Quality by design approach for novel drug delivery systems. In: NK Jain and N Bajwa, (Eds.), *Introduction to Quality by Design (QBD): From Theory to Practice.* Springer, Nature, 2024, pp. 243–61. doi:10.1007/978-981-99-8034-5_10.
175. Gu Y, Li M. Molecular modeling. In: H Ishida (Ed.), *Handbook of Benzoxazine Resins.* Elsevier, Amsterdam, 2011, pp. 103–10.
176. Casalini T. Not only in silico drug discovery: molecular modeling towards in silico drug delivery formulations. *J Control Release.* 2021;332:390–417.
177. Bera I, Payghan P V. Use of molecular dynamics simulations in structure-based drug discovery. *Curr Pharm Des.* 2019;25(31):3339–49.
178. Al-Mohaya M, Mesut B, Kurt A, Çelik YS. In silico approaches which are used in pharmacy. *J Appl Pharm Sci.* 2024;14(4):239–53.
179. Dvořák Z, Kopp F, Costello CM, Kemp JS, Li H, Vrzalová A et al. Targeting the pregnane X receptor using microbial metabolite mimicry. *EMBO Mol Med.* 2020;12(4):e11621.
180. Baker RE, Pena J-M, Jayamohan J, Jérusalem A. Mechanistic models versus machine learning, a fight worth fighting for the biological community? *Biol Lett.* 2018;14(5):20170660.
181. European Medicines Agency. *Guideline on the Reporting of Physiologically based Pharmacokinetic (PBPK) Modelling and Simulation.* European Medicines Agency, London, UK, 2018.
182. Basu S, Lien YT, Vozmediano V, Schlender J-F, Eissing T, Schmidt S et al. Physiologically based pharmacokinetic modeling of monoclonal antibodies in pediatric populations using PK-Sim. *Front Pharmacol.* 2020;11:868.
183. Trenfield SJ, Awad A, McCoubrey LE, Elbadawi M, Goyanes A, Gaisford S et al. Advancing pharmacy and healthcare with virtual digital technologies. *Adv Drug Deliv Rev.* 2022;182:114098.
184. Harrer S, Shah P, Antony B, Hu J. Artificial Intelligence for clinical trial design. *Trends Pharmacol Sci.* 2019;40(8):577–91.

8 Nucleic Acid-Based Therapeutic Delivery System

Anjali Kumari[1], *Geeta Aggarwal*[1,2], *and Amanpreet Kaur*[1,2]
[1]School of Pharmaceutical Sciences, Department of Pharmaceutics, Delhi Pharmaceutical Sciences and Research University, New Delhi, India
[2]Centre for Advanced Formulation Technology, Department of Pharmaceutics, Delhi Pharmaceutical Sciences and Research University, New Delhi, India

8.1 INTRODUCTION

The public is now more aware of nucleic acid-based therapeutics due to the rapid global response during the pandemic caused by COVID-19 [1]. Messenger RNA (mRNA) vaccine development has progressed rapidly, but it builds upon years of research and interest among scientists. This century marks the dawn of genomic medicine, driven by advancements in genetic engineering following the completion of the human genome project. These advances have revolutionized our understanding of how genes influence biological functions in both health and disease, making the entire genome accessible for targeted therapeutic interventions using nucleic acid-based molecules [2,3]. Nucleic acid-based therapeutics, therefore, offer a promising avenue to target diseases at the genetic level, expanding treatment possibilities for conditions previously deemed difficult to treat or lacking effective options in modern medicine [4–6]. Traditionally, translational research has primarily focused on small molecules and recombinant proteins, resulting in significant advancements in medicine. Small molecules offer several benefits, including cell permeability, low immunogenicity, and the capability to be tailored synthetically for specific molecular interactions with target proteins [7,8]. However, their utility is limited to proteins with known binding pockets. Recombinant proteins, on the other hand, have been pivotal in various applications like protein replacement therapy and cancer treatments using therapeutic antibodies. However, their large size and polarity hinder their ability to cross cell membranes freely and restrict their scope of application. The structural and functional diversity of proteins makes the development of broadly applicable delivery systems difficult [9,10].

Nucleic acid-based therapeutics are changing the area of medicine, offering a promising approach to improve the treatment options by targeting the genetic roots of diseases. Nucleic acid-based therapeutics are innovative platforms for introducing genetic material, such as RNA or DNA, into target cells for therapeutic effects. These strategies hold immense potential in revolutionizing medicine by enabling precise gene editing, regulating gene expression, and modulating cellular pathways [11,12]. However, the successful delivery of these therapeutics to target cells remains a significant challenge because nucleic acids have unfavorable physicochemical properties that impede cellular transmission, immune activation, and nuclease degradation, which make them challenging to use as therapeutics. Advanced delivery technologies are necessary for delivering nucleic acid therapies to the target site, thereby achieving safety and an effective response. These technologies also protect nucleic acids from degradation, ensure stability in circulation, facilitate localization to target tissues, and promote effective intracellular delivery. The important elements of delivery systems include viral vectors such as adenoviruses, adeno-associated viruses, herpes simplex viruses, retroviruses, and lentiviruses, as well as non-viral vectors such as lipid nanoparticles, polymers, cell-penetrating peptides, and engineered exosomes.

This chapter overview covers nucleic acid-based therapeutics, their types and challenges, and how to overcome those challenges. It also covers the types of delivery systems that have facilitated clinical translation of nucleic acid-based therapeutics. We have also discussed the ongoing clinical trials and FDA-approved nucleic acid-based therapeutics.

8.2 NUCLEIC ACID-BASED THERAPEUTICS

Nucleic acids serve as the fundamental genetic units responsible for encoding proteins, offering the ability to treat a broader spectrum of diseases compared to conventional drug targets. Expanding on Watson-Crick base pairing, most of the nucleic acid-based therapies are based on the five canonical bases: uracil, thymine, guanine, cytosine, and adenine [13]. Apart from the variation in substitutions at the sugar ribose backbone's 2'-site, DNA predominantly uses thymine, while RNA substitutes it with uracil. Despite sharing identical building blocks of nucleotides, modifications can be made to nucleosides to enhance stability, reduce immunogenicity, and increase diversity in nucleic acid-based therapeutics. Nucleic acids carry a negative charge because of their sugar-phosphate backbone, varying in size from 20 nucleotides to over 1,000 base pairs [14]. Furthermore, diverse nucleic acid-based therapeutics can

 DOI: 10.1201/9781003508601-8

be designed for specific purposes like genetic knockdown or stimulating protein expression.

8.2.1 Types of Nucleic Acid-Based Therapeutics

8.2.1.1 DNA-Based Therapeutics

In DNA-based therapeutics, DNA molecules are used as the primary therapeutic substance in order to treat a variety of diseases. This therapeutic technique includes oligonucleotides for antisense, plasmids for gene therapy [15], and aptamers and CRISPR-cas9 gene editing [16,17]. DNA-based therapeutics have a number of advantages over conventional therapeutics, one of which is their high specificity and selective recognition of molecular targets. They can be applied preventively or as soon as the manifest, preventing its progression and associated complications. DNA-based therapeutics ensure specificity by targeting selected genes for ablation. Although these therapies may minimize side effects compared to conventional drugs, challenges include diminished biological stability, reduced cellular uptake, short half-lives, and unpredictable pharmacokinetics, which limit their development and success. Recent advancements in understanding DNA uptake, metabolism, and trafficking have led to improved stability and reduced degradation of DNA-based therapeutics. The stability of short DNA-based therapeutics can be enhanced by chemical derivatization, while synthetic and natural platforms deliver plasmid DNA and viral vectors which transfer therapeutic genes [18]. These delivery systems improve pharmacokinetics and targeted tissue delivery. These DNA-based therapeutics have attracted a lot of interest because they can target particular genes or pathways, providing targeted and possibly curative treatments [19].

8.2.1.2 RNA-Based Therapeutics

RNA-based therapies use sophisticated delivery systems to deliver RNA molecules like Antisense Oligonucleotides (ASOs), siRNAs, miRNAs, and mRNAs into target cells to alter gene expression or protein synthesis precisely. ASOs (antisense oligonucleotides) block RNA via steric hindrance or RNAse H activity [20,21]. siRNAs (small interfering RNA) induce mRNA degradation through RISC [22]. miRNAs (micro-RNA) modulate translation or promote degradation [23]. mRNA (messenger RNA) therapies enable protein expression and vaccination [24]. Delivery systems like LNPs enhance stability and cellular uptake, showing promise in treating cancers, genetic disorders, and infectious diseases. Several clinical trials have explored nucleic acid-based therapeutics using various delivery systems to treat specific diseases, as detailed in Table 8.1 (Figure 8.1).

8.3 DELIVERY CHALLENGES

Although nucleic acid therapeutics offer potential treatments for a wide range of diseases, as shown in Table 8.1, their widespread use is hindered by numerous delivery obstacles. These therapeutics are characterized by their high molecular weight and strong negative charge, which hinders their passive diffusion through cellular membranes [26]. Due to these physicochemical properties, delivery vectors are required to transport these therapeutics to the targeted diseased cells. Apart from the cellular membrane's physical barrier, various systemic and intracellular obstacles highlight the necessity for well-designed delivery vehicles [27]. These barriers stem from the immune system's defense mechanisms, evolved to detect and neutralize viral and bacterial genetic material. In the body, endonucleases and exonucleases interact with nucleic acids to break down foreign RNA. Additionally, the innate immune system recognized nucleic acids as pathogen-associated molecular patterns (PAMP), leading to toll-like receptor (TLR) and other pattern recognition receptor (PRR) activation downstream. To tackle these issues, numerous chemical alterations have been explored, expanding on traditional nucleoside bases and backbone structures [28,29]. The main challenges in delivering nucleic acid-based therapeutics are specificity, cellular uptake, endosomal escape, avoiding immune system recognition, and clearance.

8.4 DELIVERY SYSTEMS

This chapter covers recent developments in the delivery of nucleic acids, which emphasize vectors that are viral as well as non-viral. Viral vectors possessing high transduction efficiency, such as lentiviruses, herpes simplex viruses, adenoviruses, and adeno-associated viruses, have been associated with safety concerns such as immunogenicity and insertional mutagenesis. Non-viral vectors, which are scalable, easily modifiable, and have low immunogenicity, such as lipid nanoparticles (LNPs), polyethyleneimine (PEI), poly(amidoamine) (PAMAM) dendrimers, and other common non-viral gene vectors, present promising alternatives, such as physical methods.

8.4.1 Viral Vectors

Viral vectors excel as DNA delivery tools because they naturally target specific cell types and uniquely possess the ability to transport genetic material directly into the nucleus, a crucial step for DNA-based payloads to be functional since they are inactive outside the nucleus. This feature is immensely significant in highlighting the effectiveness of viral vectors compared to other delivery methods. Its advantages are highly efficient delivery and long-lasting expression [30]. Its disadvantages are immunogenicity and potential for insertional mutagenesis. Applications are gene therapy for genetic disorders like cystic fibrosis and hemophilia. Viral vectors have been pivotal in advancing gene therapy due to their superior ability to deliver genes compared to non-viral vectors. Additionally, they offer versatility in achieving gene expression tailored to specific requirements, whether for short-term, high-level expression as beneficial in cancer gene therapy, or for sustaining long-term expression crucial in chronic conditions

TABLE 8.1
Clinical Studies Using Nucleic Acid-Based Therapeutics [25]

Delivery System	Disease	Company	Brand Name	Target	Status	Clinical Trial
			Cancer-Related Clinical Studies Using Nucleic Acid-based Therapeutics			
ASO naked (modified)	Liver cancer	Hoffmann-La Roche	RO7070179	HIF1A	Ph 1 completed	NCT02564614
ASO Naked (modified)	Solid cancer	Ionis Pharmaceuticals, Inc.	IONIS-STAT3Rx	STAT 3	Ph 2 completed	NCT01563302
ASO LNP	Myeloid leukemia	Bio-Path Holdings, Inc	BP1001	Grb2	Ph 2 completed	NCT02781883
LNP	Mycosis fungoides, cutaneous T-cell lymphoma	miRagen Therapeutics, Inc.	Cobomarsen	miR-155	Ph 1 completed	NCT02580552
siRNA	Solid neoplasm colorectal, pancreatic, and renal cell carcinoma	Wake Forest University Health Sciences	APN401	E3 ubiquitin ligase Cbl-b	Ph 1 completed	NCT03087591
siRNA	Pancreatic cancer	Monroe Dunaway Anderson Cancer Center	KRASG12D	KRASG12D	Ph 1	NCT03608631
siRNA loaded in LNP	Solid neoplasm	M.D. Anderson Cancer Center	siRNAEPhA2-DOPC	EphA2	Ph 1 recruiting	NCT01591356
siRNA loaded in a polymeric matrix	Pancreatic ductal cancer and adenocarcinoma	Silenseed Ltd	siG12D LODER	KRASG12D	Ph 1 completed	NCT01188785
SNALP	Hepatocellular carcinoma	Arbutus Biopharma Corporation	TKM-080301	PLK1	Ph 2 completed	NCT02191878
siRNA-gold nanoparticle	Recurrent glioblastoma and gliosarcoma	Northwestern University	NU-0129	Bcl2L12	Ph 1 completed	NCT03020017
RGD-siRNA conjugate	Clear cell renal cell carcinoma	Arrowhead Pharmaceuticals	ARO-HIF2	HIF-2α	Ph 1 recruiting	NCT04169711
mRNA (naked)	Melanoma	BioNTech RNA Pharmaceuticals GmbH	IVAC MUTANOM, RBL001/ RBL002	Melanoma-associated target antigens	Ph 1 completed	NCT02035956
			RBL001/RBL002	Melanoma-associated target antigens	Ph 1 completed	NCT01684241
	Stage II and III phases of colorectal cancer	BioNTech SE	RO7198457	TAA	Ph 2 recruiting	NCT04486378
Langerhans cells electroporated with CT7, MAGEA3, and WT1 mRNA	Multiple myeloma	Memorial Sloan Kettering Cancer Center	NA	NA	completed	NCT01995708
			Clinical Studies Using Nucleic Acid-Based Therapy to Treat Infectious Diseases			
Modified ASO naked	Hepatitis B	GlaxoSmithKline	IONIS-HBVRx	Hepatitis B Virus (HPV) surface antigen	Ph 2 completed	NCT02981602
Modified siRNA	Respiratory syncytial	Alnylam Pharmaceuticals	ALNRSV01	RSV nucleocapsid gene	Ph 2 completed	NCT00658086

(Continued)

TABLE 8.1 (*Continued*)
Clinical Studies Using Nucleic Acid-Based Therapeutics [25]

Delivery System	Disease	Company	Brand Name	Target	Status	Clinical Trial
siRNA cholesterol conjugate	Hepatitis B	Arrowhead HBV gene Pharmaceuticals	ARC-520	HBV gene	Ph 1 completed	NCT02535416
GalNAc–siRNA conjugate		Arrowhead Pharmaceuticals	ARO-HBV	HBV gene	Ph 2 completed	NCT03365947
Nucleotide analog therapy	Hepatitis B chronic	Arbutus Biopharma Corporation	ARB001467	HBV genome	Ph 2 completed	NCT02631096
GalNAc–siRNA conjugate NA	Hepatitis B, chronic	Dicerna Pharmaceuticals, Inc.	DCRHBVS	NA	Ph 1 Completed	NCT03772249
		Vir Biotechnology, Inc.	VIR-2218	HBV	Ph 1 recruiting	NCT03672188
mRNA-LNP	Influenza	ModernaTX, Inc	VAL506440	H10N8 antigen	Ph 1 completed	NCT03076385
	Zika Virus		mRNA1325	Antigen proteins	Ph 1 completed	NCT03014089
	Influenza		VAL339851	H7 antigen	Ph 1 completed	NCT03345043
	Cytomegalovirus		mRNA1647/1443	CMV proteins	Ph 1 completed	NCT03382405
	Chikungunya virus		VAL181388	Viral antigens	Ph 1 completed	NCT03325075
	Human parainfluenza virus		mRNA1653	Viral antigens	Ph 1 completed	NCT03392389
Nucleoside modified mRNA: LNPs	Influenza H10N8	ModernaTX, Inc.	mRNA1440	Antigen HA	Ph 1 completed	NCT03076385
	Influenza H7N9		mRNA1851/ VAL339851	Antigen HA	Ph 1 completed	NCT03345043
	Zika Virus/Influenza		mRNA1893	Viral antigens	Ph 1 completed	NCT04064905
mRNA: LNPs	COVID-19	CureVac	CVnCoV vaccine	SARS-CoV-2 Spike Protein	Ph 1 completed	NCT04449276
Naked mRNA Rabies	Rabies	CureVac	CV7201	virus glycoprotein	Ph 1 completed	NCT02241135
Naked mRNA			CV8102	virus glycoprotein	Ph 1 completed	NCT02238756
Naked mRNA	HIV infections	CureVac	AGS-004	HIV target antigens	Ph 1 completed	NCT01069809
mRNA: LNPs	Rabies		CV7202	RABV-G glycoprotein	Ph 1 completed	NCT03713086
	COVID-19	BioNTech SE	BNT162	SARS-CoV-2 Proteins	Ph 1/2 unknown status	NCT04380701
					Ph 2/3 unknown status	NCT04368728
		Clinical Studies Using Nucleic Acid-based Therapy to Treat Liver Disease				
Conjugate (GalNAc)	Elevated triglycerides/familial hypercholesterolemia		Ionis Pharmaceuticals, Inc	IONIS ANGPTL3-LRx	Ph 1 completed	NCT02709850
Naked (modified)	Familial amyloid polyneuropathy (FAP)			IONIS-TTR-Rx	Ph 2 completed	NCT01737398
LNP	Transthyretin-mediated Amyloidosis (ATTR)		Alnylam Pharmaceuticals	ALN-TTR02	Ph 2 completed	NCT01617967
				ALN-TTR02	Ph 3 completed	NCT01960348
	Hepatic fibrosis		Bristol-Myers Sqibb	ND-L02-s0201	Ph 1 completed	NCT02227459
				ND-L02-s0201	Ph 1 completed	NCT01858935
				ND-L02-s0201	Ph 2 completed	NCT03420768
	Propionic acidemia		ModernaTX, Inc	mRNA-3927	Ph 1/2 Active	NCT04159103

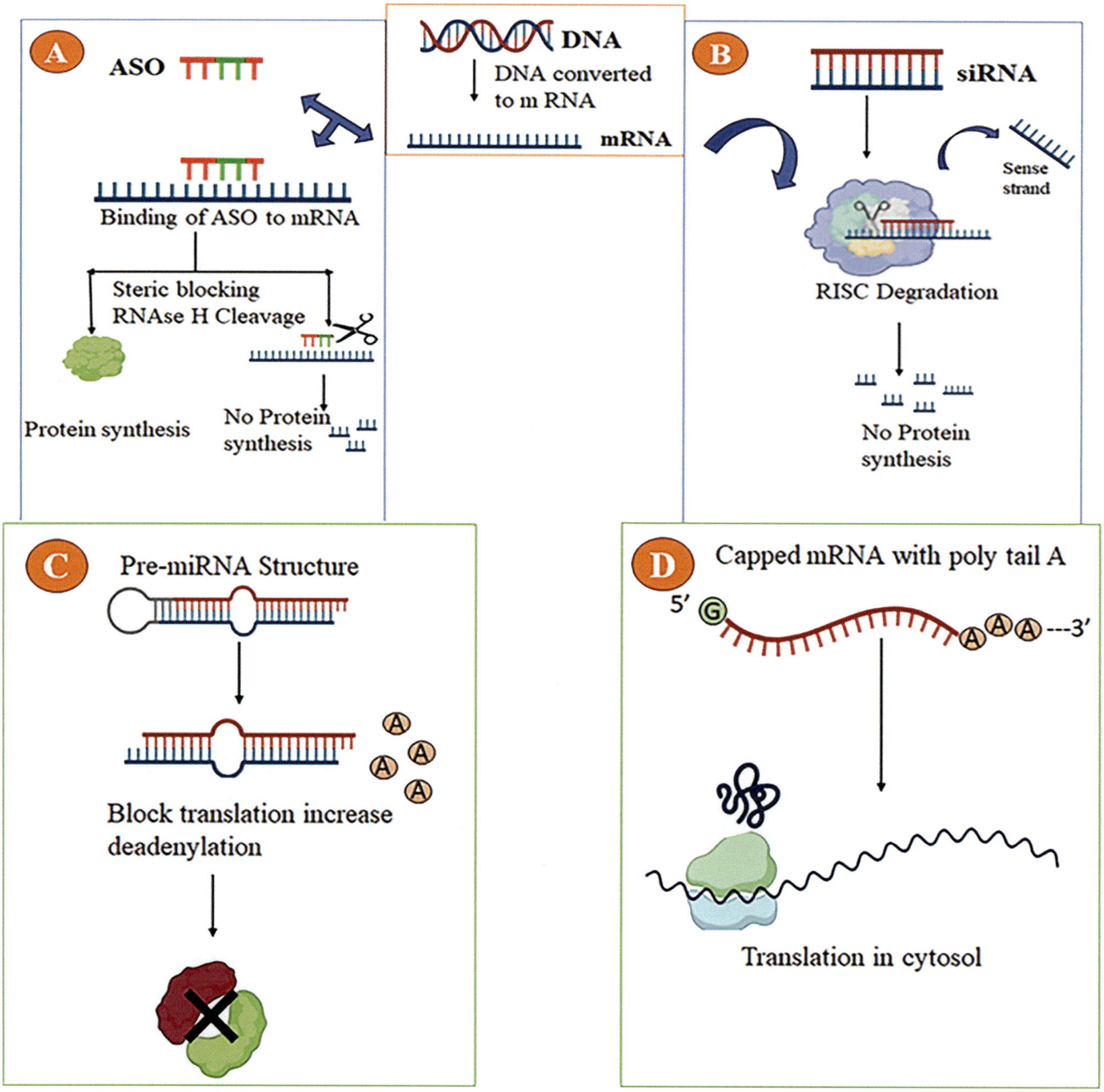

FIGURE 8.1 Types of therapeutic nucleic acids.

like hemophilia [31]. However, because there is a risk that virus progeny will spread, using viral vectors requires more stringent biosafety measures than using non-viral techniques, especially when using oncolytic or replication-competent viruses [32]. A range of viral vectors derived from DNA and RNA viruses has been designed for use in applications of gene therapy, as depicted in Figure 8.2. The type of disease being targeted has a significant impact on the choice of vector; for example, short-term expression needs are required for acute conditions like infectious diseases, whereas long-term expression needs are necessary for chronic diseases. For acute diseases, replication-deficient viral vectors can offer high-level transient expression, leading to therapeutic effectiveness. Conversely, for chronic conditions, achieving long-term expression typically involves the chromosomal integration of the viral vector/transgene or its extrachromosomal presence, ensuring prolonged therapeutic benefits. Various types of viral factors are available, such as adenovirus, Adeno-associated virus, Herpes simplex viruses, Retroviruses, Lentivirus, Alphavirus Vectors, Flavivirus Vectors, Measles Virus Vectors. Below is an extensive overview of different types of viral vectors [33].

- **Adenovirus**: Adenovirus (Ad) vectors have long been favored for gene transfer into mammalian cells [34]. The double-stranded DNA (dsDNA) genome of these non-enveloped viruses can contain

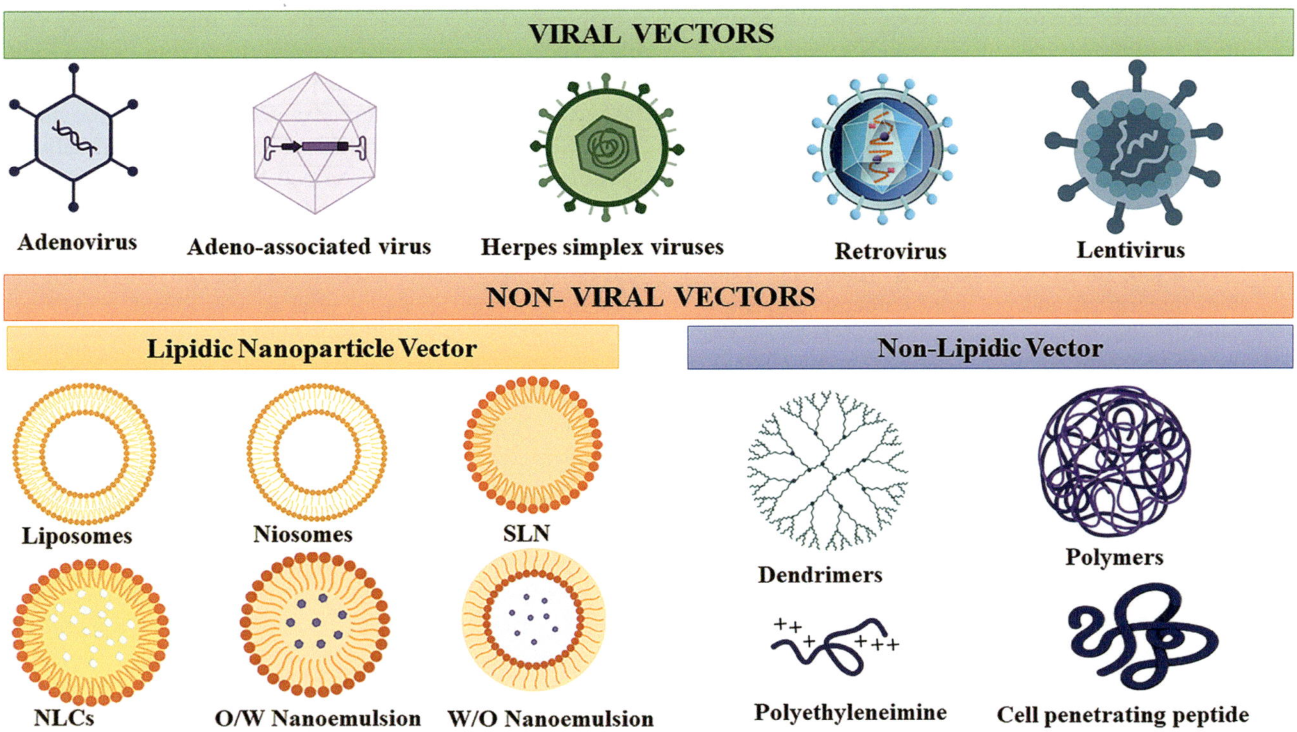

FIGURE 8.2 Viral and non-viral vectors for nucleic acid delivery.

up to 7.5 kb of foreign DNA, but modified Ad shuttle vectors can hold up to 14 kb [35]. Initial Ad vectors faced challenges due to strong immunogenicity post-E1/E3 gene removal, which was mitigated in later generations. [36]. Third-generation high-capacity adenoviral (HC-Ad) vectors are capable of carrying up to 37 kb of foreign DNA [37,38]. Oncolytic, replication-competent Ad vectors target tumor cells specifically, aiding in their elimination. Packaging cell line advancements have enabled production of good manufacturing practices (GMP)-grade Ad vectors on a large scale [39,40]. Stable extrachromosomal transgene expression is ensured by Ad vectors, which can last over a year without genomic integration. Non-human primate studies have demonstrated expression up to 7 years, with minimal negative long-term effects [41, 42].

- **Adeno-Associated Virus [AAV]**: AAV, a tiny and non-enveloped single-stranded DNA (ssDNA) virus, is usually limited to 4 kb of foreign DNA [43]. However, new developments, such as trans-splicing, overlapping, or fragmented dual AAV vectors, have enhanced this ability [44,45]. Generally safe, AAV vectors can elicit immune responses upon repeated administration, which can be mitigated by using different serotypes for re-administration or employing exosome-associated AAV (Exo-AAV) for reduced doses and lower immune responses [46]. Exo-AAV8 vectors have shown achievement in achieving prolonged gene transfer directed by the liver [47]. Although mostly extrachromosomal, genes delivered by AAV can integrate into the host genome, with enhanced integration potential using 28S ribosomal DNA homology sequences [48].
- **Herpes Simplex Viruses**: These are large enveloped dsDNA viruses known for causing latent infections in neural ganglia [49]. Traditional antivirals do not effectively treat latent Herpes Simplex Virus (HSV) in the ganglia, leading to reactivation once therapy is stopped. Engineered HSV expression vectors allow for long-lasting transgene expression [50]. These linear viruses circularize into viral episomes inside the nucleus, remaining extrachromosomal without integration [51]. HSV vectors boast a high capacity and accommodate over 30 kb of foreign genetic material [52]. They can be engineered HSV amplicons with a packaging capability of 150 kb of DNA [53]. Despite their strong cytopathogenicity, this has been mitigated by deleting non-essential genes from the HSV genome [54]. Additionally, introducing micro-RNA sequences (miR-145) into the HSV ICP27 gene has led to the development of oncolytic HSV vectors, selectively reducing cell proliferation in non-small cell lung cancer (NSCLC) cells [55]. Advanced HSV packaging systems like the helper virus-free system for HSV amplicons, using an ICP27-deleted, oversized HSV-1 DNA within a bacterial artificial chromosome (BAC), have been engineered [56].

- **Retroviruses**: Retroviruses (RVs) that are enveloped in single-stranded RNA (ssRNA) can encapsulate up to 8 kb of foreign sequence [57]. Reverse transcriptase activity, which permits RVs to create dsDNA copies from the RNA genome for integration into the host genome, is one of their distinctive characteristics [58]. Long-term transgene expression is supported by chromosomal integration; however, concerns regarding random integration and potential leukemia development in severe combined immunodeficiency-X1 (SCID-X1) patients have been raised. To address this, self-inactivating γRV (SIN-γRV) vectors have been developed, proving safe in clinical trials with no observed adverse integration or leukemia cases [59]. Interestingly, ADA-SCID patients treated with γRV show a lower risk of insertional oncogenesis compared to other inherited immunodeficiencies [60]. For example, in a clinical study, none of the ten patients developed leukemia [61], and just one of the 50 ADA-SCID patients receiving γRV exhibited leukemia-like symptoms [62]. Additionally, to enable the large-scale production of GMP-grade materials, packaging cell lines specifically designed for RV vectors are used [63]. Among the main disadvantages of using RV vectors for gene therapy is that they can only transduce cells that are dividing, not those that are not.
- **Lentivirus**: lentivirus (LV) vectors, which are also members of the retrovirus family, have the capacity to transduce mammalian cells that are dividing and those that are not [64]. Similar to RVs, LV vectors can carry up to 8 kb of foreign DNA and have a single-stranded RNA genome. Although insertional oncogenesis and rare adverse events have been reported, LV vectors are recognized for their negligible cell cytotoxicity and "semi-random" chromosomal integration, which improves their biosafety profile and applications in clinical trials [65]. For instance, modifying the integration pattern of human immunodeficiency virus-1 (HIV-1) by combining specific regions of proteins reduced the frequency of integration events [66]. LV vectors have also been engineered using expression systems derived from non-human viruses like the feline immunodeficiency virus, equine infectious anemia virus, and simian immunodeficiency virus [67,68]. In addition, producer cell lines are created to enable large-scale production [69]. But obstacles like low titers and toxicity have prevented them from being widely used.

8.4.2 Non-Viral Vectors

Non-viral gene vectors provide a number of benefits over viral vectors, including numerous raw material options, a flexible chemical composition, an easily adjustable topology, a high capacity for loading DNA, and improved safety profiles. Non-viral vectors encompass various types such as Lipid nanoparticles (LNPs), exosomes, cationic polymers (e.g., PEI, PAMAM), and polysaccharide macromolecules (like chitosan (CS), and cyclodextrin (CD)) [70–72]. Non-viral delivery systems depend on a phase of cell fission where the nuclear membrane collapses and then reforms, enabling genetic material to diffuse inside the nucleus passively. Currently, LNPs and cationic polymeric-based carriers are the most promising candidates among non-viral gene vectors [73]. For instance, LNP-formulated mRNA vaccines have also been approved by the FDA quickly to combat the COVID-19 pandemic.

8.4.2.1 Lipid Nanoparticles [LNP]

To enhance the entry of large, charged macromolecules like nucleic acids through biological membranes, lipid polymorphic phases can temporarily reduce permeability barriers. To overcome these challenges, Felgner et al.'s work on cationic lipids pioneered lipid-mediated delivery systems [74]. LNPs have been extensively used in the clinical development of non-viral gene therapy vehicles. Typically, LNPs comprise ionizable lipids like N-[1-(2,3-dioleoyloxy) propyl]-N,N,N-trimethylammonium chloride (DOTAP) and N-[1-(2,3-dioleyloxy)propyl]-N,N,N-trimethylammonium chloride (DOTMA), along with neutral lipids such as cholesterol and phospholipid, and PEG-modified lipids. Cationic lipids are composed of three components: a cationic head, a linking part, and hydrophobic tails. Lipids, especially cationic lipids, are widely used in the delivery of siRNA, DNA, and mRNA due to their high capacity for carrying genetic material, degradability, practicality, adaptable structure, and charge, as well as their scalable manufacturing process. Additionally, Different commercial transfection reagents, such as dioctadecylamido-glycylspermine (DOGS), DOTMA, DOTAP, lipofectamine 2000, and lipofectamine 3,000 have been explored by scientists [75,76]. Several research findings highlight the significant impact of factors such as charge density, composition, head and tail numbers, hydrophobicity/hydrophilicity, linking groups, and shape (rod or spherical) on gene transfection efficiency. Two COVID-19 mRNA vaccines were announced by BioNTech-Pfizer and Moderna on November 18, 2020. LNPs are essential for supplying host cells with mRNA sequences that encode the SARS-CoV-2 spike protein, which is ultimately necessary to fight SARS-CoV-2 infection [77]. For example, LNP delivery technology is used by Patisiran (marketed as Onpattro) to treat hereditary transthyretin amyloidosis (hATTR), stemming from gene mutations causing abnormal aggregation of transthyretin (TTR) protein. This condition leads to systemic pathologies affecting various organs, given TTR's secretion by hepatocytes. Like Tegsedi ASO, before Onpattro was developed, patients with hATTR could only receive a liver transplant. Onpattro capitalizes on LNP-siRNA buildup in hepatocytes (TTR source) and their gene silencing power to suppress mutated TTR expression, reducing the amount of nucleic acid needed for effective silencing [78].

8.4.2.2 Polyethyleneimine [PEI]

PEI has been recognized as one of the most promising potential vectors for the delivery of nucleic acids or genes. This is due to its unique structure, having three nitrogen atoms for every one nitrogen atom and multiple primary, secondary, and tertiary amines in branched PEI [79]. These features enable PEI to exhibit superior transfection properties, facilitate endosomal escape through proton buffering, and achieve high gene condensation due to its strong electric charge. In in vitro transfection studies, linear PEI (25 kDa) is frequently utilized as a positive control and is commonly considered the standard for transfection. Due to its strong positive charge, high molecular weight PEI exhibits excellent transfection efficiency but has high cytotoxicity. On the other hand, low-molecular-weight PEI is less toxic but has a lower transfection efficiency, which limits its application [80]. To address these issues, strategies such as introducing degradable groups, cross-linking low-weight PEI for improved transfection, grafting functionalized monomers for enhanced targeting, and composite structures made of different materials have been investigated as a way to balance the cytotoxicity and efficiency of PEI-based gene vectors. Although these strategies still face challenges in clinical trials, they hold promise for addressing the issues. Chen et al. developed a high-performance gene vector by combining PEI with graphene oxide (GO) through carboxyl-amino reactions. This branch PEI-GO vector presented enhanced efficiency of transfection and decreased toxicity compared to PEI (25 kDa) [81]. Similarly, other composite systems like PEI/nanosilicon, bPEI/quantum dots, bPEI/nanohydroxyapatite, and PEI/carbon nanotubes have been explored for diverse applications. Given the challenge of single-phase compounds meeting all needs, multi-component PEI composite systems represent an encouraging avenue for research. Additionally, Hu et al. demonstrated that conjugating indolicidin to PEI enhanced its transfection efficiency, highlighting the potential of hydrophobic modifications [82]. Numerous studies have emphasized the benefits of branched polymers in gene therapy; however, they face drawbacks such as imprecise structure and limited systematic research, which impede the clinical use of PEI-based vectors. Exploring branched polymer vectors could be highly impactful, involving tasks like precisely synthesizing functionalized branched PEI, creating organic-inorganic PEI nanoparticles, and conducting comprehensive animal and preclinical trials.

8.4.2.3 PAMAM Dendrimers

PAMAM dendrimers are highly symmetrical spherical polymers with strong gene transfection abilities due to their abundance of secondary and tertiary amines. Their transfection efficiency is influenced by the generation number, with partially degraded dendrimers often showing better performance [83]. Despite their precise structure and low polydispersity index, challenges such as complex preparation, limited scalability, poor degradation, and high toxicity impede their clinical use [84]. Dendrimer modification methods often involve terminal functionalization, creating diverse topologies, and developing composite systems like PAMAM dendrimer/nanoparticle combinations. Liu and colleagues devised a PAMAM dendrimer grafted with boronic acid for the delivery of proteins that utilizes nitrogen-borate complexation along with cation-π and ionic interplay of the polymer and protein (126). The effectiveness of the boronic acid-grafted dendrimer in delivering various proteins, including trypsin, ribonuclease A, fluorescent proteins, Cas9 ribonucleoprotein β-galactosidase, and saporin, was demonstrated in cell experiments [85]. This vector overcomes delivery challenges posed by proteins with uncertain charges, providing an adaptable platform for protein delivery and CRISPR-Cas9 gene editing. While hyperbranched polymers are mainly used for genetic materials and small molecule drugs, hyperbranched polymers grafted with boric acid may be created for protein delivery. Functional modifications like fluorine grafting or guanidinobenzoic acid grafting on PAMAM dendrimers enhance their application range, although challenges like steric hindrance and non-degradability persist. Strategies like introducing disulfide bonds aim to address the non-degradability issue in PAMAM dendrimers [86].

8.4.2.4 Polysaccharide

Developing suitable vehicles for delivering new macromolecules emerging from the biotech industry is a substantial difficulty facing scientists studying pharmaceuticals. Since oligonucleotides, proteins, peptides, and genes are all volatile substances, they must be shielded from deterioration in the biological environment. Additionally, their limited ability to get past biological barriers and reach their target sites significantly limits their efficacy. Therefore, the development of a reliable delivery mechanism for administering these molecules internally into the body is essential to realizing their potential as therapeutic agents. In the process of creating these delivery systems, a number of approaches have been looked into, with biodegradable nanoparticles drawing a lot of attention [87]. Notably, polysaccharides, such as cyclodextrin and chitosan, show considerable promise.

Cyclodextrin (CD) is a naturally occurring cyclic oligosaccharide molecule that can be divided into three groups according to the total number of connected glucose units: α-, β-, and γ-CD. CD is well known for having several reactive sites, a high cargo capacity, and high biocompatibility [88]. Modified CDs are widely used for gene delivery, with CD-based vectors employed in the first non-viral gene therapy clinical trials. However, due to their lack of charge, low transfection efficiency is frequently caused by CDs' inadequate proton buffer capacity and DNA condensation [89]. Many approaches are being studied to enhance the cyclodextrin vectors' transfection efficiency, including coating cyclodextrin surfaces with cationic oligomers to boost positive charge; using cyclodextrin as a core to produce star-shaped vectors, whose topological structure increases transfection efficiency; and cyclodextrin terminal groups being changed for particular uses, such as supramolecular self-assembly or cancer cell targeting. For example, CD-based hybrid vector outperformed its PLGA-based counterparts, Lipofectamine 2,000 and branched PEI (25 kDa), in terms of transfection efficiency and cytotoxicity. Furthermore, the hybrid vector efficiently delivered antisense oligonucleotides, which inhibited the growth of tumor cells and encouraged cell death [90]. In order to create a star-shaped CD-based vector, Loh and colleagues used CD as the core and poly [2-(dimethylamino)ethyl methacrylate] (pDMAEMA) copolymer as the arms. It is evident from this that topological structure is essential for enhancing transfection efficiency [91].

Chitosan (CS) is a positively charged, biodegradable polymer produced by deacetylation of chitin through chemical or enzymatic hydrolysis [92]. It is mainly used for nucleic acid delivery due to its advantages, which include decreased immunogenicity, low toxicity, biodegradability, and better biocompatibility [93]. Chitosan binds to nucleotides to create complexes that shield them from degradation and increase the effectiveness of transfection. Parameters like molecular weight, degree of deacetylation, pH, and N/P ratio affect these complex characteristics and efficacy. Despite these benefits, issues like aggregation, poor proton buffering, and low transfection efficiency remain. With the help of PEI and CS, Tripathi et al. created a targeted gene vector that significantly increased gene expression in mouse peritoneal macrophages [94]. Similarly, Zhao et al. linked CS and PEI via disulfide bonds, resulting in a biodegradable CS-based vector that effectively delivers the bone morphogenetic protein 2 (BMP2) gene to promote osteoblast activity and mineralization. Multifunctional CD-based vectors should be created, taking charge density, targeting, and topological structure into account in order to overcome constraints [95].

8.4.2.5 Cell-Penetrating Peptides [CPPs]

CPPs, also referred to as protein transduction domains (PTDs), were discovered several years ago. These peptides, typically 6–30 amino acids in length, can either be synthetic or naturally occurring. They can carry different payloads across biological membranes without losing their functionality. Since their initial discovery, the utilization of CPPs as vectors has experienced significant growth. These peptides can transport a diverse range of cargoes, including tiny peptides, complete proteins, liposomes, nanoparticles, viral particles, and nucleic acids such as RNA and DNA, as well as fluorescent imaging probes and radioisotopes [96,97]. Despite the few clinical trials using CPPs, a number of important difficulties must be addressed before CPP-based therapeutics can be widely used. These challenges include the possibility of CPP-containing proteins or peptides being immunogenic, particularly in long-term therapies; the lack of oral bioavailability, requiring topical or intravenous administration; the non-specificity of cationic and hydrophobic CPPs, which impacts therapeutic efficacy; and the need for rigorous toxicity testing prior to clinical trials [97]. Additionally, the high production costs and potential expense of CPP-based therapies, though offset by improved efficacy in time-limited treatments like cancer, remain a concern for chronic conditions necessitating frequent dosing [98].

8.4.3 Physical Methods

8.4.3.1 Electroporation

Electroporation is a powerful delivery method that utilizes brief, high-voltage pulses to temporarily increase cell membrane permeability, enabling the efficient uptake of nucleic acids, including DNA, RNA, and oligonucleotides. It offers high transfection efficiency, even in challenging cell types like stem and primary cells, making it valuable for gene therapy and personalized medicine. Despite concerns about cell viability and off-target effects, ongoing refinements aim to enhance safety and specificity, positioning electroporation as a promising tool for future nucleic acid-based therapeutic advancements [99–101].

8.4.3.2 Sonoporation

Sonoporation, a novel method using ultrasound waves with microbubbles or nanoparticles, enhances cell membrane permeability to deliver nucleic acids like DNA and RNA [102]. It offers high transfection efficiency, precise spatial-temporal control, and broad applicability in vitro, ex vivo, and in vivo. Though optimization challenges remain, advancements in ultrasound and nanotechnology are enhancing its safety and efficacy, making sonoporation a promising approach for gene therapy and personalized medicine [103].

8.4.3.3 Exosomes

Exosomes are nanosized, cell-secreted vesicles that naturally transport nucleic acids and proteins, making them promising carriers for nucleic acid therapeutics. They offer advantages such as low immunogenicity, biocompatibility, and protection of therapeutic cargo. Exosomes can deliver various nucleic acid, such as DNA, mRNA, miRNA, siRNA, and CRISPR/Cas9 with positive preclinical outcomes [104]. Challenges include large-scale production, low loading efficiency, and targeting precision. Advances in microfluidics, lipid nanoparticles, and imaging technologies are improving these limitations. With continued research and optimization, exosome-based delivery systems may revolutionize personalized medicine and gene therapy [105] (Table 8.2).

8.5 OUTLOOK AND CONCLUSION

Although nucleic acid-based therapeutics have been studied for many years, this technology did not receive much attention until the COVID-19 pandemic. The amount of preclinical and clinical data supporting the efficacy and safety of nucleic acid-based therapeutics has increased significantly during the last two years. Patients who

TABLE 8.2
FDA and European Medicines Agency (EMA)-Approved Nucleic Acid Therapies

Category	Product (Generic Name)	Product (Brand Name)	Gene Target	Administration	Indication	Company
				2013		
ASOs	Mipomersen	KYNAMRO	Apolipoprotein B-100	SQ	Hypercholesterolemia	Ionis Pharmaceuticals
				2016		
ASOs	Eteplirsen	EXONDYS 51	Dystrophin (exon 51)	IT	Duchenne muscular dystrophy	Sarepta Therapeutics
ASOs	Nusinersen	SPINRAZA	Survival of motor neuron 2	IT	Spinal muscular atrophy	Ionis Pharmaceuticals
				2017		
AAV vectors	Voretigene neparvovec-rzyl	LUXTURNA	RPE65 (AAV2)	SRF	Leber congenital amaurosis	Spark Therapeutics
				2018		
ASOs	Inotersen	TEGSEDI	TTR	SQ	TTR-mediated amyloidosis	Ionis Pharmaceuticals
LNP-RNA	Patisiran	ONPATTRO	TTR siRNA	IV	TTR-mediated amyloidosis	Alnylam Pharmaceuticals
				2019		
ASOs	Volanesoren	WAYLIVRA	Apolipoprotein CIII	SQ	Familial chylomicronemia syndrome	Ionis Pharmaceuticals/ Akcea
ASOs	Golodirsen	VYONDYS 53	Dystrophin (exon 53)	SQ	Duchenne muscular dystrophy	Sarepta Therapeutics
GalNAc-siRNA conjugates	Givosiran	GIVLAARI	ALAS1	SQ	Acute hepatic porphyrias	Alnylam Pharmaceuticals
AAV vectors	Onasemnogene abeparvovec	ZOLGENSMA	SMN1 (AAV9)	IV	Spinal muscular atrophy	AveXis/Novartis
				2020		
GalNAc-siRNA conjugates	Inclisiran	LEQVIO	PCSK9	SQ	Hypercholesterolemia	Novartis/Alnylam Pharmaceuticals
GalNAc-siRNA conjugates	Lumasiran	OXLUMO	Glycolate oxidase	SQ	Primary hyperoxaluria type 1	Alnylam Pharmaceuticals
LNP-RNA	Tozinameran	COMIRNATY	SARS-CoV-2 spike protein mRNA	IM	COVID-19 (FDA, emergency use; Switzerland, full approval)	BioNTech/Pfizer
LNP-RNA	mRNA-1273	mRNA-1273			COVID-19 (FDA, emergency use)	(Moderna/NIAID/ BARDA)
				2021		
ASOs	Casimersen	AMONDYS 45	Dystrophin (exon 45)	SQ	Duchenne muscular dystrophy	Sarepta Therapeutics
Adenovirus vectors	AZD1222, ChAdOx1 nCoV-19	VAXZEVRIA	SARS-CoV-2 spike protein DNA (ChAdOx1)	IM	COVID-19 (emergency use)	AstraZeneca
Adenovirus vectors	Ad26.COV2.S	Ad26.COV2.S	SARS-CoV-2 spike protein DNA (Ad26)			Johnson & Johnson
Adenovirus vectors	Ad5-nCoV	CONVIDECIA	SARS-CoV-2 spike protein DNA (Ad5)		COVID-19 (Approved in China)	CanSinoBIO

(Continued)

TABLE 8.2 (*Continued*)

FDA and European Medicines Agency (EMA)-Approved Nucleic Acid Therapies

Category	Product (Generic Name)	Product (Brand Name)	Gene Target	Administration	Indication	Company
				2022		
Cell-based gene therapy	Betibeglogene autotemcel	ZYNTEGLO	The βA-T87Q-globin gene is transduced into Hematopoietic Stem Cell (HSC) using Lentiviral Vector (LVV)	IV infusion	Patients with β-thalassemia, both adult and pediatric, who need frequent RBC transfusions	Bluebird bio Inc,
LVV	Ciltacabtagene autoleucel	CARVYKTI	BCMA on the cell surface	IV infusion	Refractory or relapsed multiple myeloma	Janssen Biotech, Inc.
Adeno-associated virus vector-based gene therapy	Etranacogene dezaparvovec-drlb	HEMGENIX	Human coagulation factor IX with a Padua variant	IV	Congenital Factor IX deficiency, or hemophilia B	CSL Behring LLC
				2023		
LVV	Lovotibeglogene autotemcel	LYFGENIA	HSCs transduced with BB305 (HbAT87Q), a non-replicating LVV	IV infusion	A medical history of vaso-occlusive events and sickle cell disease	Bluebird bio, Inc
				2024		
Cell-based gene therapy	Atidarsagene autotemcel	LENMELDY	Arylsulfatase (ARSA) gene (ARSA LVV)	IV infusion	Metachromatic leukodystrophy that presents as pre-symptomatic late infantile, early symptomatic early juvenile or pre-symptomatic early juvenile	Orchard Therapeutics (Europe) Limited

previously had few or no treatment options now have valuable options due to currently approved nucleic acid therapeutics. Advancements in delivery systems are essential to fulfill this promise and make these cutting-edge treatments available to patients everywhere. Nucleic acid-based therapeutics faced several challenges, such as off-target effects, cellular uptake, endosomal escape, and lack of specificity. In order to overcome these challenges, some modifications are required for successfully transferring the genes into the host for the treatment of the aforementioned chronic diseases.

In this chapter, we have discussed various approved nucleic acid therapeutics for the treatment of cancer, infectious diseases, and liver diseases. We have also focused on various delivery systems, such as viral and non-viral vectors, which are responsible for the transfer of genes to the target site without any interaction with biological barriers. Considerable interest has been shown in expanding potential applications, particularly for LNP as non-viral vectors and AAV vector systems as viral vectors, for improving tissue specificity (e.g., for ASO and siRNA conjugates), which gives confidence in their successful clinical application. However, the viral vector AAV has a low loading capacity of drugs, and it can also integrate with the human genome, which is why it may be difficult to accept it in clinical trials. On the other hand, non-viral vectors are ideal vectors for the delivery of genes to the target site because of their good loading capacity and safety profile. Hence, it is easily acceptable by the FDA due to these benefits. Then again, in some cases, these vectors stimulate immunogenicity *in vivo* and are also prone to improper uptake by the digestive organs; therefore, researchers have added modifications or peptides to prevent immune clearance. Therefore, exosomes derived from organisms and proteins & peptides are also used as non-viral vectors to prevent immune clearance and help deliver the drug at the target site [106].

Most of the FDA-approved nucleic acid-based therapeutics are tolerated effectively, but these therapeutics frequently require pre- or co-medication to avoid immunological reactions or unwanted infusions. For example, a patient taking Onpattro is also required to take a combination of antihistamines, dexamethasone, and acetaminophen, at least 60 min before starting treatment for effective results [106]. Nevertheless, the clinical results of these therapeutics balance the risk of possible adverse reactions because most of the approved nucleic acid-based therapeutics cure rare or monogenic disorders for which there would otherwise be few or no treatments available. Nucleic acid-based therapeutics are currently showing clinical advantages in patients due to their long-lasting or even curative effects. Delivery technologies play an important role in their clinical translation due to the promotion of internalization, enhancement of stability, and target affinity.

The clinical translation of gene-editing treatments using nuclease-induced double-strand breaks or single-base DNA mutations is facilitated by platform technologies that carry large nucleic acid payloads. Unlike traditional gene transfer techniques, gene editing can treat disorders caused by dominant-negative mutations, making these methods very promising. Additionally, gene editing enables precise engineering, likely resulting in physiologically accurate gene expression, such that gene insertion or correction is regulated by an endogenous promoter. Apart from in vitro methods, a number of *in vivo* gene editing treatments, facilitated by vector technology such as LNP or AAV, are being evaluated for their potential to treat metabolic disorders, retinal dystrophies, and hATTR in ongoing clinical trials.

Recently, more is being done in the rapidly emerging field of nucleic acid-based therapeutics delivery to address the current challenges and develop new delivery strategies. Future developments that might occur include smart nanoparticles, which will be able to precisely deliver information-based cellular signals; a bioinformatics tool using genetic information for providing a personalized delivery system; and gene-editing technologies providing accurate genome editing using CRISPR-based delivery systems.

To sum up, the platform technologies examined in this chapter have made nucleic acid therapeutics a reality in clinical settings. Originally designed to treat rare diseases, these delivery systems are now being used to develop more broadly ranging genetic medications and are also helping to accelerate the development of pandemic vaccines. They also make it possible to apply cutting-edge techniques like gene-editing medicines in clinical settings. Nucleic acid therapeutics have the potential to completely transform the way that many previously incurable diseases are treated, despite obstacles related to toxicity, manufacturing, and socioeconomic issues.

SQ, subcutaneous; IT, intrathecal; SRF, subretinal fluid; TTR, transthyretin; IV, intravenous, IM, intramuscular, HSC, hematopoietic stem cells; LVV, lentiviral vector; RBC, red blood cell; BCMA, B-cell maturation antigen.

REFERENCES

1. Thanh Le T, Andreadakis Z, Kumar A, Gómez Román R, Tollefsen S, Saville M et al. The COVID-19 vaccine development landscape. *Nat Rev Drug Discov.* 2020;19(5):305–6.
2. Patil SD, Rhodes DG, Burgess DJ. DNA-based therapeutics and DNA delivery systems: a comprehensive review. *AAPS J.* 2005;7(1):E61–77.
3. Islam MA, Reesor EKG, Xu Y, Zope HR, Zetter BR, Shi J. Biomaterials for mRNA delivery. *Biomater Sci.* 2015;3(12):1519–33.
4. Kole R, Krainer AR, Altman S. RNA therapeutics: beyond RNA interference and antisense oligonucleotides. *Nat Rev Drug Discov.* 2012;11(2):125–40.
5. Kaczmarek JC, Kowalski PS, Anderson DG. Advances in the delivery of RNA therapeutics: from concept to clinical reality. *Genome Med.* 2017;9(1):60.
6. Ahmadzada T, Reid G, McKenzie DR. Fundamentals of siRNA and miRNA therapeutics and a review of targeted nanoparticle delivery systems in breast cancer. *Biophys Rev.* 2018;10(1):69–86.
7. Fenton OS, Olafson KN, Pillai PS, Mitchell MJ, Langer R. Advances in biomaterials for drug delivery. *Adv Mater.* 2018;30(29):e1705328.
8. Tibbitt MW, Dahlman JE, Langer R. Emerging frontiers in drug delivery. *J Am Chem Soc.* 2016;138(3):704–17.
9. Tai W, Zhao P, Gao X. Cytosolic delivery of proteins by cholesterol tagging. *Sci Adv.* 2020;6(25):eabb0310.
10. Stewart MP, Langer R, Jensen KF. Intracellular delivery by membrane disruption: mechanisms, strategies, and concepts. *Chem Rev.* 2018;118(16):7409–531.
11. Ingle RG, Fang WJ. An overview of the stability and delivery challenges of commercial nucleic acid therapeutics. *Pharmaceutics.* 2023;15:522.
12. Takakura Y. Nucleic acid drug delivery and targeting. Vol. 28, *Pharm Res.* 2011. p. 691–3.
13. Limbach PA, Crain PF, McCloskey JA. Summary: the modified nucleosides of RNA. *Nucleic Acids Res.* 1994;22(12):2183–96.
14. Willem de Vries J, Schnichels S, Hurst J, Strudel L, Gruszka A, Kwak M et al. DNA nanoparticles for ophthalmic drug delivery. *Biomaterials.* 2018 Mar;157:98–106.
15. Pan X, Veroniaina H, Su N, Sha K, Jiang F, Wu Z, et al. Applications and developments of gene therapy drug delivery systems for genetic diseases. Vol. 16, *Asian J Pharm Sciences. Shenyang Pharm Univ.* 2021. p. 687–703.
16. Riccardi C, Napolitano E, Platella C, Musumeci D, Melone MAB, Montesarchio D. Anti-VEGF DNA-based aptamers in cancer therapeutics and diagnostics. *Med Res Rev.* 2021;41:464–506.
17. Pickar-Oliver A, Gersbach CA. The next generation of CRISPR–Cas technologies and applications. *Nat Rev Mol Cell Biol.* 2019;20(8):490–507.
18. Egli M, Manoharan M. Chemistry, structure and function of approved oligonucleotide therapeutics. Nucleic Acids Res. 2023;51(6):2529–73. doi: 10.1093/nar/gkad067.
19. Sussman C, Liberatore RA, Drozdz MM. Delivery of DNA-based therapeutics for the treatment of chronic diseases. *Pharmaceutics.* 2024;16(4):535.
20. Dobrowolski C, Paunovska K, Hatit MZC, Lokugamage MP, Dahlman JE. Therapeutic RNA delivery for COVID and other diseases. *Adv Healthc Mater.* 2021;10(15).
21. Roberts TC, Langer R, Wood MJA. Advances in oligonucleotide drug delivery. *Nat Rev Drug Discov.* 2020;19(10):673–94.

22. Kanasty R, Dorkin JR, Vegas A, Anderson D. Delivery materials for siRNA therapeutics. *Nat Mater.* 2013;12(11):967–77.
23. Ambros V. The functions of animal microRNAs. *Nature.* 2004;431(7006):350–5.
24. Kirschman JL, Bhosle S, Vanover D, Blanchard EL, Loomis KH, Zurla C et al. Characterizing exogenous mRNA delivery, trafficking, cytoplasmic release, and RNA–protein correlations at the level of single cells. *Nucleic Acids Res.* 2017;45(12):e113–e113.
25. ClinicalTrials.gov [Internet]. Bethesda (MD): National Library of Medicine (US).
26. Lipinski CA, Lombardo F, Dominy BW, Feeney PJ. Experimental and computational approaches to estimate solubility and permeability in drug discovery and development settings. *Adv Drug Deliv Rev.* 1997 Jan;23(1–3):3–25.
27. McMahon HT, Gallop JL. Membrane curvature and mechanisms of dynamic cell membrane remodelling. *Nature.* 2005;438(7068):590–6.
28. Jensen S, Thomsen AR. Sensing of RNA viruses: a review of innate immune receptors involved in recognizing RNA virus invasion. *J Virol.* 2012;86(6):2900–10.
29. Yin H, Kanasty RL, Eltoukhy AA, Vegas AJ, Dorkin JR, Anderson DG. Non-viral vectors for gene-based therapy. *Nat Rev Genet.* 2014;15(8):541–55.
30. Bulcha JT, Wang Y, Ma H, Tai PWL, Gao G. Viral vector platforms within the gene therapy landscape. *Signal Transduct Target Ther.* 2021;6(1):53.
31. Rothe M, Modlich U, Schambach A. Biosafety challenges for use of lentiviral vectors in gene therapy. *Curr Gene Ther.* 2014;13(6):453–68.
32. Fukuhara H, Ino Y, Todo T. Oncolytic virus therapy: a new era of cancer treatment at dawn. *Cancer Sci.* 2016;107(10):1373–9.
33. Lundstrom K. Viral vectors in gene therapy: where do we stand in 2023? *Viruses.* 2023;15(3):698.
34. Tatsis N, Ertl HCJ. Adenoviruses as vaccine vectors. *Molecular Therapy.* 2004 Oct;10(4):616–29.
35. Ehrke-Schulz E, Zhang W, Schiwon M, Bergmann T, Solanki M, Liu J et al. Cloning and large-scale production of high-capacity adenoviral vectors based on the human adenovirus type 5. *J Visual Exp.* 2016;(107).
36. Crystal RG. Adenovirus: the first effective *in vivo* gene delivery vector. *Hum Gene Ther.* 2014 Jan;25(1):3–11.
37. Wen S, Schneider DB, Driscoll RM, Vassalli G, Sassani AB, Dichek DA. Second-generation adenoviral vectors do not prevent rapid loss of transgene expression and vector DNA from the arterial wall. *Arterioscler Thromb Vasc Biol.* 2000 Jun;20(6):1452–8.
38. Ricobaraza A, Gonzalez-Aparicio M, Mora-Jimenez L, Lumbreras S, Hernandez-Alcoceba R. High-capacity adenoviral vectors: expanding the scope of gene therapy. *Int J Mol Sci.* 2020;21(10):3643.
39. Wang F, Wang Z, Tian H, Qi M, Zhai Z, Li S et al. Biodistribution and safety assessment of bladder cancer-specific recombinant oncolytic adenovirus in subcutaneous xenografts tumor model in nude mice. *Curr Gene Ther.* 2012;12(2):67–76.
40. Wei Q, Fan J, Liao J, Zou Y, Song D, Liu J et al. Engineering the rapid adenovirus production and amplification (RAPA) cell line to expedite the generation of recombinant adenoviruses. *Cell Physiol Biochem.* 2017;41(6):2383–98.
41. Ehrhardt A, Xu H, Kay MA. Episomal persistence of recombinant adenoviral vector genomes during the cell cycle in vivo. *J Virol.* 2003 Jul;77(13):7689–95.
42. Brunetti-Pierri N, Ng T, Iannitti D, Cioffi W, Stapleton G, Law M et al. Transgene expression up to 7 years in nonhuman primates following hepatic transduction with helper-dependent adenoviral vectors. *Hum Gene Ther.* 2013 Aug;24(8):761–5.
43. Samulski RJ, Muzyczka N. AAV-mediated gene therapy for research and therapeutic purposes. *Annu Rev Virol.* 2014;1(1):427–51.
44. Grieger JC, Samulski RJ. Packaging capacity of adeno-associated virus serotypes: impact of larger genomes on infectivity and postentry steps. *J Virol.* 2005 Aug;79(15):9933–44.
45. Mcclements ME, Maclaren RE. Adeno-associated virus (AAV) dual vector strategies for gene therapy encoding large transgenes. *Yale J Biol Med.* 2017;90:611–623.
46. Mingozzi F, High KA. Immune responses to AAV vectors: overcoming barriers to successful gene therapy. *Blood.* 2013;122(1):23–36.
47. Meliani A, Boisgerault F, Fitzpatrick Z, Marmier S, Leborgne C, Collaud F et al. Enhanced liver gene transfer and evasion of preexisting humoral immunity with exosome-enveloped AAV vectors. *Blood Adv.* 2017;1(23):2019–31.
48. Deyle DR, Russell DW. Adeno-associated virus vector integration. *Curr Opin Mol Ther.* 2009;11(4):442–7.
49. Singh N, Tscharke DC. Herpes simplex virus latency is noisier the closer we look. *J Virol.* 2020;94(4).
50. Epstein A, Marconi P, Argnani R, Manservigi R. HSV-1-derived recombinant and amplicon vectors for gene transfer and gene therapy. *Curr Gene Ther.* 2005;5(5):445–57.
51. Morissette G, Flamand L. Herpesviruses and chromosomal integration. *J Virol.* 2010 Dec;84(23):12100–9.
52. Holmes KD, Cassam AK, Chan B, Peters AA, Weaver LC, Dekaban GA. A multi-mutant herpes simplex virus vector has minimal cytotoxic effects on the distribution of filamentous actin, α-actinin 2, and a glutamate receptor in differentiated PC12 cells. *J Neurovirol.* 2000 Jan;6(1):33–45.
53. Kasai K, Saeki Y. DNA-based methods to prepare helper virus-free herpes amplicon vectors and versatile design of amplicon vector plasmids. *Curr Gene Ther.* 2006;6(3):303–14.
54. Wu N, Watkins SC, Schaffer PA, DeLuca NA. Prolonged gene expression and cell survival after infection by a herpes simplex virus mutant defective in the immediate-early genes encoding ICP4, ICP27, and ICP22. *J Virol.* 1996 Sep;70(9):6358–69.
55. Li JM, Kao KC, Li LF, Yang TM, Wu CP, Horng YM et al. MicroRNA-145 regulates oncolytic herpes simplex virus-1 for the selective killing of human non-small cell lung cancer cells. *Virol J.* 2013;10(1):241.
56. Saeki Y, Fraefel C, Ichikawa T, Breakefield XO, Chiocca EA. Improved helper virus-free packaging system for HSV amplicon vectors using an ICP27-deleted, oversized HSV-1 DNA in a bacterial artificial chromosome. *Mole Ther.* 2001 *Apr*;3(4):591–601.
57. Schambach A, Morgan M. Retroviral vectors for cancer gene therapy. *In* 2016. p. 17–35.
58. Lesbats P, Engelman AN, Cherepanov P. Retroviral DNA integration. *Chem Rev.* 2016;116(20):12730–57.
59. Hacein-Bey-Abina S, Pai SY, Gaspar HB, Armant M, Berry CC, Blanche S et al. A modified γ-retrovirus vector for x-linked severe combined immunodeficiency. *New England J Med.* 2014;371(15):1407–17.
60. Pai SY. Built to last: gene therapy for ADA SCID. *Blood.* 2021;138(15):1287–8.
61. Reinhardt B, Habib O, Shaw KL, Garabedian E, Carbonaro-Sarracino DA, Terrazas D et al. Long-term outcomes after gene therapy for adenosine deaminase severe combined immune deficiency. *Blood.* 2021;138(15):1304–16.

62. Kohn LA, Kohn DB. Gene therapies for primary immune deficiencies. *Front Immunol.* 2021;12.
63. Berg K, Schäfer VN, Bartnicki N, Eggenschwiler R, Cantz T, Stitz J. Rapid establishment of stable retroviral packaging cells and recombinant susceptible target cell lines employing novel transposon vectors derived from Sleeping Beauty. *Virology.* 2019 May;531:40–7.
64. Kay MA, Glorioso JC, Naldini L. Viral vectors for gene therapy: the art of turning infectious agents into vehicles of therapeutics. *Nat Med.* 2001 Jan;7(1):33–40.
65. Ciuffi A. Mechanisms governing lentivirus integration site selection. *Curr Gene Ther.* 2008;8(6):419–29.
66. Nakajima T, Nakamaru K, Ido E, Terao K, Hayami M, Hasegawa M. Development of novel simian immunodeficiency virus vectors carrying a dual gene expression system. *Hum Gene Ther.* 2000 Sep;11(13):1863–74.
67. Hartmann K. Clinical aspects of feline retroviruses: a review. *Viruses.* 2012;4(11):2684–710.
68. Olsen J. Gene transfer vectors derived from equine infectious anemia virus. *Gene Ther.* 1998;5(11):1481–7.
69. Ferreira MV, Cabral ET, Coroadinha AS. Progress and perspectives in the development of lentiviral vector producer cells. *Biotechnol J.* 2021;16(1).
70. Xiao Y, Tang Z, Huang X, Chen W, Zhou J, Liu H et al. Emerging mRNA technologies: delivery strategies and biomedical applications. *Chem Soc Rev.* 2022;51(10):3828–45.
71. Tenchov R, Bird R, Curtze AE, Zhou Q. Lipid nanoparticles—from liposomes to mRNA vaccine delivery, a landscape of research diversity and advancement. *ACS Nano.* 2021;15(11):16982–7015.
72. Dong S, Feng Z, Ma R, Zhang T, Jiang J, Li Y et al. Engineered design of a mesoporous silica nanoparticle-based nanocarrier for efficient mRNA delivery *in Vivo. Nano Lett.* 2023;23(6):2137–47.
73. Wang C, Pan C, Yong H, Wang F, Bo T, Zhao Y et al. Emerging non-viral vectors for gene delivery. *J Nanobiotechnol.* 2023;21(1):272.
74. Felgner PL, Gadek TR, Holm M, Roman R, Chan HW, Wenz M et al. Lipofection: a highly efficient, lipid-mediated DNA-transfection procedure (liposomes/cationic lipid vesides/gene transfer). *Proc Nat Acad Sci* USA. 84:1987:7413–7.
75. Mintzer MA, Simanek EE. Nonviral vectors for gene delivery. *Chem Rev.* 2009;109(2):259–302.
76. Patil, Gao, Lin, Li, Dang, Tian et al. The development of functional non-viral vectors for gene delivery. *Int J Mol Sci.* 2019;20(21):5491.
77. Nanomedicine and the COVID-19 vaccines. *Nat Nanotechnol.* 2020;15(12):963–963.
78. Suhr OB, Coelho T, Buades J, Pouget J, Conceicao I, Berk J et al. Efficacy and safety of patisiran for familial amyloidotic polyneuropathy: a phase II multi-dose study. *Orphanet J Rare Dis.* 2015;10(1):109.
79. Kircheis R, Kichler A, Wallner G, Kursa M, Ogris M, Felzmann T et al. Coupling of cell-binding ligands to polyethylenimine for targeted gene delivery. *Gene Ther.* 1997; 4(5):409–18.
80. Wang Y, Zheng M, Meng F, Zhang J, Peng R, Zhong Z. Branched polyethylenimine derivatives with reductively cleavable periphery for safe and efficient in vitro gene transfer. *Biomacromolecules.* 2011;12(4):1032–40.
81. Chen B, Liu M, Zhang L, Huang J, Yao J, Zhang Z. Polyethylenimine-functionalized graphene oxide as an efficient gene delivery vector. *J Mater Chem.* 2011;21(21):7736.
82. Hu WW, Yeh CC, Tsai CW. The conjugation of indolicidin to polyethylenimine for enhanced gene delivery with reduced cytotoxicity. *J Mater Chem B.* 2018;6(36):5781–94.
83. Pack DW, Hoffman AS, Pun S, Stayton PS. Design and development of polymers for gene delivery. *Nat Rev Drug Discov.* 2005;4(7):581–93.
84. Wang H, Jiang Y, Peng H, Chen Y, Zhu P, Huang Y. Recent progress in microRNA delivery for cancer therapy by non-viral synthetic vectors. *Adv Drug Deliv Rev.* 2015 Jan;81:142–60.
85. Liu C, Wan T, Wang H, Zhang S, Ping Y, Cheng Y. A boronic acid–rich dendrimer with robust and unprecedented efficiency for cytosolic protein delivery and CRISPR-Cas9 gene editing. *Sci Adv.* 2019 Jun;5(6).
86. Wang M, Cheng Y. Structure-activity relationships of fluorinated dendrimers in DNA and siRNA delivery. *Acta Biomater.* 2016 Dec;46:204–10.
87. Janes KA, Calvo P, Alonso MJ. Polysaccharide colloidal particles as delivery systems for macromolecules. *Adv Drug Deliv Rev.* 2001 Mar;47(1):83–97.
88. Wright KJ, Badwaik VD, Samaddar S, Hyun SH, Glauninger K, Eom T et al. Organocatalytic synthesis and evaluation of polycarbonate pendant polymer:β-cyclodextrin-based nucleic acid delivery vectors. *Macromolecules.* 2018;51(3): 670–8.
89. Mellet CO, Fernández JMG, Benito JM. Cyclodextrin-based gene delivery systems. *Chem Soc Rev.* 2011;40(3):1586–608.
90. Chen H, Liu X, Dou Y, He B, Liu L, Wei Z et al. A pH-responsive cyclodextrin-based hybrid nanosystem as a nonviral vector for gene delivery. *Biomaterials.* 2013 May;34(16):4159–72.
91. Loh XJ, Wu YL. Cationic star copolymers based on β-cyclodextrins for efficient gene delivery to mouse embryonic stem cell colonies. *Chem Commun.* 2015;51(54):10815–8.
92. Karayianni M, Sentoukas T, Skandalis A, Pippa N, Pispas S. Chitosan-based nanoparticles for nucleic acid delivery: technological aspects, applications, and future perspectives. *Pharmaceutics.* 2023;15(2):534.
93. Mao S, Sun W, Kissel T. Chitosan-based formulations for delivery of DNA and siRNA. *Adv Drug Deliv Rev.* 2010 Jan;62(1):12–27.
94. Tripathi SK, Goyal R, Kashyap MP, Pant AB, Haq W, Kumar P et al. Depolymerized chitosans functionalized with bPEI as carriers of nucleic acids and tuftsin-tethered conjugate for macrophage targeting. *Biomaterials.* 2012 Jun;33(16):4204–19.
95. Zhao X, Li Z, Pan H, Liu W, Lv M, Leung F et al. Enhanced gene delivery by chitosan-disulfide-conjugated LMW-PEI for facilitating osteogenic differentiation. *Acta Biomater.* 2013 May;9(5):6694–703.
96. Geng J, Xia X, Teng L, Wang L, Chen L, Guo X et al. Emerging landscape of cell-penetrating peptide-mediated nucleic acid delivery and their utility in imaging, gene-editing, and RNA-sequencing. *J Controll Release.* 2022 Jan;341:166–83.
97. Timotievich ED, Shilovskiy IP, Khaitov MR. Cell-penetrating peptides as vehicles for delivery of therapeutic nucleic acids. *Mech Appl Med. Biochem (Moscow).* 2023;88(11):1800–17.
98. Bottens RA, Yamada T. Cell-penetrating peptides (CPPs) as therapeutic and diagnostic agents for cancer. *Cancers (Basel).* 2022;14(22):5546.
99. Al-Dosari MS, Gao X. Nonviral gene delivery: principle, limitations, and recent progress. *AAPS J.* 2009;11(4):671.
100. Rodriguez A, del A, Angeles M. Non-viral delivery systems in gene therapy. In: Rodríguez A, editor. *Gene Therapy: Tools and Potential Applications.* Rijeka: InTech; 2013, pp. 155–176. doi:10.5772/55510.

101. C. Heller L, Heller R. Electroporation gene therapy preclinical and clinical trials for melanoma. *Curr Gene Ther.* 2010;10(4):312–7.
102. Fan Z, Kumon RE, Deng CX. Mechanisms of microbubble-facilitated sonoporation for drug and gene delivery. *Ther Deliv.* 2014;5(4):467–86.
103. Ramamoorth M. Non-viral vectors in gene therapy: an overview. *J Clin Diag Res.* 2015;9(1):GE01-6.
104. Chen H, Yao H, Chi J, Li C, Liu Y, Yang J et al. Engineered exosomes as a drug and RNA co-delivery system: new hope for enhanced therapeutics? *Front Bioeng Biotechnol.* 2023;11:1254356.
105. Zhang Y, Liu Q, Zhang X, Huang H, Tang S, Chai Y et al. Recent advances in exosome-mediated nucleic acid delivery for cancer therapy. *J Nanobiotechnol.* 2022;20(1):279.
106. Kulkarni JA, Witzigmann D, Thomson SB, Chen S, Leavitt BR, Cullis PR et al. The current landscape of nucleic acid therapeutics. *Nat Nanotechnol Nat Res.* 2021;16:630–43.

9 Understanding the Concept of Quality by Design (QbD) Strategy

Case Study for Optimization of Lyophilized Solid Lipid Nanoparticles to Boost Oral Bioavailability of Antimalarial Drugs

Neha Bajwa[1,2], *Preet Amol Singh*[2], *Narendra Kumar Jain*[3], *and Neeraj Mishra*[4]

[1]Pharma Innovation Lab, Department of Pharmaceutical Sciences & Technology, Maharaja Ranjit Singh Punjab Technical University, Bathinda, Punjab, India

[2]University Center of Excellence in Research, Baba Farid University of Health Sciences, Faridkot, India

[3]Department of Pharmaceutics, Dr. H.S. Gour University, Sagar, MP, India

[4]Department of Pharmaceutics, Amity Institute of Pharmacy, Amity University Madhya Pradesh, Gwalior, MP, India

9.1 INTRODUCTION

Parasites affect around 30% of the world's population, making them a major worldwide issue. Malaria has had a greater impact on world history than any other infectious disease. With the emergence of widespread chloroquine resistance and global scarcity of quinine, a search for newer antimalarial drugs has become imperative [1]. For effective treatment of malaria, WHO recommends the use of artemisinin, an active principle of the Chinese medicinal plant, *Artemisia annua* (Quinghaosu), and its semisynthetic derivatives, viz., arteether, artemether, artesunate, and artemotil. Treatments containing an artemisinin derivative (artemisinin-combination therapies, ACTs) are now the standard therapeutic approach worldwide for *Plasmodium falciparum* malaria [2,3]. Artemisinin and its derivatives, the key ingredient in the Chinese herb Artemisia annua L., have become the most widely prescribed antimalarial medications in recent years. The therapeutic efficacy of artemisinins is becoming more appealing as multidrug-resistant bacteria become more common [4,5].

Artemisinin derivatives with vigorous antimalarial activity include dihydroartemisinin, artemether, artesunate, and ART [6,7]. These chemicals have been used for over 20 years and are especially effective against chloroquine-resistant malaria parasites [8]. Brossi and its coauthors introduced ART, a dihydroartemisinin ethyl ether derivative, as a new semisynthetic antimalarial drug that was shown to be effective against the erythrocytic stage of chloroquine-resistant *P. falciparum* and cerebral malaria. Artemisinin derivatives such as artemether and artesunate are less effective than ART in treating malaria [9].

Currently, only intramuscular injections of this drug are available on the market, which have lower patient acceptance and compliance due to being involved with pricking, pain at the site of injection, numbness, redness, and even leading to muscle fibrosis, granuloma, tissue necrosis, and injuries. This formulation also suffers from various issues, like erratic absorption patterns and increased cost of production. No oral dosage form of arteether has yet been developed due to its low solubility (~17 μg/mL), high degradation in the stomach (~40%), and poor oral bioavailability [10,11].

Solid lipid nanoparticles (SLNs) have been introduced as an alternative drug carrier system suitable to enhance the solubility of water-insoluble drugs. SLNs have attracted significant concerns from scientists for their outstanding advantages over other colloidal carriers in terms of solubilized features, high stability, low bio-toxicity, and large-scale production [12–14]. Therefore, applications of SLNs for oral, skin, ocular, pulmonary, and parenteral administration have been widely developed, with some presenting attractive results. Disadvantages that remain for SLNs include entrapment efficiency and encapsulation of hydrophilic materials. Recently, the novel concept of combining methods from SLNs and other polymeric systems has been explored [15–18].

Overall, SLNs have been investigated for various pharmaceutical applications, including improving the bioavailability of poorly water-soluble drugs. The SLNs systems have several advantages, including the fact that they are not readily taken up by cells in the reticulo endothelial system (RES) and thus bypass liver and spleen filtering, the fact that the lipid matrix is made from physiological lipid, which reduces the risk of acute and chronic toxicity, and the fact that they are easier to manufacture than bi-polymeric nanoparticles [19–21]. The SLNs' drug delivery system enables the controlled and targeted release of encapsulated

DOI: 10.1201/9781003508601-9

drugs. By coating or attaching ligands to SLNs, a wider range of drugs can be targeted. Moreover, the stability of the drug has been enhanced, resulting in the development of SLNs with a three-year shelf life, which is a notable improvement compared to other colloidal carrier systems.

QbD analyzes the aspects that are crucial for product quality from the customer's perspective and converts them into qualities that the product should have, as well as how the different parameters may be adjusted in a limited way to obtain the intended product/drug target profile [22]. QbD identified the links between process variables, formulation, and product characteristics, as well as the elements that cause variability. This information may also create a more flexible, resilient, and dependable manufacturing process that consistently delivers high-quality products over time [23–25]. Because QbD is a broad concept that extends beyond the pharmaceutical industry, a new vocabulary, "Formulation by Design," has been developed based on the QbD integrated principle, particularly in the development of pharmaceutical dosage forms. Formulation by Design is a more precise approach to producing high-quality pharmaceuticals. The five phases are primarily responsible for achieving the objective of high-quality medication manufacture [26].

The FDA has required the industry to use a QbD-based optimization strategy (Figure 9.1), which is a cornerstone of the ICH Q8–Q11 standards [27,28].

The objective of the study was to formulate an oral ART dosage form that would enhance oral bioavailability while reducing the dose, thereby lessening the risk of toxicity. We successfully attempted to prepare ART-SLN utilizing the spontaneous nanoprecipitation approach and evaluated the physicochemical features of the ART-SLN as well as its oral bioavailability.

9.2 FORMULATION DEVELOPMENT, OPTIMIZATION, AND CHARACTERIZATION OF SLN

9.2.1 Risk Assessment Studies for Preparation of SLNs

The main objective of risk assessment was to identify the plausibility of risks/failures. To do so, the Ishikawa fishbone diagram was constructed with the aid of Minitab 17 statistical software (M/s Minitab Inc., Philadelphia, PA) to depict a cause-and-effect relationship among the possible material attributes that were going to affect the critical quality attributes (CQAs) of the formulation [29]. The obtained fishbone diagram depicted the influence of various process parameters and material specifications for the development of ART-loaded SLNs. Furthermore, a risk assessment matrix (RAM) was conducted to prioritize the selection of high-risk factors, demonstrating the potential risks associated with different material specifications and process parameters of ART-SLNs in relation to potential CQAs. The RAM assigned a low, medium, or high grade to each of the studied factors [30].

9.2.2 Outlining Quality Target Product Profile (QTPP) and CQAs for Preparation of SLNs

QTPP is a primary step toward QbD-oriented development of ART-SLNs. It is defined as 'the imminent glossary of the quality attributes of a drug product that will be achieved exemplarily to ensure the desired quality, along with safety and efficacy of the drug product'. To meet the QTPP, various CQAs were identified, such as particle size and drug entrapment. The QTPP components and CQAs that impact the performance of ART-SLNs are listed in Table 9.1.

9.3 MATERIALS AND METHODS

9.3.1 Materials

Brooks Laboratories Limited, Baddi, India, provided arteether as a free sample. SD Fine Chemicals in India provided Tween 80, sodium lauryl sulfate, and stearic acid. Merck in India provided the acetone and methanol. All of the compounds and solvents were spectroscopic or analytical in nature.

9.3.1.1 Preparation of SLNs

ART-loaded SLNs were prepared by the spontaneous nanoprecipitation method with some modifications. Briefly, ART and stearic acid (6 mg) were dissolved in an acetone/ethanol (1:1) mixture to obtain an organic phase. Then, the organic phase was added dropwise using a syringe to 120 mL with

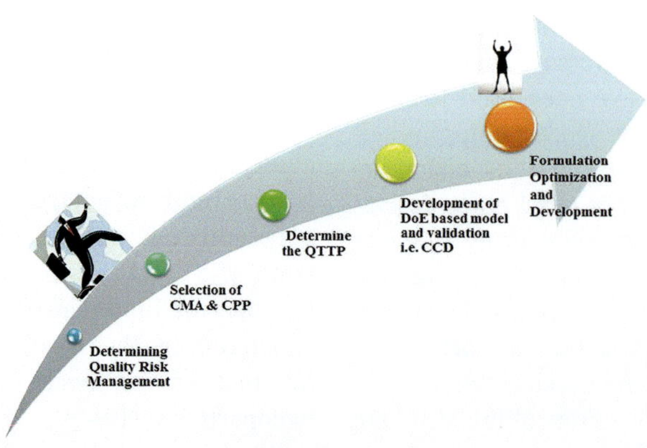

FIGURE 9.1 Steps involved in the drug product formulation based on QbD.

TABLE 9.1
QTPP Elements and CQAs for SLNs-Based Formulations and Their Justification

QTPP Elements	Target	Justification	
Dosage form	Liquid	To enhance bioavailability.	
Novel drug delivery system designed	SLNs	The selection of SLNs helps in improved drug loading and enhances bioavailability.	
Route of administration	Oral	The proposed route of delivery is to improve patient compliance.	
Strength	0.04% w/w	This strength is within the minimum effective concentration range of ART as per the literature and preclinical investigations.	
Packaging	Suitable for a dosage form	Suitable packaging is required for proper extrusion and consumer acceptability.	
Stability	At least 12 months at refrigerated storage	To maintain the optimum strength of ART for the therapeutic efficacy of the drug during the storage period.	
The Quality Attribute of the Drug Product	**Target**	**Justification**	**Is this a CQA?**
Physical appearance	Smooth and uniform No unpleasant color or odor	Physical appearance is not considered critical as it is not directly linked to safety and efficacy.	No
Total drug content	~100%	It is a vital parameter for dosage form design. Since SLNs are not a unit dose system, it is considered moderately critical.	No
Particle size	Nanometer range (50–200 nm)	It is a very critical parameter as particle size will significantly affect the bioavailability and solubility.	Yes
PDE	80%–100%	A higher value of PDE will provide good formulation stability as well as drug stability. Hence, it is a very critical attribute.	Yes
Entrapment Efficiency	More than 90%	A higher value of percentage EE will significantly affect the dose loading and dose delivery.	Yes
In vitro drug release	pH-dependent drug release	Minimum drug release in an acidic environment and maximum in basic pH.	Yes

a 70°C temperature of the aqueous phase under continuous stirring at a magnetic stirrer for half an hour [31,32]. The nanoparticle suspension was then lyophilized. The lyophilized SLN were then encapsulated in enteric-coated capsule shells [33]. Figure 9.2 illustrates the method for the preparation of lyophilized SLN formulations.

9.3.2 Optimization of SLNs by DoE

The randomized central composite design was employed to optimize ART-loaded SLN. The surfactant concentration and the acetone-to-ethanol volume ratio in the organic phase were selected as independent variables. Particle size and percentage entrapment efficiency were selected as responses. The statistical analysis of the factorial design formulations was performed using Design Expert® software (Version 13, Stat-Ease Inc.). Thirteen batches were prepared as suggested by the software using a central composite design with three levels, two factors, and two responses [34–36].

For the optimization of these factors, a central composite design (CCD) was employed with the aid of Design Expert software. CCD was selected because of its capability to give fewer experimental runs in comparison to other experimental designs.

Two factors, i.e., concentration of surfactant and amount of organic phase, were selected as the independent variables, and CCD gave 13 experimental runs. Entrapment efficiency and particle size were selected as responses for the experimental runs. Table 9.2 depicts the 13 experimental runs confirmed by CCD.

9.3.2.1 Physicochemical Characterization of ART-Loaded SLNs

9.3.2.1.1 Particle Size Analysis and Polydispersity Index Studies of ART-Loaded SLNs

Particle size and polydispersity index (PDI) of SLNs were determined by the zeta sizer nano series (Malvern, USA). SLNs were subsequently redispersed in Milli-Q water, and the particle size was recorded at 25°C. All determinations were carried out in triplicate.

9.3.2.1.2 Morphology of ART-Loaded SLN

Transmission electron microscopy (TEM). The particle shape of ART-loaded SLN was studied using TEM (FEI Tecnai F20, USA). One drop of the aqueous dispersion of SLN was loaded onto a carbon-coated copper grid and stained with a 2% w/v aqueous solution of phosphotungstic acid. The specimens were viewed under the microscope at an accelerating voltage of 120 kV.

9.3.2.1.3 Entrapment Efficiency (EE) of ART-Loaded SLNs

The nanoparticle dispersion was centrifuged for 15 min at 18,000 rpm using a centrifuge (Remi, India). From the supernatant, 0.1 mL was withdrawn and diluted up to 10 mL

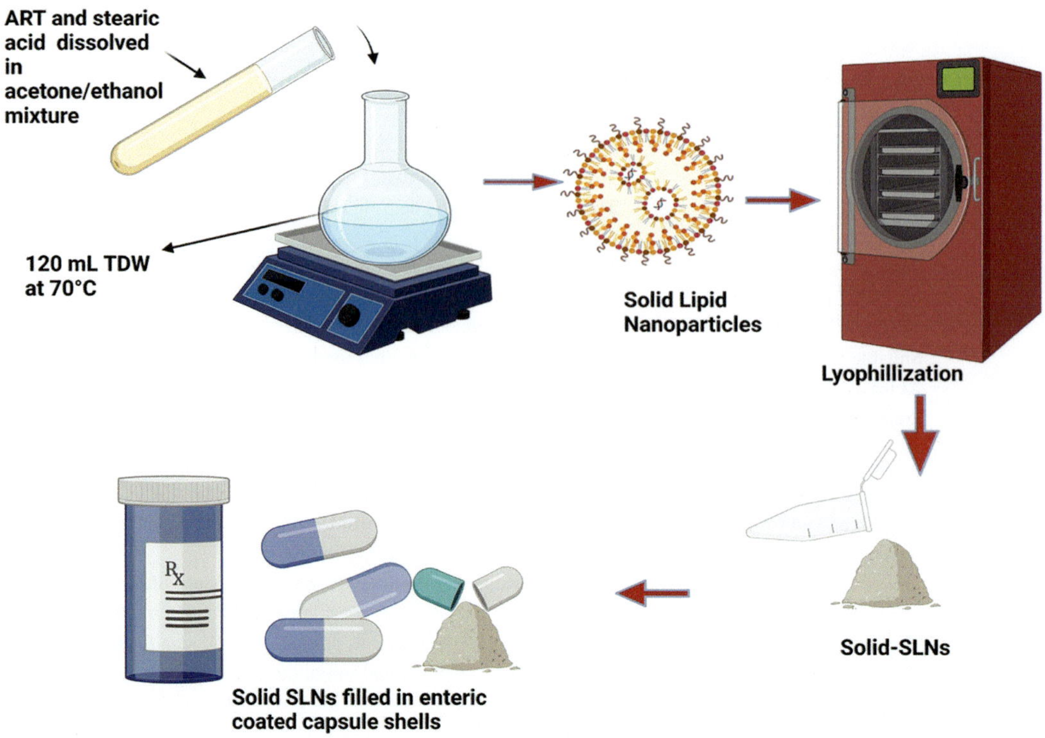

FIGURE 9.2 Method of preparation of solid ART-loaded SLN.

TABLE 9.2
Experimental Runs Suggested by Central Composite Design

Run	Factor 1: A: Surfactant Concertation (mg)	Factor 2: B: Organic Phase (mL)
1	100	3
2	50	3
3	75	7.5
4	75	7.5
5	75	3
6	75	7.5
7	75	12
8	50	12
9	100	7.5
10	75	7.5
11	100	12
12	50	7.5
13	75	7.5

using methanol, which was then analyzed using a UV spectrophotometer at 254 nm after acid degradation.

9.3.2.1.4 Determination of Total Drug Content (TDC) of ART-Loaded SLNs

Similarly, for the estimation of TDC, the SLN dispersion (1 mL) was disrupted using a suitable volume of chloroform: methanol mixture (2:1). The samples were vortexed to ensure complete extraction of the drug. The TDC was determined spectrophotometrically at λ_{max} of 254 nm. The studies were conducted in triplicate.

9.3.2.1.4.1 Determination of pH of ART-loaded SLN The pH of SLNs was determined to check whether it would irritate oral administration or not.

The prepared ART-SLNs were lyophilized to form powdered SLNs, which were then further evaluated using various parameters. A variety of parameters, including size, shape, flow properties (angle of repose, Hausner's ratio, and Carr's index), porosity, friability, disintegration time, drug content, and in vitro drug release, were also assessed for the powdered SLNs.

9.3.2.1.5 Particle Size Analysis of Lyophilized ART-Loaded SLNs

The particle size of the powder SLNs was analyzed by the microscopic technique and was calculated directly by using a motic microscope.

9.3.2.1.6 Flow Properties of Lyophilized Solid ART-Loaded SLN

Hausner's ratio, compressibility index, and angle of repose may all be used to evaluate powder flow.

9.3.2.1.7 Angle of Repose of Lyophilized ART-Loaded SLN

The angle of repose was measured using the fixed funnel technique. A funnel was positioned above graph paper that was laid out horizontally and fastened with its tip at a certain height (h). The mixture was gently poured down the funnel until the peak of the conical pile touched the funnel's tip. The conical pile's base radius was calculated.

The angle of repose (θ) was calculated using the following equation:

$$\tan\theta = \frac{h}{r} \tag{9.1}$$

where θ=angle of repose, h=height of the cone, r=radius of the cone base. Values for angle of repose ≤30° usually indicate a free-flowing material, and angles ≥40° suggest a poorly flowing material, 25–30 shows excellent flow properties, 31–35 shows good flow properties, 36–40 show fair flow properties, and 41–45 shows passable flow properties.

9.3.2.1.8 Carr's Index of Lyophilized ART-Loaded SLN

The ability of a powder to be put into capsules is gauged by the compressibility index (Carr's index) as given in Table 9.3. From the bulk and tapped densities, it is calculated. Theoretically, a material is more flowable if it is less compressible. As a result, it serves as a gauge for the relative significance of particle interactions. Such interactions are often less pronounced in a free-flowing powder, and the values of the bulk and tapped densities will be more similar. There are usually more particle interactions in poorly flowing materials, which results in a larger discrepancy between the bulk and tapped densities [37]. These differences are reflected in the Carr's index, which is calculated using the following formula:

$$\text{Carr's index} = \frac{\rho_{\text{tap}} - \rho_b}{\rho_{\text{tap}}} \times 100 \tag{9.2}$$

where ρ_b=bulk density, ρ_{tap}=tapped density

9.3.2.1.9 Hausner's Ratio of Lyophilized ART-Loaded SLN

The Hausner's ratio is a proximate indicator of powder flow simplicity. The formula used to compute it is as follows.

$$\text{Hausner' ratio} = \frac{t}{b} \tag{9.3}$$

TABLE 9.3
Compressibility Index Values

Compressibility Index (%)	Flow Character
≤10	Excellent
11–15	Good
16–20	Fair
21–25	Passable
26–31	Poor
32–37	Very poor
>38	Very very poor

where b is the bulk density and t is the tapped density. Hausner's ratios between 1.25 and 1.5 exhibit moderate flow qualities, while those above 1.5 suggest poor flow. Lower Hausner's ratios (1.25) imply better flow properties than higher ones.

9.3.2.1.10 Porosity of Lyophilized ART-Loaded SLN

Lyophilized ART-loaded SLN percentage porosity was calculated using the following equation:

$$\%\ \varepsilon = \frac{(\rho_t - \rho_b)}{\rho_b} \times 100 \tag{9.4}$$

where ε=effective porosity, ρ_t=true density and ρ_b=bulk density.

9.3.2.1.11 In vitro Release Studies of Lyophilized ART-Loaded SLN

In vitro, a drug release study of ART-SLNs was performed using the dialysis bag method. ART-SLNs (equivalent to 50 mg of ART) were dispersed in Milli-Q water and filled in a dialysis bag (Dialysis membrane, Hi media; MWCO 12,000e14,000 Da). The dialysis bag was closed from both sides and incubated in 250 mL of 0.1 N hydrochloric acid for the first 2 h, followed by phosphate buffer pH 6.8 for the remaining period on a magnetic stirrer (50 rpm, 37°C). Tween-80 (1%, w/v) was added to the release medium to maintain the sink condition during the dissolution study. A sample (2 mL) was withdrawn at pre-determined time points for 24 h and was replaced by an equal amount of fresh dissolution medium. Each sample after acid degradation was analyzed by UV-visible spectrophotometer at 254 nm [38]. The study was carried out in triplicate.

9.3.2.1.12 Permeability Studies of Prepared Formulations

The permeability of the intestine of a pig was measured in triplicate using the Franz diffusion cell procedure. The medium was phosphate buffer (6.8 pH). The benefactor chamber of the Franz cell was supplemented with an excess of ART and lyophilized SLN. The testing was performed by withdrawing 1 mL of the test and substituting the same amount of fresh media to ensure sink conditions at specific time points (15, 30, 60, 90, 120, 240, and 360 min). The acquired sample was adequately diluted and introduced into the HPLC to examine the drug contained in the selection. After 6 h, the Franz diffusion cell was deconstructed, and the skin (from the diffusion research) was gently removed from the cell [39]. The formulation was swabbed on the skin's surface with phosphate buffer. The treatment was administered twice to ensure that no formulation residue remained on the skin's surface. After cutting the skin into tiny pieces, it was maintained in a buffer of pH 6.8 to extract the medication contained in the skin [40]. After appropriate dilution and filtering, an HPLC-based method [41] was used to quantify the amount of drugs deposited in the skin.

9.3.2.1.13 *In-vivo Studies of Prepared Formulations*

Male New Zealand rabbits weighing 2.5–3 kg were procured from the Lala Lajpat Rai University of Veterinary and Animal Sciences, Hisar, and housed in ventilated rabbit houses maintained at 242°C and 50% relative humidity. The animals were acclimated to the necessary temperature and humidity three days prior to the experiment. Before starting studies, the Institutional Animal Ethics Committee authorized the study procedures approved by Institutional Animal Ethics Committee (IAEC) (Approval number—MRSPTU/IAEC/2019/16). To determine and compare bioavailability, ART aqueous suspension (ART dispersed in water), IV ART, IM ART, and enteric-coated capsules filled with lyophilized SLN were administered separately. All the formulations were delivered orally at a 6 mg/kg dosage (except *i.m./i.v.*). The marginal ear vein was used for blood collection using microfuge tubes with spray-dried Ethylenediaminetetraacetic Acid (EDTA). After centrifuging the blood for 15 min at 2,500 rpm, the plasma was collected, stored at 20°C, and analyzed using High-Performance Liquid Chromatography (HPLC).

9.3.2.1.14 *Sample Preparation*

Plasma was drawn from the rabbit's marginal ear vein and kept in EDTA-free tubes. The liquid was vortexed for 3 min after adding 0.4 mL acetonitrile and centrifuged at 25,000 rpm for 15 min at 10°C. The sample was followed by transferring the organic phase to fresh tubes and autosampler vials, and injecting it into the HPLC for analysis.

9.3.2.1.15 *Pharmacokinetics and Biostatistics*

Pharmacokinetics and statistical analysis were also conducted to produce ART plasma concentration-time data. The maximal plasma concentration (C_{max}) and the time required to attain that concentration (T_{max}) were assessed using acquired data. The area under the concentration-time curve (AUC_{0-t}) was analyzed using the trapezium technique. To determine C_{max} and T_{max}, individual animal concentrations were plotted against time. The absolute bioavailability of ART aqueous solution (ART dispersed in water), IM ART, and enteric-coated capsules filled with lyophilized SLNs was calculated by dividing the respective Analytical Ultracentrifugation (AUC) of each formulation by the AUC of *i.v.* data. The value $p < 0.01$ denotes statistical significance.

9.4 RESULT AND DISCUSSION

9.4.1 Formulation Development, Optimization, and Characterization of SLNs

9.4.1.1 Analysis of SLNs Using QbD Approach

The primary objective of risk assessment was to determine the likelihood of risks/failures. To do so, the Ishikawa fishbone diagram was constructed with the aid of Minitab 17 statistical software (M/s Minitab Inc., Philadelphia, PA) to depict a cause-and-effect relationship among the possible material attributes that were going to affect the CQAs of the formulation. The obtained fishbone diagram depicted the influence of various process parameters and material specifications for the development of ART-CD complex-loaded SLNs. Furthermore, a RAM was carried out for prioritization of selecting high-risk factors, to demonstrate the potential risks involved with different material specifications and process parameters of ART-CD complex-loaded SLNs on potential CQAs by designating a low, medium, and high grade to each of the studied factors.

9.4.1.2 Outlining Quality Target Product Profile (QTPP) and CQAs of SLN

QTPP is a primary step toward QbD-oriented development of ART-loaded SLNs. It is defined as 'the imminent glossary of the quality attributes of a drug product that will be achieved exemplarily to ensure the desired quality, along with safety and efficacy of the drug product'. To meet the QTPP, various CQAs were allocated, such as particle size and drug entrapment. Table 9.4 enlists the QTPP elements and CQAs for products/processes affecting the performance of ART-SLNs.

9.4.1.3 Preparation of SLNs

ART-loaded SLNs were prepared by the spontaneous nanoprecipitation method with some modifications. Briefly, ART (6 mg) and stearic acid (6 mg) were dissolved in an acetone/ethanol mixture (1:1) to obtain an organic phase. Then, the organic phase was added dropwise using a syringe to 120 mL with 70°C of the aqueous phase under continuous stirring at a magnetic stirrer for half an hour. The nanoparticle suspension was then lyophilized. After lyophillization, the SLNs were redispersed in water by sonication, which was used to determine the particle size, and the supernatant was collected to evaluate entrapment efficiency (%).

9.4.1.4 Selection of Lipids of SLNs

The solid lipid selected was stearic acid (290 ± 0.9 mg/mL), as ART showed maximum solubility in it. Equilibrium solubility studies were done in various organic solvents, such as ethanol and acetone. The solubility of ART was maximum in ethanol (232 ± 0.8 mg/mL) and acetone (301 ± 0.7 mg/mL). Hence, based on the results of solubility studies, stearic acid, ethanol, and acetone were selected as lipid and organic solvents, respectively, for the development of SLNs.

9.4.1.5 Risk Assessment Studies of SLNs

Figure 9.3 demonstrates the Ishikawa fishbone diagram depicting the various causes that affect the CQAs of SLNs formulations. The 6 Ms, probably affecting the CQAs like machines, men, material (API & excipients), measurements, methods, and milieu/environment were depicted by the fishbone diagram. The RAM was also constructed to observe the possible causes affecting the CQAs. The risk matrix for SLNs, enlightening the low, medium, and high-risk levels related to each of the process parameters/material specifications, is illustrated in Table 9.5. RAM suggested that the concentration of solid lipid, liquid lipid, and surfactant, PL, was a high-risk level as assigned with red color, and the aqueous

TABLE 9.4
QTPP Elements and CQAs for SLNs-Based Formulations and Their Justification

QTPP Elements	Target	Justification	
Dosage form	Liquid	To enhance bioavailability.	
Novel drug delivery system designed	SLNs	The selection of SLNs helps in improved drug loading and enhances bioavailability.	
Route of administration	Oral	The proposed route of delivery is to improve patient compliance.	
Strength	0.04% w/w	This strength is within the minimum effective concentration range of ART as per the literature and preclinical investigations.	
Packaging	Suitable for a dosage form	Suitable packaging is required for proper extrusion and consumer acceptability.	
Stability	At least 12 months at refrigerated storage	To maintain the optimum strength of ART for the therapeutic efficacy of the drug during the storage period.	
Quality Attribute of the Drug Product	**Target**	**Justification**	**Is this a CQA?**
Physical appearance	Smooth and uniform no unpleasant color or odor	Physical appearance is not considered critical as it is not directly linked to safety and efficacy.	No
Total drug content	100%	It is a vital parameter for dosage form design. Since SLNs are not a unit dose system it is considered moderately critical.	No
Particle size	Nanometer range (50–200 nm)	It is a very critical parameter as particle size will significantly affect the bioavailability and solubility.	Yes
PDE	80%–100%	Higher value of PDE will provide good formulation stability as well drug stability. Hence, it is a very critical attribute.	Yes
Entrapment efficiency	More than 90%	A higher value of percentage EE will significantly effect the dose loading and dose delivery.	Yes
In vitro drug release	pH-dependent drug release	Minimum drug release in an acidic environment and maximum in a basic pH.	Yes

TABLE 9.5
Initial RAM for ART-Loaded SLNs

	Risk Estimation Matrix						
CQAs of SLNs	Conc. of Solid Lipid	Conc. of Smix	Conc. of Organic Solvent	The Ratio of Solid: Liquid Lipid	Stirring Time	Stirring Speed	Temp.
Particle size	High	High	High	Medium	Medium	Medium	Low
PDE	High	High	High	High	Low	Low	Low
% EE	High	High	High	High	Low	Low	Low

Note: Red color = high-risk parameter; yellow color = medium-risk parameter; green = low-risk parameter.

phase was associated with medium risk as assigned with brown color. In contrast, the stirring speed, stirring time, and temperature were associated with medium to low levels of risk, as assigned with green color.

9.4.2 Design of Experiments (DoE) for SLNs

The randomized CCD was employed to optimize ART-loaded SLNs. Surfactant concentration and acetone to-ethanol volume ratio in the organic phase were selected as independent variables. Particle size and entrapment efficiency were selected as responses. The statistical analysis of the factorial design formulations was performed using Design Expert® software (Version 13, Stat-Ease Inc.). Thirteen batches were prepared as suggested by the software using a CCD with three levels, two factors, and two responses.

For optimization of these factors, CCD was used with the help of Design Expert software. CCD was selected because of its capability to give a smaller number of experimental runs in comparison to other experimental designs. Table 9.6 shows the low high values of the factors selected by DoE.

9.4.3 Optimization and Development

9.4.3.1 DoE-based Optimization of Selected Factors of SLNs

DoE was employed to optimize the selected independent variables, i.e., the concentration of surfactant and the

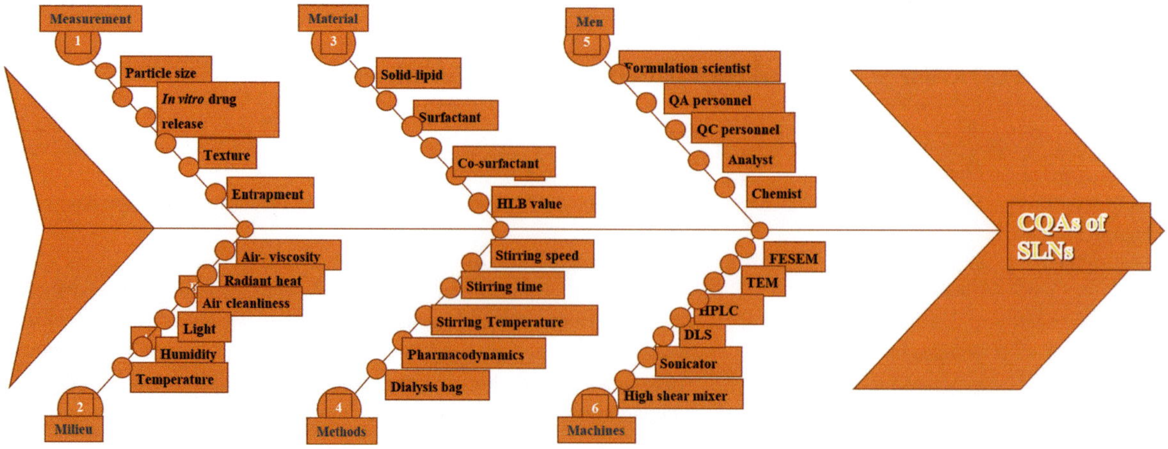

FIGURE 9.3 Ishikawa fishbone diagram depicting various causes that affect the CQAs of SLNs formulation.

TABLE 9.6
Low and High Values of Selected Factors

A	Surf	mg	50	100	−1 ↔ 50	+1 ↔ 100	75	17.68
B	Organic phase	mL	3	12	−1 ↔ 3	+1 ↔ 12	7.50	3.18

TABLE 9.7
Experimental Runs and Value of Response as Per Experimental Runs

Std	Run	*A*: Surfactant Concentration (mg)	*B*: Organic Phase (mL)	Entrapment Efficiency (%)	Particle Size (nm)
2	1	100	3	88.75	340.4
1	2	50	3	81.01	293
10	3	75	7.5	84	402.3
11	4	75	7.5	79.58	443.4
7	5	75	3	85.98	461.8
13	6	75	7.5	86.52	465.3
8	7	75	12	95.5	513.6
3	8	50	12	92.37	519.6
6	9	100	7.5	80.7	211.2
9	10	75	7.5	87.65	366.7
4	11	100	12	98.55	334.3
5	12	50	7.5	86.03	479.9
12	13	75	7.5	84.45	436.8

amount of organic phase, with respect to entrapment efficiency (EE) and particle size (PS) as the response variables. CCD gave 13 experimental runs for different values of independent variables, and the value of the response is shown in Table 9.7.

9.4.3.2 Statistical Analysis of Data of SLNs

ANOVA was involved in the statistical data analysis of the information generated with the help of software for better optimization of selected variables for the evaluation of the

TABLE 9.8
ANOVA-Based Statistical Parameters for SLNs

	EE (%)		PS		
Source	*F*-value	*p*-Value	*F*-Value	*p*-Value	Inference
Model	4.65	0.0343	5.17	0.0265	Significant
A-Surfactant	0.9636	0.3590	8.99	0.0200	
B-Organic phase	12.29	0.0099	4.03	0.0846	
AB	0.0477	0.8334	4.42	0.0737	
A^2	0.1495	0.7105	8.30	0.0236	
B^2	9.27	0.0187	1.92	0.2080	
Residual					
Lack of Fit	1.77	0.2925	3.44	0.1319	Not significant
R^2	0.7687		0.7868		
Adjusted R^2	0.6035		0.6345		
Adeq Precision	5.822		7.585		

significance of the model. A significant model will have a *p*-value less than 0.05. The adequacy of our defined model was calculated based on lack of fit or *p*-value and R^2 value. The R^2 value exhibited closeness of predicted data and came closer to experimental data, and the value should be closer to 1 [38]. ANOVA statistics shown in Table 9.8 suggested that our *p*-value for both responses. i.e., EE and PS were approximately 0.0343 and 0.0265, respectively (much less than 0.05), showing the significance of the developed model. The same result was confirmed by the value of R^2 for both responses, which were found to be 0.9767 and 0.9198, respectively, which is much nearer to 1 [35].

9.4.3.3 Effect of the Variables on Entrapment Efficiency of SLNs

Entrapment efficiency, defined as the amount of drug entrapped in SLNs, is one of the critical parameters that can affect the final quality target profile of SLNs. The design of the experiment gave a quadratic mathematical model to depict the effect of a different variable on EE.

$$\text{EE} = +84.37 + 1.43\,A + 5.11B - 0.3900AB - 0.8312A^2 + 6.54\,B^2 \tag{9.5}$$

where A and B are the selected variables, i.e. surfactant concentration and organic phase.

The relationship between the independent variable and entrapment efficiency was determined by plotting a 2D contour plot and a 3D response surface plot, as shown in Figure 9.4.

As per the above, *EE* started increasing with an increasing amount of organic phase, but it decreased when the value of the organic phase and surfactant continuously increased. The value of entrapment efficiency again starts increasing with the increasing value of the organic phase after a certain point. Thus, EE was more in the higher and lower values of the organic solvent.

9.4.3.4 Effect of the Variable on Particle Size of SLNs

PS is one of the important parameters that can affect the final quality profile of SLN.

The equation below shows the relationship between the PS and the selected independent variable, as shown in Figure 9.5.

$$\text{PS} = 428.21 - 67.77A - 45.38B - 58.17AB - 95.95\,A^2 + B^2 46.20 \tag{9.6}$$

where A and B are the selected two variables.

According to Figure 9.5, particle size continuously increased with the increase in the concentration of the organic phase. But with an increase in the concentration of surfactant, PS increased to a certain extent, then started decreasing if the concentration of surfactant further increased.

9.4.4 Overlay Plot and Validation of Model

It is essential to define the design space across all experimental areas, allowing for the easy achievement of the desired quality profile of the final product. An overlay plot, generated with the help of Design Expert software, was created, illustrating the design space in the yellow region and the entire experimental region in the gray background, as shown in Figure 9.6.

Validation of the model is an important aspect of QbD-based formulation development, as it ensures that the design is capable of replicating the experimental region. So validation of the above-developed model was done by comparing the result of responses at the lab experimental level to the value suggested by the design itself. The above design was 92.4% (197.90 nm predicted and 109 observed) and 93.8% (96.23% predicted and 93.4% observed) nearest in predicting both responses, i.e., PS and entrapment efficiency.

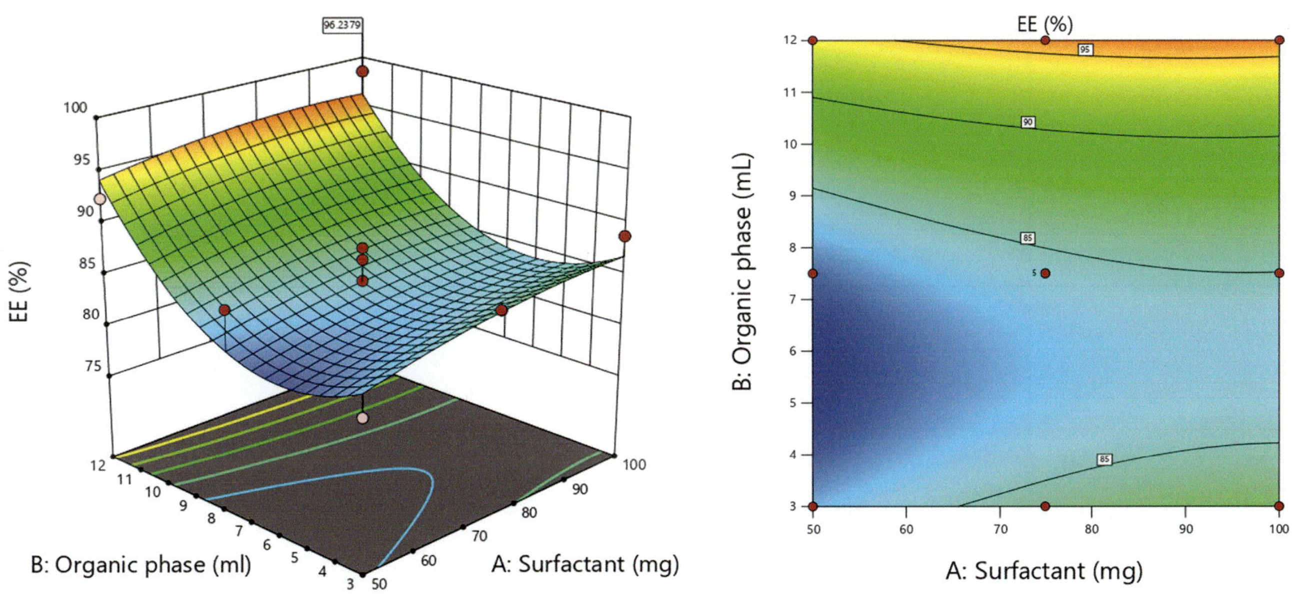

FIGURE 9.4 The 2-dimensional contour plot and 3-dimensional response surface plot for entrapment efficiency.

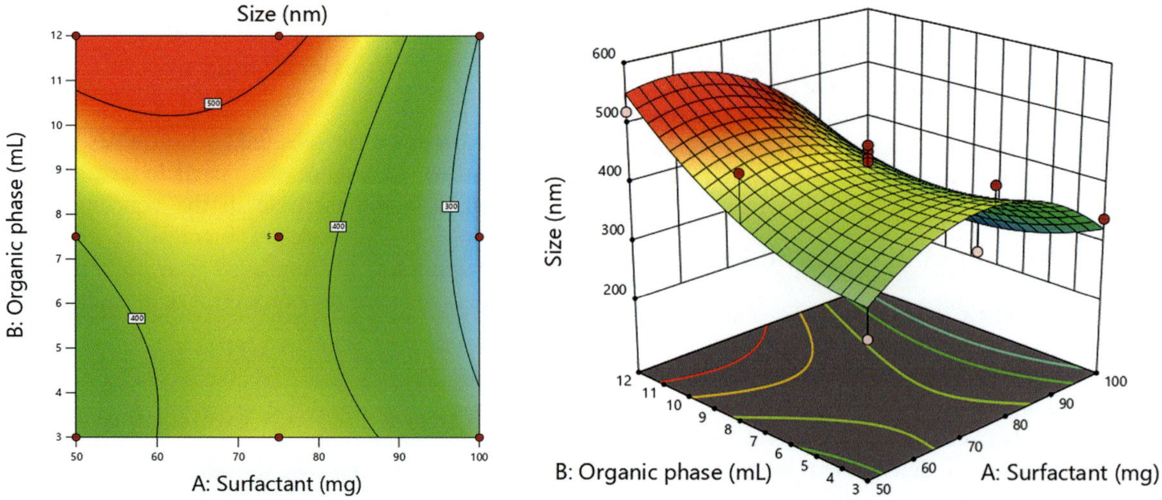

FIGURE 9.5 The 2-dimensional contour plot and the 3-dimensional response surface plot for particle size.

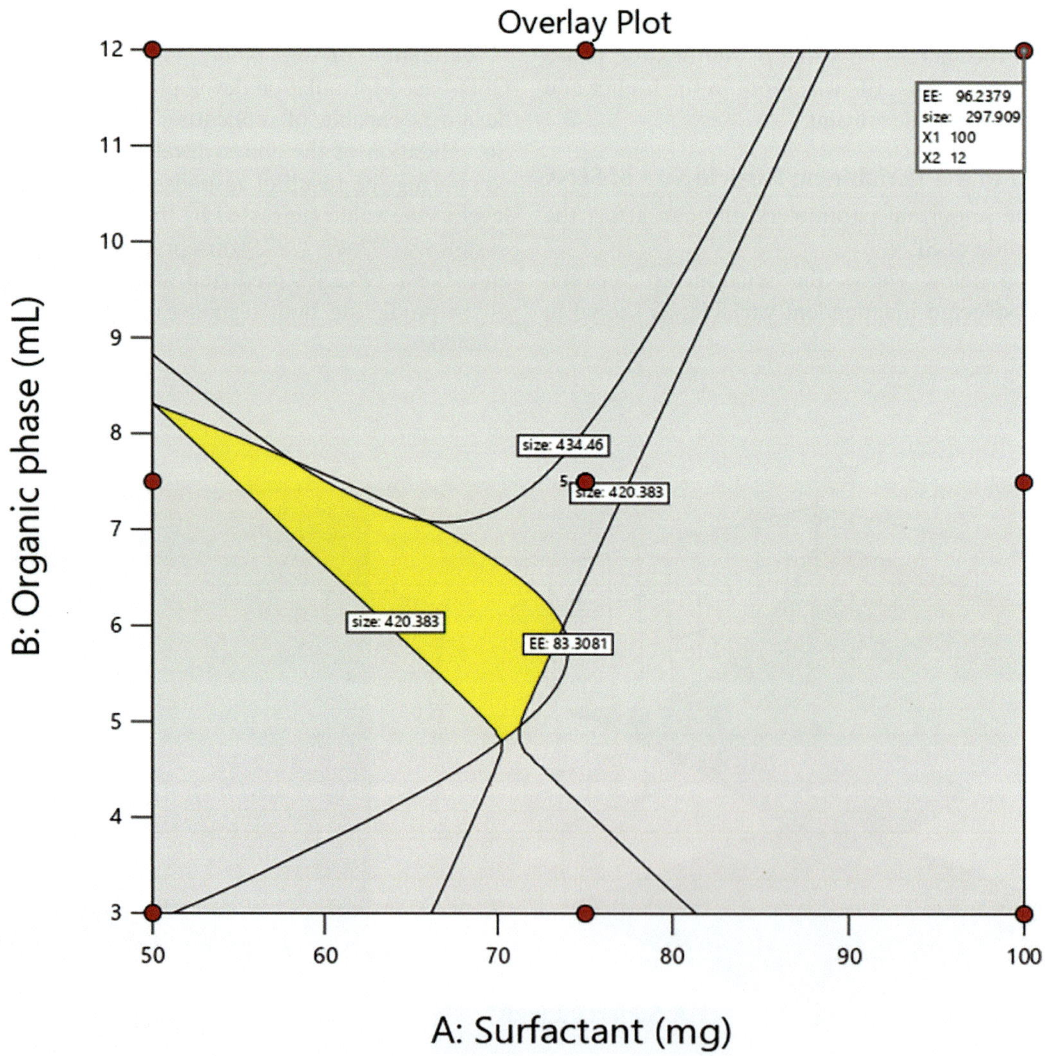

FIGURE 9.6 Overlay plot showing the design space in the yellow region and the whole experimental region in the gray background.

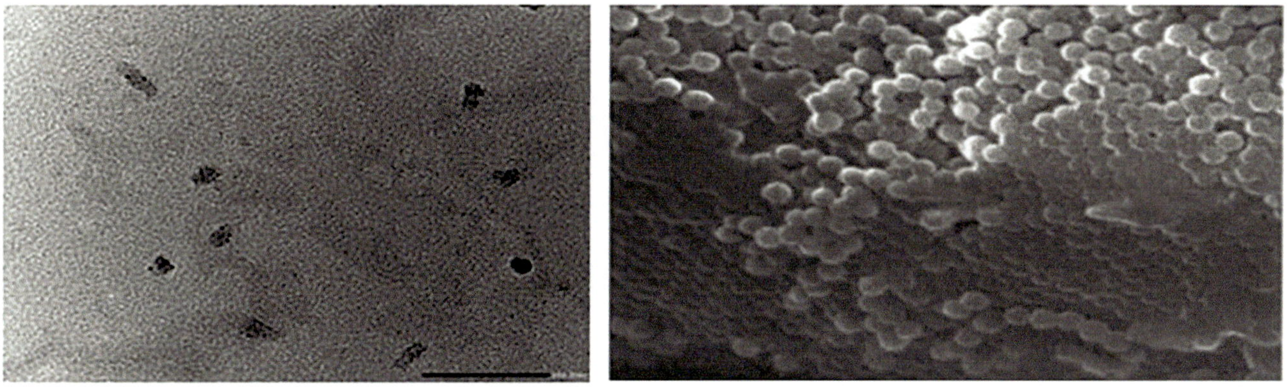

FIGURE 9.7 TEM image and SEM image of prepared SLNs.

9.4.5 Preparation of SLNs

ART-loaded SLN was prepared by the spontaneous nanoprecipitation method with some modifications. Briefly, ART (6mg) and stearic acid (50mg) were dissolved in 12mL of an acetone/ethanol mixture (1:1) to obtain an organic phase. Then, the organic phase (12mL) was added dropwise using a syringe to the 120mL aqueous phase at a temperature of 70°C under continuous stirring for half an hour with a magnetic stirrer. The nanoparticle suspension was then lyophilized under certain conditions. After lyophillization, the SLNs were redispersed in water by sonication, which was used to determine the PS, and the supernatant was collected to evaluate percentage entrapment efficiency.

9.4.6 Physicochemical Characterization of ART-Loaded SLN

The key advantage of ART-SLN is the protection of ART from enzymatic degradation, which delays its metabolism. The endoperoxide ring of ART, which is mainly responsible for its antimalarial activity, has been reported to be cleaved in the acidic environment of the stomach, which can decrease its activity and absorption. The incorporation of ART in SLN can reduce its exposure to the acidic pH of the stomach, thereby preventing degradation and increasing its bioavailability and activity.

9.4.7 Particle Size Distribution

The result shows the formation of a homogeneous ART-SLN having a PDI of less than 0.2. Figure 9.5 illustrates the PS and its distribution of ART-SLN measured by Dynamic Light Scattering (DLS). The results demonstrate that ART-SLNs have been prepared successfully, with a mean diameter of ART-SLN of 109.2±15nm. The zeta potential of the SLN (–29.1) was determined so that the contribution of charge distribution can also be assessed, as shown in Figure.

The PS of ART-SLN was less than 200nm, the PDI was found to be <0.5, which provides a large surface area for the absorption of ART.

9.4.8 Surface Morphology of SLNs

The PS of the ART-SLN was further confirmed by HR-TEM, and the results revealed that the PS was less than 200nm (Figure 9.7). FESEM photomicrograph revealed spherical nanoparticles with a smooth and uniform surface. The TEM photomicrograph also indicated the presence of spherical nanoparticles.

9.4.9 XRD of SLNs

The Powder X-Ray Diffraction (PXRD) spectra of lyophilized optimized ART-loaded SLN formulations showed the absence of all the ART characteristic peaks, as shown in Figure 9.8. The absence of characteristic ART peaks demonstrated the successful incorporation of ART into the SLNs.

9.4.10 Entrapment Efficiency Percentage of SLNs

The maximum entrapment efficiency (% EE) in ART-SLN was found to be 93.7%±6.2%. The drug loading of the optimized SLNs was found to be 78.54%±0.62%.

9.4.11 pH

The pH of ART-SLNs was found to be 7.2, which is somewhat neutral. This is important because substances that are too acidic or too basic can cause irritation or damage to the skin or other tissues. Therefore, the fact that the ART-SLNs have a neutral pH confirms that they are not likely to cause any irritation or harm to the gastrointestinal tract.

9.4.12 Particle Size Distribution of Lyophilized SLNs

The prepared SLN was lyophilized to get powder SLN, and the flow properties of the powdered SLN were further analyzed. The PS distribution of powdered SLN is as shown in Table 9.9.

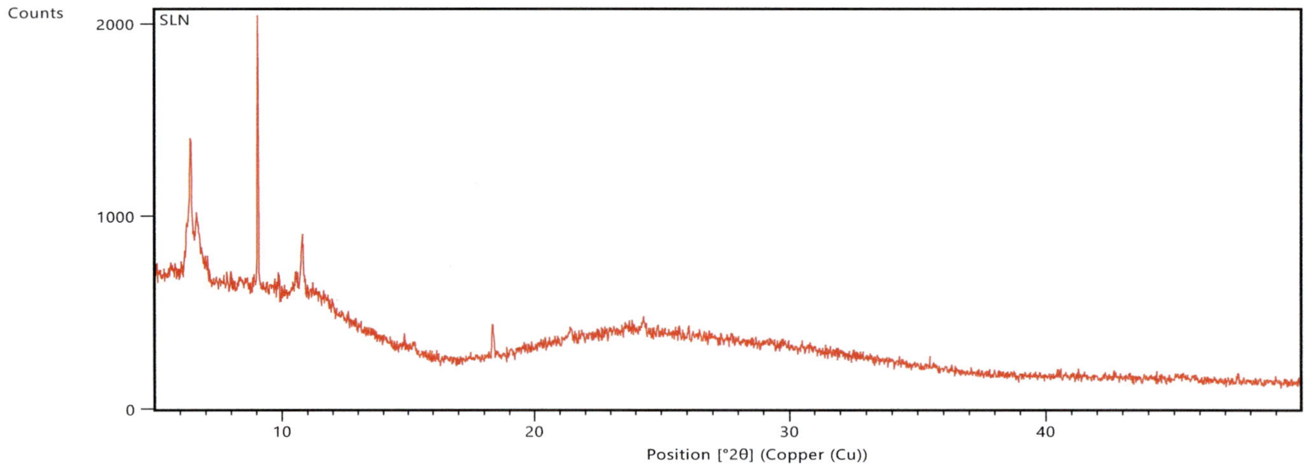

FIGURE 9.8 The PXRD spectra of prepared SLNs.

TABLE 9.9
Particle Size Distribution

S.No.	Sieve No	Percentage Weight Retained (Uncoated)
1	18	0
2	22	1.2
3	25	5.4
4	30	89

9.4.13 Shape of Lyophilized SLNs

The shape of solid SLN was found to be spherical when seen under a motic microscope. The spherical nature of the spheroid was calculated by pellips (which is a characteristic of pellet shape). A powder with pellips equal to 1.0 is considered good for pharmaceutical processing, as shown in Figure 9.9.

$$P_{\text{ellips}} = p/\pi d \max = 1(\text{spherical}) \qquad (9.7)$$

where p=perimeter, π=27/7, d_{max}=maximum diameter of SLNs.

9.4.14 Bulk Density of Lyophilized SLNs

Bulk density and tapped density were used for the calculation of Hausner's ratio and compressibility index. These help in determining the flow property and the compressibility of ingredients. A negligible variation in bulk density was observed. The bulk density of solid SLNs was found to be 0.81.

9.4.15 Tapped Density of Lyophilized SLNs

A negligible difference was experimentally observed in the tapped density and was found to be 0.89.

9.4.16 Carr's Index (%) of Lyophilized SLNs

Results of the compressibility study showed that the compressibility index of the solid SLN had excellent flowability, as the results fell in the range of ≤10%. The compressibility index was found to be 7.2, which showed excellent flow.

9.4.17 Hausner's Ratio of Lyophilized SLNs

According to USP 2000, powders with a Hausner's ratio between 1.12 and 1.18 exhibited good flow, while those above 1.35 showed poor flow properties. The result was 1.14.

9.4.18 Angle of Repose of Lyophilized SLNs

The angle of repose of freeze-dried SLNs was found to be 29. According to I.P., this is an excellent flow.

9.4.19 *In vitro* Drug Release of Lyophilized SLNs

Figure 9.10 depicts the cumulative ART release from ART-SLN, at 37°C. Around 40% of the drug was released from ART-SLN within the first 12 h. Following that, the ART release rate increased to 30 h, with a maximum drug release of 79.1%. The results show that the percentage drug release of all formulations was less than 10% in the acidic pH, while at 50 h, the maximum drug release was attained.

9.5 PERMEABILITY STUDIES

9.5.1 Studies on Permeability

The Franz diffusion cell technique was used to study the permeability of the produced SLNs of ART, which may result in higher bioavailability due to increased immersion dissolvability. The diffusion limit restricts the absorption of medicine into the lipophilic apical layer of the colon, and the drug must be dissolved to enhance final bioavailability. Furthermore, robust state properties such as molecule size, internal energy, glasslike or nebulous structure, atomic mobility, and wettability impact the medication's solubility. Using ART-loaded SLNs *vs.* pure ART [40].

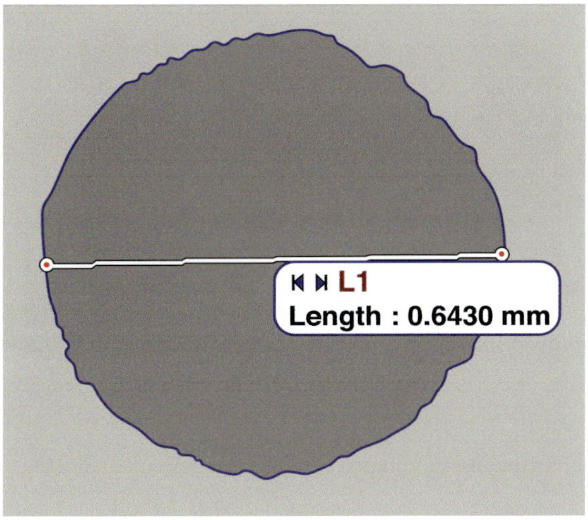

FIGURE 9.9 Shape of *lyophilized* SLN.

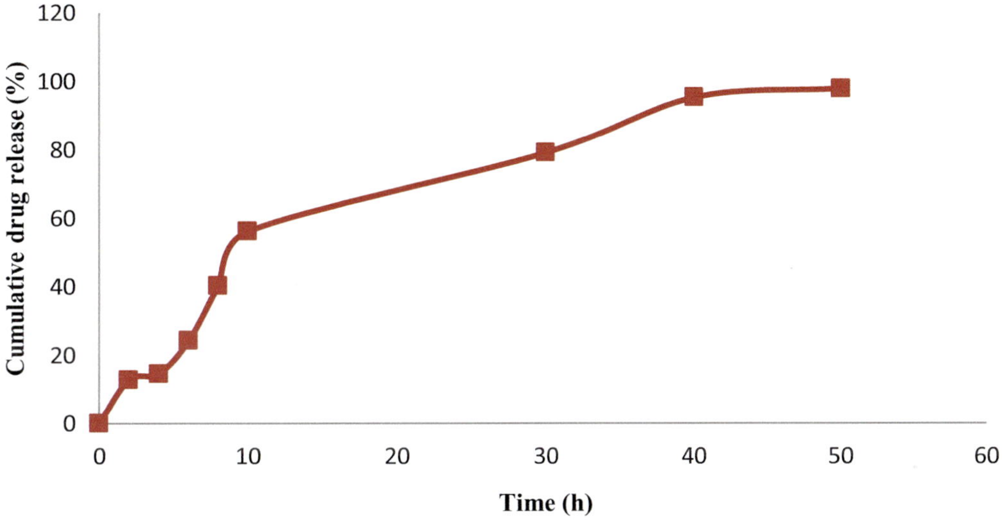

FIGURE 9.10 The percentage drug release of ART-SLN.

The results showed a considerable improvement in the penetration measure of ART, with SLNs having a 7.82-fold higher permeability than ART in its pure form. The results back up prior discoveries of equilibrium dissolvability data. SLNs with improved penetrability can be utilized to confirm the presence of microcrystalline and amorphous ART fragments (Figure 9.11).

9.6 PHARMACOKINETIC STUDIES

Non-compartmental analysis was performed on the data [41]. The C_{max} and $AUC_{0\text{-}\infty}$ values for powder SLN formulations are substantially higher than those for the ART aqueous solution, according to the findings.

The area under the curve for ordinary arteether (AUC_{0-t} 58.25 h.g/mL) was much higher (AUC). The C_{max} of SLN is 123.6 g/mL, and the $t_{1/2}$ is 7.037 h. In comparison to *iv* data, the SLNs had an absolute bioavailability of 27.64%. Figure 9.12 compares the plasma drug concentrations of SLNs and the simple drug throughout time.

9.7 CONCLUSION

The increased oral bioavailability of ART-SLN is justified by the higher AUC and C_{max} values for ART-SLN. These parameters, which are characteristics of the drug formulation and are crucial for comparative bioavailability studies, describe the amount of drug absorption, and C_{max} is a crucial component of the blood level profile. This might be related to the ART-strong SLN's lipid shielding, which prevents enzymatic and chemical destruction of the ART. Hence, SLNs formulation's enhanced oral bioavailability, which may be used to treat multidrug-resistant malaria with higher patient compliance.

ETHICS APPROVAL AND CONSENT TO PARTICIPATE

Yes, the research work was approved by the Institutional Animal Ethics Committee authorized by IAEC (Approval number-MRSPTU/IAEC/2019/16).

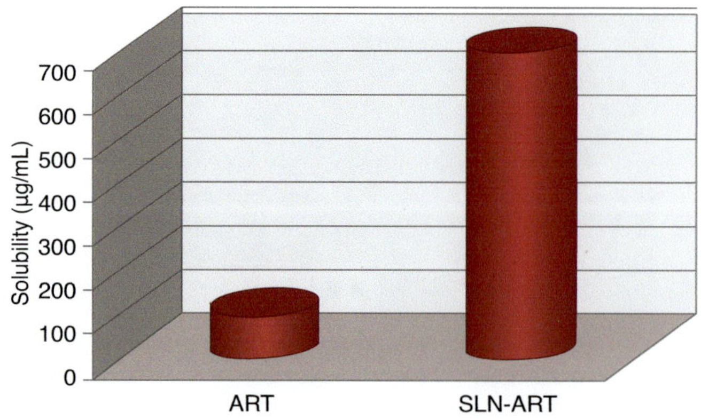

FIGURE 9.11 Intestinal permeability of formulations compared with arteether.

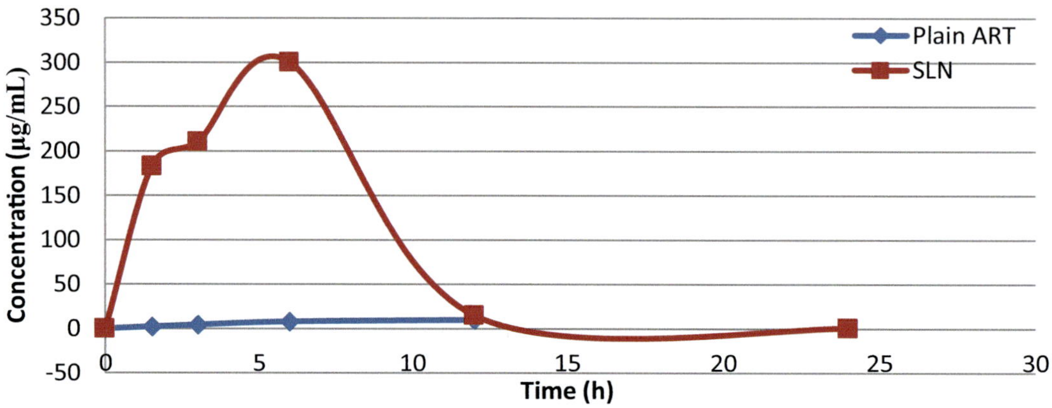

FIGURE 9.12 Plasma concentrations time plot of plain drug with formulated SLNs.

FUNDING

The author thanks the Indian Council of Medical Research (ICMR), New Delhi, for funding her work as a senior research fellow under the ICMR-SRF scheme (F.No. 53/2019/-ECD-II dated 19/07/2019).

CONFLICT OF INTEREST

The authors declare no conflicts of interest, financial or otherwise.

ACKNOWLEDGMENTS

The author highly acknowledges ICMR, New Delhi, for providing the ICMR-SRF fellowship to one of the author, Neha Bajwa, for providing ICMR-SRF fellowship to carry out research work.

CONFLICT OF INTEREST

The author has no conflict of interest to declare. There are no financial interests to disclose, and all coauthors have reviewed and approved the manuscript's contents. We attest to the submission's accuracy. We certify that the submission is original work and is not under submission to any other journal.

REFERENCES

1. Chan K, Tusting LS, Bottomley C, Saito K, Djouaka R, and Lines J. Malaria transmission and prevalence in rice-growing versus non-rice-growing villages in Africa: a systematic review and meta-analysis. *Lancet Planet Heal.* 2022;6(3):e257–e269.
2. Shmuklarsky MJ, Klayman DL, Milhous WK, Kyle DE, and Rossan RN. *Comparison of Beta-Artemether and Beta-Arteether against Malaria Parasites In Vitro and In Vivo.* Walter Reed Army Institute of Research, Washington, DC, 1993.
3. Oleinikov AV. Malaria parasite plasmodium falciparum proteins on the surface of infected erythrocytes as targets for novel drug discovery. *Biochem.* 2022;87(1): S192–S202.
4. Thakur S, Bajwa N, and Baldi A. New analytical methods for estimation of arteether by uv and fluorescence spectrophotometry: development and validation. *J Drug Deliv Ther.* 2018;8(5):151–158.
5. Tripathi R, Sen M, and Dutta GP. *The Use of Artemisinin Derivative Suppositories as Life-Saving Remedy for*

Critical Malaria Patients. Cambridge Scholars Publishing, Cambridge, 2021.
6. Müller RH, and Souto EB. Lipid nanoparticles (SLN and NLC) for drug delivery. In: Kumar R, Tabata Y, Domb A (Eds.), *Nanoparticles for Pharmaceutical Applications*. American Scientific Publishers, Stevenson Ranch, CA, 2005, pp. 369–393.
7. Müller RH, Mäder K, and Gohla S. Solid lipid nanoparticles (SLN) for controlled drug delivery–a review of the state of the art. *Eur J Pharm Biopharm*. 2000;50(1):161–177.
8. Gunjan S, Sharma T, Yadav K, Chauhan BS, Singh SK, Siddiqi MI, and Tripathi R. Artemisinin derivatives and synthetic trioxane trigger apoptotic cell death in asexual stages of plasmodium. *Front Cell Infect Microbiol*. 2018;8:256.
9. Brossi A, Venugopalan B, Dominguez Gerpe L, Yeh HJC, Flippen-Anderson JL, Buchs P, Luo XD, Milhous W, and Peters W. Arteether, a new antimalarial drug: synthesis and antimalarial properties. *J Med Chem*. 1988;31(3):645–650.
10. Mirchandani Y, Patravale VB, and Brijesh S. Solid lipid nanoparticles for hydrophilic drugs. *J Control Release* 2021;335:457–464.
11. Rajpoot K. Solid lipid nanoparticles: a promising nanomaterial in drug delivery. *Curr Pharm Des*. 2019;25(37):3943–3959.
12. Fichera M, Keck C, and Müller RH. Nanopure® technology–drug nanocrystals for the delivery of poorly soluble drugs. In: Müller RH, Keck CM, Bushrab NF (Eds.), *Particles 2004–Particle Synthesis, Characterization and Particle-Based Advanced Materials*. American Scientific Publishers, Orlando, FL, 2004.
13. Battaglia L, and Gallarate M. Lipid nanoparticles: state of the art, new preparation methods and challenges in drug delivery. *Expert Opin Drug Deliv*. 2012;9(5):497–508.
14. Cunha S, Costa CP, Moreira JN, Lobo JMS, and Silva AC. Using the quality by design (QbD) approach to optimize formulations of lipid nanoparticles and nanoemulsions: a review. *Nanomed Nanotechnol Biol Med*. 2020;28:102206.
15. Zur Mühlen A, Schwarz C, and Mehnert W. Solid lipid nanoparticles (SLN) for controlled drug delivery–drug release and release mechanism. *Eur J Pharm Biopharm*. 1998;45(2):149–155.
16. Schwarz C. Solid lipid nanoparticles (SLN) for controlled drug delivery II. Drug incorporation and physicochemical characterization. *J Microencapsul*. 1999;16(2):205–213.
17. Dingler A, and Gohla S. Production of solid lipid nanoparticles (SLN): scaling up feasibilities. *J Microencapsul*. 2002;19(1):11–16.
18. Dwivedi P, Khatik R, Khandelwal K, Shukla R, Paliwal SK, Dwivedi AK, and Mishra PR. Preparation and characterization of solid lipid nanoparticles of antimalarial drug arteether for oral administration. *J Biomater Tissue Eng*. 2014;4(2):133–137.
19. Gohla SH, and Dingler A. Scaling up the feasibility of the production of solid lipid nanoparticles (SLN). *Pharmazie*. 2001;56(1):61–63.
20. Kaur R, Sharma N, Tikoo K, and Sinha VR. Development of mirtazapine-loaded solid lipid nanoparticles for topical delivery: optimization, characterization, and cytotoxicity evaluation. *Int J Pharm*. 2020;586:119439.
21. Scioli Montoto S, Muraca G, and Ruiz ME. Solid lipid nanoparticles for drug delivery: pharmacological and biopharmaceutical aspects. *Front Mol Biosci*. 2020;7:587997.
22. Raza K, Singh B, Singal P, Wadhwa S, and Katare OP. Systematically optimized biocompatible isotretinoin-loaded solid lipid nanoparticles (SLNs) for topical treatment of acne. *Colloids Surfaces B Biointerfaces*. 2013;105:67–74.
23. Beg S, Jain S, Kushwah V, Bhatti GK, Sandhu PS, Katare OP, and Singh B. Novel surface-engineered solid lipid nanoparticles of rosuvastatin calcium for low-density lipoprotein-receptor targeting: a quality by design-driven perspective. *Drug Deliv*. 2017;12(4):333–356.
24. Mishra V, Thakur S, Patil A, and Shukla A. Quality by design (QbD) approaches in the current pharmaceutical set-up. *Expert Opin Drug Deliv*. 2018;15(8):737–758.
25. Gurumukhi VC, and Bari SB. Quality by design (QbD)–based fabrication of atazanavir-loaded nanostructured lipid carriers for lymph targeting: bioavailability enhancement using chylomicron flow block model and toxicity studies. *Drug Deliv Transl Res*. 2022;12(5):1230–1252.
26. Zielińska A, Ferreira NR, Durazzo A, Lucarini M, Cicero N, Mamouni S El, Silva AM, Nowak I, Santini A, and Souto EB. Development and optimization of alpha-pinene-loaded solid lipid nanoparticles (SLN) using experimental factorial design and dispersion analysis. *Materials* 2019;24(15):2683.
27. Mukherjee S, Ray S, and Thakur RS. Solid lipid nanoparticles: a modern formulation approach in drug delivery systems. *Indian J Pharm Sci*. 2009;71(4):349.
28. Bajwa N, Naryal S, Mahal S, Singh PA, and Baldi A. Quality-by-design strategy for the development of arteether-loaded solid self-microemulsifying drug delivery systems. *J Drug Deliv Sci Technol*. 2022;103707.
29. Naman S, Madhavi N, Singh B, Madan J, and Baldi A. Implementing risk-based quality by design for development and optimization of flavored oral disintegrating mini tablets. *J Drug Deliv Sci Technol*. 2021;66:102799.
30. Bajwa N, Singh PA, Jyoti K, Baldi A, Chandra R, and Madan J. Scale-up, Preclinical, and clinical status of poly (lactide-co-glycolide) and its copolymers-based drug delivery systems. *Nanomater Evol Adv Towar Ther Drug Deliv (Part II)*. 2021;13:246.
31. Glaubitt K, Ricci M, and Giovagnoli S. Exploring the nano spray-drying technology as an innovative manufacturing method for solid lipid nanoparticle dry powders. *AAPS PharmSciTech*. 2019;20(1):1–14.
32. Kathe N, Henriksen B, and Chauhan H. Physicochemical characterization techniques for solid lipid nanoparticles: principles and limitations. *Drug Dev Ind Pharm*. 2014;40(12):1565–1575.
33. Rohit B, and Pal KI. A method to prepare solid lipid nanoparticles with improved entrapment efficiency of hydrophilic drugs. *Curr Nanosci*. 2013;9(2):211–220.
34. Ali H, and Singh SK. Preparation and characterization of solid lipid nanoparticles of furosemide using quality by design. *Part Sci Technol*. 2018;36(6):695–709.
35. Naseri N, Valizadeh H, and Zakeri-Milani P. Solid lipid nanoparticles and nanostructured lipid carriers: structure, preparation and application. *Adv Pharm Bull*. 2015;5(3):305.
36. Olbrich C, Kayser O, and Müller RH. Lipase degradation of Dynasan 114 and 116 solid lipid nanoparticles

(SLN)—effect of surfactants, storage time, and crystallinity. *Int J Pharm.* 2002;237(1–2):119–128.
37. Thakur R, and Arora V. Comprehensive review on polymeric and metal nanoparticles: possible therapeutic avenues. *Int J Polym Mater Polym Biomater.* 2022;1–21.
38. Bajwa N, Mahal S, Madan J, and Baldi A. Implementation of the QbD approach to the development and validation of an analytical method for alpha-beta arteether. *Lett Drug Des Discov.* 2022;19:123–130.
39. Bajwa N, Singh M, Naryal S, Mahal S, Mehta S, Madan J, and Baldi A. Formulation Development and assessment of solid dispersion and hydrotropy for BCS Class II drug solubility enhancement. *Lett Drug Des Discov.* 2022;19. doi: 10.2174/1570180819666220822122016.
40. Bajwa N, Mahal S, Naryal S, Singh PA, and Baldi A. Development of novel solid nanostructured lipid carriers for bioavailability enhancement using a quality by design approach. *AAPS PharmSciTech.* 2022;23(7):1–14.
41. Bajwa N, Saroch P, Baldi A. Implementation of analytical quality by design methodology to develop a UV spectrometric technique for arteether quantification. *Ind J Pharm Edu Res.* 2023;57(3):1–7.

10 Nanomedicine Scale-Up Technologies

Feasibilities and Challenges

Jayaraman Rajangam[1], *G. S. N. Koteswara Rao*[2], *Narahari N. Palei*[3], *Haja Bava Bakrudeen*[4], *Dinesh Kumar Upadhyay*[5], *Sabareesh Mekala*[1], *and G. Sabareesh*[6]

[1] Department of Pharmacology, Shri Venkateshwara College of Pharmacy, Ariyur, Pondicherry, India

[2] Department of Pharmaceutics, Shobhaben Pratapbhai Patel School of Pharmacy & Technology Management, SVKM's NMIMS, Vile Parle (W), Mumbai, India

[3] Department of Pharmaceutics, AMITY Institute of Pharmacy, AMITY University, Lucknow, Uttar Pradesh, India

[4] Pharmaceutical Technology Unit, AIMST University, Jalan Bedong—Semeling, Bedong, Kedah, Malaysia

[5] School of Pharmaceutical Sciences, Jaipur National University, Jaipur, Rajasthan, India

[6] Department of Pharmaceutical Sciences, School of Biotechnology and Pharmaceutical Sciences, Vignan's Foundation for Science, Technology and Research, Vadlamudi, Guntur, India

10.1 INTRODUCTION TO NANOPARTICLES AS NANOMEDICINE

Nanotechnology is the study of manipulating matter at the atomic and molecular scales, as well as the science that deals with matter at the scale of one billionth of a meter (10^{-9} m = 1 nm). The most basic element in creating a nanostructure is a nanoparticle, which is larger than an atom or a simple molecule that is subject to quantum mechanics but much smaller than the universe of common things that are characterized by Newton's laws of motion [1,2].

Any particulate element having at least one dimension within the range of 1–100 nm is commonly referred to as a nanoparticle. They can take on a variety of forms, including spheres, wires, rods, planes, stars, cages, and multipods. In contrast to their bulk counterparts, nanoparticles exhibit several unique characteristics, including a high surface-to-volume ratio, high surface energy, and distinct mechanical, thermal, electrical, magnetic, and optical properties. Their characteristics render them suitable for a wide range of applications, including electronics, energy extraction and storage, communications, biology, and medicine. By adding desirable functional groups to their surface, their chemical reactivity and dispersibility in different solvents can be adjusted to suit a certain application [3].

10.1.1 Nanoparticles Technology and Nano Medicine

In general, spherical polymeric particles made of synthetic or natural polymers are called nanoparticles. Their sizes vary from 10 to 500 nm. These particles have an extensive range of possible uses due to their spherical form and high surface area to volume ratio [4].

The field of nanoparticle technology is rapidly expanding, offering cutting-edge and efficient therapies for a wide range of diseases. Because they can effectively carry and distribute biological materials, medicinal medicines, or imaging probes to specific locations, such as a particular organ, tissue, or even an underlying cell, nanoparticles are essential to nanomedicine. Furthermore, a few of them have active properties that make them useful as nanoprobes for sensing and imaging or as agents for cutting-edge treatments.

Molecular-level functional systems engineering is a component of nanotechnology. Disciplines ranging from materials science to biology are drawn to these systems due to their distinctive physical, optical, and electrical properties. Nanomedicine, which combines nanotechnology with highly specialized medical treatments for illness prevention, diagnosis, and therapy, is one of the most active fields of nanotechnology research. Drug delivery systems currently account for more than 75% of all sales in nanomedicine [5]. Nanomedicine uses nanoparticles for disease detection, monitoring, control, prevention, and therapy, utilizing nanotechnology. Definition developed by EC, involving natural, incidental, or manufactured materials [6]. Nanomedicine utilizes nanosized tools for disease diagnosis, prevention, and treatment, thereby improving the quality of life. It comprises polymer therapies, bioships, imaging agents, and medication delivery using nanoparticles [7].

The National Institutes of Health of the United States provided the following definition of nanomedicine: "the science that uses nanomaterials for the development of diagnosis, treatment, and prevention of specific medical applications". The development of this technology has sparked improvements in a

DOI: 10.1201/9781003508601-10

number of medical domains, including molecular imaging, biomarkers, controlled medication delivery, and biosensing. Dental, pharmaceutical, and medical fields, among others, stand to gain much from the application of nanomedicine. These developments, which demonstrate the interdisciplinary nature of this field of study, are the outcome of collaborations between businesses and research institutions, including universities [8]. The branch of medicine known as "nanomedicine" uses small particles, or "nanoparticles," to diagnose and treat medical conditions. The goal of this branch of advanced medicine is to address the numerous problems and shortcomings in conventional medicine, including low target specificity, systemic and organ toxicity, and inadequate bioavailability [9,10]. Some selected nanomaterials and their biomedical applications are depicted in Table 10.1, and a summary of the composition and fabrication methods of functional nano-systems is shown in Table 10.2.

Examples of nanocarrier-based drugs on the market [5] (Table 10.3):

10.2 SCALE-UP TECHNOLOGIES OF NANOMEDICINE

In recent years, nanomedicine has made remarkable strides, transforming the way we approach drug delivery, diagnostics, and therapeutics. Creating nanoparticles has become a key focus, where achieving the optimal size, shape, and composition can lead to groundbreaking treatments. With this progress comes a growing need for ways to mass-produce these nanoscale materials. In this chapter, we'll explore the world of scaling-up nanomedicine technologies. It's where creativity and innovation come together to push the boundaries of what's possible. We'll dive into the various methods used to transform lab experiments into market-ready products, discussing the hurdles and opportunities associated with each approach. Today, with technology evolving rapidly, there's a wide range of scaling-up techniques available. From using supercritical fluids to crafting intricate microfluidic systems and employing advanced membrane extrusion methods, the toolbox for scaling-up has never been more diverse or promising. We are witnessing a surge in the adoption of these cutting-edge techniques due to a significant demand for scalable manufacturing processes to meet the growing need for nanomedicine products. Whether it's big pharmaceutical companies or innovative startups, everyone in the industry is turning to these methods to speed up the development and launch of the next generation of treatments. In this chapter, we'll take a deep dive into three standout scaling-up technologies that are reshaping nanomedicine: the supercritical fluid method, microfluidics, and membrane extrusion, giving you a glimpse into how they're shaping the future of healthcare.

TABLE 10.1
Nanomaterials and Their Biomedical Applications

Nanomaterials	Biomedical Applications	Challenges and Future Perspectives
Semiconductor nanoparticles (quantum dots, rods)	Fluorescence bioimaging, multiphoton bioimaging, fluorescence and ratiometric biosensing, in vitro diagnostics.	Development of heavy metal-free quantum dots/rods, surface modification prevent degradation/erosion, detailed toxicological analysis, two-photon photodynamic therapy (pdt).
Rare-earth-doped upconversion nanoparticles	Deep tissue bioimaging, sensitive biosensing (oxygen, metal ions, DNA, etc.), multimodal bioimaging (MRI, PET, CT), PDT.	Scaling-up synthesis, surface modification to prevent leaching of rare-earth ions, detailed toxicological analysis, light-regulated drug release, intraoperative light guidance, theranostics.
Plasmonic nanoparticles (metallic nanostructures, self-doped semiconductor)	Dark field optical imaging, multiphoton bioimaging, photoacoustic (PA) bioimaging, photothermal therapy (PTT), plasmonic sensors, and gene therapy.	Development of core–shell nanoparticles, multimodal bioimaging, light-regulated drug release, and theranostics.
Silica-based nanoparticles (ormosil, doped silica)	Fluorescence bioimaging, multiphoton bioimaging, multimodal bioimaging, drug release, PDT, and gene therapy.	Detailed toxicological analysis, theranostics.
Carbon nanostructures (carbon dots, nanotubes, graphene)	Fluorescence bioimaging, multiphoton bioimaging, amperometric biosensing, drug release, gene therapy, tissue engineering.	Detailed toxicological analysis, externally activated therapies, theranostics.
Polymeric and lipid nanocarriers (polymeric micelles, hydrogels, dendrimers, liposomes, solid–lipid nanoparticles)	Sustained and stimulus-responsive drug release, gene delivery, PDT, photothermal therapy (PTT), gene therapy, and tissue engineering.	Scaled-up synthesis, externally activated drug release, theranostics.
Magnetic nanoparticles (iron oxide nanoparticles, cobalt ferrites)	Magnetic-resonance imaging (MRI), magnetically guided drug delivery, magnetic biosensing and cell separation, gene therapy.	Multimodal bioimaging, theranostics.
Nanoscale metal–organic frameworks (nmofs)	Multimodal bioimaging, drug release, gene therapy, tissue engineering.	Scaled-up synthesis, controlled drug release, detailed toxicological analysis, theranostics.

TABLE 10.2
Composition and Fabrication Methods of Some Nano-Systems

Nano-Systems	Composition	Characterization	Fabrication Methods
SLNs	Solid lipids, surfactants,co-surfactants	Shape: spherical Size: 50–1,000 nm	Thin film hydration; hand shaking; reverse-phase evaporation; sonication; ether injection; pro-niosome method; freeze and thaw; dehydration; rehydration; bubble method; solvent injection-homogenization technique; microfluidics.
NLCs	Solid lipids, liquid lipids (oils), surfactants	Shape: spherical Size: 120.9–185.2 nm	High-pressure homogenization (hot/cold); ultrasonication (probe/bath); solvent evaporation; solvent emulsification diffusion; microemulsion; supercritical fluid; spray drying; hot-melt extrusion (HME) methods.
Liposomes	Cholesterol, phospholipids	Shape: spherical Size: 25–5,000 nm	Reverse-phase evaporation; membrane extrusion; thin film hydration; sonication (probe/bath); ethanol injection; ether injection; freeze-thaw; micro-emulsification; microfluidics; and heating methods.
Ethosomes	Ethanol, phospholipids	Shape: spherical	Thin film hydration; vortex/sonication; ethanol injection; hot and cold methods.
Transfersomes	Phospholipids, surfactant/ edge activators, alcohol buffering agent	Shape: spherical Size: 90–140 nm	Thin film hydration.
Niosomes	Cholesterol, nonionic surfactant	Shape: spherical Size: 100–2,000 nm	Cold and hot method.
ISAsomes	Inverse bicontinuous (Q2) and discontinuous hexagonal (H2) phases	Shape: nonlamellar Size: 100–200 nm	High-energy emulsification methods, microfluidization, high-pressure homogenization.
Polymeric NPs	Natural/synthesis polymers	Shape: spherical Size: 10–1,000 nm	Emulsification/solvent diffusion; salting; nanoprecipitation; polymerization; dialysis; solvent evaporation; supercritical fluidization
Polymeric Micelles	Polymers preferably, polyethylene glycol, aqueous solution, ligands (target molecules).	Shape: spherical Size: 113.0±34.0 nm, 498.7±35.4 nm.	Chemical conjugation; physical entrapment through dialysis; emulsification; poly-ionic complexation.
Dendrimers	Core molecule, surface units, monomers	Shape: spherical Size: 20 nm	Divergent; convergent method.
Metallic NPs	Metal atom, ligands	Shape: rod, spherical, triangular, spherical Size: 2–100 nm	Microfluidics.
Metallic oxide NPs	Metal oxide	Shape: non-spherical Size: <100 nm	Green synthesis methods.
QDs	Metalloid crystalline core, surface groups/polymer/ peptide ligand/antibody	Shape: spherical Size: 20 nm	Oxidative cleavage, hydrothermal/solvothermal, microwave/ ultrasonic-assisted, electrochemical, pulsed laser ablation, chemical vapor deposition, electron beam irradiation, controllable synthesis, carbonization
Silica NPs	Silica	Shape: spherical Size: 0.05–2 nm	A system chemical method.

10.2.1 Supercritical Fluid Technology (SCF)

10.2.1.1 Principles and Applications of SCF Technology

In SCF technology, a substance is brought to a state where its temperature and pressure are above its critical point, allowing it to exhibit properties of both liquids and gases. This unique state enables the SCF to act in various roles in nanoparticle production, such as a solvent, swelling and plasticizing agent, antisolvent, and a medium for polymerization [11,12]. SCF technology enhances nanoparticle production by breaking down agglomerates, thereby improving solubility, dissolution rates, controlled release, and drug absorption, while ensuring minimal residual solvent concentration [13] (Figure 10.1).

10.2.1.2 Process Optimization for Scale-Up

Scaling up supercritical fluid (SCF) technology for nanomedicine production involves careful optimization of several key factors to ensure that the process remains efficient,

TABLE 10.3
Some Nano Carrier-Based Drugs on the Market

Types of Nanoformulations	Brand Name	Active Ingredient	Indications
Nanocrystalline drugs	Rapamune®	Rapamycin	Immunosuppressive
	Emend®	Aprepitant	Anti-emetic
	Tricor®	Fenofibrate	Hypercholesterolemia
	Megace®	Megestrol	Anti-anorexia
Liposomes	AmBisome®	Amphotericin B	Fungalinfections
	Doxil®	Doxorubicin	Ovarian cancer, Kaposi's sarcoma, and breast cancer
	Caelyx®	Doxorubicin	Ovarian cancer, Kaposi's sarcoma, and breast cancer
	Depocyt®	Cytarabine	Lymphomatous meningitis
	Daunoxome®	Daunorubicin	Kaposi's sarcoma
Polymer–drug conjugates	Adagen®	Adenosine deaminase	Adenosine deaminase enzyme deficiency
	Onscaspar®	L-asparaginase	Acute lymphoblastic leukemia
	Pegasys®	PEGylated IFN-α-2a	Hepatitis C
Polymeric micelles	Genexol-PM®	Paclitaxel	Cancer chemotherapy
Protein (albumin) nanoparticles	Abraxane®	Paclitaxel	Metastatic breast cancer
Lipid colloidal dispersion	Amphotec®	Amphotericin B	Fungal infections

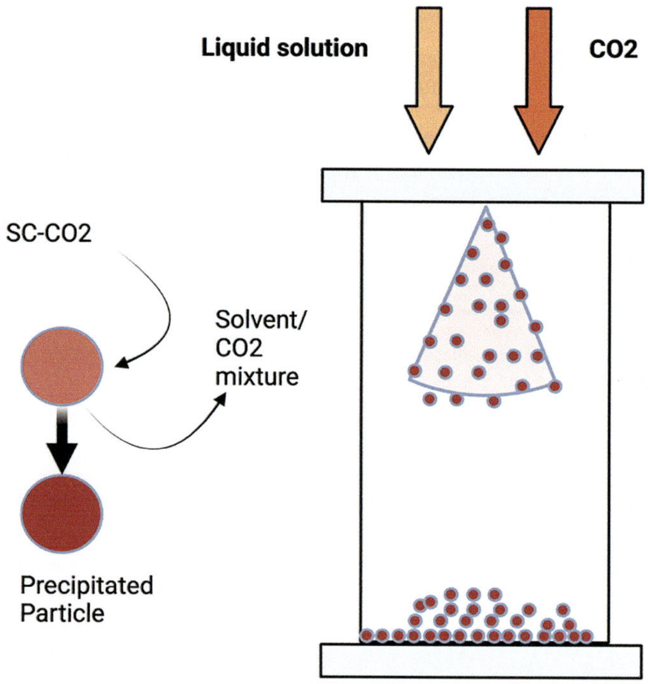

FIGURE 10.1 The above figure explains the general procedure of supercritical fluid technology. The liquid solution and CO_2 are entered into the chamber, and it gets converted to a precipitated particle. As per the type of SCF technology implemented, there are some minor modifications, but this general system remains the same.

consistent, and economical. This section explores the key aspects of process optimization, including pilot-scale studies, equipment selection, and the implementation of process analytical technology (PAT) [14].

10.3 PILOT-SCALE STUDIES

Pilot-scale studies are a crucial step in scaling up SCF technology from laboratory research to full-scale industrial production. These studies help identify potential scale-up issues and optimize process parameters to ensure consistent product quality and process efficiency.

10.3.1 Steps in Pilot-Scale Studies

1. **Initial Scale-Up**:
    - Transition from small-scale laboratory experiments to larger volumes, typically ranging from liters to tens of liters.

- This initial scale-up tests the feasibility of the SCF process at a larger scale and identifies any preliminary issues.

2. **Parameter Optimization**:
   - Fine-tune critical parameters such as pressure, temperature, flow rate, and solvent composition to achieve optimal solubility, particle size, and product yield.
   - Conduct experiments to determine the best operating conditions for achieving the desired product characteristics.
3. **Consistency Testing**:
   - Perform repeated runs under optimized conditions to assess the consistency and reproducibility of the process.
   - Evaluate critical quality attributes (CQAs) such as particle size distribution, encapsulation efficiency, and product stability across multiple batches.
4. **Data Analysis**:
   - Collect and analyze data from pilot-scale runs to identify trends and correlations between process parameters and product quality.
   - Use statistical tools and process modeling to interpret the data and make informed decisions about further optimization.
5. **Optimization**:
   - Use the insights gained from pilot-scale studies to refine process parameters and ensure robust scalability.
   - Implement process adjustments as needed to address any identified issues and improve overall efficiency and consistency [14].

10.3.2 Scale-Up Considerations

The scale-up of SCF formulation processes presents a formidable challenge in the realm of chemical engineering. While fundamental principles for SCF equipment design exist, the dearth of data for large-scale SCF plants hinders comprehensive calculations for their main components. Moreover, the literature on scaling up SCF particle design processes remains sparse, with only a handful of papers addressing this crucial aspect. This chapter aims to elucidate the intricacies of scaling-up SCF formulation processes, focusing on three critical steps: particle generation, particle collection, and fluid purification and recycling.

10.3.3 Particle Generation

Atomization stands as the cornerstone of particle generation in most SCF formulation processes. The ideal approach to scaling-up involves maintaining all parameters constant, which is achieved by replacing a single nozzle with multiple similar nozzles. However, practical challenges arise in achieving perfect fluid distribution and preventing droplet coalescence. An alternative method, applicable to the Rapid Expansion of Supercritical Solutions (RESS) process, utilizes a sintered disk as a fluid distributor. Despite its potential, this method faces limitations due to continuous pore plugging by generated nanoparticles.

Nucleation of solid particles from liquid droplets necessitates careful consideration of droplet size, which is correlated with dimensionless numbers such as the Reynolds and Weber numbers. Scaling up while maintaining constant parameters proves challenging, with liquid velocity emerging as a key determinant. The post-nucleation growth of particles depends on maintaining a consistent residence time before collection, thereby ensuring reproducible particle size and morphology.

10.3.4 Particle Collection

Particle collection poses significant challenges, especially for micronic or submicronic particles. Various methods, including inertia, electrostatic forces, liquid scrubbing, and filtration, are employed to collect particles from gas flows. However, fine particles necessitate filtration for dry particles or liquid scrubbing for suspension. Despite filtration's efficiency, challenges such as batch processing variability and nonhomogeneous properties persist. Innovative techniques, including semibatch processes and in situ harvesting, address these challenges on a case-by-case basis [15].

10.3.5 Residual Solvent Stripping

Processes employing organic solvents require meticulous elimination of residual solvent from formed particles. Large-scale stripping poses challenges, with fluid bypass and aggregate formation being notable concerns. Strategies involving percolation by supercritical carbon dioxide prove effective, necessitating careful control to prevent solvent concentration anomalies.

10.3.6 Fluid Purification and Recycle

For certain SCF processes, fluid purification and recycling are imperative. While some processes permit venting of carbon dioxide, others mandate fluid recycling. Challenges arise in recompressing fluid from atomization pressure to dissolution pressure, particularly for RESS processes. Organic solvent recycling presents additional complexities, with challenges in eliminating residual solvent and maintaining constant solvent concentration.

10.3.7 Scale-Up Results

Successful scale-up from laboratory to commercial scale underscores the feasibility of SCF formulation processes. Pilot plants and commercial-scale facilities demonstrate consistent particle size distributions across various scales. Safety considerations remain paramount throughout the design, installation, and operation of equipment to mitigate hazards associated with handling SCFs and liquefied gases.

The scale-up of SCF formulation processes represents a complex endeavor, requiring meticulous attention to detail and innovative solutions to overcome inherent challenges. By addressing critical aspects of particle generation, collection, solvent stripping, and fluid recycling, successful scale-up can be achieved, paving the way for the widespread application of SCF technologies in pharmaceutical and industrial settings [14].

10.3.8 Microfluidics Technology

Microfluidizer technology is a relatively new and rapidly evolving method for nanoparticle production. This technique, exemplified by products like the Microfluidizer® from Microfluidics Inc., utilizes high-pressure fluid dynamics to create nanoparticles with precise control over size and distribution [16].

10.3.9 Principles of Microfluidizer Technology

Microfluidizer technology operates by forcing two fluid streams to collide under extremely high pressure, up to 1,700 bar. This high-pressure collision generates intense shear forces and cavitation, which are crucial for breaking down the fluids into nanoparticles. The simultaneous action of these forces leads to the efficient production of uniformly sized nanoparticles.

1. **High-Pressure Collision**: The core mechanism involves the frontal collision of two fluid streams under high pressure. This collision induces several physical changes, including the generation of shear and cavitation forces, which are essential for nanoparticle formation.
2. **Shear and Cavitation Forces**: These forces play a significant role in reducing particle size. Shear forces help in breaking down larger particles, while cavitation creates microbubbles that collapse and further break down the particles.
3. **Surfactant Addition**: To stabilize the nanoparticles formed during the process, surfactants are added to the solution. Surfactants help prevent agglomeration, ensuring that the nanoparticles remain discrete and stable [16] (Figure 10.2).

10.3.10 Scale-Up Considerations

The choice of materials for microfluidic devices is crucial for the fabrication of Nanoparticulate Drug Delivery Systems (NDDSs) as it impacts the wetting properties and contact angles of solutions, thereby affecting the properties of the NDDSs themselves. Various materials have been employed for microfluidic devices, including silicon, glass, paper, and PDMS (polydimethylsiloxane).

Silicon and Glass: These are among the earliest materials used for microfluidic devices. Glass offers advantages such as transparency, insulation, high thermal conductivity, and stable electro-osmotic mobility. Silicon, on the other hand, exhibits chemical compatibility, high thermal stability, design flexibility, and semiconductor properties. Both materials are suitable for microfluidic device fabrication, although silicon is more prone to opacity and fragility.

PDMS (Polydimethylsiloxane): In recent years, there has been a shift toward PDMS-based microfluidic devices due to their cost-effectiveness, high optical transparency, excellent biocompatibility, and elasticity. However, PDMS can be susceptible to deformation and corrosion by organic solvents, which may affect the physicochemical properties

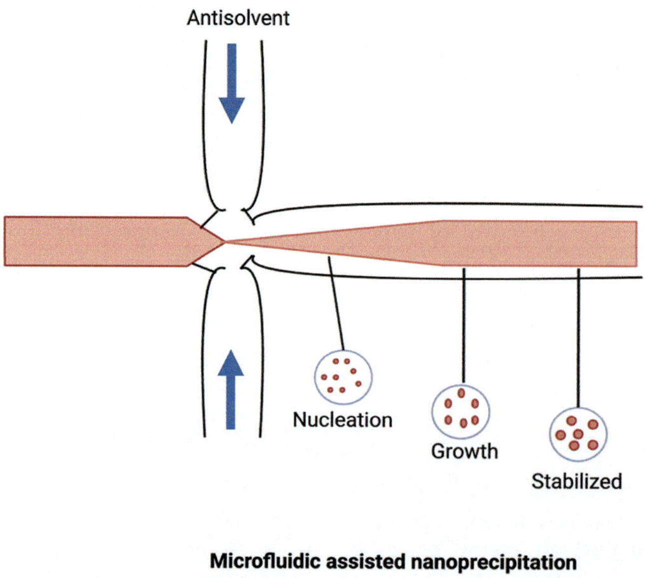

FIGURE 10.2 The above figure explains how the microfluidic process works. This also depicts that there are 3 stages of particle formation here. The entry for the solution and the antisolvent is different, so they collide at high pressure, which causes the breakdown of fluids into nanoparticles.

of NDDSs. Despite these drawbacks, PDMS remains a popular choice for microfluidic devices due to its favorable properties.

When selecting materials for microfluidic devices, factors such as chemical compatibility, durability, and transparency must be carefully considered to effectively meet the requirements for fabricating NDDSs.

In microfluidic devices, the fluidic regime is typically categorized into laminar and turbulent flow. Laminar flow occurs when a fluid flows in parallel layers with a constant velocity, whereas turbulent flow lacks distinct streamlines. The Reynolds number (Re) is a crucial parameter for reflecting the fluid regime. At low Re (<1,800), viscous forces dominate, resulting in laminar flow, whereas at high Re (>2,000), inertia forces dominate, leading to turbulent flow. The transition state (1,800<Re<2,000) occurs between laminar and turbulent flow.

In microfluidic devices, laminar flow is predominant due to the low Reynolds numbers typically encountered (Re<100). This steady flow regime enables precise fluid manipulation and efficient mass transfer dominated by passive molecular diffusion. Molecular diffusion, driven by Brownian motion, facilitates the gradual mixing of precursors in microfluidic devices.

Additionally, the Peclet number (Pe) reflects molecular diffusion, with lower Pe values indicating dominance of molecular diffusion. Pe is positively related to fluid velocity and diffusion distance. In microfluidic channels, molecular diffusion is relatively complex due to the low Pe, resulting in the formation of liquid-liquid interfaces and minimized transfer of molecules between liquids.

Microfluidic mixing efficiency, critical for controlling the properties of NDDSs, can be achieved through various strategies, including active and passive mixing. Active microfluidic mixing utilizes external energy sources, such as pressure, ultrasonics, or temperature, to perturb precursors and enhance mixing efficiency. In contrast, passive microfluidic mixing relies on fluid pump and channel design to increase the contact surface area and reduce the diffusion distance, without requiring external energy input.

Microfluidic liquid manipulation techniques, such as continuous-flow and segmented-flow microfluidics, enable precise control over fluid behaviors to fabricate NDDSs with narrow particle size distributions, higher drug-loading capacities, and excellent dispersion properties. Continuous-flow microfluidics focuses on miscible fluids, while segmented-flow microfluidics introduces immiscible phases to produce regularly spaced segmented interfaces, such as those in droplet microfluidics.

Droplet microfluidics, a segmented-flow technique, allows for the production of monodisperse droplets with controlled sizes and structures, making it suitable for fabricating NDDSs. Droplets can be classified based on their structures, including single-emulsion, double-emulsion, multi-emulsion, and multiple core' droplets, each offering specific advantages for NDDS synthesis.

Overall, microfluidics offers a versatile platform for fabricating NDDSs with precise control over particle size, morphology, and drug-loading capacity, making it a promising technology for pharmaceutical and biomedical applications [17,18].

10.3.11 Membrane Extrusion

Extrusion is a direct method for nanoparticle production, particularly for hydrophobic drugs. It is a scalable and cost-effective technique that involves converting hydrophobic drugs directly into nanoparticles. A common approach is manual extrusion using gastight syringes and polycarbonate membranes. However, this method can lead to heterogeneity, particularly when pore sizes are smaller than 100 nm, due to variations in the manual pressure applied [19].

Recently, Guo et al. (2013) [52] introduced a nanoporous membrane extrusion method specifically for producing nanoparticles loaded with hydrophobic drugs. This process enhances the precipitation of drug-loaded nanoparticles at the nanopore exits, increasing uniformity and consistency. Similarly, Khinast et al. (2019) developed a novel one-step process that combines hot-melt extrusion with internal devolatilization to convert a liquid-stabilized nano-suspension into a solid formulation. In this method, a polymer is melted in an extruder, and a stable nano-suspension is fed into it via side-feeding devices. The water is then removed through devolatilization, and the polymer solidifies at the outlet [20].

Membrane extrusion involves using a polycarbonate membrane with a well-defined pore diameter to prepare vesicles from a lipid suspension. The diameter of these vesicles closely matches the pore diameter of the membrane. By passing the suspension through the membrane multiple times, homogeneous large unilamellar vesicles or nanoliposomes are produced, which depend on the pore size. This technique offers significant advantages, including the elimination of the need to remove detergents or organic solvents from the final product. Moreover, it ensures reproducibility in the average liposomal size and size distribution [21].

One study compared two liposomal preparation methods for scale-up objectives. Liposomes composed of cholesterol/sphingomyelin containing vinblastine-N-oxide prepared by the extrusion method exhibited a lower polydispersity index (PDI) of 0.1 and a mean size of 64.26 nm. In contrast, a state-of-the-art microfluidics method yielded a PDI of 0.2 and an average size of 56.02 nm, demonstrating the efficacy and reproducibility of the extrusion method. Extrusion is not only a simple but also an effective technique for producing nanoparticles from hydrophobic drugs in a scalable and inexpensive manner. This method utilizes high pressure, shear, and heat to alter the structure of compounds, leading to processes like gelatinization, melting, and fragmentation [19]. Hot-melt extrusion, in particular, converts liquid-stabilized nano-suspensions into solid nanocrystals through a combination of heat and pressure, producing uniform-sized nanoparticles continuously.

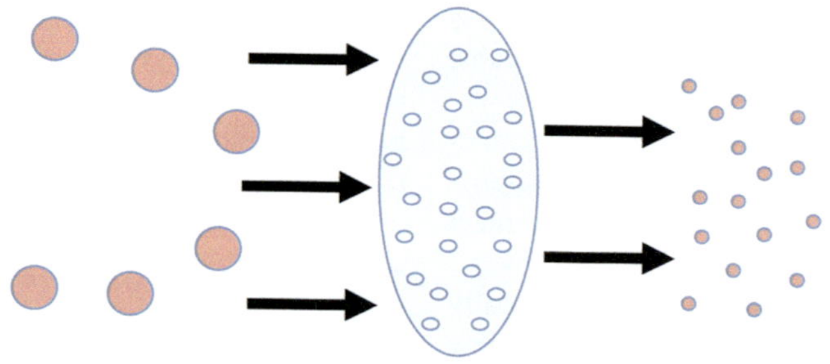

FIGURE 10.3 The above figure explains the principle of membrane extrusion technology. The pore size decides the final particle size.

Nanoporous membrane extrusion can be further employed to prepare hydrophobic nanoparticles using various strategies such as vesicle extrusion, membrane emulsification, precipitation extrusion, and biological membrane extrusion. These strategies are applicable in fabricating different nanocarriers, including liposomes, emulsions, nanoparticles, nanotubes, and nanofibers. Despite its advantages, membrane extrusion faces challenges, particularly during scale-up. Issues such as the preparation of nanomaterials using small amounts of raw materials and the accumulation of filtration residue at the channel's feeding site, which can lead to blockages, need to be addressed [22] (Figure 10.3).

10.3.12 Process Optimization for Scale-Up

Scaling up membrane extrusion for nanomedicine production involves careful optimization of several key factors to ensure that the process remains efficient, consistent, and economical. This section explores the key aspects of process optimization, including pilot-scale studies, equipment selection, and the implementation of PAT.

Pilot-scale studies serve as an intermediate step between laboratory-scale research and full-scale commercial production. They are essential for identifying and resolving potential scale-up issues and for optimizing process parameters.

10.4 REGULATORY GUIDELINES FOR NANOMEDICINES

Nanotechnology has emerged as a disruptive medical force, providing unprecedented prospects for targeted drug administration, improved imaging tools, and personalized treatments. Nanomedicines are the cutting edge of this innovation; they are materials at the nanoscale designed to interact molecularly with biological systems [23]. While promising, the research and marketing of nanomedicines are subject to severe regulatory requirements designed to ensure their safety, efficacy, and quality. Regulatory authorities worldwide, including the FDA (Food and Drug Administration) in the United States and the EMA (European Medicines Agency) in Europe, have developed detailed frameworks tailored to the properties and potential risks of nanoscale materials [24,25]. These recommendations cover all stages of nanomedicine development, from preliminary research and preclinical testing to clinical trials, manufacturing, and post-market surveillance.

The regulatory landscape for nanomedicines is dynamic, developing along with advances in nanotechnology and our understanding of nanoparticle behavior in biological environments. A detailed study of nanoparticle safety profiles, including biocompatibility, biodistribution, and potential toxicity, is crucial for regulatory approval [26]. Furthermore, strict production processes and quality control requirements ensure nanoparticle qualities remain consistent and reliable from batch to batch. Manufacturing standards play an important role in ensuring uniformity and quality control in nanoparticle manufacturing. Guidelines mandate robust protocols to maintain uniformity across batches, safeguarding product integrity and patient safety [27]. Post-market surveillance continues after approval, evaluating long-term safety and efficacy and responding quickly to any emergent issues.

Moreover, global harmonization efforts seek to align regulatory standards across different regions, facilitating efficient approval processes while maintaining robust safety and efficacy standards. This alignment not only accelerates the translation of nanomedicine discoveries from research laboratories to clinical applications but also promotes international cooperation and commercial access for these groundbreaking therapeutics (Figure 10.4).

10.5 FEASIBILITY AND CHALLENGES IN SCALE-UP TECHNOLOGY OF NANOMEDICINES

Advancements in nanotechnology and materials science have revolutionized the design and manufacturing of nanomedicines. These advancements have led to the development of new nanomaterials with enhanced drug delivery, targeted specificity, and improved biocompatibility. The loading and release kinetics in nanoparticles have also been improved, leading to increased therapeutic efficacy. Multifunctional nanocarriers have been developed to target

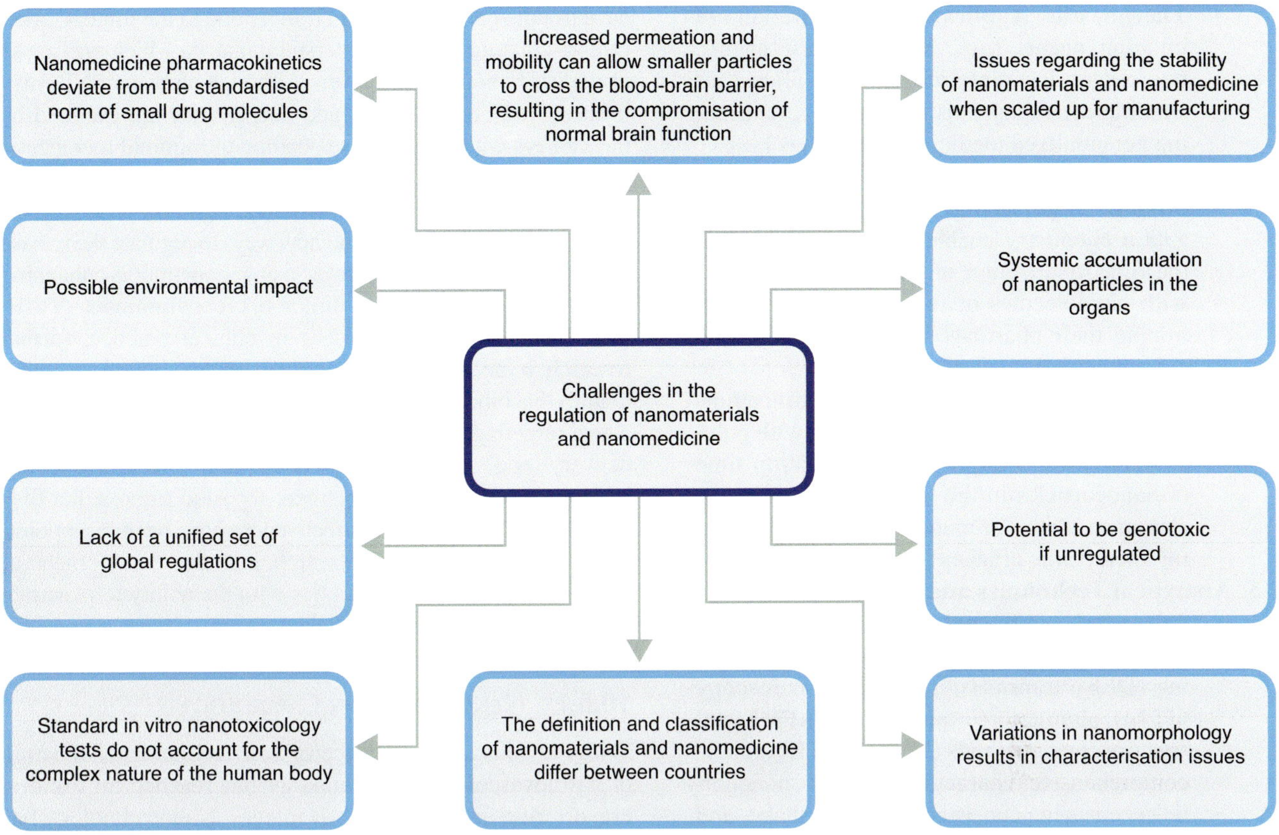

FIGURE 10.4 Challenges in the regulation of nanomaterials and nanomedicine.

specific cells or tissues, transport multiple medications, and respond to external stimuli. Bioconjugation techniques have also been enhanced, enabling the coupling of targeted ligands to nanoparticles, thereby improving selectivity and minimizing off-target effects. Improved analytical techniques have also been developed for quality control and scalability. Some of the notable advancements in nanotechnology and materials science of nanomedicine, and their therapeutic potential are listed here.

1. **Innovative Nanomaterials:**
   - **Polymeric Nanoparticles:** Polymeric nanoparticles, such as PLGA (poly(lactic-co-glycolic acid)) and PEGylated nanoparticles, offer controlled release capabilities and improved biocompatibility. These materials are widely used for delivering various drugs, including small molecules and biologics [27].
   - **Lipid-Based Nanocarriers:** Liposomes and lipid nanoparticles provide excellent biocompatibility and the ability to encapsulate both hydrophilic and hydrophobic drugs. They are used for targeted delivery and controlled release of therapeutic agents [28].
   - **Inorganic Nanoparticles:** Gold nanoparticles, quantum dots, and magnetic nanoparticles have unique optical, magnetic, or electronic properties useful for imaging and targeted therapy applications [29].
2. **Improved Drug Loading and Release:**
   - **Controlled Release Systems:** Advances in nanotechnology have enabled the development of stimuli-responsive nanoparticles that release drugs in response to environmental cues such as pH, temperature, or enzymatic activity [30].
   - **Enhanced Encapsulation Efficiency:** Techniques such as nanoprecipitation, emulsion-solvent evaporation, and microfluidics have been optimized to improve drug loading efficiency and ensure uniform distribution within nanoparticles [31].
3. **Multifunctional Nanocarriers:**
   - **Targeting Specificity:** Surface modification of nanoparticles with targeting ligands (e.g., antibodies, peptides, aptamers) enhances their specificity for diseased cells or tissues, reducing systemic toxicity and improving therapeutic efficacy [32].
   - **Combination Therapy:** Nanocarriers capable of delivering multiple drugs simultaneously enable combination therapies, which can synergistically enhance therapeutic outcomes and overcome drug resistance [32].

- **Theranostic Applications:** Integration of imaging agents (e.g., MRI contrast agents, fluorescent dyes) into nanocarriers allows for simultaneous diagnosis and therapy, facilitating personalized medicine approaches [33].

4. **Bioconjugation Techniques:**
   - **Surface Engineering:** Advances in bioconjugation chemistry enable precise control over the functionalization of nanoparticle surfaces with biomolecules or targeting ligands, optimizing their pharmacokinetics and targeting efficiency [34].
   - **Enhanced Biocompatibility:** Modifications such as PEGylation (surface coating with polyethylene glycol) improve the circulation time of nanoparticles in the bloodstream, reducing clearance by the immune system and enhancing therapeutic efficacy [35].
5. **Analytical Techniques and Quality Control:**
   - **Advanced Characterization Methods:** Techniques such as dynamic light scattering (DLS), transmission electron microscopy (TEM), atomic force microscopy (AFM), and spectroscopic methods (UV-Vis, FTIR) enable comprehensive characterization of nanoparticles, ensuring consistency in size, shape, and surface properties [36].
   - **Safety Assessment:** Detailed studies on pharmacokinetics, biodistribution, and potential toxicity using in vitro and in vivo models provide critical data for regulatory approvals and clinical translation of nanomedicines [37].

These advancements underscore the transformative potential of nanotechnology in overcoming challenges in drug delivery and improving therapeutic outcomes. Continued research and innovation in this field hold promise for addressing complex medical conditions with greater precision and effectiveness [38].

10.6 FUTURE PERSPECTIVE ON NANOMEDICINE

Incorporating nanotechnology into conventional pharmaceuticals has the potential to increase their therapeutic effectiveness. For the next wave of nanotherapeutics, developing very efficient and tailored drug delivery systems that avoid the accumulation of toxic substances in healthy tissues is a top goal. Modern technology has allowed for the creation of smart medicine delivery devices with active targeting and stimulus response. These smart drug delivery systems provide both a significant problem and an exciting opportunity for the field of nanomedicine as it continues to evolve; they show promise as a method of eliminating diseased tissues while sparing healthy ones [39].

In contrast to tumor pathological aspects that are difficult to manipulate, the circulation half-lives of nanotherapeutics may be controlled by modifying the surface of the drug carrier and choosing the appropriate drug carrier. Large nanodrugs may have trouble penetrating solid tumors due to the tumors' aberrant vasculature and thick interstitial matrix, which may cause drug distributions to be uneven. Although ultrasmall NPs may infiltrate tumors uniformly and deeply, they are excreted by the kidneys with relative ease, leading to minimal therapeutic outcome [40]. Hence, methods need to be devised to increase the intratumoral penetration of nanodrugs and prolong their half-life. It is important to find new ways to regulate the release of drugs into tumors, such as using penetration-enhancing ligands or influencing the tumor microenvironment (TME) [41]. To implement these ideas in clinical practice, further research is necessary. By carrying surface-bound targeting ligands that bind to receptors overexpressed in diseased tissues, active-targeting nanotherapeutics can accomplish site-specific delivery and selective drug absorption. Nanotherapeutics have great promise as a treatment for solid tumors, but thus far, tumor absorption and effectiveness have been disappointing [42]. Hence, further research into the development of active-targeting nanotherapeutics with the ability to penetrate solid tumors should be undertaken.

10.6.1 Nanomedicine in Cardiology

With the constant emergence of new technologies, the area of cardiovascular nanotechnology has reached an exciting era, offering much-needed advantages. Nanotechnology has a lot of potential in new technological developments. It is believed that wearable cardiac monitors will soon play a more significant role in clinical practice, complementing their current diagnostic uses. Potential future applications of nanotechnology in healthcare include screening high-risk individuals for specific indications and initiating early preventive pathways to reduce health and financial costs. Engineered nanoparticles and nanofluidic devices have the potential to initiate interactions between electronics and biological systems, similar to how small-molecule signaling controls cellular activity [43]. There may be more room for nanocomposites and nanocoatings in implanted cardiac electrical devices. Stents, artificial heart valves, and embolic spheres are several medical devices that are now under investigation for potential applications of nanocoatings.

Nanotechnology has the potential to impact significantly device sensitivity and performance due to its exceptional structural manipulation capabilities at the nanoscale. Advanced sensors made of nanomaterials will enable better early detection and ongoing symptom monitoring. Switches or remote-controlled manipulation at the nanoscale will enhance illness prevention. Additionally, nanotechnology will aid in the treatment and restoration of heart tissue by delivering molecules or agents with specific properties in a sustained and precise manner [44].

10.6.2 Nanomedicine in Neurology

Nanomedicine is unquestionably advancing at an inexorable rate, aided by the combined efforts of brilliant minds throughout the world, in the fight against neurodegenerative diseases and their terrible effects. Designing efficient nanocarriers to transport cargo and discharge them in a spatiotemporally regulated manner across the BBB requires a more comprehensive understanding of the processes underlying BBB regulation and malfunction in healthy and diseased brains [45]. Many have stated that nanomedicine has a restorative impact on various neurological illnesses in animal models; nonetheless, this subject requires further clarification [46]. Excitingly, neurodegenerative diseases (NDs) may now be effectively bioimaged, diagnosed, and treated using multifunctional nanomaterials. Clearly, further research is needed to describe improved ND and NP cause and movement monitoring.

10.6.3 Nanotechnology in Ophthalmology

The use of nanotechnology in medicine has completely altered the landscape of eye disease prevention [47]. Despite the clear benefit compared to conventional ocular administration methods, researchers are still in the early stages of their investigations and face numerous obstacles, such as the need for large-scale production and late-phase clinical trials, before they can produce strong results and evidence. It is essential to consider the various aspects of the disease, including its origins and progression, when designing a nanoscale therapeutic drug delivery vehicle for the future. It is necessary to ascertain how the pharmacokinetic and pharmacotoxic profiles are impacted by factors such as particle size, surface charge, content, and aggregation.

10.6.4 Nanomedicine in Dermatology

Topical delivery has several advantages, but the biggest problem is that it's challenging to determine the correct dosage and duration of treatment. The majority of topical preparations are given directly to the affected area, which often harbors beneficial microorganisms. When beneficial microorganisms in the surroundings are subjected to large doses of antibiotics, drug-resistance phenotypes may develop. Limiting antibiotics' impact on beneficial microorganisms is a very challenging task. Consequently, there is a need for a surge in research into nanotherapeutics to validate their promise, with a focus on clinical trials to assess their efficacy, safety, and skin distribution [48]. It is crucial to thoroughly understand how nanotherapeutics interact with biological barriers during transport, as well as how transport efficiency and targeting affect the safety and effectiveness of these nanotherapeutics. One of the challenges that need particular attention for future study is the stability of lipid-based nanoformulations delivered topically [49].

10.6.5 Nanotechnology in Regenerative Medicine

A growing number of experts are beginning to acknowledge the importance of nanotechnology in regenerative medicine [50]. By utilizing nanotechnology, biomaterials can have their physical and biological characteristics controlled with greater precision than with traditional methods. The use of nanostructures with the potential to resemble natural tissue has led to the creation of more durable and practical scaffolds, making it one of the most prominent applications in this area. To create scaffolds with exceptional biomimicking capacity, which can have their chemical composition and structure optimized, it is necessary to choose the new polymer matrix/nanomaterial combination with extreme care. To overcome these limitations, current scaffolds need to be enhanced. Research and development on nanofibers, which may mimic the characteristics of real cardiac tissue, is now underway. There has been a surge of interest in the therapeutic use of growth factors for wound healing. The best treatment option for skin tissue regeneration seems to be biodegradable scaffolds combined with various growth agents. Very little research has focused on the application of regenerative nanomedicine in ophthalmology. Still, research into nano-ophthalmology is ongoing, with the hope of one day preserving or restoring eyesight, which it promises.

10.6.6 Safety Concerns Regarding Nanomedicine

Nanotherapeutics, once within the body, interact with a variety of biological settings, including the blood, extracellular matrix, cytoplasm, and organelles of cells. Concerns about the safety of nanotherapeutics are multifaceted. There is a present paucity of guidelines and methodologies for evaluating the safety of nanodrugs, even though clinical translation relies on thorough assessments of nanotherapeutics' safety. The evaluation of nanotherapeutics' safety cannot be done using the same methods as conventional medications [51]. There could be significant impacts on toxicity from nanocarrier coatings and ligands. To improve the chances of successfully translating innovative nanotherapeutics into clinical practice, it is necessary to understand the toxicity of existing medications, the physicochemical characteristics of NPs, and the characteristics of nanocarriers.

When it comes to translating nanotherapeutics for clinical use, quality and safety are paramount. It is possible to remove a nanotherapeutic from sale if safety concerns arise, even after it has received clinical approval. Ensuring patient safety and improving the quality of medication are the primary goals of Good Laboratory Practice (GLP). Unfortunately, there are currently no GLPs available that are tailored to nanotherapeutics. To increase the success rate of clinical application and foster the development of safe and effective nanotherapeutics, it is crucial to construct GLPs for nanotherapeutics.

10.7 CONCLUSION

Nanomedicine is a revolutionary approach in the healthcare sector, utilizing nanoparticles to enhance the safety, delivery, and effectiveness of medications, as well as diagnostic imaging tools and therapeutic interventions. These nanoparticles, with their unique features, have enabled advancements in various medical fields, including oncology, cardiology, neurology, and regenerative medicine.

They target specific cells or tissues during drug administration, thereby increasing bioavailability and reducing systemic adverse effects, particularly in the treatment of complex diseases such as cancer. Advances in nanoparticle composition, such as liposomes, lipid-based systems, and polymeric carriers, have improved efficiency and safety in drug delivery processes. However, scaling up nanomedicine from laboratory research to commercial production presents several challenges, including cost, repeatability, and process optimization. Safety and toxicity are critical factors in the development and application of nanomedicines, and thorough safety evaluations are necessary to identify and mitigate potential risks. International regulatory agencies, such as the FDA and EMA, are working to develop guidelines that take into account the unique properties of nanomedicines. Nanomedicine has the potential to revolutionize illness management and treatment paradigms with sustained advancements in nanotechnology and strict respect for safety and regulatory norms.

REFERENCES

1. M. Benelmekki, An introduction to nanoparticles and nanotechnology, *Des. Hybrid Nanoparticles.* (2015). https://doi.org/10.1088/978-1-6270-5469-0ch1.
2. S. Horikoshi, N. Serpone, Introduction to nanoparticles. In: S. Horikoshi, N. Serpone (Eds.), *Microwaves in Nanoparticle Synthesis: Fundamentals and Applications*. Wiley, 2013, pp. 1–24. https://doi.org/10.1002/9783527648122.ch1.
3. G. Chen, I. Roy, C. Yang, P.N. Prasad, Nanochemistry and nanomedicine for nanoparticle-based diagnostics and therapy, *Chem. Rev.* 116 (2016) 2826–2885. https://doi.org/10.1021/acs.chemrev.5b00148.
4. M. Aschner, Nanoparticles: transport across the olfactory epithelium and application to the assessment of brain function in health and disease, *Prog. Brain Res.* 180 (2009) 141–152. https://doi.org/10.1016/S0079-6123(08)80008-8.
5. S. Bamrungsap, Z. Zhao, T. Chen, L. Wang, C. Li, T. Fu, W. Tan, Nanotechnology in therapeutics: a focus on nanoparticles as a drug delivery system, *Nanomedicine.* 7 (2012) 1253–1271. https://doi.org/10.2217/nnm.12.87.
6. D. Astruc, Introduction to nanomedicine, *Molecules.* 21 (2016) 4. https://doi.org/10.3390/molecules21010004.
7. J. Cancino-Bernardi, V.S. Marangoni, V. Zucolotto, Nanomedicine. In: Y. Dahman, T. Truong, M. Jawaid, S. Widjaja (Eds.), *Nanotechnology and Functional Materials for Engineers*. Elsevier (2017), pp. 229–249. https://doi.org/10.1016/B978-0-323-49780-0.00003-X.
8. Y. Dahman, Nanomedicine. In: T. Truong, M. Jawaid, S. Widjaja (Eds.), *Nanotechnology and Functional Materials for Engineers: Micro and Nano Technologies*. Elsevier (2017), pp. 229–249. https://doi.org/10.1016/b978-0-323-51256-5.00011-3.
9. M. Sabareesh, J. Pedda Yanadaiah, K.B. Chandrasekhar, Novel nanoproniosomal vesicular carriers for the effective and efficient drug delivery: fundamentals, recent advancements and applications. *Res. J. Pharm. and Tech.* 14(11) (2021) 6155–6165. https://doi.org/10.52711/0974-360X.2021.01067.
10. I.P. Ejidike, O. Ogunleye, M.O. Bamigboye, O.M. Ejidike, A. Ata, M.O. Eze, H.S. Clayton, V.U. Nwankwo, J.O. Fatokun, Role of nanotechnology in medicine: opportunities and challenges, *Environ. Sci. Eng. Part* F 2070 (2024) 353–375. https://doi.org/10.1007/978-3-031-45956-6_14.
11. D. Sanli, S.E. Bozbag, C. Erkey, Synthesis of nanostructured materials using supercritical CO 2: part I, *Physical transformations, J. Mater. Sci.* 47 (2012) 2995–3025. https://doi.org/10.1007/s10853-011-6054-y.
12. M. Herrero, J.A. Mendiola, A. Cifuentes, E. Ibáñez, Supercritical fluid extraction: recent advances and applications, *J. Chromatogr. A.* 1217 (2010) 2495–2511. https://doi.org/10.1016/j.chroma.2009.12.019.
13. P. Girotra, S.K. Singh, K. Nagpal, Supercritical fluid technology: a promising approach in pharmaceutical research, *Pharm. Dev. Technol.* 18 (2013) 22–38. https://doi.org/10.3109/10837450.2012.726998.
14. M. Perrut, Supercritical fluid applications: industrial developments and economic issues, *Ind. Eng. Chem. Res.* 39 (2000) 4531–4535. https://doi.org/10.1021/ie000211c.
15. J. Jung, M. Perrut, Particle design using supercritical fluids: literature and patent survey, *J. Supercrit. Fluids.* 20 (2001) 179–219. https://doi.org/10.1016/S0896-8446(01)00064-X.
16. P. Shrimal, G. Jadeja, S. Patel, A review on novel methodologies for drug nanoparticle preparation: microfluidic approach, *Chem. Eng. Res. Des.* 153 (2020) 728–756. https://doi.org/10.1016/j.cherd.2019.11.031.
17. I. V. Zhigaltsev, N. Belliveau, I. Hafez, A.K.K. Leung, J. Huft, C. Hansen, P.R. Cullis, Bottom-up design and synthesis of limit size lipid nanoparticle systems with aqueous and triglyceride cores using millisecond microfluidic mixing, *Langmuir.* 28 (2012) 3633–3640. https://doi.org/10.1021/la204833h.
18. A. Gdowski, K. Johnson, S. Shah, I. Gryczynski, J. Vishwanatha, A. Ranjan, Optimization and scale up of microfluidic nanolipomer production method for preclinical and potential clinical trials, *J. Nanobiotechnology.* 16 (2018) 1–10. https://doi.org/10.1186/s12951-018-0339-0.
19. S. Jain, S.K. Cherukupalli, A. Mahmood, S. Gorantla, V.K. Rapalli, S.K. Dubey, G. Singhvi, Emerging nanoparticulate systems: preparation techniques and stimuli responsive release characteristics, *J. Appl. Pharm. Sci.* 9 (2019) 130–143. https://doi.org/10.7324/JAPS.2019.90817.
20. J. Khinast, R. Baumgartner, E. Roblegg, Nano-extrusion: a one-step process for manufacturing of solid nanoparticle formulations directly from the liquid phase, *AAPS PharmSciTech.* 14 (2013) 601–604. https://doi.org/10.1208/s12249-013-9946-0.
21. H. Zhang, J. Yang, R. Sun, S. Han, Z. Yang, L. Teng, Microfluidics for nano-drug delivery systems: from fundamentals to industrialization, *Acta Pharm. Sin. B.* 13 (2023) 3277–3299. https://doi.org/10.1016/J.APSB.2023.01.018.
22. U.M. Musazzi, S. Franzè, F. Condorelli, P. Minghetti, P. Caliceti, Feeding next-generation nanomedicines to Europe: regulatory and quality challenges, *Adv. Healthc. Mater.* 12 (2023) 2301956. https://doi.org/10.1002/adhm.202301956.
23. J. Allan, S. Belz, A. Hoeveler, M. Hugas, H. Okuda, A. Patri, H. Rauscher, P. Silva, W. Slikker, B. Sokull-Kluettgen, W. Tong, E. Anklam, Regulatory landscape of nanotechnology and nanoplastics from a global perspective, *Regul. Toxicol. Pharmacol.* 122 (2021). https://doi.org/10.1016/j.yrtph.2021.104885.
24. C.S. Auletta, Regulatory toxicology. In M. J. Derelanko, C. S. Auletta (Eds.), *Handbook of Toxicology,* 3rd ed. CRC Press, 2014, pp. 811–833. https://www.taylorfrancis.com/chapters/mono/10.1201/b16632-31.
25. J.A. Damasco, S. Ravi, J.D. Perez, D.E. Hagaman, M.P. Melancon, Understanding nanoparticle toxicity to direct

a safe-by-design approach in cancer nanomedicine, *Nanomaterials.* 10 (2020) 1–41. https://doi.org/10.3390/nano10112186.
26. W. Oualikene-Gonin, V. Sautou, E. Ezan, H. Bastos, E. Bellissant, L. Belgodère, P. Maison, J. Ankri, Regulatory assessment of nano-enabled health products in public health interest. Position of the scientific advisory board of the French National Agency for the Safety of Medicines and Health Products, *Front. Public Heal.* 11 (2023). https://doi.org/10.3389/fpubh.2023.1125577.
27. E.B. Souto, G.F. Silva, J. Dias-ferreira, A. Zielinska, F. Ventura, A. Durazzo, M. Lucarini, E. Novellino, A. Santini, Nanopharmaceutics: part I—clinical trials legislation and good manufacturing practices (GMP) of nanotherapeutics in the EU, *Pharmaceutics.* 12 (2020) 146. https://doi.org/10.3390/pharmaceutics12020146.
28. M.G. Tirumala, P. Anchi, S. Raja, M. Rachamalla, C. Godugu, novel methods and approaches for safety evaluation of nanoparticle formulations: a focus towards in vitro models and adverse outcome pathways, *Front. Pharmacol.* 12 (2021) 612659. https://doi.org/10.3389/fphar.2021.612659.
29. Bhavna, S. Bhargava, A. Ojha, International Council for Harmonisation (ICH) Guidelines, *Regul. Aff. Pharm. Ind.* (2022) 47–74. https://doi.org/10.1016/B978-0-12-822211-9.00008-3.
30. T.A. Adekola, *Regional Cooperation, Intellectual Property Law and Access to Medicines: A Holistic Approach for Least Developed Countries.* Taylor & Francis (2023). https://doi.org/10.4324/9781003342519.
31. D. Bobo, K.J. Robinson, J. Islam, K.J. Thurecht, S.R. Corrie, Nanoparticle-based medicines: a review of FDA-approved materials and clinical trials to date, *Pharm. Res.* 33 (2016) 2373–2387. https://doi.org/10.1007/s11095-016-1958-5.
32. V.P. Torchilin, Multifunctional, stimuli-sensitive nanoparticulate systems for drug delivery, *Nat. Rev. Drug Discov.* 13 (2014) 813–827. https://doi.org/10.1038/nrd4333.
33. T.H. Kim, S. Lee, X. Chen, Nanotheranostics for personalized medicine, *Expert Rev. Mol. Diagn.* 13 (2013) 257–269. https://doi.org/10.1586/erm.13.15.
34. R. Singh, S.P. Srinivas, M. Kumawat, H.K. Daima, Ligand-based surface engineering of nanomaterials: trends, challenges, and biomedical perspectives, *Open Nano.* 15 (2024) 100194. https://doi.org/10.1016/j.onano.2023.100194.
35. J.S. Suk, Q. Xu, N. Kim, J. Hanes, L.M. Ensign, PEGylation as a strategy for improving nanoparticle-based drug and gene delivery, *Adv. Drug Deliv. Rev.* 99 (2016) 28–51. https://doi.org/10.1016/j.addr.2015.09.012.
36. P.C. Lin, S. Lin, P.C. Wang, R. Sridhar, Techniques for physicochemical characterization of nanomaterials, *Biotechnol. Adv.* 32 (2014) 711–726. https://doi.org/10.1016/j.biotechadv.2013.11.006.
37. D.T. Savage, J.Z. Hilt, T.D. Dziubla, In vitro methods for assessing nanoparticle toxicity, *Methods Mol. Biol.* 1894 (2019) 1–29. https://doi.org/10.1007/978-1-4939-8916-4_1.
38. M. Chehelgerdi, M. Chehelgerdi, O.Q.B. Allela, R.D.C. Pecho, N. Jayasankar, D.P. Rao, T. Thamaraikani, M. Vasanthan, P. Viktor, N. Lakshmaiya, M.J. Saadh, A. Amajd, M.A. Abo-Zaid, R.Y. Castillo-Acobo, A.H. Ismail, A.H. Amin, R. Akhavan-Sigari, Progressing nanotechnology to improve targeted cancer treatment: overcoming hurdles in its clinical implementation, *Mol. Cancer.* 22 (2023). https://doi.org/10.1186/s12943-023-01865-0.
39. R. Paliwal, R.J. Babu, S. Palakurthi, Nanomedicine scale-up technologies: feasibilities and challenges, *Ageing Int.* 15 (2014) 1527–1534. https://doi.org/10.1208/s12249-014-0177-9.
40. N. Desai, Challenges in development of nanoparticle-based therapeutics, *AAPS J.* 14 (2012) 282–295. https://doi.org/10.1208/s12248-012-9339-4.
41. S. Das Mandal, S. Mandal, Y. V. Pathak, J.K. Patel, Current challenges and future directions in nanomedicine, *Emerg. Technol. Nanoparticle Manuf.* (2021) 575–583. https://doi.org/10.1007/978-3-030-50703-9_27.
42. J. Liu, M. Li, Z. Luo, L. Dai, X. Guo, K. Cai, Design of nanocarriers based on complex biological barriers in vivo for tumor therapy, *Nano Today.* 15 (2017) 56–90. https://doi.org/10.1016/j.nantod.2017.06.010.
43. M.U. Munir, Nanomedicine penetration to tumor: challenges, and advanced strategies to tackle this issue. *Cancers (Basel).* 14 (2022). https://doi.org/10.3390/cancers14122904.
44. C. Paus, R. van der Voort, A. Cambi, Nanomedicine in cancer therapy: promises and hurdles of polymeric nanoparticles, *Explor. Med.* 2 (2021) 167–185. https://doi.org/10.37349/emed.2021.00040.
45. S. Mukherjee, V.S. Madamsetty, D. Bhattacharya, S. Roy Chowdhury, M.K. Paul, A. Mukherjee, Recent advancements of nanomedicine in neurodegenerative disorders theranostics, *Adv. Funct. Mater.* 30 (2020) 2003054. https://doi.org/10.1002/adfm.202003054.
46. T. Bhattacharya, G.A.B.E. Soares, H. Chopra, M.M. Rahman, Z. Hasan, S.S. Swain, S. Cavalu, Applications of phyto-nanotechnology for the treatment of neurodegenerative disorders, *Materials (Basel).* 15 (2022). https://doi.org/10.3390/ma15030804.
47. M.G. Sharaf, S. Cetinel, L. Heckler, K. Damji, L. Unsworth, C. Montemagno, Nanotechnology-based approaches for ophthalmology applications, *Asia-Pacific J. Ophthalmol.* 3 (2014) 172–180. https://doi.org/10.1097/apo.0000000000000059.
48. F. Farjadian, A. Ghasemi, O. Gohari, A. Roointan, M. Karimi, M.R. Hamblin, Nanopharmaceuticals and nanomedicines currently on the market: challenges and opportunities, *Nanomedicine.* 14 (2019) 93–126. https://doi.org/10.2217/nnm-2018-0120.
49. Sabareesh M, Yanadaiah JP, Sekhar KC. Formulation development, Ex vivo evaluation and In vivo antihypertensive study of losartan potassium loaded nanoproniosomal gel: A novel vesicular approach for transdermal delivery. Research Journal of Pharmacy and Technology. (2021) Mar 1 14(3) 1423–30. http://dx.doi.org/10.5958/0974-360X.2021.00254.
50. M.E. Gomes, M.T. Rodrigues, R.M.A. Domingues, R.L. Reis, Tissue engineering and regenerative medicine: new trends and directions: a year in review, *Tissue Eng. B Rev.* 23 (2017) 211–224. https://doi.org/10.1089/ten.teb.2017.0081.
51. Q. Liu, J. Zou, Z. Chen, W. He, W. Wu, Current research trends of nanomedicines, *Acta Pharm. Sin. B.* 13 (2023) 4391–4416. https://doi.org/10.1016/j.apsb.2023.05.018.
52. Guo P, Hsu TM, Zhao Y, Martin CR, Zare RN. Preparing amorphous hydrophobic drug nanoparticles by nanoporous membrane extrusion. *Nanomedicine* (2013) Mar 1 8(3) 333–41. https://doi.org/10.2217/nnm.12.119.

11 Role of Metallic Nanoparticles in Controlled and Targeted Drug Delivery

Ahmet Dogan Ergin[1,2] *and Burcu Uner*[3,4,5]

[1]Department of Pharmaceutical Technology, Faculty of Pharmacy, Trakya University, Edirne, Turkey

[2]Department of Pharmaceutical Nanotechnology, Institute of Health Sciences, Trakya University, Edirne, Turkey

[3]Department of Pharmaceutical and Administrative Sciences, University of Health Science and Pharmacy, St. Louis, Missouri

[4]Department of Anesthesiology, Center for Clinical Pharmacology, Washington University School of Medicine, St. Louis, Missouri

[5]Department of Pharmaceutical Technology, Faculty of Pharmacy, Istanbul Kent University, Istanbul, Turkey

11.1 INTRODUCTION

Drug delivery systems are essential for the effective and targeted delivery of pharmaceutical agents to specific sites within the body. These systems encompass a variety of technologies and formulations aimed at optimizing drug administration. Different routes of drug administration are utilized to ensure the delivery of the correct drug amount to the precise site at the appropriate time, achieving optimal therapeutic concentrations while minimizing side effects [1]. Formulation technologies play a crucial role in drug delivery systems, ensuring that drugs are delivered in a manner that maximizes their pharmacological activity [2]. Lipid-based vesicular systems have been studied as drug carriers, with a focus on stability and regulatory considerations to enhance drug delivery efficiency [3]. Novel drug delivery systems, such as nanoparticles, microemulsions, and liposomes, are being developed to meet scientific requirements and improve drug delivery by addressing factors like pharmacokinetics and site-specific action [4–6]. Injectable in situ depot-forming drug delivery systems offer advantages in controlled drug delivery, providing ease of application and localized drug release [7]. Transdermal drug delivery systems have become essential in modern pharmaceutical dosage forms, offering unique benefits in drug administration [8].

Innovative drug delivery systems, such as vesicular systems, aim to deliver drugs at a controlled rate tailored to the body's needs during treatment, directing the active compounds to the intended site of action [6,9]. Ocular drug delivery systems have been designed to enhance drug persistence near target cells while minimizing exposure to non-target cells, thereby improving therapeutic effects and reducing toxic effects [10,11]. Floating drug delivery systems are advantageous for drugs absorbed in the upper gastrointestinal tract, facilitating controlled drug release in specific segments of the digestive system [12]. Implantable drug delivery systems enable the manipulation of drug release kinetics using external stimuli, providing a tailored approach to drug administration [13]. Liposomal drug delivery systems have demonstrated potential in enhancing drug efficacy by improving bioavailability and enabling controlled release [14]. Novel drug delivery systems, including nanogels, lignin-based systems, and DQAsomes, are being explored for their potential in drug and gene delivery applications [15–17].

The development of nanoparticle drug delivery systems with controlled size and geometry has enabled passive drug enrichment in tumor regions through size- and shape-dependent effects, thereby enhancing drug delivery efficiency and targeting in cancer therapy [18]. Various innovative systems, including micro-reservoir implants, transdermal patches, nanoparticles, antibody-drug conjugates, and microneedles, have been developed for controlled drug delivery, offering new opportunities for therapeutic interventions [19]. Advancements in drug delivery technologies, including DNA origami-based systems, magnetosomes, and polymeric micelles, have shown significant potential for targeted drug delivery in cancer therapy [20–22]. These systems can overcome drug resistance mechanisms, increase drug accumulation at specific sites, and enable multifunctional approaches that combine imaging and therapy for improved treatment outcomes.

Metal oxide nanoparticles have garnered significant interest across scientific disciplines due to their unique properties and versatile applications. Nanoparticles like aluminum oxide (Al_2O_3), zinc oxide (ZnO), and cerium oxide (CeO_2) are known for their high surface-to-volume ratio, optical properties, and catalytic activity [23]. These nanoparticles have been extensively researched for their roles in catalysis, optoelectronics, bioimaging, and biodiagnosis, leveraging their distinct electronic and optical properties [23]. Various methods, including green synthesis and

DOI: 10.1201/9781003508601-11

thermal decomposition of metal organic frameworks, have been explored for the synthesis of metal oxide nanoparticles like copper oxide (CuO) and titanium dioxide (TiO_2), offering diverse applications in nanotechnology [24]. Metal oxide nanoparticles have shown promise in various applications, including antimicrobial hydrogels, environmental remediation, and biomedical applications [24,25]. To fully utilize their potential in scientific and technological fields, it is essential to comprehend the physicochemical properties and biological interactions of metal oxide nanoparticles.

11.2 SYNTHESIS METHODS OF METAL OXIDE NANOPARTICLES

The synthesis of metal oxide nanoparticles (MONPs) involves various methodologies that can be broadly categorized into physical, chemical, and biological techniques. Each method has distinct advantages and drawbacks, making them suitable for different applications based on the desired properties of the nanoparticles.

One of the most versatile and widely used methods is the *sol-gel process*. This chemical method involves transitioning from a liquid sol to a solid gel phase, often using metal alkoxides or metal chlorides as precursors (Figure 11.1). The sol-gel method is highly valued for its ability to control the composition and morphology of the resulting nanoparticles with high precision, resulting in highly pure and homogenous products. However, it requires long processing times and high-temperature calcination, which can be a significant drawback in terms of energy consumption and time [26,27]. Various nanoparticles, including calcium oxide [28], iron (III) oxide [29], zirconia [30], CuO [31], Ce_2O [32], AgO [33], TiO_2 [34], etc.

Hydrothermal and solvothermal synthesis methods involve crystallizing substances from high-temperature aqueous solutions or organic solvents under high vapor pressures. These methods are advantageous for their precise control over particle size and morphology, making them ideal for large-scale production. The hydrothermal method uses water as the solvent, while the solvothermal method employs organic solvents (Figure 11.2). Despite their benefits, these methods require specialized high-pressure equipment and involve long reaction times, which can limit their practicality for certain applications. In hydrothermal synthesis, metal oxide hollow spheres can be produced by incorporating metal ions into carbon spheres during the treatment [36]. On the other hand, solvothermal synthesis enables the production of nanoparticles, such as Cu2ZnSnSe4, by injecting sulfur and selenium precursors into a solution of metal salts [37].

Chemical vapor deposition (CVD) is another prominent technique used for synthesizing MONPs. In CVD, a thin film is deposited on a substrate through the chemical reaction of vapor-phase precursors. This method produces high-quality, crystalline nanoparticles with controlled dimensions, making it highly suitable for applications requiring precise structural properties. However, the high cost and complexity of the equipment, along with potential safety hazards associated with the use of toxic gases, are significant challenges to this method [39]. CVD has been extensively employed in the growth of carbon nanotubes (CNTs) by leveraging the catalytic activity of transition metal nanoparticles, highlighting the use of CVD in synthesizing CNTs, where transition metal nanoparticles deposited on a support facilitate the dissociation-diffusion-precipitation mechanism for CNT growth [40]. Additionally, the controlled synthesis of carbon nanotubes was demonstrated by utilizing nearly monodisperse iron nanoclusters to define the diameters of the nanotubes grown via CVD [41].

Furthermore, CVD has played a crucial role in the synthesis of various other nanomaterials. For example, they discussed the promising nature of CVD in producing copper nanoparticles on large-area graphene [42]. Additionally, the synthesis of Ni/NiO nanoparticles using metal organics CVD of nickel acetylacetonate in a tube flow reactor was reported, showcasing the versatility of CVD in nanoparticle synthesis [43].

Laser ablation is a physical method that utilizes high-energy laser pulses to vaporize a target material, thereby forming nanoparticles in a gas or liquid medium. This method is known for producing highly pure nanoparticles without the need for chemical reagents, making it an environmentally friendly option (Figure 11.3). However, laser ablation is limited in terms of scalability and involves

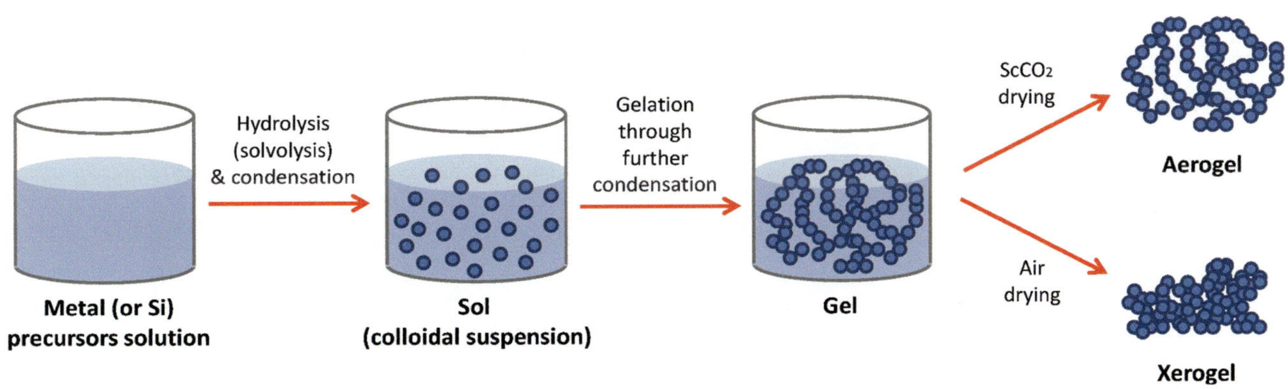

FIGURE 11.1 Sol-gel synthesis method [35]. (Direct permission granted by MDPI.)

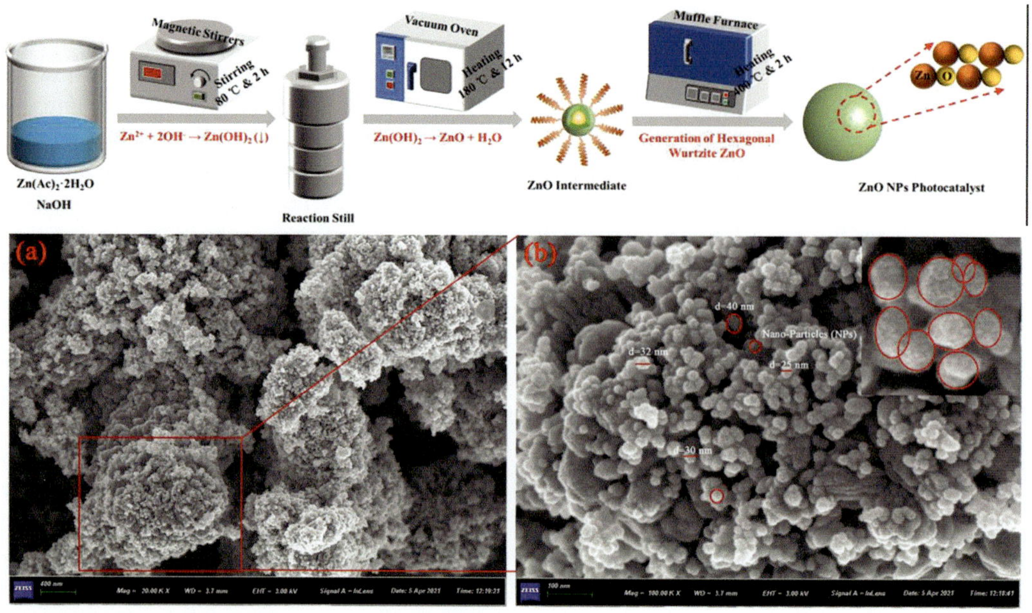

FIGURE 11.2 Preparation of ZnO NPs by the solvothermal method and SEM images of ZnO NPs. (a) Agglomeration and (b) nanoparticles [38]. (Direct permission granted by MDPI.)

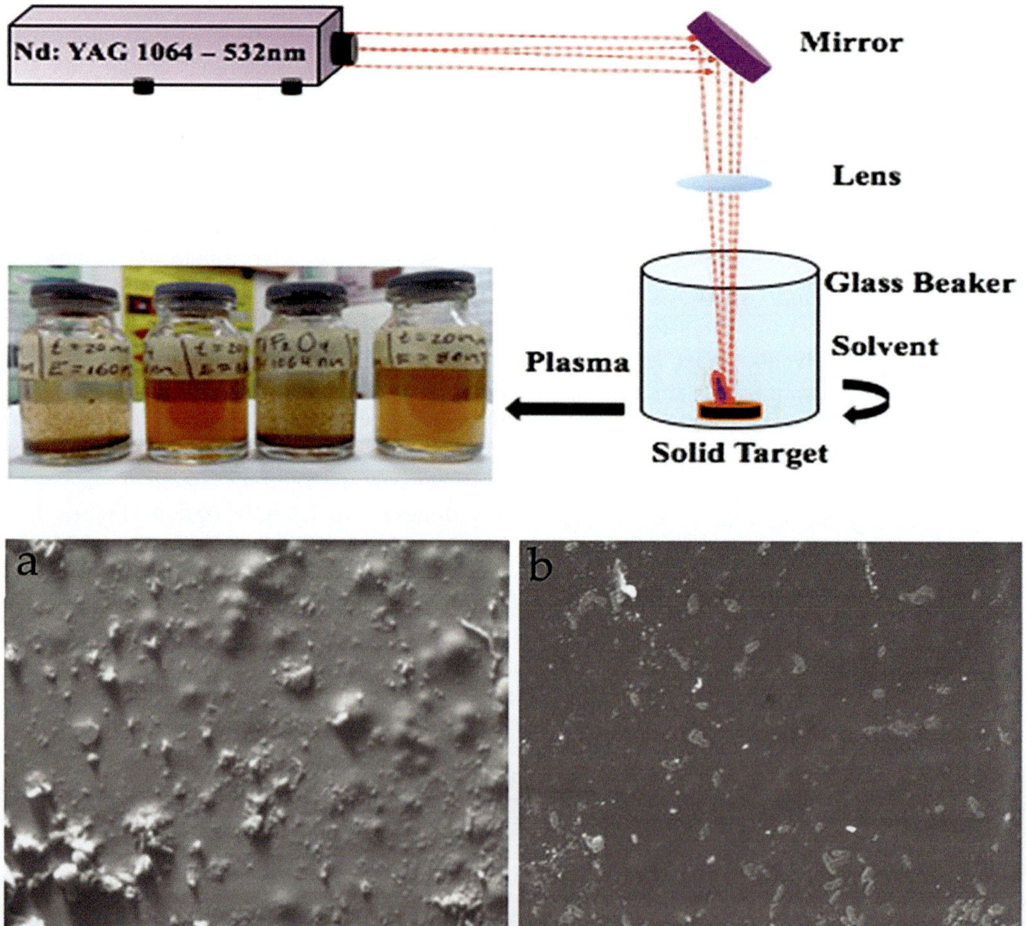

FIGURE 11.3 Laser ablation-based nanoparticle production and SEM images [47]. (Direct permission granted by MDPI.)

high energy consumption, which can be a constraint for large-scale production [39]. Different studies have demonstrated the effectiveness of laser ablation in producing nanoparticles of various materials. For instance, Salman et al. (2016) utilized laser ablation to prepare TiO_2 nanoparticles by ablating a titanium target in deionized water using Nd:YAG lasers [44]. Similarly, Muniz-Miranda et al. (2011) employed laser ablation to produce copper nanoparticles in aqueous solutions, which were characterized using various spectroscopic techniques [45]. Furthermore, Al-Ogaidi & Al-Ogaidi (2019) used laser ablation to synthesize silver nanoparticles and silver nanoparticle-graphene oxide nanocomposites, showcasing the versatility of this method in producing composite materials [46].

Experimental configuration employed for the preparation of colloidal nanoparticles (NPs) using laser ablation in a liquid medium and scanning electron microscopy (SEM) images of (a) $TiFe_2O_4$ and (b) $ZnFe_2O_4$ nanoparticles deposited on a silica substrate. $TiFe_2O_4$ and $ZnFe_2O_4$ are solid targets and are ablated for 15 min in 6 mL of liquid [47].

Microwave-assisted synthesis utilizes microwave radiation to rapidly heat the reaction mixture, resulting in the formation of nanoparticles. This technique offers the advantage of short reaction times and high yields due to the enhanced reaction rates and uniform heating provided by microwaves (Figure 11.4). Despite these benefits, microwave-assisted synthesis is generally limited to small-batch production and can suffer from issues related to uneven heating. Various studies have demonstrated the effectiveness of microwave-assisted techniques in synthesizing different metallic nanoparticles. For instance, Bilecka et al. (2008) reported the rapid synthesis of highly crystalline metal oxide nanoparticles, including CoO, ZnO, Fe_3O_4, MnO, Mn_3O_4, and $BaTiO_3$, by reacting metal alkoxides, acetates, or acetylacetonates with benzyl alcohol under microwave heating. This study showcases the efficiency of microwave-assisted synthesis in producing crystalline metal oxide nanoparticles in a short period [48].

Additionally, Nishida et al. (2017) detailed a technique for quickly producing uniformly sized nanoparticles of Ru, Rh, Pd, Ir, and Pt, stabilized by poly(N-vinyl-2-pyrrolidone), using a microwave-assisted chemical reduction process involving ethanol. This approach highlights the versatility of microwave-assisted techniques in synthesizing noble metal nanoparticles with controlled sizes and properties [50].

Green synthesis of metal and metal oxide nanoparticles has emerged as a promising alternative to conventional chemical and physical synthesis methods, addressing the environmental and toxicity concerns associated with these traditional approaches. This eco-friendly method utilizes natural sources, including plant extracts, fungi, algae, and bacteria, which serve as reducing and capping agents to convert metal ions into nanoparticles (Figure 11.5). The use of biomass not only reduces the production of hazardous by-products but also enhances the biocompatibility and stability of the nanoparticles. Green synthesis is characterized by its simplicity, cost-effectiveness, and scalability, making it suitable for large-scale production. Additionally, the biogenic nanoparticles exhibit unique properties, such as enhanced antimicrobial activity, which are advantageous for various biomedical applications, including drug delivery, tissue engineering, and diagnostic imaging. The integration of green synthesis principles into nanotechnology not only aligns with the Sustainable Development Goals but also opens new avenues for safer and more effective therapeutic solutions.

Each of these synthesis methods offers unique benefits and faces specific challenges, making the choice of method highly dependent on the intended application and desired properties of the metal oxide nanoparticles. Table 11.1 shows examples of green-synthesized nanoparticles.

11.3 METAL/METAL OXIDE NANOPARTICLE TYPES

11.3.1 Gold Nanoparticles

Gold nanoparticles are a versatile class of nanomaterials with unique properties that make them valuable for various applications. These nanoparticles have been extensively studied and utilized in different fields due to their biocompatibility, ease of functionalization, and distinctive optical and electronic characteristics [68]. Gold nanoparticles have been utilized in the medical field for various applications, including drug delivery systems, cancer treatment, and diagnostic materials [69]. They have shown promise in enhancing orthodontic anchorage and bone tissue regeneration, highlighting their potential in the field of medicine [70]. Furthermore, gold nanoparticles have been utilized in cancer therapy, particularly in photothermal therapy, due to their ability to serve as a simple platform for thermal ablation while maintaining biocompatibility in vivo [71]. The synthesis of gold nanoparticles using green methods has garnered attention due to its eco-friendly approach and potential biomedical applications. These green-synthesized gold nanoparticles have been explored for their anti-inflammatory activity and cytotoxic potential against cancer cells, showcasing their diverse applications [72].

Gold nanoparticles have been utilized as carriers for targeted drug delivery, enhancing drug efficacy and reducing side effects (Figure 11.6) [73]. They have also been employed in developing nanoplatforms for the co-delivery of multiple agents, such as doxorubicin and siRNA, demonstrating their ability to deliver combination therapies effectively [74]. Additionally, gold nanoparticles have been used in drug vectors for photodynamic therapy, enabling the efficient delivery of hydrophobic drugs to specific sites [75].

Furthermore, gold nanoparticles have been linked with drugs such as doxorubicin to aid in intracellular drug delivery, addressing multidrug resistance in cancer cells [76]. These nanoparticles have also been integrated into drug delivery systems that respond to near-infrared light

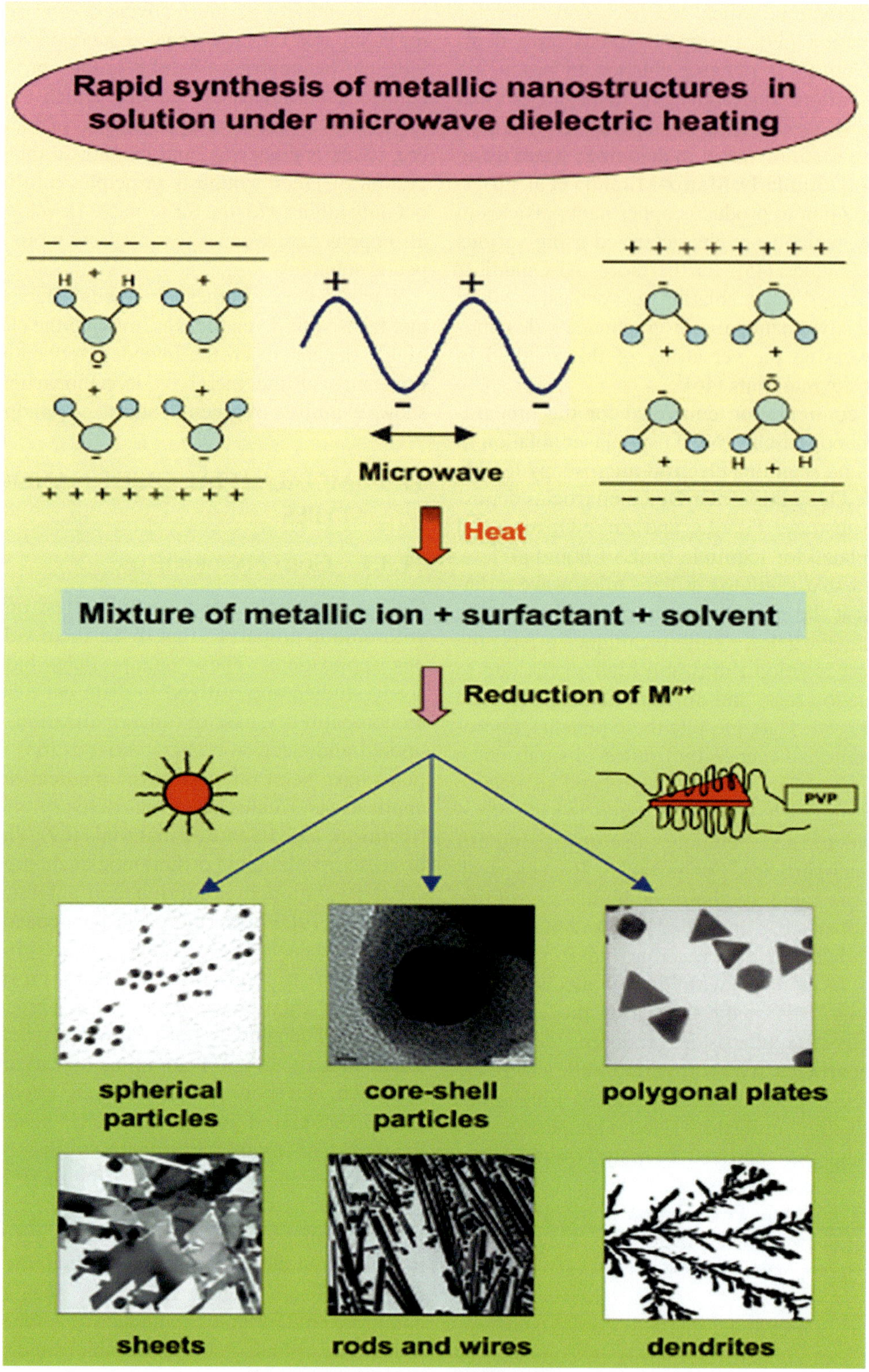

FIGURE 11.4 Microwave-assisted synthesis of metallic nanostructures [49]. (Permission granted by Copyright Clearance Center (CCC).)

irradiation, allowing for controlled drug release mechanisms [77]. Moreover, gold nanoparticles have been conjugated with pectin, a biopolymer, to function as drug delivery systems, emphasizing their potential in targeted drug delivery applications [78].

Gold nanoparticles have demonstrated efficacy in drug delivery systems for targeted therapy across various diseases, including cancer. These nanoparticles can be linked with targeting agents, such as antibodies or peptides, to actively target solid tumors and improve drug absorption [80].

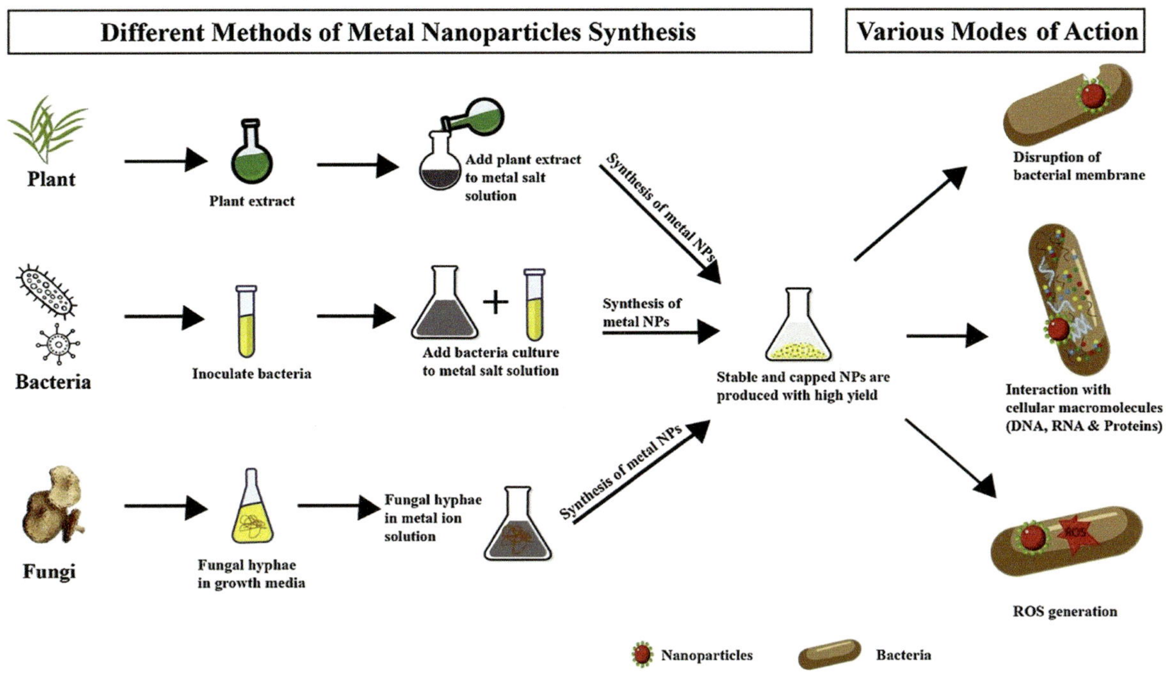

FIGURE 11.5 Green synthesis of metallic nanoparticles [51]. (Permission granted by MDPI.)

TABLE 11.1
Green-Synthesized Metallic Nanoparticles

Biological Source	Type of Nanoparticle	Applications	References
Selaginella species	Ag	Antibacterial, antifungal, anticoagulant, and antiplatelet properties	[52]
Vernonia amygdalina extract		Antibacterial	[53]
Fusarium chlamydosporum and *Penicillium chrysogenum*		Antifungal and anti-mycotoxin	[54]
Endophytic fungi isolated from *Calotropis procera (Ait) latex*		Antibacterial	[55]
Saccharomyces cerevisiae		*Antibacterial*	[56]
golden rod (*Solidago canadensis*) plant extract	Au	Anticancer	[57]
Paracoccus haeundaensis BC74171T (marine-based bacteria)		Antioxidant activity and antiproliferative effect	[58]
Croton Caudatus Geisel extract		Antibacterial, antifungal, anticancer	[59]
Aerva javanica plant leaf extract	CuO	Antibacterial, antifungal	[60]
Microalgae (*Chlorella kessleri, Dunaliella tertiolecta,* and *Tetraselmis suecica*)		–	[61]
Laurus nobilis L. leaves	α-Fe_2O_3	Antifungal And Antibacterial	[62]
Seaweed Kappaphycus alvarezii (K. alvarezii)	Fe_3O_4	–	[63]
Plantain peel extract		Magnetic effect	[64]
Extracts of two red algae (*Bostrychia tenella* and *Laurencia obtusa*), a green alga (*Halimeda tuna*), and a brown alga (*Sargassum filipendula*)+marine sponge	TiO_2	Antifouling effect	[65]
Extracts of black pepper (*Piper nigrum*), coriander (*Coriandrum sativum*) and clove (*Syzygium aromaticum*)		Enhanced encapsulation efficiency	[66]
Juglans regia L. leaf extract	ZnO	Antimicrobial effect	[67]
Veronica multifida leaf extract		Antimicrobial effect on Escherichia coli ATCC 43895, Staphylococcus aureus ATCC 29213, Bacillus subtilis, Bacillus licheniformis, Pseudomonas aeruginosa, and Salmonella typhimurium	

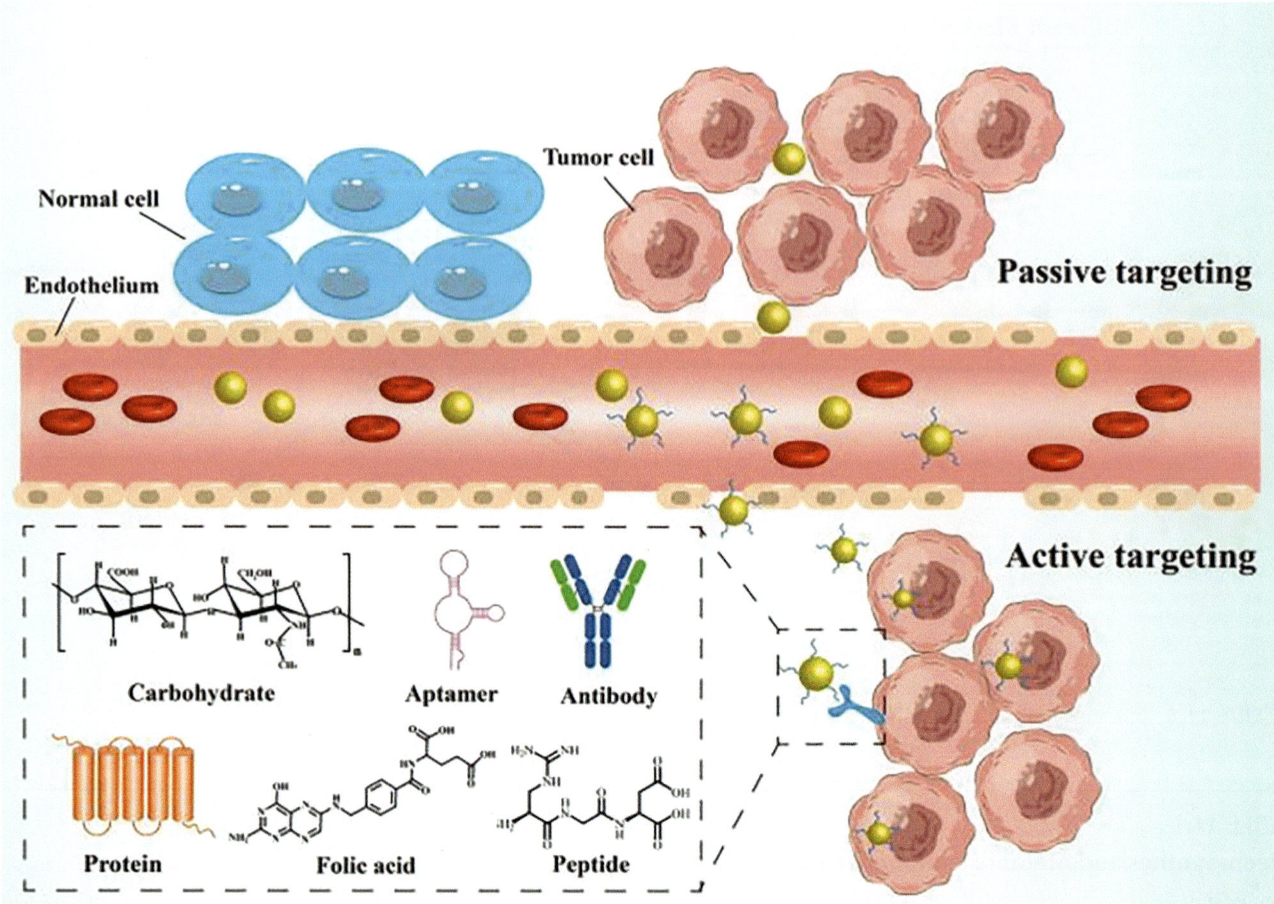

FIGURE 11.6 Targeting gold nanoparticles [79]. (Direct permission granted by MDPI.)

Additionally, the uptake of gold nanoparticles can be associated with specific biomarkers, like epidermal growth factor receptor (EGFR) expression, allowing for precise delivery to cancer cells [81]. The distinctive physical and chemical properties of gold nanoparticles make them well-suited for targeted drug delivery systems. By combining gold nanoparticles with therapeutic agents, such as tamoxifen or paclitaxel, one can achieve selective targeting and enhanced drug effectiveness in specific cell types, including breast cancer cells [82,83]. Moreover, the combination of gold nanoparticles with targeted medications can offer advantages in targeting specific areas, such as the bone, brain, or microorganisms [84].

11.3.2 Silver Nanoparticles

Silver nanoparticles, which are tiny particles of silver, have garnered considerable interest due to their distinctive properties and potential applications in various fields, including medicine, electronics, and environmental science. Various synthesis methods are available to produce silver nanoparticles, each providing unique advantages and specific characteristics [85].

Silver nanoparticles (AgNPs) can be synthesized using various methods, each offering unique advantages. The chemical reduction method, widely used due to its simplicity, involves reducing a silver salt like silver nitrate in an aqueous solution with reducing agents such as sodium borohydride, ascorbic acid, or citrate, and stabilizing agents like polyvinyl alcohol, polyvinylpyrrolidone, or citrate to prevent aggregation [86]. Green synthesis, an eco-friendly approach, uses biological entities such as plant extracts, microorganisms, or enzymes, leveraging the reducing and stabilizing properties of biomolecules in plant extracts from sources like neem, tea, and aloe vera [87]. Physical methods, including laser ablation and evaporation-condensation, offer alternatives that do not require chemical reagents. Laser ablation involves irradiating a silver target in a liquid medium with a high-intensity laser, while evaporation-condensation heats silver to produce vapor that condenses into nanoparticles [85]. Electrochemical methods, which reduce silver ions at the cathode in an electrochemical cell, allow precise control over nanoparticle size and morphology by adjusting current density, voltage, and electrolyte composition [46,87]. Each synthesis method can be customized to yield silver nanoparticles with specific properties, aligning them with particular applications based on the required particle size, shape, purity, and intended purpose.

Silver nanoparticles have been explored for their potential in delivering a wide range of therapeutic agents, including anticancer substances like doxorubicin [88]. The use of silver nanoparticles in drug delivery systems allows for the targeted delivery of drugs to specific cells or tissues, enhancing the therapeutic efficacy of the treatment [89].

Silver nanoparticles offer several advantages as drug delivery vehicles. They possess antimicrobial properties, making them suitable for combating infections at the site of drug delivery [89]. Briefly, the application of drugs to silver nanoparticles for drug delivery involves strategies such as functionalization, coating, conjugation, and specific design of carrier systems to enhance the efficacy and stability of the drug delivery system. The conjugation approach involves the synthesis of a silver nanoparticle-based drug delivery system that enables the simultaneous intracellular delivery of drugs like doxorubicin and improves their therapeutic efficacy [88].

Functionalization of silver nanoparticles is crucial for drug delivery applications. Coating silver nanoparticles can enhance their properties and applications, making them suitable for drug delivery systems [90]. Additionally, the conjugation of drugs onto silver nanoparticles can protect the drugs from the body's immune system, thereby extending their circulation time in the bloodstream [91].

Moreover, the method of drug loading onto silver nanoparticles is essential for effective drug delivery. For instance, the ability of alkaline degradation products of certain drugs to reduce Ag+ ions to silver nanoparticles has been utilized for drug delivery applications, especially analysis methods [92]. Furthermore, the design of degradable polymeric nanoparticles specifically for silver loading has been explored as potential carriers for silver-based antimicrobials, highlighting the importance of the carrier system in drug delivery [93].

11.3.3 Iron Nanoparticles

Iron oxide nanoparticles are a group of nanoparticles composed of iron and oxygen atoms, exhibiting unique magnetic properties that make them valuable in various applications. Different types of iron oxide nanoparticles include magnetite (Fe_3O_4), hematite (Fe_2O_3), and maghemite (γ-Fe_2O_3), each with distinct characteristics and applications (Figure 11.7) [94,95]. These nanoparticles have been extensively studied for their potential in magnetic nanodevices, biomagnetic applications, hyperthermia therapy, and drug delivery systems [96,97].

The synthesis of iron oxide nanoparticles can be achieved through various methods, such as biomimetic synthesis using Bacillus megaterium, green synthesis using plant extracts like Hordeum vulgare and Rumex acetosa, and chemical methods like co-precipitation and hydrothermal synthesis [99–101]. Characterization techniques like X-ray diffraction (XRD), transmission electron microscopy (TEM), and Raman spectroscopy are commonly used to analyze the properties of iron oxide nanoparticles [94].

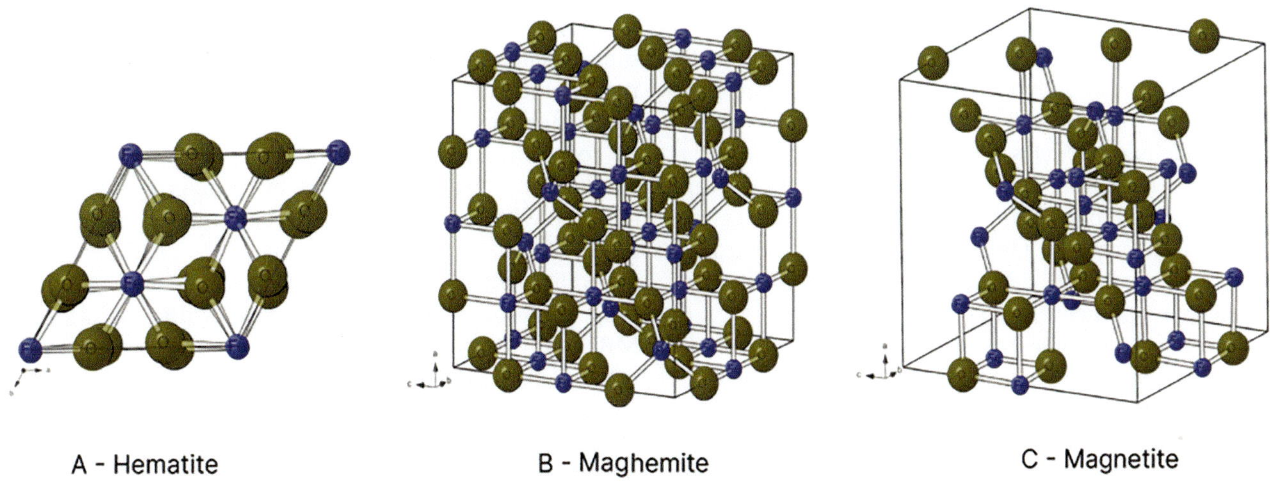

FIGURE 11.7 Appearance and crystalline structures of hematite (A), maghemite (B), and magnetite (C). Oxygen atoms are shown in yellow and iron atoms in blue [98].

Iron oxide nanoparticles have shown promise in applications such as antibacterial activity, antioxidant activity, magnetic resonance imaging (MRI) contrast enhancement, and water purification (Madhu et al., 2019; Osborne et al., 2012; Palchoudhury & Lead, 2014) [130–132]. The biocompatibility and stability of these nanoparticles are crucial factors in their potential use for biomedical applications, such as drug delivery and hyperthermia therapy (Li et al., 2016; Hajiali et al., 2022) [133,134].

Characterization studies of iron oxide nanoparticles involve various techniques to understand their properties and behavior. These studies are crucial for applications in diverse fields such as medicine, catalysis, and environmental remediation. Techniques used for characterization include spectroscopic methods, microscopy, thermal analysis, and magnetic measurements. One common method mentioned in the references is the use of SEM and energy-dispersive X-ray spectroscopy (EDS) to characterize the morphology and elemental composition of iron oxide nanoparticles [102]. Additionally, high-resolution transmission electron microscopy (HRTEM) is employed to investigate the structural properties of these nanoparticles [103]. These techniques provide valuable information on the size, shape, and distribution of nanoparticles. Moreover, studies utilize techniques like XRD to analyze the crystalline structure of iron oxide nanoparticles [103]. This helps in determining the phase composition and crystallographic properties of the nanoparticles. Mössbauer spectroscopy is another method used to investigate the magnetic properties of iron oxide nanoparticles [103].

Thermal analysis techniques such as thermogravimetric analysis (TGA) and differential scanning calorimetry (DSC) are employed to study the thermal stability and behavior of iron oxide nanoparticles [104]. These analyses provide insights into the thermal properties and stability of the nanoparticles. Furthermore, magnetic measurements are conducted to evaluate the magnetic behavior of iron oxide nanoparticles [105]. Understanding the magnetic properties is crucial for applications such as magnetic separation and targeted drug delivery systems.

Iron oxide nanoparticles have received significant attention in the field of drug delivery due to their unique properties. These nanoparticles can be utilized as carriers for various drugs, enabling targeted delivery and controlled release. Several studies have explored the application of iron oxide nanoparticles in drug delivery systems.

Pham et al. demonstrated the potential of chitosan-coated iron oxide nanoparticles for curcumin drug delivery, showing promising results in magnetically targeted drug delivery [106]. Similarly, the use of doxorubicin-loaded carboxymethylated PST001-coated iron oxide nanoparticles (DOX@CM-PST-IONPs) for cancer therapy is highlighted, leveraging their high drug encapsulation, protection against degradation, and controlled release properties. Derived from Tamarindus indica, PST001 exhibits excellent biocompatibility and stability, enabling efficient fabrication of these biogenic nanoparticles. The research demonstrates that DOX@CM-PIONPs effectively penetrate both 2D and 3D tumor models, induce reactive oxygen species (ROS)-mediated cell death, and exhibit higher cytotoxicity in cancer cells compared to free doxorubicin. In vivo studies in mice reveal significant tumor reduction and increased survival rates, indicating their potential for clinical application in targeted cancer treatment [107].

Furthermore, the development and surface modification of superparamagnetic iron oxide nanoparticles for applications in chemotherapy have gained attention due to their role as drug delivery agents [108]. The review study by Kievit and Zhang explained well the use of superparamagnetic iron oxide nanoparticles (SPIONs) for targeted cancer therapy, emphasizing their dual functionality as therapeutic delivery systems and MRI contrast agents. SPIONs consist of an iron oxide core, a biocompatible coating, and a therapeutic payload, often enhanced with targeting ligands for the specific recognition of cancer cells. The core's superparamagnetic properties enable noninvasive MRI, while the coating ensures stability and biocompatibility, helping the SPIONs navigate physiological barriers like the blood-brain barrier, liver, kidneys, and spleen. Active targeting is achieved by conjugating SPIONs with ligands such as folic acid, transferrin, or peptides like RGD and chlorotoxin, which recognize and bind to receptors overexpressed on cancer cells. The study explores various therapeutic applications, including gene therapy, where SPIONs deliver DNA or RNA molecules to silence or replace genes in cancer cells; chemotherapy, where SPIONs enhance the delivery of drugs like doxorubicin by overcoming multidrug resistance and ensuring higher intracellular concentrations; protein therapy, where SPIONs protect and deliver therapeutic proteins to target cells; and magnetic hyperthermia, where SPIONs generate heat under an alternating magnetic field to induce apoptosis in cancer cells or trigger drug release from thermosensitive materials. In vivo studies in animal models show that SPIONs can effectively target tumors, enhance imaging contrast, and achieve significant therapeutic outcomes. They emphasize the need for rigorous characterization and optimization of SPION synthesis for clinical translation, highlighting their potential to revolutionize cancer treatment by combining therapeutic delivery with real-time imaging and monitoring capabilities [109].

Additionally, IONPs can be used for the dual functional properties of magnetic nanoparticles for drug delivery and MRI, showcasing the versatility of iron oxide nanoparticles in biomedical applications. Jain et al. investigate the dual functional properties of magnetic nanoparticles (MNPs) for drug delivery and MRI. Oleic acid-coated iron oxide nanoparticles stabilized with Pluronic F-127 exhibited high drug incorporation efficiency (74%–95%) and sustained drug release for doxorubicin and paclitaxel, both alone and in combination. These MNPs maintained their magnetic properties, exhibiting higher T2 relaxivity than Feridex IV, which indicates superior sensitivity for T2-weighted imaging. In MCF-7 breast cancer cells, the drug-loaded MNPs exhibited synergistic antiproliferative effects. In vivo, the circulation half-life of these MNPs in mice was significantly longer than Feridex IV, suggesting potential clinical applications for real-time monitoring of drug distribution and tumor response to therapy [110].

11.3.4 Zinc Oxide Nanoparticles

Zinc oxide nanoparticles have gained significant attention due to their unique properties and diverse applications, including in the fields of medicine, environmental remediation, and agriculture. Zinc oxide nanoparticles exhibit remarkable antimicrobial properties, making them effective agents against various pathogens, both gram-positive and gram-negative bacteria [111–114].

Zinc oxide nanoparticles have demonstrated significant potential in drug delivery and targeting applications due to their unique properties. These nanoparticles can be customized for specific drug delivery systems to improve therapeutic outcomes. By functionalizing zinc oxide nanoparticles, targeting ligands such as antibodies or peptides can be attached to enable precise binding to target cells or tissues, facilitating targeted drug delivery [115]. In the field of cancer therapy, zinc oxide nanoparticles have exhibited promise in effectively delivering anticancer drugs. These nanoparticles, either alone or when loaded with anticancer agents such as doxorubicin, have demonstrated antitumor activities, highlighting their potential for use in targeted drug delivery systems for cancer treatment [116]. The increased cytotoxic potential of anticancer drug-loaded zinc oxide nanoparticles is attributed to pH-dependent drug release and receptor-mediated endocytosis, which contribute to their efficacy in cancer therapy [116,117].

Furthermore, zinc oxide nanoparticles have been investigated for their antibacterial properties, rendering them suitable for pharmaceutical applications. By incorporating these nanoparticles into drug delivery systems, the efficacy of antimicrobial agents can be enhanced, potentially targeting specific pathogens and overcoming drug resistance in microbial infections [118].

11.3.5 Cerium Oxide Nanoparticles

Cerium oxide nanoparticles, also known as nanoceria, have been utilized in various fields such as polishing, solar cells, fuel oxidation, corrosion protection, and automobile exhaust treatment [119]. In the realm of drug delivery, cerium oxide nanoparticles have shown promise in therapeutic applications due to their antioxidative properties and potential to deliver drugs to target cell types [120].

Cerium oxide nanoparticles can catalyze the dismutation of superoxide anion radicals into oxygen and hydrogen peroxide, a process similar to that of the enzyme superoxide dismutase (SOD). This reaction is crucial for protecting cells from the damage caused by highly reactive superoxide anion radicals, which can harm DNA, proteins, and lipids. The SOD-like reaction of cerium oxide nanoparticles can be represented as follows:

$$O^{-2} + Ce^{+4} \rightarrow O_2 + Ce^{+3}$$

$$O^{-2} + Ce^{+4} \rightarrow O_2 + Ce^{+3}$$

$$O^{-2} + Ce^{+3} + 2H^{+} \rightarrow H_2O_2 + Ce^{+4}$$

Cerium oxide nanoparticles may exhibit catalase-like activity as well. In this reaction, cerium oxide nanoparticles can break down hydrogen peroxide into water and oxygen. Thus, cerium oxide nanoparticles can reduce oxidative stress and help protect cells from oxidative damage. The catalase-like reaction of cerium oxide nanoparticles can be represented as:

$$2H_2O_2 + Ce^{+4} \rightarrow 2H_2 + 2O_2 + Ce^{+3}$$

Cerium oxide nanoparticles may also demonstrate peroxidase-like activity. In this reaction, cerium oxide nanoparticles can oxidize hydrogen peroxide to produce oxygen and water. The oxidase-like reaction of cerium oxide nanoparticles can be represented by the following equation:

$$H_2O_2 + R \rightarrow ROH + H_2O$$

$$O_2 + R \rightarrow ROH$$

In this reaction, cerium oxide nanoparticles act as catalysts, facilitating the reaction between O_2 and R without being consumed, thereby aiding in the reaction. R can be any organic compound, such as glucose, lipids, or proteins, among others. The oxidase-like reaction of cerium oxide nanoparticles is pH-dependent, and due to its maximum activity in acidic conditions at pH 4, it holds potential for a range of applications [121,122].

Moreover, cerium oxide nanoparticles have been explored for their potential in targeted drug delivery systems. By modifying these nanoparticles with specific ligands, they can be tailored for precise binding to target cells or tissues, enhancing drug delivery efficiency [123]. Additionally, cerium oxide nanoparticles have been incorporated into protective microcapsules for enzyme delivery, preventing oxidation and enhancing therapeutic efficacy [124]. In conclusion, cerium oxide nanoparticles hold promise in drug delivery and targeting applications. Their antioxidative properties, enzyme-like activities, and potential for targeted drug delivery make them valuable tools in advancing biomedical treatments.

11.3.6 Titanium Dioxide Nanoparticles

Titanium dioxide nanoparticles, also known as TiO_2 nanoparticles, have been investigated for their potential applications in drug delivery and targeted therapy. Research has shown that these nanoparticles can enhance the anticancer efficacy of drugs like doxorubicin while reducing their side effects [125]. Furthermore, titanium dioxide nanoparticles have been utilized in targeted drug delivery systems, such as those incorporating paclitaxel attached to hydroxyapatite and titanium dioxide nanoparticles, to enhance drug delivery efficiency [126]. Studies have demonstrated that titanium dioxide nanoparticles synthesized using Mucuna pruriens exhibit antioxidant and anti-inflammatory activities, indicating their potential in therapeutic applications [127]. Additionally, titanium dioxide nanoparticles coated with polyethylene glycol have shown promise as carriers for drugs like 2-methoxyestradiol, highlighting their effective drug adsorption capabilities [128].

Moreover, titanium dioxide nanoparticles have been investigated for their various properties and applications, including their photocatalytic properties, biocompatibility, and potential use in water pollution treatment [129]. In conclusion, titanium dioxide nanoparticles offer a wide range of potential applications in drug delivery and targeting. Their unique properties make them valuable in enhancing drug efficacy, reducing side effects, and improving therapeutic outcomes in various biomedical and pharmaceutical settings.

11.4 COMPARATIVE ANALYSIS OF DIFFERENT METAL OXIDE NANOPARTICLES

Nanoparticles have become a cornerstone of modern nanotechnology, offering unique properties that make them invaluable in various applications, particularly in the biomedical field. Among the various types of nanoparticles, gold, silver, iron, zinc oxide, cerium oxide, and titanium dioxide nanoparticles have garnered significant attention due to their distinct characteristics and potential uses. Table 11.2 aims to evaluate these nanoparticles based on their efficacy in drug delivery, biocompatibility and toxicity, and scalability and cost-effectiveness.

11.5 CHALLENGES AND LIMITATIONS

Metal and metal oxide nanoparticles offer numerous advantages due to their unique properties, such as high surface area, low concentration requirements, and ease of separation after treatment. These nanoparticles have been explored for various applications, including wastewater treatment, the diagnosis of viral infections, and therapeutic tools. However, their use presents challenges and limitations in several key areas, including toxicity and biocompatibility, scale-up and reproducibility, and regulatory and ethical considerations.

One of the primary challenges with metal and metal oxide nanoparticles is their potential toxicity to biological systems. This toxicity can arise from the generation of ROS, the release of metal ions, and interactions with cellular components, which can lead to oxidative stress, inflammation, and cell damage. For instance, silver nanoparticles, while effective as antimicrobial agents, can be toxic to mammalian cells at high concentrations.

Ensuring that nanoparticles are biocompatible, meaning they do not elicit an adverse immune response, is crucial for their application in medical treatments. Biocompatibility varies significantly depending on the type of nanoparticle, its surface chemistry, size, shape, and dose. Gold nanoparticles are generally considered biocompatible, but the presence of surface coatings or contaminants can compromise their biocompatibility.

Nanoparticles intended for medical use must undergo rigorous testing and approval by regulatory bodies such as the FDA (Food and Drug Administration) or EMA (European Medicines Agency). The complex nature of nanoparticles introduces additional complexity to the regulatory process, as their behavior can differ significantly from that of bulk materials. For example, titanium dioxide nanoparticles used in sunscreens must demonstrate safety and efficacy, requiring extensive testing and documentation.

The use of nanoparticles, particularly in medical applications, raises ethical questions regarding their long-term impact on human health and the environment. There is a need for transparent risk assessment and management strategies to address public concerns. The potential for nanoparticles to enter and persist in the environment poses ethical questions about their ecological impact and the long-term effects on human health.

By addressing these challenges and limitations, researchers and manufacturers can more effectively harness the potential of metal and metal oxide nanoparticles, ensuring their safe and effective application in various fields.

TABLE 11.2
Comparison of Metallic Nanoparticles

Nanoparticles	Efficacy in Drug Delivery	Biocompatibility and Toxicity	Scalability and Cost-Effectiveness
Gold	High due to surface functionalization	High biocompatibility, low toxicity	Moderate to high cost, scalable
Silver	Moderate antimicrobial properties	Moderate biocompatibility, can be toxic at high concentrations	Moderate cost, scalable
Iron	High magnetic properties for targeted delivery	High biocompatibility, low toxicity	Low cost, highly scalable
Zinc oxide	Moderate, used in sunscreens and wound healing	High biocompatibility, low toxicity	Low cost, highly scalable
Cerium oxide	High antioxidant properties	High biocompatibility, low toxicity	Moderate cost, scalable
Titanium dioxide	Moderate UV protection applications	High biocompatibility, low toxicity	Low cost, highly scalable

11.6 FUTURE PERSPECTIVES

The future of metal nanoparticles is poised to be shaped by several emerging trends and advancements. One significant trend is the design and functionalization of nanoparticles to achieve greater control over their size, shape, and surface chemistry, enabling the creation of multifunctional nanoparticles with capabilities for targeting, imaging, and therapy. For instance, stimuli-responsive nanoparticles that react to specific environmental cues, such as pH or temperature, are being developed to release drugs or alter their behavior in a controlled manner. Additionally, the development of bio-inspired and fully biodegradable nanoparticles aims to reduce long-term toxicity and environmental impact.

In the realm of personalized medicine and theranostics, metal and metal oxide nanoparticles are expected to play a crucial role. Targeted drug delivery systems that use functionalized nanoparticles to deliver drugs directly to specific cells can minimize side effects and improve treatment efficacy. Theranostic nanoparticles, which combine therapeutic and diagnostic functions, enable real-time monitoring and adjustment of treatments, thereby enhancing precision in medical interventions. Furthermore, the development of patient-specific nanoparticles tailored to individual genetic and molecular profiles holds promise for highly personalized treatment regimens, improving outcomes and reducing adverse effects.

The integration of nanotechnology with other therapeutic modalities is also anticipated to enhance treatment efficacy and broaden the scope of treatable conditions. For example, combination therapies that utilize nanoparticles to deliver multiple drugs simultaneously can target different disease pathways in a synergistic manner. Moreover, the use of nanoparticles in photothermal and photodynamic therapies, where nanoparticles convert light into heat or ROS to kill cancer cells, exemplifies the potential of combining nanotechnology with existing treatments. Additionally, nanoparticles are being explored as carriers for gene editing tools, such as CRISPR/Cas9, offering precise genetic modifications for the treatment of genetic disorders and cancers. By continuing to address these emerging trends, the integration of nanotechnology into medical practice promises to revolutionize treatment paradigms and improve patient outcomes.

11.7 CONCLUSION

In conclusion, metal and metal oxide nanoparticles represent a promising frontier in nanomedicine, with the potential to revolutionize drug delivery, diagnostics, and therapeutic applications. Their unique physical and chemical properties enable targeted and efficient drug delivery, enhancing treatment outcomes while minimizing side effects. However, challenges related to toxicity, biocompatibility, scalability, and regulatory compliance must be addressed to harness their potential fully. Future research should focus on optimizing the design and functionalization of these nanoparticles to enhance their therapeutic efficacy and safety. By overcoming these challenges, metal and metal oxide nanoparticles can significantly advance the field of personalized medicine and integrated therapeutic modalities, ultimately improving patient care and outcomes.

REFERENCES

1. Jain, K.K., *Drug Delivery Systems: An Overview*. Humana Press, Totowa, NJ, 2008, pp. 1–50.
2. Pund, A. et al., Recent Patents on Modified Release Oral Dosage Forms. *Journal of Drug Delivery and Therapeutics*, 2021. **11**(4-S): 195–211.
3. Shrestha, H., R. Bala, and S. Arora, Lipid-based Drug Delivery Systems. *Journal of Pharmacy (Cairo)*, 2014. 2014. **2014**: 801–820.
4. Uner, B., P. Dwivedi, and A.D. Ergin, Effects of Arginine on Coenzyme-Q10 Micelle Uptake for Mitochondria-Targeted Nanotherapy in Phenylketonuria. *Drug Delivery and Translational Research*, 2023. 14: 8901–8920.
5. Ergin, A.D. et al., Improving the Bioavailability and Efficacy of Coenzyme Q10 on Alzheimer's Disease through the Arginine-based Proniosomes. *Journal of Pharmaceutical Sciences*, 2023. **112**(11): 2921–2932.
6. Sezgin-Bayindir, Z. et al., Evaluation of Various Block Copolymers for Micelle Formation and Brain Drug Delivery: In Vitro Characterization and Cellular Uptake Studies. *Journal of Drug Delivery Science and Technology*, 2016. **36**: 120–129.
7. Chitkara, D. et al., Biodegradable Injectable in Situ Depot-Forming Drug Delivery Systems. *Macromolecular Bioscience*, 2006. **6**(12): 977–990.
8. Alkilani, A.Z., M.T. McCrudden, and R.F. Donnelly, Transdermal Drug Delivery: Innovative Pharmaceutical Developments based on Disruption of the Barrier Properties of the Stratum Corneum. *Pharmaceutics*, 2015. **7**(4): 438–470.
9. Uner, B. and A.D. Ergin, Enhanced Mitochondrial Co-Localization of β-Escin Micelle and Pancreatic Tumor Accumulation Relation. *Journal of Drug Delivery Science and Technology*, 2023. **89**: 104994.
10. Uner, B. et al., Timolol-Loaded Ethosomes for Ophthalmic Delivery: Reduction of High Intraocular Pressure In Vivo. *International Journal of Pharmaceutics*, 2023. **640**: 123021.
11. Özdemir, S. et al., Preparation and Characterization Studies of Dorzolamide-Loaded Ophthalmic Implants for Treating Glaucoma. *Turkish Journal of Pharmaceutical Sciences*, 2023. **20**(3): 149–156.
12. Özsoy, Y., N.S. Başarır, and B. Mesut, A Review on Current Perspective of Gastroretentive Drug Delivery Systems Prioritising Floating Dosage Forms. *Sağlık Bilimlerinde İleri Araştırmalar Dergisi*, 2021. **4**(2): 43–64.
13. Rajgor, N.B., M.P. Patel, and B. Veldurthy, Implantable Drug Delivery Systems: An Overview. *Systematic Reviews in Pharmacy*, 2011. **2**(2): 91.
14. Ergin, A.D. and B. Uner, Characterization, Optimization, and In Vitro Evaluation of Cholesterol-Free Liposomes. *Journal of Drug Delivery Science and Technology*, 2023. **84**: 104468.
15. Jain, S. et al., An Overview of Nanogel–Novel Drug Delivery System. *Asian Journal of Pharmaceutical Research and Development*, 2019. **7**(2): 47–55.

16. Kumar, R. et al., Lignin: Drug/Gene Delivery and Tissue Engineering Applications. *International Journal of Nanomedicine*, 2021. **16**: 2419–2441.
17. Uner, B. et al., Assessing the Performance of Coenzyme Q10 loaded DQAsomes to Treat Leigh Syndrome Caused by NDUFS4 Knockout. *Journal of Drug Delivery Science and Technology*, 2024. **97**: 105809.
18. Zhang, Q. et al., DNA Origami as an In Vivo Drug Delivery Vehicle for Cancer Therapy. *ACS Nano*, 2014. **8**(7): 6633–6643.
19. Sanjay, S.T. et al., Recent Advances of Controlled Drug Delivery Using Microfluidic Platforms. *Advanced Drug Delivery Reviews*, 2018. **128**: 3–28.
20. Puri, R. et al., Magnetosomes: A Tool for Targeted Drug Delivery in the Management of Cancer. *Journal of Nanomaterials*, 2022. **2022**: 1–12.
21. Nasongkla, N. et al., Multifunctional Polymeric Micelles as Cancer-Targeted, Mri-Ultrasensitive Drug Delivery Systems. *Nano Letters*, 2006. **6**(11): 2427–2430.
22. Halley, P.D. et al., Daunorubicin-Loaded DNA Origami Nanostructures Circumvent Drug-Resistance Mechanisms in a Leukemia Model. *Small*, 2015. **12**(3): 308–320.
23. Abdelrahman, M.S. et al., Studies of Polylactic Acid and Metal Oxide Nanoparticles-based Composites for Multifunctional Textile Prints. *Coatings*, 2020. **10**(1): 58.
24. Vijayaram, S. et al., Applications of Green Synthesized Metal Nanoparticles: A Review. *Biological Trace Element Research*, 2024. **202**(1): 360–386.
25. Wahid, F. et al., Recent Advances in Antimicrobial Hydrogels Containing Metal Ions and Metals/Metal Oxide Nanoparticles. *Polymers*, 2017. **9**(12): 636.
26. Bokov, D.O. et al., Nanomaterial by Sol-Gel Method: Synthesis and Application. *Advances in Materials Science and Engineering*, 2021. **2021**: 1–21.
27. Niederberger, M., Nonaqueous Sol–Gel Routes to Metal Oxide Nanoparticles. *Accounts of Chemical Research*, 2007. **40**(9): 793–800.
28. Habte, L. et al., Synthesis of Nano-Calcium Oxide from Waste Eggshell by Sol-Gel Method. *Sustainability*, 2019. **11**(11): 3196.
29. Gash, A.E. et al., Use of Epoxides in the Sol–Gel Synthesis of Porous Iron(III) Oxide Monoliths From Fe(III) Salts. *Chemistry of Materials*, 2001. **13**(3): 999–1007.
30. Joo, J. et al., Multigram Scale Synthesis and Characterization of Monodisperse Tetragonal Zirconia Nanocrystals. *Journal of the American Chemical Society*, 2003. **125**(21): 6553–6557.
31. Dörner, L. et al., Cost-Effective Sol-Gel Synthesis of Porous CuO Nanoparticle Aggregates with Tunable Specific Surface Area. *Scientific Reports*, 2019. **9**(1): 11758.
32. Hosseini, M. et al. Sol–Gel Synthesis, Physico-Chemical and Biological Characterization of Cerium Oxide/Polyallylamine Nanoparticles. *Polymers*, 2020. **12**, doi:10.3390/polym12071444.
33. Mohammad, S. et al., Synthesis and Characterization of Silver Nanoparticles by Sol-Gel Technique. *Nanoscience and Nanometrology*, 2017. **3**(1): 34–39.
34. Priyanka, K.P. et al., Microbicidal Activity of TiO_2 Nanoparticles Synthesised by Sol-Gel Method. *IET Nanobiotechnology*, 2016. **10**(2): 81–6.
35. Tao, Y. and P.P. Pescarmona Nanostructured Oxides Synthesised via scCO2-Assisted Sol-Gel Methods and their Application in Catalysis. *Catalysts*, 2018. **8**, doi:10.3390/catal8050212.
36. Titirici, M.M., M. Antonietti, and A. Thomas, A Generalized Synthesis of Metal Oxide Hollow Spheres Using a Hydrothermal Approach. *Chemistry of Materials*, 2006. **18**(16): 3808–3812.
37. Wu, L., Y. Sun, and Y. Xiang, Synthesis of $Cu_2ZnSnSe_4$ Nanoparticles via Solvothermal Route. *Journal of Alloys and Compounds*, **682** (2016) 1–7.
38. Wang, Y. et al. Solvothermal Synthesis of ZnO Nanoparticles for Photocatalytic Degradation of Methyl Orange and p-Nitrophenol. *Water*, 2021. **13**, doi:10.3390/w13223224.
39. Baig, N., I. Kammakakam, and W. Falath, Nanomaterials: A Review of Synthesis Methods, Properties, Recent Progress, and Challenges. *Materials Advances*, 2021. **2**(6): 1821–1871.
40. Ostadhossein, A. et al., Do Nickel and Iron Catalyst Nanoparticles Affect the Mechanical Strength of Carbon Nanotubes? *Extreme Mechanics Letters*, 2018. **20**: 29–37.
41. Cheung, C.L. et al., Diameter-Controlled Synthesis of Carbon Nanotubes. *The Journal of Physical Chemistry B*, 2002. **106**(10): 2429–2433.
42. Ünlü, C.G. et al., The Production of Cu Nanoparticles on Large Area Graphene by Sputtering and In-Flight Sintering. *Crystal Research and Technology*, 2017. **52**(11), doi:10.1002/crat.201700149.
43. Moravec, P. et al., NiOx Nanoparticle Synthesis by Chemical Vapor Deposition from Nickel Acetylacetonate. *Materials Sciences and Applications*, 2011. **2**(4): 258–264.
44. Salman, A.A., Synthesis and Characterization of TiO_2 Nanoparticles by Laser Ablation in Liquid. *Engineering and Technology Journal*, 2016. **34**(1): 100–106.
45. Muniz-Miranda, M., C. Gellini, and E. Giorgetti, Surface-Enhanced Raman Scattering From Copper Nanoparticles Obtained by Laser Ablation. *The Journal of Physical Chemistry C*, 2011. **115**(12): 5021–5027.
46. Al-Ogaidi, M. and I. Al-Ogaidi, Evaluating the Antibacterial Activity of AgGO Nanocomposite Against Clinical Isolate Bacteria. *Journal of Southwest Jiaotong University*, 2019. **54**(6).
47. Duque, J.S. et al. Colloidal Metal Oxide Nanoparticles Prepared by Laser Ablation Technique and Their Antibacterial Test. *Colloids and Interfaces*, 2019. **3,** doi:10.3390/colloids3010025.
48. Bilecka, I., I. Djerdj, and M. Niederberger, One-Minute Synthesis of Crystalline Binary and Ternary Metal Oxide Nanoparticles. *Chemical Communications*, 2008(7): 886–888.
49. Tsuji, M. et al., Microwave-Assisted Synthesis of Metallic Nanostructures in Solution. *Chemistry – A European Journal*, 2005. **11**(2): 440–452.
50. Nishida, Y. et al., Facile Synthesis of Size-Controlled Rh Nanoparticles via Microwave-Assisted Alcohol Reduction and Their Catalysis of CO Oxidation. *Chemistry Letters*, 2017. **46**(8): 1254–1257.
51. Singh, A. et al., Green Synthesis of Metallic Nanoparticles as Effective Alternatives to Treat Antibiotics-Resistant Bacterial Infections: A Review. *Biotechnology Reports*, 2020. **25**: e00427.
52. Dakshayani S, S. et al., Antimicrobial, Anticoagulant, and Antiplatelet Activities of Green Synthesized Silver Nanoparticles Using Selaginella (Sanjeevini) Plant Extract. *International Journal of Biological Macromolecules*, 2019. **131**: 787–797.
53. Tesfaye, M. et al., Green Synthesis of Silver Nanoparticles Using Vernonia Amygdalina Plant Extract and its Antimicrobial Activities. *Heliyon*, 2023. **9**(6): e17356.

54. Khalil, N.M., M.N. Abd El-Ghany, and S. Rodríguez-Couto, Antifungal and Anti-Mycotoxin Efficacy of Biogenic Silver Nanoparticles Produced by Fusarium chlamydosporum and Penicillium chrysogenum at Non-Cytotoxic Doses. *Chemosphere*, 2019. **218**: 477–486.
55. Mohamed, N.H. et al., Antimicrobial Activity of Green Silver Nanoparticles from Endophytic Fungi Isolated from Calotropis procera (Ait) latex. *Microbiology (Reading)*, 2019. **165**(9): 967–975.
56. Niknejad, F. et al., Green Synthesis of Silver Nanoparticles: Advantages of the Yeast Saccharomyces Cerevisiae Model. *Current Medical Mycology*, 2015. **1**(3): 17–24.
57. Botha, T.L. et al., Cytotoxicity of Ag, Au and Ag-Au Bimetallic Nanoparticles Prepared Using Golden Rod (Solidago canadensis) Plant Extract. *Scientific Reports*, 2019. **9**(1): 4169.
58. Patil, M.P. et al., Extracellular Synthesis of Gold Nanoparticles Using the Marine Bacterium Paracoccus Haeundaensis BC74171T and Evaluation of their Antioxidant Activity and Antiproliferative Effect on Normal and Cancer Cell Lines. *Colloids and Surfaces B: Biointerfaces*, 2019. **183**: 110455.
59. Vijaya Kumar, P., S. Mary Jelastin Kala, and K.S. Prakash, Green Synthesis of Gold Nanoparticles Using Croton Caudatus Geisel leaf extract and their biological studies. *Materials Letters*, 2019. **236**: 19–22.
60. Amin, F. et al., Green Synthesis of Copper Oxide Nanoparticles Using Aerva javanica Leaf Extract and Their Characterization and Investigation of In Vitro Antimicrobial Potential and Cytotoxic Activities. *Evidence-Based Complementary and Alternative Medicine*, 2021. **2021**: 5589703.
61. Salas-Herrera, G. et al., Impact of Microalgae Culture Conditions on the Capacity of Copper Nanoparticle Biosynthesis. *Journal of Applied Phycology*, 2019. **31**(4): 2437–2447.
62. Jamzad, M. and M. Kamari Bidkorpeh, Green Synthesis of Iron Oxide Nanoparticles by the Aqueous Extract of Laurus Nobilis L. Leaves and Evaluation of the Antimicrobial Activity. *Journal of Nanostructure in Chemistry*, 2020. **10**(3): 193–201.
63. Yew, Y.P. et al., Green Synthesis of Magnetite (Fe_3O_4) Nanoparticles Using Seaweed (Kappaphycus alvarezii) Extract. *Nanoscale Research Letters*, 2016. **11**(1): 276.
64. Venkateswarlu, S. et al., Biogenic Synthesis of Fe_3O_4 Magnetic Nanoparticles Using Plantain Peel Extract. *Materials Letters*, 2013. **100**: 241–244.
65. Alarif, W.M. et al., Green Synthesis of TiO(2) Nanoparticles Using Natural Marine Extracts for Antifouling Activity. *Marine Drugs*, 2023. **21**(2).
66. Bhullar, S., N. Goyal, and S. Gupta, Rapid green-synthesis of TiO(2) nanoparticles for therapeutic applications. *RSC Advances*, 2021. **11**(48): 30343–30352.
67. Darvishi, E., D. Kahrizi, and E. Arkan, Comparison of Different Properties of Zinc Oxide Nanoparticles Synthesized by the Green (Using Juglans regia L. leaf extract) and Chemical Methods. *Journal of Molecular Liquids*, 2019. **286**: 110831.
68. Ahmed, M., Z. Deng, and R. Narain, Study of Transfection Efficiencies of Cationic Glyconanoparticles of Different Sizes in Human Cell Line. *ACS Applied Materials & Interfaces*, 2009. **1**(9): 1980–1987.
69. Gotov, O., G. Battogtokh, and Y.T. Ko, Docetaxel-Loaded Hyaluronic Acid–Cathepsin B-Cleavable-Peptide–Gold Nanoparticles for the Treatment of Cancer. *Molecular Pharmaceutics*, 2018. **15**(10): 4668–4676.
70. Harikrishnan, S., R. Navaneethan, and K.S. Rajesh, Green Synthesis and Characterization of Bisphosphonate Conjugated Gold Nanoparticle With Cissus Quadrangularis Extract to Enhance Orthodontic Anchorage. *International Journal of Health Sciences*, 2022: 7388–7398.
71. Fazal, S. et al., Green Synthesis of Anisotropic Gold Nanoparticles for Photothermal Therapy of Cancer. *ACS Applied Materials & Interfaces*, 2014. **6**(11): 8080–8089.
72. Kumar, P.A., Biosynthesis and Characterization of Cotton Seed Oil Cake Gold Nanoparticles With Reference to Its Anti-Inflammatory Activity. *International Journal of Health Sciences*, 2022: 14288–14303.
73. Syed, A. et al., Extracellular Biosynthesis of Monodispersed Gold Nanoparticles, Their Characterization, Cytotoxicity Assay, Biodistribution and Conjugation With the Anticancer Drug Doxorubicin. *Journal of Nanomedicine & Nanotechnology*, 2012. **04**(01).
74. Shrestha, B. et al., Gold Nanoparticles Mediated Drug-Gene Combinational Therapy for Breast Cancer Treatment. *International Journal of Nanomedicine*, 2020. **15**: 8109–8119.
75. Cheng, Y. et al., Highly Efficient Drug Delivery With Gold Nanoparticle Vectors for in Vivo Photodynamic Therapy of Cancer. *Journal of the American Chemical Society*, 2008. **130**(32): 10643–10647.
76. Wang, F. et al., Doxorubicin-Tethered Responsive Gold Nanoparticles Facilitate Intracellular Drug Delivery for Overcoming Multidrug Resistance in Cancer Cells. *ACS Nano*, 2011. **5**(5): 3679–3692.
77. Haine, A.T. and T. Niidome, Drug Delivery Systems Controlled by Irradiation of Near Infrared Light. *Journal of Photopolymer Science and Technology*, 2015. **28**(5): 705–710.
78. Khodashenas, B. et al., Conjugation of Pectin Biopolymer With Au-nanoparticles as a Drug Delivery System: Experimental and DFT Studies. *Applied Organometallic Chemistry*, 2020. **34**(6).
79. Huang, H. et al., Gold Nanoparticles: Construction for Drug Delivery and Application in Cancer Immunotherapy. *Pharmaceutics*, 2023. **15**(7).
80. Brown, S.D. et al., Gold Nanoparticles for the Improved Anticancer Drug Delivery of the Active Component of Oxaliplatin. *Journal of the American Chemical Society*, 2010. **132**(13): 4678–4684.
81. Patra, C.R. et al., Targeted Delivery of Gemcitabine to Pancreatic Adenocarcinoma Using Cetuximab as a Targeting Agent. *Cancer Research*, 2008. **68**(6): 1970–1978.
82. Dreaden, E.C. et al., Tamoxifen–Poly(ethylene Glycol)–Thiol Gold Nanoparticle Conjugates: Enhanced Potency and Selective Delivery for Breast Cancer Treatment. *Bioconjugate Chemistry*, 2009. **20**(12): 2247–2253.
83. Paciotti, G.F. et al., Synthesis and Evaluation of Paclitaxel-Loaded Gold Nanoparticles for Tumor-Targeted Drug Delivery. *Bioconjugate Chemistry*, 2016. **27**(11): 2646–2657.
84. Eissa, D.M. et al., Gold Nanoparticles' Concentration Dependent Reactivity to Alendronate, Memantine, Andtobramycin. *Nanomaterials*, 2022. **12**(4), 505.
85. Rai, M., A. Yadav, and A. Gade, Silver Nanoparticles as a New Generation of Antimicrobials. *Biotechnol Adv*, 2009. **27**(1): 76–83.
86. Sau, T.K. and C.J. Murphy, Room Temperature, High-Yield Synthesis of Multiple Shapes of Gold Nanoparticles in

Aqueous Solution. *Journal of the American Chemical Society*, 2004. **126**(28): 8648–8649.

87. Sharma, V.K., R.A. Yngard, and Y. Lin, Silver Nanoparticles: Green Synthesis and their Antimicrobial Activities. *Advances in Colloid and Interface Science*, 2009. **145**(1–2): 83–96.
88. Benyettou, F. et al., Synthesis of Silver Nanoparticles for the Dual Delivery of Doxorubicin and Alendronate to Cancer Cells. *Journal of Materials Chemistry B*, 2015. **3**(36): 7237–7245.
89. Dhayalan, M. et al., Biomedical Applications of Silver Nanoparticles. In: G. Sharma (Ed.), *Silver Nanoparticles*. IntechOpen, 2021, pp. 1–21.
90. Fahmy, H.M. et al., Coated Silver Nanoparticles: Synthesis, Cytotoxicity, and Optical Properties. *RSC Advances*, 2019. **9**(35): 20118–20136.
91. Akbari, M., R. Morad, and M. Mâaza, First Principle Study of Silver Nanoparticles Interactions with Antimalarial Drugs Extracted From Artemisia Annua Plant. *Journal of Nanoparticle Research*, 2020. **22**: 331.
92. El-Alfy, W. et al., Application of Silver Nanoparticles for the Spectrophotometric Determination of Cefdinir and Cefepime HCl in Pharmaceutical Preparations and Human Urine Samples. *Nano Biomedicine and Engineering*, 2019. **11**(4): 381–390
93. Lim, Y.H. et al., Preparation and In Vitro Antimicrobial Activity of Silver-Bearing Degradable Polymeric Nanoparticles of Polyphosphoester-block-Poly(l-Lactide). *ACS Nano*, 2015. **9**(2): 1995–2008.
94. Schwaminger, S.P., C. Syhr, and S. Berensmeier, Controlled Synthesis of Magnetic Iron Oxide Nanoparticles: Magnetite or Maghemite? *Crystals*, 2020. **10**(3): 214.
95. Jubb, A.M. and H.C. Allen, Vibrational Spectroscopic Characterization of Hematite, Maghemite, and Magnetite Thin Films Produced by Vapor Deposition. *ACS Applied Materials & Interfaces*, 2010. **2**(10): 2804–2812.
96. Sun, S. et al., Monodisperse MFe_2O_4 (M=Fe, Co, Mn) *Nanoparticles. Journal of the American Chemical Society*, 2003. **126**(1): 273–279.
97. Hajiali, S., S. Daneshjou, and S. Daneshjoo, Biomimetic Synthesis of Iron Oxide Nanoparticles From Bacillus Megaterium to Be Used in Hyperthermia Therapy. *AMB Express*, 2022. **12**(1): 145.
98. de Souza, T.C. et al. Synthesis of Iron Oxides and Influence on Final Sizes and Distribution in Bacterial Cellulose Applications. *Polymers*, 2023. **15**, doi:10.3390/polym15153284.
99. Makarov, V.V. et al., Biosynthesis of Stable Iron Oxide Nanoparticles in Aqueous Extracts of Hordeum Vulgare and Rumex Acetosa Plants. *Langmuir*, 2014. **30**(20): 5982–5988.
100. Khan, M.I. et al., Green Synthesis of Magnetic Iron Oxide Nanoparticle for Antibacterial Activity: A Review. *Biological Sciences - PJSIR*, 2021. **64**(2): 202–210.
101. Hyeon, T. et al., Synthesis of Highly Crystalline and Monodisperse Maghemite Nanocrystallites Without a Size-Selection Process. *Journal of the American Chemical Society*, 2001. **123**(51): 12798–12801.
102. Kiyani, M.M. et al., A Histological Study for the Evaluation of Potential Harmful Effects of Orally Ingested Iron Oxide Nanoparticles in Mice. *The Annals of Clinical and Analytical Medicine*, 2021. **12**(3): 238–241.
103. Kim, K.-C. and Y.-S. Kim, Post Annealing Effects on Iron Oxide Nanoparticles Synthesized by Novel Hydrothermal Process. *Journal of Magnetics*, 2010. **15**(4): 179–184.
104. Can, H.K. et al., Preparation, Characterization and Dynamical Mechanical Properties of Dextran-Coated Iron Oxide Nanoparticles (DIONPs). *Artificial Cells Nanomedicine and Biotechnology*, 2017. **46**(2): 421–431.
105. Hufschmid, R. et al., Ex Situ and in Situ (S)TEM of Iron Oxide Nanoparticles Synthesized by Decomposition of an Organometallic Precursor. *Microscopy and Microanalysis*, 2015. **21**(S3): 965–966.
106. Pham, T. T. et al., Chitosan-coated iron oxide nanoparticles for magnetically targeted drug delivery of curcumin. *Colloids and Surfaces B: Biointerfaces*, 2011. **86**(2): 421–428.
107. Unnikrishnan, B.S., G.U. Preethi, and T.T. Sreelekha, A Comprehensive Study on 2D, 3D and Solid Tumor Environment to Explore a Multifunctional Biogenic Nanoconjugate. *Scientific Reports*, 2021. **11**(1): 8721.
108. Mahmoudi, M. et al., Superparamagnetic Iron Oxide Nanoparticles (SPIONs): Development, Surface Modification and Applications in Chemotherapy. *Advanced Drug Delivery Reviews*, 2011. **63**(1–2): 24–46.
109. Kievit, F.M. and M. Zhang, Surface Engineering of Iron Oxide Nanoparticles for Targeted Cancer Therapy. *Accounts of Chemical Research*, 2011. **44**(10): 853–862.
110. Jain, T.K. et al., Magnetic Nanoparticles with Dual Functional Properties: Drug Delivery and Magnetic Resonance Imaging. *Biomaterials*, 2008. **29**(29): 4012–21.
111. Siddiqi, K.S. et al., Properties of Zinc Oxide Nanoparticles and their Activity Against Microbes. *Nanoscale Research Letters*, 2018. **13**(1): 141.
112. Singh, J. et al., 'Green' Synthesis of Metals and Their Oxide Nanoparticles: Applications for Environmental Remediation. *Journal of Nanobiotechnology*, 2018. **16**(1): 84.
113. Nivethitha, R. et al., Antimicrobial Activity of Zinc Oxide Nanoparticles Synthesized Using Leaves Extract of Abies Webbiana. *Journal of Pharmaceutical Research International*, 2021. **33**: 3702–3710.
114. Chaudhary, A. et al., Antimicrobial Activity of Zinc Oxide Nanoparticles Synthesized From Aloe Vera Peel Extract. *Sn Applied Sciences*, 2018. **1**(1): 136.
115. Rasmussen, J.W. et al., Zinc Oxide Nanoparticles for Selective Destruction of Tumor Cells and Potential for Drug Delivery Applications. *Expert Opinion on Drug Delivery*, 2010. **7**(9): 1063–1077.
116. Swidan, M.M., F. Marzook, and T.M. Sakr, pH-Sensitive Doxorubicin Delivery Using Zinc Oxide Nanoparticles as A Rectified Theranostic Platform: In Vitro Anti-Proliferative, Apoptotic, Cell Cycle Arrest and In Vivo Radio-Distribution Studies. *Journal of Materials Chemistry B*, 2024. 12: 6257–6274.
117. Murali, M. et al., Plant-Mediated Zinc Oxide Nanoparticles: Advances in the New Millennium Towards Understanding Their Therapeutic Role in Biomedical Applications. *Pharmaceutics*, 2021. **13**(10): 1662.
118. Younas, Z. et al., Mechanistic Approaches to the Application of Nano-Zinc in the Poultry and Biomedical Industries: A Comprehensive Review of Future Perspectives and Challenges. *Molecules*, 2023. **28**(3): 1064.
119. Miri, A. et al., In Vitro Cytotoxic Performance of Pure and Co-Doped CeO_2 Nanoparticles on Breast Cancer Cell, Colon Cancer Cell, and Mouse Embryo Fibroblast Cell Lines. *Research Journal of Medical Sciences*, 2022. **24**(14): 1056–1062.
120. Lord, M.S. et al., Hyaluronan Coated Cerium Oxide Nanoparticles Modulate CD44 and Reactive Oxygen

Species Expression in Human Fibroblasts. *Journal of Biomedical Materials Research Part A*, 2016. **104**(7): 1736–1746.
121. Bai, Y. et al., Advanced Biological Applications of Cerium Oxide Nanozymes in Disease Related to Oxidative Damage. *ACS Omega*, 2024. **9**(8): 8601–8614.
122. Yadav, N., Cerium Oxide Nanostructures: Properties, Biomedical Applications and Surface Coatings. *3 Biotech*, 2022. **12**(5): 121.
123. Wang, M. et al., Preparation, Characterization and Multiple Biological Properties of Peptide-Modified Cerium Oxide Nanoparticles. *Biomolecules*, 2022. **12**(9): 1277.
124. Popov, A.I. et al., Ceria Nanoparticles-Decorated Microcapsules as a Smart Drug Delivery/Protective System: Protection of Encapsulated P. Pyralis Luciferase. *ACS Applied Materials & Interfaces*, 2018. **10**(17): 14367–14377.
125. Chen, Y. et al., Anticancer Efficacy Enhancement and Attenuation of Side Effects of Doxorubicin with Titanium Dioxide Nanoparticles. *International Journal of Nanomedicine*, 2011. **6**: 2321–2331.
126. Venkatasubbu, G.D. et al., Acute and Subchronic Toxicity Analysis of Surface Modified Paclitaxel Attached Hydroxyapatite and titanium Dioxide Nanoparticles. *International Journal of Nanomedicine*, 2015. **62A**: 137.
127. Aditya, B. et al., Antioxidant and Antiinflammatory Activity of Titanium Dioxide Nanoparticles Synthesised Using Mucuna Pruriens. *Journal of Pharmaceutical Research International*, 2021. **10**: 414–422.
128. León, A. et al., FTIR and Raman Characterization of TiO2 Nanoparticles Coated With Polyethylene Glycol as Carrier for 2-Methoxyestradiol. *Applied Sciences*, 2017. **7**(1): 49.
129. Ahmed, N.M. and A.M. Taha, Histopathological Effects of Titanium Dioxide Nanoparticles on the Liver of Japanese Quail Coturnix Coturnix Japonica. *Iraqi Journal of Veterinary Sciences*, 2022. **36**(2): 349–358.
130. Madhu, A. et al., One-pot green synthesis of yttrium and zirconium doped iron oxide nanoparticles using Simarouba Glauca leaf extract. *Materials Today: Proceedings*, 2023.
131. Osborne, E. A. et al., Rapid microwave-assisted synthesis of dextran-coated iron oxide nanoparticles for magnetic resonance imaging. *Nanotechnology*, 2012. **23**(21): 215602.
132. Palchoudhury, S et al., Platinum attachments on iron oxide nanoparticle surfaces. *Journal of Applied Physics*, 2010. **107**(9).
133. Li, X et al., Current investigations into magnetic nanoparticles for biomedical applications. *J Biomed Mater Res Part A*, 2016. **104A**: 1285–1296.
134. Hajiali, S et al., Biomimetic synthesis of iron oxide nanoparticles from Bacillus megaterium to be used in hyperthermia therapy. *AMB Expr*, 2022. **12**: 145. https://doi.org/10.1186/s13568-022-01490-y

12 Nanoparticle Strategies for Enhanced Protein and Peptide Drug Delivery

Burcu Uner[1,2,3], *Ahmet Dogan Ergin*[4,5], *and Pankaj Dwivedi*[1,2]

[1]Department of Pharmaceutical and Administrative Sciences, University of Health Science and Pharmacy, St. Louis, Missouri

[2]Department of Anesthesiology, Center for Clinical Pharmacology, Washington University School of Medicine, St. Louis, Missouri

[3]Department of Pharmaceutical Technology, Faculty of Pharmacy, Istanbul Kent University, Istanbul, Turkey

[4]Department of Pharmaceutical Technology, Faculty of Pharmacy, Trakya University, Edirne, Turkey

[5]Department of Pharmaceutical Nanotechnology, Institute of Health Sciences, Trakya University, Edirne, Turkey

12.1 INTRODUCTION

The development of advanced drug delivery systems has revolutionized modern medicine by enhancing the efficacy and safety of therapeutic agents. Among these advancements, protein and peptide drug delivery systems have emerged as critical tools in targeting specific cellular pathways with high precision. These biologics offer unique advantages over traditional small-molecule drugs, including higher specificity, reduced side effects, and the ability to modulate complex biological functions. As a result, protein and peptide therapeutics have become indispensable in treating a wide range of diseases, from chronic conditions like diabetes to complex disorders such as cancer. This section explores the significance of these drug delivery systems, emphasizing their potential to address unmet medical needs and enhance patient care.

12.1.1 Importance of Protein/Peptide Drug Delivery Systems

Protein and peptide drug delivery systems have gained substantial attention due to their therapeutic potential and high specificity. Unlike traditional small-molecule drugs, proteins and peptides can precisely target cellular processes, enhancing therapeutic effects and reducing side effects. Their complex structures enable interactions with biological pathways that small molecules cannot achieve, offering a higher degree of specificity in modulating biological functions [1–3]. Proteins and peptides play a crucial role in numerous physiological processes, making them promising candidates for treating diseases such as cancer, diabetes, cardiovascular disorders, and infections. Insulin, a peptide hormone, revolutionized diabetes treatment, significantly improving patient outcomes. Similarly, monoclonal antibodies, a class of protein-based drugs, have revolutionized cancer therapy by specifically targeting tumor cells, thereby minimizing damage to healthy tissues [1,3,4].

However, the development of protein and peptide drugs is challenging due to their instability and susceptibility to enzymatic degradation. Effective delivery systems are crucial for protecting these molecules, ensuring they reach their target sites [5, 6], and maintaining therapeutic concentrations. Advances in drug delivery technologies have enhanced the stability, bioavailability, and controlled release of protein and peptide drugs. Nanotechnology-based delivery systems, such as liposomes, polymeric nanoparticles, and micelles, have shown promise in protecting and delivering these drugs. These nanocarriers can be designed for sustained and controlled release, targeted delivery, and evasion of the immune system, thereby increasing therapeutic efficacy and reducing dosage requirements.

Conjugation strategies, in which proteins or peptides are linked to polymers or other molecules, enhance stability and prolong the circulation time of these molecules in the bloodstream. PEGylation, which involves attaching polyethylene glycol (PEG) chains to therapeutic proteins, is widely used to increase the half-life, reduce immunogenicity, and enhance the pharmacokinetic profiles of these drugs. Advanced formulation techniques, like lyophilization and microencapsulation, have enhanced the storage stability and shelf life of protein and peptide drugs. These methods protect active molecules from environmental stress, maintaining therapeutic efficacy over time [7,8].

The importance of protein and peptide drug delivery systems lies in their potential to address unmet medical needs and provide new treatment options for challenging diseases. As molecular biology and disease mechanisms are better understood, the role of these biologics in modern medicine will likely expand, offering hope for patients and driving innovative therapeutic developments [9]. Therefore,

DOI: 10.1201/9781003508601-12

effective delivery to target sites is vital for maximizing clinical benefits and improving patient outcomes. Ongoing advancements in drug delivery technologies and formulation strategies are set to overcome existing challenges, leading to the next generation of protein and peptide-based therapies.

12.1.2 Challenges in Delivering Protein/Peptide Drugs

Delivering protein and peptide drugs effectively presents several significant challenges due to the unique properties of these biomolecules. One of the primary issues is their inherent instability, as proteins and peptides are prone to denaturation and degradation when exposed to environmental conditions such as heat, pH changes, and enzymatic activity. This instability can significantly reduce their therapeutic efficacy. Another major challenge is the rapid clearance of these molecules from the bloodstream [1,10,11]. Due to their relatively large size and susceptibility to proteolytic enzymes, proteins and peptides often have short half-lives in the circulatory system. This rapid clearance necessitates frequent dosing, which can be inconvenient and uncomfortable for patients.

The delivery of protein and peptide drugs also faces significant barriers at the cellular level. These molecules often struggle to cross cellular membranes due to their size and hydrophilicity. Effective delivery to intracellular targets thus requires innovative strategies to facilitate cellular uptake [11,12].

Additionally, immune responses can pose a challenge. Proteins and peptides can be immunogenic, leading to potential adverse reactions and the formation of neutralizing antibodies. This immunogenicity can not only reduce the efficacy of the treatment but also pose safety concerns for patients. Formulating these drugs for optimal delivery is another complex task. The formulation must protect the drug from degradation while ensuring its release at the target site [13,14]. Achieving this balance requires sophisticated delivery systems, such as nanoparticles, liposomes, or polymer conjugates, which can add to the complexity and cost of development.

Overcoming these challenges requires ongoing research and innovation in drug delivery technologies. Advances in nanotechnology, conjugation strategies, and formulation techniques are pivotal in enhancing the stability, bioavailability, and efficacy of protein and peptide therapeutics, paving the way for more effective treatments.

12.2 TYPES OF NANOPARTICLES IN PROTEIN/PEPTIDE DRUG DELIVERY

12.2.1 Polymeric Nanoparticles

Polymeric nanoparticles, particularly those made from biodegradable polymers such as PLGA (poly(lactic-co-glycolic acid)), are widely used due to their biocompatibility, ability to encapsulate proteins/peptides, and controlled release properties. Recent advancements have focused on enhancing the stability and bioavailability of the encapsulated agents through surface modifications and novel polymer combinations [15,16]. Xu et al. conducted a study to address the challenge of low bioavailability in orally delivered treatments aimed at regulating glucose levels in diabetic rats [17]. The nanoparticles protected insulin from gastrointestinal degradation and improved mucosal penetration (Figure 12.1a). Han et al. conducted a study to optimize a microfluidic method for encapsulating a small hexapeptide into PLGA-lipid shell nanoparticles using a two-stage microfluidic chip (Figure 12.1b). The researchers aimed to improve the process to control the size and surface charge of the nanoparticles effectively [18]. Chitosan-based nanoparticles have demonstrated significant potential for peptide delivery, as evidenced by Shrestha et al.'s study [19]. These nanoparticles, loaded with the peptide GLP-1 for diabetes treatment, exhibited enhanced stability and sustained release. This resulted in prolonged glucose regulation in animal models, indicating their effectiveness in maintaining consistent therapeutic levels and improving overall treatment outcomes for diabetes.

12.2.2 Lipid Nanoparticles

Lipid nanoparticles, including liposomes and solid lipid nanoparticles (SLNs), offer advantages in terms of biocompatibility, drug loading capacity, and the ability to protect encapsulated proteins/peptides from enzymatic degradation. Recent studies have focused on optimizing lipid composition to improve stability and targeting capabilities [20–22]. Zhang et al. developed SLNs that encapsulate both paclitaxel and siRNA. Their thorough research unequivocally demonstrated that these SLNs are engineered to target cancer cells, release the drugs gradually, and significantly inhibit tumor growth (Figure 12.1c) [21]. Another study is about a new delivery system for siRNA targeting hepatocarcinoma cells. In this study, SLNs were conjugated with cRGD peptide for targeted siRNA delivery. The authors designed these SLNs to be biocompatible and have high transfection efficiency. They achieved this by incorporating a reduction-sensitive polymer and DSPE-PEG2000-cRGD. In vitro and in vivo studies showed that the NPs were successful in delivering siRNA and inhibiting tumor growth [23].

12.2.3 Gold Nanoparticles

Gold nanoparticles (GNPs) have been utilized for their unique optical properties, ease of functionalization, and ability to enhance the stability of bound proteins/peptides. Advances in surface modification techniques have enabled targeted delivery and controlled release [24,25]. Guerrero et al. discussed that the enhanced brain delivery could be attributed to the reduced negative charge and increased lipophilicity of the GNPs, which diminished their retention by the reticuloendothelial system. The conjugation

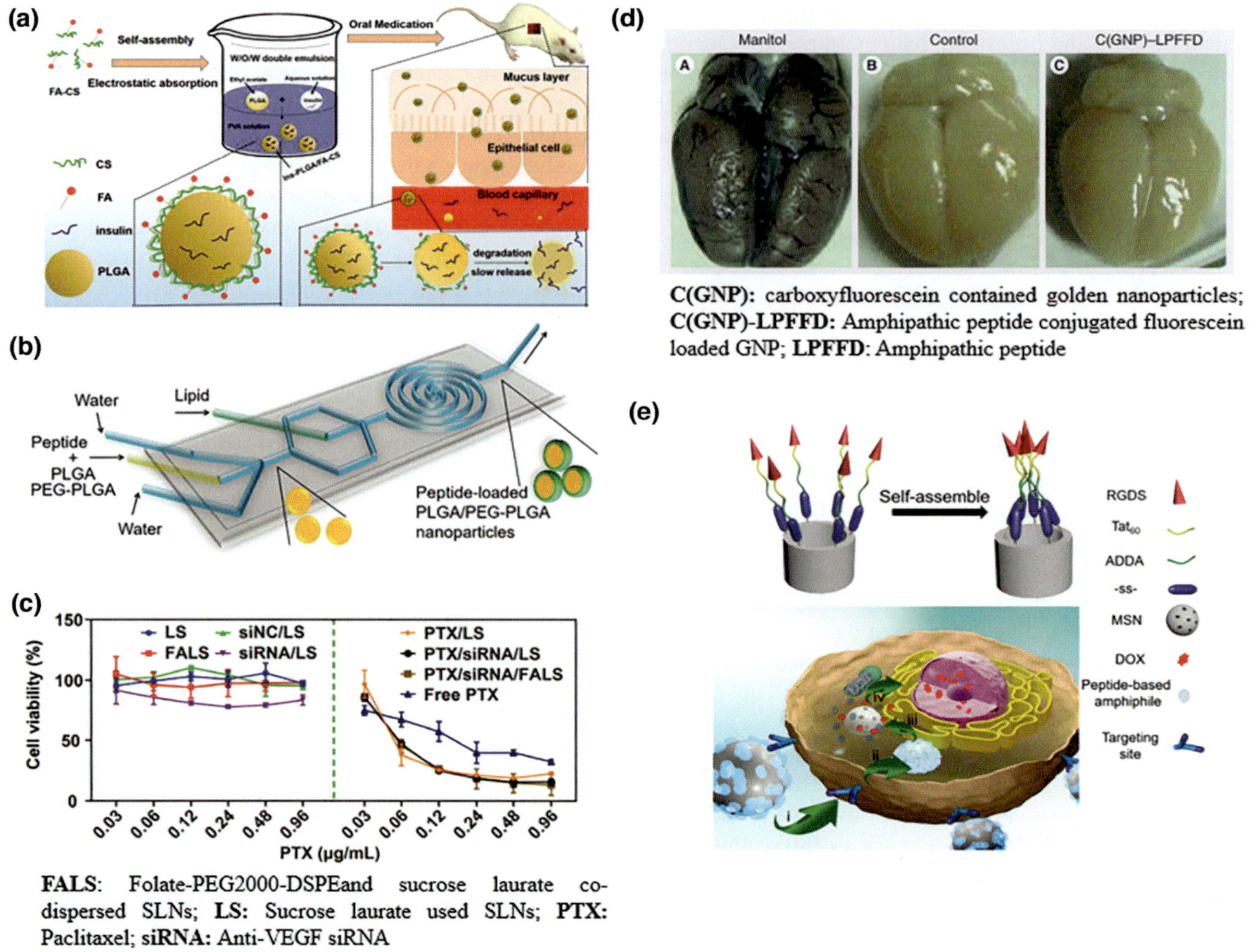

FIGURE 12.1 Protein-peptide nanocarrier systems-I. (a) Insulin-loaded PLGA nanoparticles [17]. (Permission granted by Copyright Clearance Center (CCC).) (b) C5aR1 antagonist peptide-loaded lipid shell-PLGA two-stage microfluidic chip [18]. (Permission granted by MDPI.) (c) Paclitaxel and/or siRNA-loaded SLNs cell viability study on lung cancer cell line [21]. (Permission granted by CCC.) (d) GNPs conjugation with the amphipathic peptide CLPFFD on BBB integrity [26]. (Permission granted by CCC.) (e) Self-assembling peptide-based amphiphile for MSNs delivery [30]. (Permission granted by ACS Publication.)

likely facilitated transport across the blood-brain barrier (BBB) through mechanisms that potentially involve interactions with specific receptors, such as RAGE (receptor for advanced glycation end-products). The findings from brain dissections suggest that GNPs conjugated with amphipathic peptides can cross the BBB efficiently while maintaining their integrity, making them a promising tool for targeted brain therapies without inducing adverse effects on BBB function (Figure 12.1d) [26]. In other multifunctional approaches, cRGD-polydopamine @GNPs loaded with chelator-free iodine-125 (I125] and RGD-5 poly(amidoamine) dendrimer@GNPs loaded with gadolinium (Gd) chelator/(Gd III) have been utilized as dual-mode nanoprobes for high-resolution photoacoustic (PA) and CT/MR imaging in tumor-bearing mice models. High-resolution PA imaging demonstrates that cRGD-polydopamine@AuNRs loaded with chelator-free iodine-125 specifically target tumor angiogenic vessels, showcasing their potential as effective agents for tumor therapy [27].

12.2.4 Silica Nanoparticles

Mesoporous silica nanoparticles (MSNs) are characterized by their high surface area, tunable pore size, and biocompatibility, making them suitable for loading and delivering proteins and peptides. Recent research has focused on improving drug loading efficiency and release profiles through surface functionalization [28–30]. Cheng et al. developed an MSN, which is coated with a peptide-based amphiphile called ADDA-TCPP. This coating acts as a nanovalve, preventing the drug from leaking out prematurely. The nanovalve is designed to be redox-responsive, meaning it can be triggered to open by the high levels of glutathione (GSH) found inside tumor cells. Once inside the tumor cells, the nanovalve opens, releasing the doxorubicin (DOX) and killing the cells (Figure 12.1e). The study found that the MSNs were effective in targeting and killing tumor cells while minimizing harm to healthy cells [30]. On the other hand, Pan et al. found that MSN-TAT peptide conjugated systems with a diameter of 50 nm or smaller could efficiently target the nucleus and deliver the DOX into the targeted nucleus, killing cancer cells with enhanced efficiency. The study also highlights the importance of the TAT peptide in facilitating the nuclear localization of MSNs. The authors conclude that this nuclear-targeted drug delivery system based on MSNs-TAT could be a promising strategy for cancer therapy [31].

12.2.5 Dendrimers

Dendrimers, with their highly branched, tree-like structures, offer numerous surface functional groups for drug conjugation. Advances in dendrimer chemistry have focused on enhancing biocompatibility and drug delivery efficiency [32,33]. Ren et al. developed a polymer that can bind to peptides and facilitate their entry into cells. The polymer is based on manganese (II) ions and a molecule called terpyridine (TPy). The researchers tested the G5 conjugated polymer with different peptides and found that it could efficiently deliver them into various types of cells [33]. Zhang et al. developed a PEGylated lysine peptide dendrimer-gemcitabine conjugate (dendrimer-GEM) nanoparticle. This nanoparticle utilizes the enzyme Cathepsin B, which is overexpressed in tumor cells, to trigger the release of the chemotherapeutic drug gemcitabine (GEM). The dendrimer-GEM nanoparticle was designed to self-assemble into a size suitable for enhanced permeability and retention (EPR) effect, allowing it to accumulate in tumor tissues. In vitro and in vivo experiments demonstrated the effectiveness of this drug delivery system. The in vitro drug release studies showed that the nanoparticle could rapidly release GEM in the presence of Cathepsin B. The in vivo studies in a 4T1 murine breast cancer model further confirmed the enhanced antitumor efficacy of dendrimer-GEM nanoparticles compared to free GEM, with significantly suppressed tumor growth and minimal systemic toxicity (Figure 12.2a) [34].

12.2.6 Carbon Nanotubes

Carbon nanotubes (CNTs) are notable for their high aspect ratio, large surface area, and ability to penetrate cell membranes. Recent studies have emphasized functionalization strategies to enhance biocompatibility and reduce toxicity [12,35,36]. Ohta et al. [35] explore the use of single-walled carbon nanotubes (SWCNTs) for drug and gene delivery, focusing on improving their dispersion stability in aqueous media. Despite the promising properties of SWCNTs, their hydrophobic nature limits their biomedical applications. The PEG modification, ranging from 4.1% to 13.3%, significantly improved dispersion stability in cell culture medium. Cellular uptake studies using A549 cells revealed increased uptake with higher PEG modification, confirmed

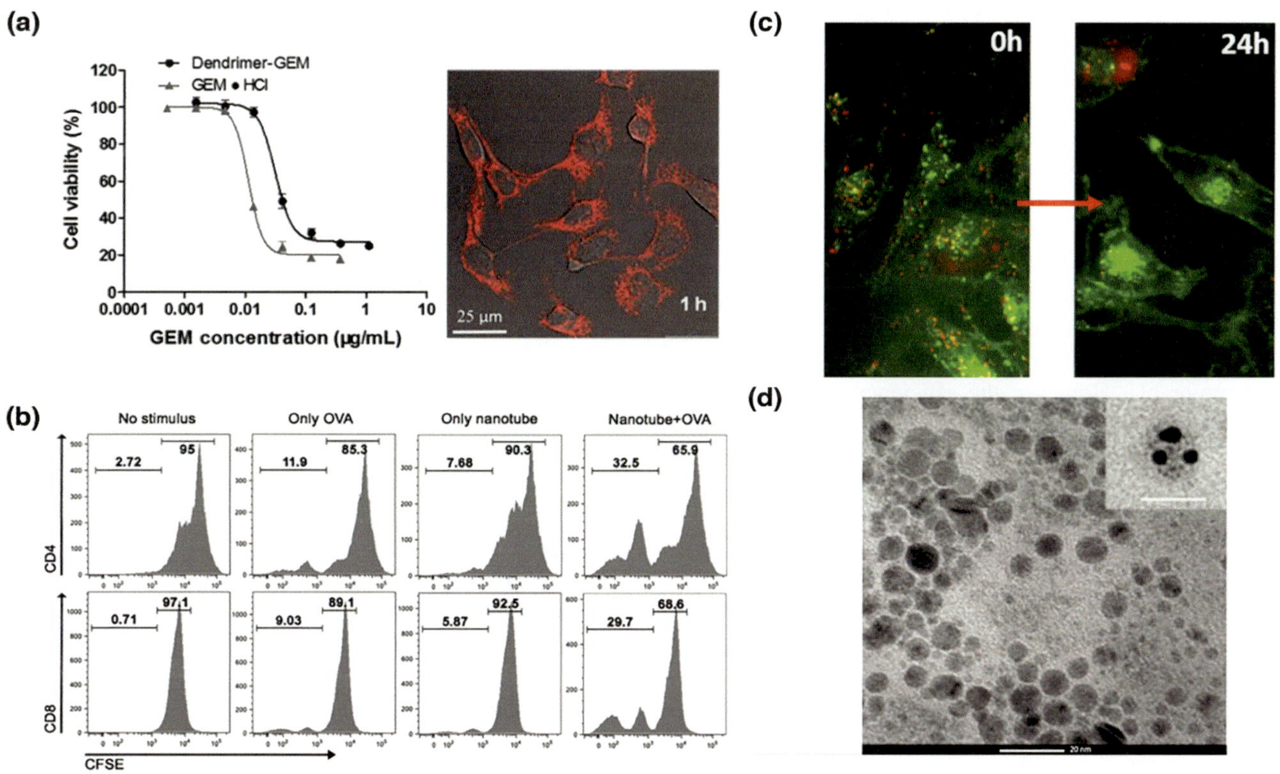

FIGURE 12.2 Protein-peptide nanocarrier systems-II. (a) Cellular uptake of dendrimer-GEM and free GEM in 4T1 cells [34]. (Permission granted by CCC.) (b) Proliferation of CD4+ and CD8+ T cells in response to different stimulation conditions using fluorescence labeling [37]. (Permission granted by ACS Publication.) (c) Confocal microscopy images show HeLa-705 cells incubated for 0 and 24 h with PF14-Alexa dyed MNPs complexes, with cell membranes stained using Fast-dio stain [41]. (Adapted figure's permission granted by Creative Commons.) (d) TEM image of APNPs at a scale of 20 nm [46]. (Permission granted by CCC.)

by fluorescence microscopy and FACS analysis. This research demonstrates that PEGylated SWCNT-peptide complexes provide a stable and functional platform for drug and gene delivery, offering enhanced dispersion stability, improved cellular uptake, and reduced cytotoxicity, which highlights their potential in biomedical applications [36]. Faria et al. developed a hybrid supramolecule consisting of MWCNTs, the cancer testis antigen NY-ESO-1, and a synthetic Toll-Like Receptor 9 (TLR9] agonist, CpG oligonucleotides. This formulation aimed to enhance antigen delivery and stimulate robust immune responses [37]. The addition of nanotubes enhances the proliferation of both CD4 and CD8 T cells, especially when combined with OVA. This indicates that nanotubes may act as adjuvants, strengthening the immune response to antigens (Figure 12.2b).

12.2.7 Magnetic Nanoparticles

Magnetic nanoparticles (MNPs) can be directed to specific sites using an external magnetic field, providing targeted delivery of proteins and peptides. Recent research has explored novel coatings and functionalizations to improve biocompatibility and drug loading [38–40]. By incorporating MNPs into complexes with cell-penetrating peptides (CPPs) and gene therapeutic agents like plasmids, splice-correcting oligonucleotides (SCOs), and small interfering RNA (siRNA), Dowaidar et al. achieved higher transfection efficiencies compared to non-magnetic complexes. The study demonstrated that these MNP-assisted complexes exhibit low cytotoxicity, high biocompatibility, and improved gene delivery performance both in vitro and in vivo (Figure 12.2c) [41]. The findings suggest that conjugating CPPs and oligonucleotides with MNPs could be a promising strategy for efficient and targeted gene therapy, offering new possibilities for treating various genetic disorders. The study by Pershina et al. also utilized 3D tumor spheroids and in vivo mouse models, demonstrating that the accumulation of MNPs in tumors correlated with the measured tumor pH. However, the efficiency of pH-triggered delivery varied significantly due to inter-tumor pH heterogeneity. These findings suggest that while pH-triggered delivery using low-pH-insertion peptides-modified MNPs holds promise for targeted cancer therapy, its effectiveness may be limited by the variability in tumor pH profiles [42].

12.2.8 Protein Nanoparticles

Protein nanoparticles, made from natural proteins like albumin, are biocompatible and biodegradable, making them ideal carriers for drug delivery. Advances have focused on improving drug loading and targeting through surface modifications [43–45]. By genetically fusing influenza hemagglutinin (HA) to ferritin, a protein that naturally forms nanoparticles, researchers created a structure that mimics the viral spike using viral proteins to form the nanoparticle's structure. This design significantly enhanced the immune response compared to traditional vaccines, generating antibodies that target highly conserved regions of HA. The vaccine demonstrated strong efficacy in animal models, protecting a wide range of H1N1 strains [43]. Yu et al.'s study demonstrated that activatable protein nanoparticles (APNPs) (Figure 12.2d) effectively delivered the neuroprotective peptide NR2B9c to the brain for stroke treatment and the anticancer peptide melittin to tumors [46]. This method showed enhanced therapeutic efficacy and specificity, with minimal systemic toxicity, making APNPs a promising platform for the targeted delivery of various therapeutic peptides.

12.3 ENCAPSULATION AND RELEASE MECHANISMS

12.3.1 Techniques for Encapsulating Proteins and Peptides

12.3.1.1 Nanoparticles

- **Polymeric Nanoparticles**: Utilized for their biodegradability, polymers like PLGA are commonly employed. Encapsulation techniques include emulsification-solvent evaporation, which involves forming an emulsion and evaporating the solvent to create nanoparticles [47,48]. Nanoprecipitation, where polymers and drugs precipitate from a solvent into nanoparticles, and spray drying, which involves spraying a polymer-drug solution into a hot chamber to form nanoparticles, are also used [48].
- **Lipid Nanoparticles**: These include structures like liposomes and SLNs. Liposomes are spherical vesicles composed of a lipid bilayer, which encapsulates drugs in their aqueous core or within the lipid layer [49,50]. SLNs and NLCs can encapsulate drugs within their solid lipid core or a blend of liquid and solid lipids. Methods such as thin-film hydration, where a thin film of lipids is hydrated to form liposomes, and high-pressure homogenization, where lipids are forced through a narrow space under high pressure to form nanoparticles, are commonly used [51,52].
- **Inorganic Nanoparticles**: Gold, silica, and magnetic nanoparticles can encapsulate proteins or peptides via physical adsorption, where molecules adhere to the nanoparticle surface, covalent attachment, involving chemical bonds between the drug and nanoparticle, or entrapment during nanoparticle synthesis [28,53].

12.3.1.2 Hydrogels

These are networks of hydrophilic polymers that can swell in water, creating a gel-like structure. Proteins and peptides can be loaded into hydrogels by simple mixing, where the drug is blended with the hydrogel precursor before gelation, or through in situ polymerization, where the drug is added to the polymerization mixture and becomes encapsulated as the hydrogel forms [54,55].

- **Microspheres**: Similar to nanoparticles but larger, these spherical particles encapsulate proteins and peptides within polymeric matrixes. Methods like solvent evaporation, where a polymer-drug solution is emulsified in water and the solvent is evaporated to form microspheres, or spray drying, where the solution is sprayed into a hot chamber to create microspheres, are used [56,57]. Zhou et al. explored the development of composition-equivalent PLGA microsphere formulations to Lupron Depot (LD), a commercial 1-month controlled-release product of leuprolide acetate [57]. They found that the encapsulation efficiency of leuprolide was primarily influenced by the intensity of the second homogenization step and the choice of polymer and gelatin. The particle size of the microspheres was affected by factors such as polymer concentration, homogenization speed, and the bloom number of gelatin. While the composition-equivalent formulations exhibited higher porosity than the reference LD product, their in vitro release kinetics were similar, suggesting that the release mechanism is primarily governed by polymer erosion. The study offers valuable insights into the development of generic or new long-acting release microsphere products, highlighting the critical role of manufacturing process variables in achieving the desired product attributes and release performance.
- **Dendrimers**: Highly branched, tree-like macromolecules that can encapsulate drugs within their internal cavities. Encapsulation occurs through electrostatic interactions, hydrogen bonding, or hydrophobic interactions, depending on the nature of the dendrimer and the drug [34,58,59]. The study of Jain et al. highlighted the synthesis of 4.5-generation (G) PPI dendrimers modified to carboxyl-terminated versions and conjugated with monoclonal antibodies targeting mesothelin, a protein frequently overexpressed in ovarian cancer cells. In vitro studies demonstrated that paclitaxel-loaded immuno dendrimers exhibited significant cytotoxicity against mesothelin-positive OVCAR-3 cells, while in vivo experiments in tumor-bearing mice showed substantial tumor-volume reduction and prolonged survival (Figure 12.3a) [59]. Biodistribution studies

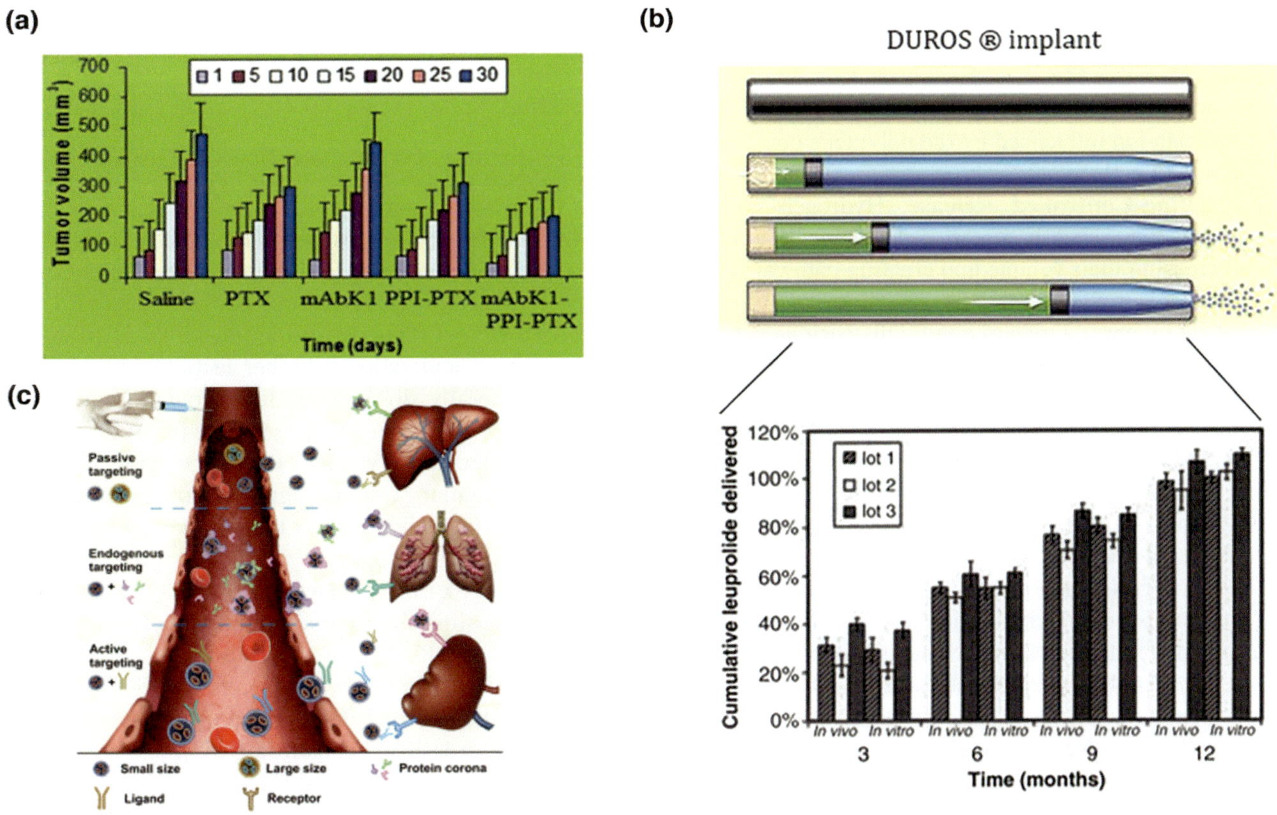

FIGURE 12.3 Administration and targeting strategies for protein-peptide NPs. (a) Tumor-volume measurements over a 30-day period for various treatment groups in an ovarian cancer study. The groups include saline (control), paclitaxel (PTX), monoclonal antibody K1 (mAbK1), poly(propylene imine) dendrimers loaded with paclitaxel (PPI-PTX), and a combination of mAbK1 conjugated with PPI-PTX (mAbK1-PPI-PTX) [59]. (Permission granted by CCC.) (b) Cross-sectional diagram of the Duros® implants and its in vivo/in vitro performance comparison for cumulative drug delivery in rats [63,64]. (Permission granted by CCC and Creative Common.) (c) Targeting strategies and administration routes of mRNA NPs) [91]. (Permission granted by CCC.)

confirmed the higher accumulation of immuno dendrimers in tumor tissues compared to nontargeted formulations. This research underscores the potential of antibody-conjugated dendrimers for targeted cancer therapy, improving drug delivery and therapeutic outcomes while minimizing side effects [59].

- **Nanoshells and Nanocapsules**: Comprising a core-shell structure, these can encapsulate proteins and peptides either within the core or adsorbed onto the shell. Techniques like layer-by-layer assembly, where alternate layers of materials are deposited to form the shell, or interfacial polymerization, where polymerization occurs at the interface between two immiscible liquids, are used [60,61].

12.3.2 Release Mechanisms

Diffusion-Controlled Release: Proteins and peptides diffuse out of the delivery system through pores or the polymer matrix. This process depends on the size of the drug molecules, the porosity of the delivery system, and the degree of cross-linking in hydrogels, which affects how easily molecules can pass through [62].

1. **Degradation-Controlled Release**: The encapsulating material degrades over time, releasing the encapsulated proteins or peptides. Biodegradable polymers like PLGA undergo hydrolysis, breaking down into lactic and glycolic acids, with the rate of degradation influenced by the polymer's composition and molecular weight.
2. **Swelling-Controlled Release**: Hydrogels swell when they absorb water or biological fluids, increasing their volume and creating pathways for the drugs to diffuse out. The swelling can be triggered by environmental conditions such as pH or temperature.
3. **Stimuli-Responsive Release**: These systems release their cargo in response to specific stimuli:
    - **pH-Responsive**: In acidic environments, such as tumor tissues or endosomes, specific polymers degrade or swell, releasing the encapsulated drugs.
    - **Thermo-Responsive**: Polymers like poly (N-isopropylacrylamide) (PNIPAAm) change their physical state at specific temperatures, leading to the release of the encapsulated drugs.
    - **Enzyme-Responsive**: Specific enzymes present in target tissues can degrade certain bonds in the polymer, causing the release of the encapsulated drugs.
4. **Osmotic Pressure-Driven Release**: Some encapsulation systems are designed to absorb water, creating osmotic pressure within the delivery system that drives the release of the encapsulated proteins or peptides. The review by Keraliya et al. discusses the use of osmotically controlled systems, specifically in the context of non-biodegradable implants. It provides information about systems like Duros®, ALZET®, and Viadur® implants, which utilize osmotic pressure to regulate the release of therapeutic proteins and peptides consistently over prolonged periods, ranging from months to a year [63]. Figure 12.3b illustrates the design and performance of the Duros® implant. The top section shows a schematic of the implant, which is comprised of a cylindrical body with an osmotic engine at one end. The implant's mechanism involves the osmotic agent drawing in water through a semipermeable membrane, generating pressure that moves a piston to push the drug out through a delivery port steadily. The lower section of the figure presents a graph comparing the cumulative release of leuprolide over time, both in vitro and in vivo, across three different lots. The data indicate a consistent and controlled release of the drug over a 12-month period, demonstrating the implant's capability to maintain therapeutic levels of leuprolide with minimal variation between different batches [63,64].
5. **Electrostatic Interaction**: Changes in the electrostatic environment, such as fluctuations in ionic strength or pH, can weaken the electrostatic bonds between the drug and the carrier, facilitating the release of the encapsulated drugs.

12.3.3 Factors Influencing Release Profiles

1. **Material Properties**:
    - **Polymer Composition**: The choice of polymer significantly impacts the release rate of encapsulated drugs. For example, PLGA degrades at different rates depending on its lactic to glycolic acid ratio. A higher lactic acid content generally slows down the degradation, resulting in a prolonged drug release [65,66].
    - **Molecular Weight**: Polymers with higher molecular weights tend to degrade more slowly, which extends the duration over which the drug is released.
    - **Cross-Linking Density**: The density of cross-linking in hydrogels or polymer networks affects the diffusion rate of drugs. Higher cross-linking results in tighter networks, slowing down the drug release due to reduced diffusion pathways [55].
2. **Particle Size and Surface Area**:
    - **Nanoparticles and Microspheres**: Smaller particles have a higher surface area relative to their volume, which can lead to faster drug release due to shorter diffusion paths and increased surface interactions.
    - **Surface Modification**: The release profile can be influenced by surface modifications such as

adding hydrophilic or hydrophobic coatings, which can alter interactions between the particle and the surrounding biological environment.

3. **Drug-Polymer Interactions**:
    - **Electrostatic Interactions**: Strong electrostatic interactions between the drug and the polymer matrix typically result in slower release rates as the drug is held more firmly within the matrix.
    - **Hydrophobic/Hydrophilic Interactions**: Hydrophobic drugs within hydrophobic matrixes or hydrophilic drugs within hydrophilic matrixes may exhibit slower release due to stronger affinity and interactions with the surrounding material [67,68].
4. **Environmental Conditions**:
    - **pH Sensitivity**: The release rate can be highly dependent on the pH of the surrounding environment. For example, certain pH-sensitive hydrogels swell or degrade at specific pH levels, such as in the acidic environment of tumor tissues, leading to accelerated drug release.
    - **Temperature Sensitivity**: Thermo-responsive polymers undergo physical changes at certain temperatures, affecting the release rate. Higher temperatures can increase polymer degradation and drug diffusion rates, leading to a faster release.
    - **Enzyme Sensitivity**: The presence of specific enzymes can accelerate the degradation of enzyme-sensitive polymers, leading to an increased release rate of the encapsulated drugs [34,69].
5. **Encapsulation Method**: The method used to encapsulate the drug, such as emulsification, nanoprecipitation, or spray drying, can influence the distribution of the drug within the delivery system and the initial release rate [1,70].
6. **Drug Properties**:
    - Smaller drug molecules typically diffuse more quickly through the polymer matrix, leading to faster release. The solubility of the drug in the polymer matrix and in the surrounding medium can affect the release rate. Poorly soluble drugs tend to release more slowly [71].
7. **Delivery System Design**:
    - Systems with a core-shell architecture, such as nanoshells and nanocapsules, provide controlled release by protecting the drug within the core and modulating its release through the shell [72–74].
8. **External Stimuli**:
    - Magnetic nanoparticles can be designed for controlled release under the influence of external magnetic fields, allowing for targeted and on-demand drug release [1,75,76].

By thoroughly understanding and manipulating these factors, researchers can tailor drug delivery systems to achieve desired release profiles, enhancing therapeutic effectiveness and improving patient outcomes.

12.3.4 Key Considerations

- **Stability of Proteins/Peptides**: It is crucial to maintain the stability and bioactivity of encapsulated proteins and peptides throughout the encapsulation and release processes. Techniques such as lyophilization (freeze-drying) and the use of stabilizing excipients can help protect the bioactivity of the drugs [77].
- **Controlled-Release Profiles**: Designing delivery systems to achieve specific release profiles, whether sustained, pulsatile, or targeted, is essential for maximizing therapeutic efficacy and improving patient compliance [54,78].

Understanding these encapsulation and release mechanisms enables the design of advanced drug delivery systems that enhance the therapeutic potential of protein and peptide drugs, resulting in more effective and patient-friendly treatments.

12.4 NANOPARTICLE DESIGN FOR STABILITY AND BIOAVAILABILITY

Creating nanoparticles for protein and peptide delivery involves a multifaceted approach to ensure both stability and enhanced bioavailability. Here's an in-depth look at the factors and strategies essential for effective nanoparticle design:

12.4.1 Choice of Materials

12.4.1.1 Biodegradable Polymers

Like PLGA; PLA, PEG [45].

12.4.1.2 Lipid-based Materials

Like liposomes, SLNs, NLCs [79].

12.4.1.3 Inorganic Nanoparticles

Like gold nanoparticles, silica nanoparticles [28,80].

12.4.2 Encapsulation Techniques

12.4.2.1 Emulsification-Solvent Evaporation

- The polymer and drug are dissolved in a volatile organic solvent and emulsified in an aqueous phase containing a surfactant. The solvent is then evaporated, resulting in the formation of nanoparticles [81].

12.4.2.2 Nano Precipitation

- Rapid mixing of a polymer-drug solution with a non-solvent (typically water) causes the polymer to precipitate, forming nanoparticles. This method is simple, does not require high temperatures, and is suitable for sensitive biomolecules like proteins and peptides.

12.4.2.3 Spray Drying

- A polymer-drug solution is atomized into a hot chamber, causing rapid solvent evaporation and the formation of nanoparticles [16].

12.5 SURFACE MODIFICATION

12.5.1 PEGylation

- PEG chains are attached to the nanoparticle surface, forming a hydrophilic barrier that reduces protein adsorption and immune recognition.

12.5.2 Ligand Attachment

- Targeting ligands (e.g., antibodies, peptides, small molecules) are conjugated to the nanoparticle surface to improve targeting specificity [82,83].

12.6 STABILIZATION STRATEGIES

12.6.1 Lyophilization [Freeze-Drying]

- The nanoparticle suspension is frozen and then subjected to a vacuum to sublimate the water, leaving behind a dry powder.

12.6.2 Cryoprotectants

- Sugars like trehalose and sucrose are added during lyophilization to protect proteins and peptides from damage caused by freezing and drying.

12.6.3 pH Buffers

- Buffers are included in formulations to maintain a stable pH, which is crucial for the stability of pH-sensitive proteins and peptides.

12.7 OPTIMIZATION OF PARTICLE CHARACTERISTICS

12.7.1 Particle Size

- Smaller nanoparticles (typically 50–200 nm) exhibit better cellular uptake and bioavailability.

12.7.2 Surface Charge

- The surface charge affects nanoparticle stability and interactions with biological membranes. Slightly negative or neutral charges help avoid rapid clearance by the reticuloendothelial system (RES) [84,85].

12.7.3 Polydispersity Index [PDI]

- A low PDI indicates uniform particle size distribution, which is crucial for consistent drug release and bioavailability.

12.8 CONTROLLED-RELEASE MECHANISMS

12.8.1 Degradation-Controlled Release

- Biodegradable polymers gradually break down in the body, releasing the encapsulated drug over time.

12.8.2 Diffusion-Controlled Release

- The drug diffuses out of the nanoparticle through pores or the polymer matrix [63,86].

12.8.3 Stimuli-Responsive Release

- Nanoparticles are designed to respond to specific stimuli such as pH changes, temperature shifts, or the presence of certain enzymes [87].

12.9 IMPROVING BIOAVAILABILITY

12.9.1 Mucoadhesive Properties

- Enhancing the mucoadhesive properties of nanoparticles increases their residence time at mucosal surfaces, improving absorption [88,89].

12.9.2 Transcytosis Pathways

- Designing nanoparticles to exploit transcytosis pathways (e.g., through M cells in the gut) enhances oral bioavailability [90]

12.9.3 Enhanced Permeability and Retention [EPR] Effect

- Leveraging the EPR effect in tumors allows nanoparticles to accumulate in cancerous tissues due to the leaky vasculature and poor lymphatic drainage [91].

12.9.4 In Vivo Stability

Protection from Enzymatic Degradation:

- Encapsulating proteins and peptides in nanoparticles shields them from proteolytic enzymes in the bloodstream and target tissues [92].

- Long Circulation Times: Surface modifications such as PEGylation help nanoparticles evade the immune system, prolonging their circulation time [93].

12.10 ADMINISTRATION ROUTES AND TARGETING STRATEGIES

Protein and peptide nanoparticles can be administered via various routes, each offering distinct advantages and challenges. Oral administration is a desirable route due to its convenience and patient compliance, but it faces significant barriers such as enzymatic degradation and poor absorption in the gastrointestinal tract. Strategies like using enteric coatings and mucoadhesive polymers can enhance stability and absorption. Parenteral routes, including intravenous, subcutaneous, and intramuscular injections, provide direct delivery into the systemic circulation, bypassing the gastrointestinal barriers. Intravenous administration ensures rapid distribution and high bioavailability, although it requires careful formulation to avoid rapid clearance by the reticuloendothelial system. Subcutaneous and intramuscular injections offer sustained release profiles, which are beneficial for chronic conditions requiring long-term therapy [94–96]. Inhalation is another effective route, particularly for targeting respiratory diseases. Nanoparticles can be delivered directly to the lungs, achieving high local concentrations while minimizing systemic side effects. Intranasal administration can also facilitate direct delivery to the central nervous system via the olfactory and trigeminal nerves, bypassing the BBB and enhancing drug bioavailability in the brain [22,97,98]. Targeting strategies for protein and peptide nanoparticles aim to improve therapeutic efficacy while minimizing off-target effects. Passive targeting exploits the EPR effect, where nanoparticles accumulate in tumor tissues due to their leaky vasculature. Active targeting involves modifying nanoparticle surfaces with ligands such as antibodies, peptides, or small molecules that bind specifically to receptors overexpressed on target cells. This approach increases the precision of drug delivery, improving therapeutic outcomes. Stimuli-responsive targeting leverages environmental triggers such as pH, temperature, or specific enzymes to release the drug at the target site, ensuring controlled and localized delivery [99,100]. mRNA nanodelivery systems are popular protein-based delivery systems that have emerged as a promising strategy for mRNA therapeutics delivery. These systems can protect mRNA from degradation, promote cellular uptake, and target specific cell types. A variety of targeting strategies have been developed for mRNA nanodelivery systems, including ligand-receptor targeting, aptamer-based targeting, and cell-penetrating peptide-based targeting (Figure 12.3a). In addition, various administration routes can be used for mRNA nanodelivery systems, including intravenous, intramuscular, and intranasal administration. The choice of administration route depends on the target tissue and the desired therapeutic effect. mRNA nanodelivery systems have the potential to revolutionize the treatment of a wide range of diseases, including cancer, infectious diseases, and genetic disorders. However, further research is needed to optimize the design and development of these systems [91,101]. By combining these administration routes and targeting strategies, protein and peptide nanoparticles can achieve optimal therapeutic efficacy and patient outcomes.

12.11 EVALUATION OF NANOPARTICLE SYSTEMS

Evaluation of protein and peptide nanoparticles involves several critical parameters to ensure their efficacy, safety, and stability. First, the physical characteristics, including particle size, size distribution, and surface charge (zeta potential), are assessed using techniques like dynamic light scattering (DLS) and electron microscopy. Lim et al. provided a comprehensive review on the use of DLS to characterize MNPs, focusing on their size distribution and colloidal stability. It delves into the mathematical analysis used to interpret DLS data, emphasizing the importance of variables like surface coating, particle size differences, and concentration. The review compares DLS with other sizing techniques such as transmission electron microscopy (TEM) and dark-field to ensure that the nanoparticles retain their integrity and functionality over time. Finally, in vivo studies are crucial to evaluate the pharmacokinetics, biodistribution, efficacy, and potential toxicity of the nanoparticles [102,103]. These studies help determine the nanoparticles' therapeutic potential and safety profile, providing comprehensive insights into their suitability for clinical applications.

12.12 CLINICAL APPLICATIONS OF NANOPARTICLE DELIVERY SYSTEMS

Protein and peptide nanoparticles have significant potential in various clinical applications due to their ability to enhance drug stability, bioavailability, and targeted delivery. In oncology, these nanoparticles are used to deliver chemotherapeutic agents directly to tumor cells, minimizing systemic toxicity and enhancing therapeutic efficacy. They are also employed in the treatment of chronic diseases such as diabetes, where insulin-loaded nanoparticles provide sustained and controlled release, improving glycemic control and patient compliance [104–106]. Pan et al. developed a post-formulation strategy using a linker peptide that integrates cargo into lipidic membranes through a self-assembling process. This approach was demonstrated by creating VCAM-1-targeted perfluorocarbon nanoparticles for in vivo targeting of atherosclerosis and breast cancer models. In the atherosclerosis model, targeted nanoparticles showed a 4.1-fold increase in binding to affected aortas compared to nontargeted nanoparticles ($P<0.0298$], while in the breast cancer model, targeted nanoparticles exhibited a 4.9-fold increase in signal from tumor vasculature compared to nontargeted nanoparticles ($P < 0.0216$]. This peptide linker post-formulation strategy is broadly applicable, potentially benefiting more than half of the lipid-coated nanosystems currently in clinical trials or use, including liposomes and micelles. This approach enhances the targeting capabilities and therapeutic efficacy of nanocarriers, offering significant clinical benefits [104] (Figure 12.4a).In the field of infectious diseases, protein and peptide nanoparticles are utilized for the delivery of vaccines and antimicrobial agents, enhancing immune responses and reducing pathogen resistance [6,22,107]. Rao et al. developed nanodecoys to address COVID-19 [108]. These nanodecoys are created by fusing cell membrane nanovesicles from genetically modified cells that express ACE2 receptors and human monocytes with abundant cytokine receptors. The nanodecoys' dual functionality allows them to neutralize SARS-CoV-2 by competing with host cells for virus binding while simultaneously adsorbing inflammatory cytokines like IL-6 and GM-CSF (Figure 12.4b). This dual action prevents viral infection and mitigates the cytokine storm associated with severe COVID-19, reducing immune dysfunction and lung injury. Clinically, these protein-peptide-loaded nanodecoys offer a promising therapeutic approach by providing strong antiviral defense and anti-inflammatory effects, potentially improving outcomes for COVID-19 patients and serving as a versatile platform for future epidemic responses. Additionally, in cardiovascular diseases, these nanoparticles can deliver therapeutic proteins that promote tissue regeneration and repair. Their use in gene therapy is also noteworthy, where nanoparticles facilitate the delivery of nucleic acids to target cells, enabling the correction of genetic disorders [85,109]. Furthermore, in neurological disorders, nanoparticles can cross the BBB, delivering neuroprotective agents directly to the brain [98]. Crook et al. identified and optimized a transferrin receptor (TfR)-binding cystine-dense peptide, demonstrating its ability to deliver neurotensin, a neuropeptide, into the mouse brain and achieve modulation of CREB signaling [110]. The flow cytometry plot on the right shows three conditions: cells expressing the surface display GFP FasL vector (SDGF)-Machupo virus glycoprotein (MaCV) construct stained with TfR (red line), elastase (pink line), and cells expressing the SDGF-Elafin construct stained with TfR (blue line). The significant rightward shift in the red line indicates strong binding of TfR to SDGF-MaCV expressing cells. In contrast, the pink and blue lines remain near the origin, indicating no significant binding. This demonstrates that the SDGF-MaCV construct specifically binds to TfR, highlighting its potential for targeted delivery of protein or peptide therapeutics through TfR-mediated targeting (Figure 12.4c). This strategy holds clinical benefits by providing an efficient method for delivering bioactive molecules to the CNS, potentially improving treatments for brain cancer, neurodegenerative diseases, and other CNS disorders through enhanced BBB penetration using protein-based therapeutics [110]. Overall, the versatility and targeted delivery capabilities of protein and peptide nanoparticles make them a promising tool for advancing personalized medicine and improving clinical outcomes across a broad range of diseases.

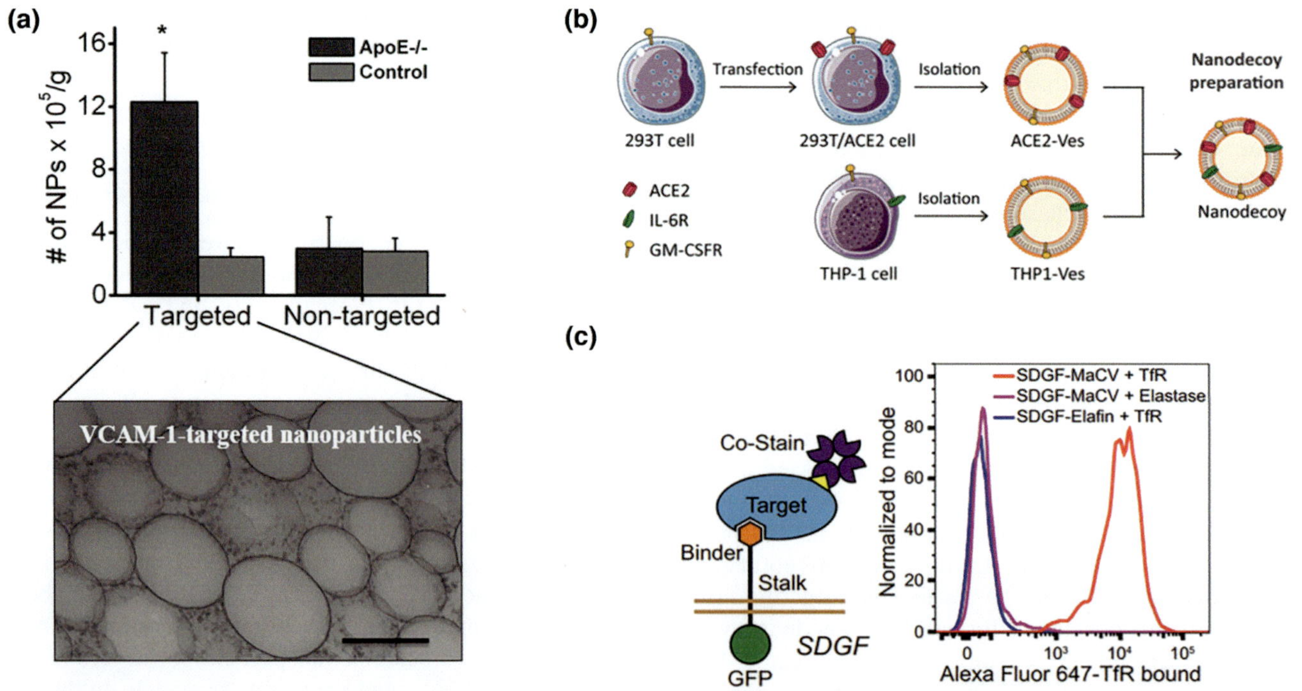

FIGURE 12.4 Clinical applications of protein-peptide NPs (a) Aortas from ApoE−/− mice given VCAM-1-targeted nanoparticles displayed significantly higher signals than other groups [104]. (Adapted figures' permission granted by Creative Commons.) (b) The process of creating nanodecoys designed to combat COVID-19 [108]. (Permission granted by CCC.) (c) The specificity of a protein-loaded system targeted to TfR using a SDGF [110]. (Adapted figures' permission granted by Creative Commons.)

12.13 REGULATORY AND MANUFACTURING CONSIDERATIONS

Regulatory and manufacturing considerations for protein and peptide nanoparticles are critical to ensuring their safety, efficacy, and quality for clinical use. Regulatory agencies, such as the FDA and EMA, require a comprehensive characterization of nanoparticles, including their physicochemical properties, stability, and biocompatibility. This involves detailed documentation and adherence to Good Manufacturing Practices (GMP) to ensure consistent production and quality control [111,112]. Manufacturers must also provide robust data from preclinical and clinical studies to demonstrate the safety and therapeutic effectiveness of the nanoparticles [113]. Scaling up the production process from laboratory to industrial scale presents challenges in maintaining the uniformity and reproducibility of nanoparticles. This necessitates the development of scalable and cost-effective manufacturing techniques, such as advanced microfluidics and high-pressure homogenization [114]. Additionally, stringent sterilization processes must be implemented to prevent contamination, and rigorous testing for endotoxins and residual solvents is required. Post-manufacturing, the stability of the nanoparticles during storage and transport must be assessed to ensure they remain effective until the point of use [53,115]. Overall, navigating the regulatory landscape and addressing manufacturing complexities are essential for the successful commercialization of protein and peptide nanoparticles.

12.14 FUTURE PERSPECTIVES AND CHALLENGES

The future of protein and peptide nanoparticles holds great promise, yet it also presents several challenges that must be addressed to realize their potential fully. Advances in nanotechnology are expected to enhance the precision and efficacy of drug delivery systems, enabling more targeted and controlled release of therapeutic agents. Innovations in materials science and engineering could lead to the development of novel nanoparticles with improved biocompatibility, stability, and functionality. However, significant challenges remain, including the need for scalable and cost-effective manufacturing processes that ensure consistent quality and reproducibility. Regulatory hurdles must also be overcome, as rigorous testing and compliance with regulatory standards are crucial for clinical approval. Additionally, understanding the long-term biocompatibility and potential toxicity of these nanoparticles is essential for their safe application. There is also a need for more comprehensive studies on the interactions between nanoparticles and biological systems to mitigate any adverse effects. Despite these challenges, the integration of interdisciplinary research and collaboration among scientists, engineers, and regulatory bodies will be key to advancing the field and bringing innovative protein and peptide nanoparticle therapies to market.

12.15 CONCLUSION

In conclusion, protein and peptide nanoparticles represent a transformative advancement in drug delivery systems, offering remarkable benefits in terms of targeted therapy, controlled release, and enhanced bioavailability. These nanoparticles have the potential to revolutionize the treatment of a wide array of diseases, including cancer, diabetes, infectious diseases, cardiovascular disorders, and neurological conditions. By providing precise delivery of therapeutic agents, they minimize side effects and improve patient outcomes. However, the journey from research to clinical application involves overcoming significant challenges related to manufacturing scalability, regulatory compliance, and ensuring long-term safety and efficacy. Continued interdisciplinary research, coupled with advancements in nanotechnology and materials science, will be critical in addressing these challenges. As we advance, the successful integration of protein and peptide nanoparticles into mainstream medical practice holds the promise of significantly improving therapeutic strategies and patient care.

REFERENCES

1. Oliveira CBP, Gomes V, Ferreira PMT, Martins JA, Jervis PJ. Peptide-based supramolecular hydrogels as drug delivery agents: recent advances. *Gels*. 2022;8(11):706. doi:10.3390/gels8110706.
2. Rong L, Lei Q, Zhang X-Z. Recent advances in peptide-based theranostic nanomaterials. *View*. 2020;1(4):20200050. doi:10.1002/VIW.20200050.
3. Fetse J, Kandel S, Mamani U-F, Cheng K. Recent advances in the development of therapeutic peptides. *Trends in Pharmacological Sciences*. 2023;44(7):425–41. doi:10.1016/j.tips.2023.04.003.
4. Pasalar M, Daneshfard B, Bagheri Lankarani K. Jollab: a safe remedy for functional dyspepsia in persian medicine. *Advanced Pharmaceutical Bulletin*. 2021;11(1):8–9. doi:10.34172/apb.2021.001.
5. Renukuntla J, Vadlapudi AD, Patel A, Boddu SH, Mitra AK. Approaches for enhancing oral bioavailability of peptides and proteins. *International Journal of Pharmaceutics*. 2013;447(1–2):75–93. doi:10.1016/j.ijpharm.2013.02.030.
6. Elzoghby AO, Elgohary MM, Kamel NM. Implications of protein- and peptide-based nanoparticles as potential vehicles for anticancer drugs. In: Donev R (Ed.), *Advances in Protein Chemistry and Structural* Biology. Academic Press, 2015, pp. 169–221.
7. Blanco E, Shen H, Ferrari M. Principles of nanoparticle design for overcoming biological barriers to drug delivery. *Nature Biotechnology*. 2015;33(9):941–51. doi:10.1038/nbt.3330.
8. Hamimed S, Jabberi M, Chatti A. Nanotechnology in drug and gene delivery. *Naunyn-Schmiedeberg's Archives of Pharmacology*. 2022;395(7):769–87. doi:10.1007/s00210-022-02245-z.
9. Fan D, Cao Y, Cao M, Wang Y, Cao Y, Gong T. Nanomedicine in cancer therapy. *Signal Transduction and Targeted Therapy*. 2023;8(1):293. doi:10.1038/s41392-023-01536-y.
10. Mitragotri S, Burke PA, Langer R. Overcoming the challenges in administering biopharmaceuticals: formulation and delivery strategies. *Nature Reviews Drug Discovery*. 2014;13(9):655–72. doi:10.1038/nrd4363.

11. Tibbitt MW, Dahlman JE, Langer R. Emerging frontiers in drug delivery. *Journal of the American Chemical Society*. 2016;138(3):704–17. doi:10.1021/jacs.5b09974.
12. Patel A, Cholkar K, Mitra AK. Recent developments in protein and peptide parenteral delivery approaches. *Therapeutic Delivery*. 2014;5(3):337–65. doi:10.4155/tde.14.5.
13. Jawa V, Terry F, Gokemeijer J, Mitra-Kaushik S, Roberts BJ, Tourdot S et al. T-cell dependent immunogenicity of protein therapeutics pre-clinical assessment and mitigation-updated consensus and review 2020. *Frontiers in Immunology*. 2020;11:1301. doi:10.3389/fimmu.2020.01301.
14. Ratanji KD, Derrick JP, Dearman RJ, Kimber I. Immunogenicity of therapeutic proteins: influence of aggregation. *Journal of Immunotoxicology*. 2014;11(2):99–109. doi:10.3109/1547691x.2013.821564.
15. Zielińska A, Carreiró F, Oliveira AM, Neves A, Pires B, Venkatesh DN et al. Polymeric nanoparticles: production, characterization, toxicology and ecotoxicology. *Molecules*. 2020;25(16):3731.
16. Zhang Y, Henry Harker A, Luo CJ, Parhizkar M, Edirisinghe M. Co-delivery of saxagliptin and dapagliflozin by electrosprayed trilayer poly (D, l-lactide-co-glycolide) nanoparticles for controlled drug delivery. *International Journal of Pharmaceutics*. 2022;628:122279. doi:10.1016/j.ijpharm.2022.122279.
17. Xu B, Jiang G, Yu W, Liu D, Liu Y, Kong X et al. Preparation of poly (lactic-co-glycolic acid) and chitosan composite nanocarriers via electrostatic self-assembly for oral delivery of insulin. *Materials Science and Engineering: C*. 2017;78:420–8.
18. Han FY, Xu W, Kumar V, Cui CS, Li X, Jiang X et al. Optimisation of a microfluidic method for the delivery of a small peptide. *Pharmaceutics*. 2021;13(9):1505.
19. Shrestha N, Araújo F, Shahbazi M-A, Mäkilä E, Gomes MJ, Airavaara M et al. Oral hypoglycaemic effect of GLP-1 and DPP4 inhibitor-based nanocomposites in a diabetic animal model. *Journal of Controlled Release*. 2016;232:113–9. doi:10.1016/j.jconrel.2016.04.024.
20. Kadari A, Pooja D, Gora RH, Gudem S, Kolapalli VRM, Kulhari H et al. Design of multifunctional peptide-collaborated and docetaxel-loaded lipid nanoparticles for antiglioma therapy. *European Journal of Pharmaceutics and Biopharmaceutics*. 2018;132:168–79.
21. Zhang C, Zhao Y, Zhang E, Jiang M, Zhi D, Chen H et al. Co-delivery of paclitaxel and anti-VEGF siRNA by tripeptide lipid nanoparticle to enhance the anti-tumor activity for lung cancer therapy. *Drug Delivery*. 2020;27(1):1397–411.
22. Xu L, Wang X, Liu Y, Yang G, Falconer RJ, Zhao C-X. Lipid nanoparticles for drug delivery. *Advanced NanoBiomed Research*. 2022;2(2):2100109. doi:10.1002/anbr.202100109.
23. Yang S, Wang D, Zhang X, Sun Y, Zheng B. cRGD peptide-conjugated polyethylenimine-based lipid nanoparticle for intracellular delivery of siRNA in hepatocarcinoma therapy. *Drug Delivery*. 2021;28(1):995–1006. doi:10.1080/10717544.2021.1928794.
24. Kalimuthu K, Lubin B-C, Bazylevich A, Gellerman G, Shpilberg O, Luboshits G et al. Gold nanoparticles stabilize peptide-drug-conjugates for sustained targeted drug delivery to cancer cells. *Journal of Nanobiotechnology*. 2018;16(1):34. doi:10.1186/s12951-018-0362-1.
25. Velasco-Aguirre C, Morales F, Gallardo-Toledo E, Guerrero S, Giralt E, Araya E et al. Peptides and proteins used to enhance gold nanoparticle delivery to the brain: preclinical approaches. *International Journal of Nanomedicine*. 2015;10:4919–36. doi:10.2147/IJN.S82310.
26. Guerrero S, Araya E, Fiedler JL, Arias JI, Adura C, Albericio F et al. Improving the brain delivery of gold nanoparticles by conjugation with an amphipathic peptide. *Nanomedicine*. 2010;5(6):897–913. doi:10.2217/nnm.10.74.
27. Rai A, Ferreira L. Biomedical applications of the peptide decorated gold nanoparticles. *Critical Reviews in Biotechnology*. 2021;41(2):186–215. doi:10.1080/07388551.2020.1853031.
28. Manzano M, Vallet-Regí M. Mesoporous Silica Nanoparticles for Drug Delivery. *Advanced Functional Materials*. 2020;30(2):1902634. doi:10.1002/adfm.201902634.
29. Kumar P, Tambe P, Paknikar KM, Gajbhiye V. Folate/N-acetyl glucosamine conjugated mesoporous silica nanoparticles for targeting breast cancer cells: a comparative study. *Colloids and Surfaces B: Biointerfaces*. 2017;156:203–12. doi:10.1016/j.colsurfb.2017.05.032.
30. Cheng Y-J, Zhang A-Q, Hu J-J, He F, Zeng X, Zhang X-Z. Multifunctional peptide-amphiphile end-capped mesoporous silica nanoparticles for tumor targeting drug delivery. *ACS Applied Materials & Interfaces*. 2017;9(3):2093–103. doi:10.1021/acsami.6b12647.
31. Pan L, He Q, Liu J, Chen Y, Ma M, Zhang L et al. Nuclear-targeted drug delivery of TAT peptide-conjugated monodisperse mesoporous silica nanoparticles. *Journal of the American Chemical Society*. 2012;134(13):5722–5. doi:10.1021/ja211035w.
32. Liu J, Gray WD, Davis ME, Luo Y. Peptide- and saccharide-conjugated dendrimers for targeted drug delivery: a concise review. *Interface Focus*. 2012;2(3):307–24. doi:doi:10.1098/rsfs.2012.0009.
33. Ren L, Gao Y, Cheng Y. A manganese (II)-based coordinative dendrimer with robust efficiency in intracellular peptide delivery. *Bioactive Materials*. 2022;9:44–53. doi:10.1016/j.bioactmat.2021.08.006.
34. Zhang C, Pan D, Li J, Hu J, Bains A, Guys N et al. Enzyme-responsive peptide dendrimer-gemcitabine conjugate as a controlled-release drug delivery vehicle with enhanced antitumor efficacy. *Acta Biomaterialia*. 2017;55:153–62. doi:10.1016/j.actbio.2017.02.047.
35. Ohta T, Hashida Y, Yamashita F, Hashida M. Development of novel drug and gene delivery carriers composed of single-walled carbon nanotubes and designed peptides with pegylation. *Journal of Pharmaceutical Sciences*. 2016;105(9):2815–24. doi:10.1016/j.xphs.2016.03.031.
36. Antonucci A, Kupis-Rozmysłowicz J, Boghossian AA. Noncovalent protein and peptide functionalization of single-walled carbon nanotubes for biodelivery and optical sensing applications. *ACS Applied Materials & Interfaces*. 2017;9(13):11321–31. doi:10.1021/acsami.7b00810.
37. Faria PCBd, Santos LId, Coelho JP, Ribeiro HB, Pimenta MA, Ladeira LO et al. Oxidized multiwalled carbon nanotubes as an antigen delivery system to promote superior CD8+ T cell response and protection against cancer. *Nano Letters*. 2014;14(9):5458–70. doi:10.1021/nl502911a.
38. Subhan MA. Advances with metal oxide-based nanoparticles as MDR metastatic breast cancer therapeutics and diagnostics. *RSC Advances*. 2022;12(51):32956–78. doi:10.1039/d2ra02005j.
39. Dowaidar M, Nasser Abdelhamid H, Hällbrink M, Langel Ü, Zou X. Chitosan enhances gene delivery of oligonucleotide complexes with magnetic nanoparticles–cell-penetrating peptide. *Journal of Biomaterials Applications*. 2018;33(3):392–401. doi:10.1177/0885328218796623.
40. Mehta RV. Synthesis of magnetic nanoparticles and their dispersions with special reference to applications in

biomedicine and biotechnology. *Materials Science and Engineering: C, Materials for Biological Applications*. 2017;79:901–16. doi:10.1016/j.msec.2017.05.135.

41. Dowaidar M, Abdelhamid HN, Hällbrink M, Freimann K, Kurrikoff K, Zou X et al. Magnetic nanoparticle-assisted self-assembly of cell-penetrating peptides-oligonucleotides complexes for gene delivery. *Scientific Reports*. 2017;7(1):9159. doi:10.1038/s41598-017-09803-z.
42. Pershina AG, Brikunova OY, Demin AM, Abakumov MA, Vaneev AN, Naumenko VA et al. Variation in tumor pH affects pH-triggered delivery of peptide-modified magnetic nanoparticles. *Nanomedicine: Nanotechnology, Biology and Medicine*. 2021;32:102317. doi:10.1016/j.nano.2020.102317.
43. Kanekiyo M, Wei C-J, Yassine HM, McTamney PM, Boyington JC, Whittle JRR et al. Self-assembling influenza nanoparticle vaccines elicit broadly neutralizing H1N1 antibodies. *Nature*. 2013;499(7456):102–6. doi:10.1038/nature12202.
44. Iqbal I, Saqib F, Mubarak Z, Latif MF, Wahid M, Nasir B et al. Alzheimer's disease and drug delivery across the blood-brain barrier: approaches and challenges. *European Journal of Medical Research*. 2024;29(1):313. doi:10.1186/s40001-024-01915-3.
45. Hong S, Choi DW, Kim HN, Park CG, Lee W, Park HH. Protein-based nanoparticles as drug delivery systems. *Pharmaceutics*. 2020;12(7). doi:10.3390/pharmaceutics12070604.
46. Yu X, Gou X, Wu P, Han L, Tian D, Du F et al. Activatable protein nanoparticles for targeted delivery of therapeutic peptides. *Advanced Materials*. 2018;30(7):1705383. doi:10.1002/adma.201705383.
47. Lalan MS, Shah P, Barve K, Jha R, Jha J. Polymeric and lipid nanocarriers in glaucoma. In: M. Rai, M. L. Occhiutto, and S. Talegaonkar (Eds.), *Nanotechnology in Ophthalmology*. Elsevier, 2023, pp. 127–149.
48. Lalu L, Tambe V, Pradhan D, Nayak K, Bagchi S, Maheshwari R et al. Novel nanosystems for the treatment of ocular inflammation: current paradigms and future research directions. *Journal of Controlled Release*. 2017;268:19–39. doi:10.1016/j.jconrel.2017.07.035.
49. Amoabediny G, Haghiralsadat F, Naderinezhad S, Helder MN, Akhoundi Kharanaghi E, Mohammadnejad Arough J et al. Overview of preparation methods of polymeric and lipid-based (niosome, solid lipid, liposome) nanoparticles: a comprehensive review. *International Journal of Polymeric Materials and Polymeric Biomaterials*. 2018;67(6):383–400.
50. Chantaburanan T, Teeranachaideekul V, Chantasart D, Jintapattanakit A, Junyaprasert VB. Effect of binary solid lipid matrix of wax and triglyceride on lipid crystallinity, drug-lipid interaction and drug release of ibuprofen-loaded solid lipid nanoparticles (SLN) for dermal delivery. *Journal of Colloid and Interface Science*. 2017;504:247–56.
51. Uner B, Ozdemir S, Yıldırım E, Ucar AY, Tas C, Uner M et al. Loteprednol -loaded nanoformulations for corneal delivery: ex-vivo permeation study, ocular safety assessment and stability studies. *Journal of Drug Delivery Science and Technology*. 2023:104252. doi:10.1016/j.jddst.2023.104252.
52. Kasongo KW, Müller RH, Walker RB. The use of hot and cold high pressure homogenization to enhance the loading capacity and encapsulation efficiency of nanostructured lipid carriers for the hydrophilic antiretroviral drug, didanosine, for potential administration to pediatric patients. *Pharmaceutical Development and Technology*. 2012;17(3):353–62.
53. Fadeel B, Garcia-Bennett AE. Better safe than sorry: understanding the toxicological properties of inorganic nanoparticles manufactured for biomedical applications. *Advanced Drug Delivery Reviews*. 2010;62(3):362–74. doi:10.1016/j.addr.2009.11.008.
54. Wei D, Pu N, Li SY, Wang YG, Tao Y. Application of iontophoresis in ophthalmic practice: an innovative strategy to deliver drugs into the eye. *Drug Delivery*. 2023;30(1):2165736. doi:10.1080/10717544.2023.2165736.
55. Xu Y, Qiu Y, Lin Q, Huang C, Li J, Chen L et al. miR-126-3p-loaded small extracellular vesicles secreted by urine-derived stem cells released from a phototriggered imine crosslink hydrogel could enhance vaginal epithelization after vaginoplasty. *Stem Cell Research & Therapy*. 2022;13(1):1–15.
56. Thundimadathil J. Formulations of microspheres and nanoparticles for peptide delivery. In: V Srivastava (Ed.), *Peptide Therapeutics: Strategy and Tactics for Chemistry, Manufacturing and Controls*. The Royal Society of Chemistry, 2019.
57. Zhou J, Walker J, Ackermann R, Olsen K, Hong JKY, Wang Y et al. Effect of manufacturing variables and raw materials on the composition-equivalent PLGA microspheres for 1-month controlled release of leuprolide. *Molecular Pharmaceutics*. 2020;17(5):1502–15. doi:10.1021/acs.molpharmaceut.9b01188.
58. Chis AA, Dobrea C, Morgovan C, Arseniu AM, Rus LL, Butuca A et al. Applications and limitations of dendrimers in biomedicine. *Molecules*. 2020;25(17). doi:10.3390/molecules25173982.
59. Jain NK, Tare MS, Mishra V, Tripathi PK. The development, characterization, and in vivo anti-ovarian cancer activity of poly(propylene imine) (PPI)-antibody conjugates containing encapsulated paclitaxel. *Nanomedicine: Nanotechnology, Biology and Medicine*. 2015;11(1):207–18. doi:10.1016/j.nano.2014.09.006.
60. Corrêa ACNTF, Vericimo MA, Dashevskiy A, Pereira PR, Paschoalin VMF. liposomal taro lectin nanocapsules control human glioblastoma and mammary adenocarcinoma cell proliferation. *Molecules*. 2019;24(3):471.
61. Chauhan MK, Sharma PK. Optimization and characterization of rivastigmine nanolipid carrier loaded transdermal patches for the treatment of dementia. *Chemistry and Physics of Lipids*. 2019;224:104794. doi:10.1016/j.chemphyslip.2019.104794.
62. Patil P, Patil V, Paradkar A. Formulation of a self-emulsifying system for oral delivery of simvastatin: in vitro and in vivo evaluation. *Acta Pharmaceutica*. 2007;57(1):111–22. doi:10.2478/v10007-007-0009-5.
63. Keraliya RA, Patel C, Patel P, Keraliya V, Soni TG, Patel RC, Patel MM. Osmotic drug delivery system as a part of modified release dosage form. *International Scholarly Research Notices*. 2012;2012(1):528079.
64. Staples M, Daniel K, Cima MJ, Langer R. Application of micro- and nano-electromechanical devices to drug delivery. *Pharmaceutical Research*. 2006;23(5):847–63. doi:10.1007/s11095-006-9906-4.
65. Wu J, Zhang Z, Gu Jg, Zhou W, Liang X, Zhou G et al. Mechanism of a long-term controlled drug release system based on simple blended electrospun fibers. *Journal of Controlled Release*. 2020;320:337–46. doi:10.1016/j.jconrel.2020.01.020.
66. Cun D, Jensen DK, Maltesen MJ, Bunker M, Whiteside P, Scurr D et al. High loading efficiency and sustained release of siRNA encapsulated in PLGA nanoparticles: quality by design optimization and characterization. *European*

Journal of Pharmaceutics and Biopharmaceutics. 2011;77(1):26–35. doi:10.1016/j.ejpb.2010.11.008.
67. Kalam MA, Iqbal M, Alshememry A, Alkholief M, Alshamsan A. Development and evaluation of chitosan nanoparticles for ocular delivery of tedizolid phosphate. *Molecules*. 2022;27(7). doi:10.3390/molecules27072326.
68. Patidar K, Panwar U, Vuree S, Sweta J, Sandhu MK, Nayarisseri A et al. An in silico approach to identify high affinity small molecule targeting m-TOR inhibitors for the clinical treatment of breast cancer. *Asian Pacific Journal of Cancer Prevention.*. 2019;20(4):1229–41. doi:10.31557/apjcp.2019.20.4.1229.
69. Kiptoo P, Calcagno AM, Siahaan TJ. Physiological, biochemical, and chemical barriers to oral drug delivery. *Drug Delivery*. 2016;19:19–34.
70. Zheng Y, Li M, Weng B, Mao H, Zhao J. Exosome-based delivery nanoplatforms: next-generation theranostic platforms for breast cancer. *Biomaterials Science*. 2022;10(7):1607–25. doi:10.1039/D2BM00062H.
71. Toussaint B, Hillaireau H, Cailleau C, Ambroise Y, Fattal E. Stability, pharmacokinetics, and biodistribution in mice of the EPAC1 inhibitor (R)-CE3F4 entrapped in liposomes and lipid nanocapsules. *International Journal of Pharmaceutics*. 2021;610:121213. doi:10.1016/j.ijpharm.2021.121213.
72. Min Y, Suminda GGD, Heo Y, Kim M, Ghosh M, Son YO. Metal-based nanoparticles and their relevant consequences on the cytotoxicity cascade and induced oxidative stress. *Antioxidants (Basel)*. 2023;12(3). doi:10.3390/antiox12030703.
73. Chen ER, Kaiser PK. Therapeutic potential of the ranibizumab port delivery system in the treatment of AMD: evidence to date. *Clinical Ophthalmology*. 2020;14:1349–55. doi:10.2147/opth.S194234.
74. Nagai N, Kaji H, Onami H, Katsukura Y, Ishikawa Y, Nezhad ZK et al. A platform for controlled dual-drug delivery to the retina: protective effects against light-induced retinal damage in rats. *Advanced Healthcare Materials*. 2014;3(10):1555–60. doi:10.1002/adhm.201400114.
75. Jo DH, Song DW, Cho CS, Kim UG, Lee KJ, Lee K et al. CRISPR-Cas9-mediated therapeutic editing of Rpe65 ameliorates the disease phenotypes in a mouse model of Leber congenital amaurosis. *Science Advances*. 2019;5(10):eaax1210. doi:10.1126/sciadv.aax1210.
76. Boenigk J, Beisser D, Zimmermann S, Bock C, Jakobi J, Grabner D et al. Effects of silver nitrate and silver nanoparticles on a planktonic community: general trends after short-term exposure. *PlOs ONE*. 2014;9(4):e95340. doi:10.1371/journal.pone.0095340.
77. Zhang L, Yao L, Zhao F, Yu A, Zhou Y, Wen Q et al. Protein and peptide-based nanotechnology for enhancing stability, bioactivity, and delivery of anthocyanins. *Advanced Healthcare Materials*. 2023;12(25):2300473. doi:10.1002/adhm.202300473.
78. Thomas MA, Morlock R, Dart C, Howard B. Sexual satisfaction results with the vaginal pH modulator from the phase 3 AMPOWER study. *Journal of Sexual Medicine*. 2022;19(6):975–82. doi:10.1016/j.jsxm.2022.03.221.
79. Göppert TM, Müller RH. Protein adsorption patterns on poloxamer-and poloxamine-stabilized solid lipid nanoparticles (SLN). *European Journal of Pharmaceutics and Biopharmaceutics*. 2005;60(3):361–72.
80. Kim K, Choi H, Choi ES, Park MH, Ryu JH. Hyaluronic acid-coated nanomedicine for targeted cancer therapy. *Pharmaceutics*. 2019;11(7). doi:10.3390/pharmaceutics11070301.
81. Jiao M, Jing L, Wei X, Liu C, Luo X, Gao M. The Yin and Yang of coordinating co-solvents in the size-tuning of Fe3O4 nanocrystals through flow synthesis. *Nanoscale*. 2017;9(47):18609–12.
82. Wu B, Sun X, Gupta HB, Yuan B, Li J, Ge F et al. Adipose PD-L1 Modulates PD-1/PD-L1 checkpoint blockade immunotherapy efficacy in breast cancer. *Oncoimmunology*. 2018;7(11):e1500107. doi:10.1080/2162402x.2018.1500107.
83. Liu C, Liu XN, Wang GL, Hei Y, Meng S, Yang LF et al. A dual-mediated liposomal drug delivery system targeting the brain: rational construction, integrity evaluation across the blood-brain barrier, and the transporting mechanism to glioma cells. *International Journal of Nanomedicine*. 2017;12:2407–25. doi:10.2147/ijn.S131367.
84. Ali RO, Quinn GM, Umarova R, Haddad JA, Zhang GY, Townsend EC et al. Longitudinal multi-omics analyses of the gut-liver axis reveal metabolic dysregulation in hepatitis C infection and cirrhosis. *Nature Microbiology*. 2023;8(1):12–27. doi:10.1038/s41564-022-01273-y.
85. Fattore L, Cafaro G, Di Martile M, Campani V, Sacconi A, Liguoro D et al. Oncosuppressive miRNAs loaded in lipid nanoparticles potentiate targeted therapies in BRAF-mutant melanoma by inhibiting core escape pathways of resistance. *Oncogene*. 2023;42(4):293–307. doi:10.1038/s41388-022-02547-9.
86. Park W, Nguyen VP, Jeon Y, Kim B, Li Y, Yi J et al. Biodegradable silicon nanoneedles for ocular drug delivery. *Science Advances*. 2022;8(13):eabn1772. doi:doi:10.1126/sciadv.abn1772.
87. Jain A, Jain SK. Stimuli-responsive smart liposomes in cancer targeting. *Current Drug Targets*. 2018;19(3):259–70. doi:10.2174/1389450117666160208144143.
88. Limpongsa E, Tabboon P, Tuntiyasawasdikul S, Sripanidkulchai B, Pongjanyakul T, Jaipakdee N. Formulation and in vitro evaluation of mucoadhesive sustained release gels of phytoestrogen diarylheptanoids from curcuma comosa for vaginal delivery. *Pharmaceutics*. 2023;15(1):264.
89. Kompella UB, Kadam RS, Lee VH. Recent advances in ophthalmic drug delivery. *Therapeutic Delivery*. 2010;1(3):435–56. doi:10.4155/tde.10.40.
90. Jones AR, Shusta EV. Blood–brain barrier transport of therapeutics via receptor-mediation. *Pharmaceutical Research*. 2007;24(9):1759–71. doi:10.1007/s11095-007-9379-0.
91. Yuan M, Han Z, Liang Y, Sun Y, He B, Chen W et al. mRNA nanodelivery systems: targeting strategies and administration routes. *Biomaterials Research*. 2023;27(1):90. doi:10.1186/s40824-023-00425-3.
92. Arowosegbe MA, Amusan OT, Adeola SA, Adu OB, Akinola IA, Ogungbe BF et al. kaempferol as a potential PAK4 inhibitor in triple negative breast cancer: extra precision glide docking and free energy calculation. *Current Drug Discovery Technologies*. 2020;17(5):682–95. doi:10.2174/1570163816666190823135948.
93. Rohan LC. *Long Acting Film Technology for Contraception and HIV Prevention (LATCH)*. NIH/NIAID R61 Grant, 2022.
94. Dholakia J, Prabhakar B, Shende P. Strategies for the delivery of antidiabetic drugs via intranasal route. *International Journal of Pharmaceutics*. 2021;608:121068. doi:10.1016/j.ijpharm.2021.121068.
95. Ugur Yilmaz C, Emik S, Orhan N, Temizyurek A, Atis M, Akcan U et al. Targeted delivery of lacosamide-conjugated gold nanoparticles into the brain in temporal lobe epilepsy in rats. *Life Sciences*. 2020;257:118081. doi:10.1016/j.lfs.2020.118081.

96. Song D, Liu P, Shang K, Ma Y. Application and mechanism of anti-VEGF drugs in age-related macular degeneration. *Frontiers in Bioengineering and Biotechnology*. 2022;10:943915. doi:10.3389/fbioe.2022.943915.
97. Ajdary M, Moosavi MA, Rahmati M, Falahati M, Mahboubi M, Mandegary A et al. Health concerns of various nanoparticles: a review of their in vitro and in vivo toxicity. *Nanomaterials*. 2018 doi:10.3390/nano8090634.
98. Duan L, Li X, Ji R, Hao Z, Kong M, Wen X et al. Nanoparticle-based drug delivery systems: an inspiring therapeutic strategy for neurodegenerative diseases. *Polymers (Basel)*. 2023;15(9). doi:10.3390/polym15092196.
99. Mostafa M, Al Fatease A, Alany RG, Abdelkader H. Recent advances of ocular drug delivery systems: prominence of ocular implants for chronic eye diseases. *Pharmaceutics*. 2023;15(6). doi:10.3390/pharmaceutics15061746.
100. Bowman C, Ma F, Mao J, Plise E, Chen E, Liu L et al. Evaluation of bottom-up modeling of the blood–brain barrier to improve brain penetration prediction via physiologically based pharmacokinetic modeling. *Biopharmaceutics & Drug Disposition*. 2023;44(1):60–70. doi:10.1002/bdd.2344.
101. Kauffman KJ, Dorkin JR, Yang JH, Heartlein MW, DeRosa F, Mir FF et al. Optimization of lipid nanoparticle formulations for mRNA delivery in vivo with fractional factorial and definitive screening designs. *Nano Letters*. 2015;15(11):7300–6. doi:10.1021/acs.nanolett.5b02497.
102. Lim J, Yeap SP, Che HX, Low SC. Characterization of magnetic nanoparticle by dynamic light scattering. *Nanoscale research letters*. 2013;8(1):3.
103. Sýkora D, Kašička V, Mikšík I, Řezanka P, Záruba K, Matějka P et al. Application of gold nanoparticles in separation sciences. *Journal of Separation Science*. 2010;33(3):372–87.
104. Pan H, Myerson JW, Hu L, Marsh JN, Hou K, Scott MJ et al. Programmable nanoparticle functionalization for in vivo targeting. *Faseb Journal*. 2013;27(1):255–64. doi:10.1096/fj.12-218081.
105. Birnbaum Y, Bajaj M, Yang HC, Ye Y. Combined SGLT2 and DPP4 Inhibition Reduces the activation of the Nlrp3/ASC inflammasome and attenuates the development of diabetic nephropathy in mice with type 2 diabetes. *Cardiovascular Drugs and Therapy*. 2018;32(2):135–45. doi:10.1007/s10557-018-6778-x.
106. Jamshidnejad-Tosaramandani T, Kashanian S, Karimi I, Schiöth HB. Synthesis of an insulin-loaded mucoadhesive nanoparticle designed for intranasal administration: focus on new diffusion media. *Frontiers in Pharmacology*. 2023;14:123456.
107. Camussi G, Deregibus MC, Bruno S, Grange C, Fonsato V, Tetta C. Exosome/microvesicle-mediated epigenetic reprogramming of cells. *American Journal of Cancer Research*. 2011;1(1):98–110.
108. Rao L, Xia S, Xu W, Tian R, Yu G, Gu C et al. Decoy nanoparticles protect against COVID-19 by concurrently adsorbing viruses and inflammatory cytokines. *Proceedings of the National Academy of Sciences*. 2020;117(44):27141–7. doi:doi:10.1073/pnas.2014352117.
109. Huang L, Li Y, Du J, Li H, Lu M, Wang Y et al. The prognostic value of retraction clefts in chinese invasive breast cancer patients. *Pathology & Oncology Research*. 2021;27:1609743. doi:10.3389/pore.2021.1609743.
110. Crook ZR, Girard E, Sevilla GP, Merrill M, Friend D, Rupert PB et al. A TfR-binding cystine-dense peptide promotes blood–brain barrier penetration of bioactive molecules. *Journal of Molecular Biology*. 2020;432(14):3989–4009. doi:10.1016/j.jmb.2020.04.002.
111. Farokhzad OC, Langer R. Impact of nanotechnology on drug delivery. *ACS Nano*. 2009;3(1):16–20. doi:10.1021/nn900002m.
112. Sainz V, Conniot J, Matos AI, Peres C, Zupancic E, Moura L et al. Regulatory aspects on nanomedicines. *Biochemical and Biophysical Research Communications*. 2015;468(3):504–10. doi:10.1016/j.bbrc.2015.08.023.
113. Shukla AA, Thömmes J. Recent advances in large-scale production of monoclonal antibodies and related proteins. *Trends Biotechnology*. 2010;28(5):253–61. doi:10.1016/j.tibtech.2010.02.001.
114. Mazayen ZM, Ghoneim AM, Elbatanony RS, Basalious EB, Bendas ER. Pharmaceutical nanotechnology: from the bench to the market. *Future Journal of Pharmaceutical Sciences*. 2022;8(1):12. doi:10.1186/s43094-022-00400-0.
115. Shah RM, Malherbe F, Eldridge D, Palombo EA, Harding IH. Physicochemical characterization of solid lipid nanoparticles (SLNs) prepared by a novel microemulsion technique. *Journal of Colloid and Interface Science*. 2014;428:286–94. doi:10.1016/j.jcis.2014.04.057.

Index

Note: **Bold** page numbers refer to tables and *italic* page numbers refer to figures.

A

AAV, *See* Adeno-associated virus (AAV)
Absorption, distribution, metabolism, and excretion (ADME)
  on drug levels, 117
  of emerging nanomaterials, 55
  properties of drugs, 34
Active efflux transporters (AETs), 87
Active pharmaceutical ingredient (API), 11–13, 98
Acute myeloid leukemia with myelodysplasia-related changes (AML-MRC), 54–55
ADCs, *See* Antibody-drug conjugates (ADCs)
Adeno-associated virus (AAV), 129
Adenoviral systems, 89
Adenovirus (Ad) vectors, 128–129
ADME, *See* Absorption, distribution, metabolism, and excretion (ADME)
AETs, *See* Active efflux transporters (AETs)
Albumin, 99–100
Albumin-based nanoparticles, 99, 100
Alginate, 99
Alternating copolymer, 95
Amphiphilic compounds, 9
Antibodies, 25, **26**, **27**, 115
Antibody-based targeting, 67
Antibody-drug conjugates (ADCs), 25, 26, 49
Antibody-guided avidin-nucleic-acid nanoassemblies (ANANAS), 49
Antibody-loaded magnetic nanoparticles, 41
Anti-epidermal growth factor receptor (EGFR) antibody, 49
Anti-retroviral drugs, conventional, 39
Aperiodic copolymer, 95
API, *See* Active pharmaceutical ingredient (API)
Aptamers, 50
Arginyl-glycyl-aspartic acid (RGD) peptide, 50
ART-loaded SLNs, 140
  angle of repose of lyophilized, 142–143
  Carr's index of lyophilized, 143
  entrapment efficiency of, 141–142
  flow properties of lyophilized solid, 142
  Hausner's ratio of lyophilized, 143
  initial RAM for, **145**
  in-vivo studies of prepared formulations, 144
  morphology of, 141
  particle size analysis of lyophilized, 142
  particle size distribution, 149
  permeability studies of prepared formulations, 143
  pharmacokinetics and biostatistics, 144
  pH of, 149
    determination, 142
  physicochemical characterization of, 141–144
  porosity of lyophilized, 143
  sample preparation, 144
  in vitro release studies of, 143
Astaxanthin (AST), 99
Azidothymidine 50-triphosphate (AZTTP) liposomes, 71
Azobenzene groups, reversible photoisomerization of, 52

B

BBB, *See* Blood-brain barrier (BBB)
BCS, *See* Biopharmaceutics Classification System (BCS)
Bergamot essential oil (BEO), 39
Bevacizumab, 56
Biocompatibility, 102
Biocompatible polymers, 46, 98
Biocompatible therapies, 26
Bioconjugation techniques, 164
Biodegradable polymer, 46
  nanoparticles, 41, 102
Biodegradable synthetic polymer, 42
Biodegradable systems, 102
Biodegradation, 102–103
Biological ligands, for targeting nanodrug carriers, *50*
Biomarkers, for drug delivery systems, **44**
Biomaterials, for control release, 41–42
  hydrogels, 43
  polymer, 42
  types, 42–43
Biopharmaceutics Classification System (BCS), 112
Block copolymer, 95
Blood-brain barrier (BBB), 30, 49, 65
  physical disruption of, 68–72
Brain drug delivery, 73, 86–88, *88*
Brain-invasive medicine delivery, 87
Brain-targeted delivery systems, 74
Branched polymer, 94–95

C

Cabazitaxel (CTX) loaded nanoparticles, 100
Camptothecin (CPT), 98–99
Cancer, 38
Carbon-based nanoparticles, 2, **3**
Carbon nanostructures, **156**
Carbon nanotubes (CNTs), 30, 169, 187–188
Cardiac fibrosis, 39
Cardiology, nanomedicine in, 164
Cardiovascular illnesses, 39
Carr's index
  of lyophilized ART-loaded SLNs, 143
  of lyophilized SLNs, 150
Cationic lipids, 81, 130
Cavitation-induced forces, 51
CDDs, advanced, 82–83
CED systems, *See* Convection-enhanced delivery (CED) systems
Cell-penetrating peptides (CPPs), 51, 67, 132
Cell-surface protein, 90
Cell therapy, nanotechnology-mediated, 39
Central nervous system (CNS)
  homeostasis, 65
  noval intranasal drug delivery to target, 84–88
Ceramic nanoparticles, **3**
Cerium oxide nanoparticles, 177
Chemokine receptors (CKR), 90
Chimeric antigen receptors (CARs), T cells with, 25
Chitosan (CS), 99, 132
Cholesterol, 5
CKR, *See* Chemokine receptors (CKR)
CNTs, *See* Carbon nanotubes (CNTs)
Collagen, 100
Colon, drug absorption mechanism in, 82
Colonic assessment
  anatomy, 110
  colonic epithelium, 111–112
  microbiota, 112–113
  mucosal fluid and gas, 112
  transit time, 111
  vascular and lymphatic system, 111
Colonic drug delivery systems, 81–82, 108, 113–116
  factors influencing oral
    gastric emptying and emptying time, 110
    gastrointestinal mucosal barrier, 110
    pH, 109–110
  route of administration for, 109, *109*
Colonic epithelium, 111–112
Colonic fluid, 112
Colonic fluid time, 82
Colonic luminal viscosity, 82
Colonic microbial ecosystem, 110
Colonic transit time, 82
Colon-targeted drugs, 113
Colorectal cancer, 108
Combined drug delivery approach, 24
Complement activation-related pseudoallergy (CARPA), 54–55
Computed tomography (CT), 44–45
Confocal laser scanning microscope (CLSM), 15
Controlled drug delivery systems
  advantages, 105
  disadvantages, 105
  immunogenicity, 103
  natural polymers in, 98–100
  particle size, 104
  polymers and polymeric nanoparticles for, 98
  solubility, 103
  stability, 103–104
  surface charge, 104
  surface modification, 104–105
  synthetic polymers in, 100–101
Controlled-release drug delivery system, 21
Controlled-release mechanisms, proteins and peptides nanoparticles, 192
Controlled-release profiles, 191
Convection-enhanced delivery (CED) systems, *68*, 68–69

Convection-enhanced diffusion (CED), 87
Conventional anti-retroviral drugs, 39
Conventional drug delivery systems, 1
Copolymer, 95
  mesoporous block, 48
  MMP-cleavable peptide-linked triblock, 47
COVID-19, 38–39
CPPs, *See* Cell-penetrating peptides (CPPs)
Cremophor EL, 100
Critical packing parameter (CPP), 9
Critical quality attributes (CQAs), 140
Crohn's disease (CD), 81, 108
Crosslinked polymer, 95
Cryoprotectants, 192
CTDDS, 82
Cubosomes, polymeric, 48
^{64}Cu-radiolabeled liposomes, 46
Curcumin-SLN microneedles, 8
Cyclodextrin (CD), 98–99, 131
Cytotoxic chemical DM1, 25

D

DDSs, *See* Drug distribution strategies (DDSs)
Degradation-controlled release, 190, 192
Dementia, 39
Dendrimers, *4*, 5, 14, 39, 47–48, 88, 189
  applications of, 6
  characteristics and applications, **6**
  morphological characterization of, 14
  PAMAM, **6**, 131
  phosphorus, **6**
  protein/peptide drug delivery systems, 187
  structural characterization of, 14
Dermatology, nanomedicine in, 165
Design of experiments (DoE), 117
  for SLNs, 145
Diffusion-controlled release, 190, 192
Dipalmitoylphosphatidylcholine (DPPC), 5
DLS, *See* Dynamic light scattering (DLS)
DNA-based therapeutics, 125
Donezepyl (DNZ) loaded nanoparticles, 101
Doxil, 22
Doxorubicin (DOX), 50, 99
  PEGylated liposomal, 22
Droplet microfluidics, 161
Drug absorption, 71, 81
  mechanism in colon, 82
Drug administration, different routes of, 168
Drug delivery, 28–29
  ophthalmic, 10
  routes of, 83
  traditional systems of, 1
Drug delivery principles, 25
  drug modifications, 25–28
  microenvironmental modifications, 28
Drug delivery system (DDS), 1, 34–35
  biomarkers for, **44**
  challenges in, 29–30, 35
  classes of therapeutic and delivery challenges, 21–22
  conventional, 1
  evolution of, 21, *22*, 78
  nanotechnology in, 35, **36–37**, *41*
    cancer, 38
    cardiovascular, 39
    COVID-19, 38–39
    dementia, 39
    eye disease, 40
    human immunodeficiency virus, 39
    lung disease, 38
    obesity, 40
    orthopedic and prosthetic, 40
    pain control, 40
    reproductive disorders, 40
    urology, 40
  physicochemical properties of, 7
  polymeric, 46–48
  regulatory considerations in, 53–54
  small molecules, 22
  targeted, 49–53
Drug distribution strategies (DDSs), 48
Drug loading and release, 163
Drug modifications, 25–28
Drug nanocarriers drug delivery systems, hyaluronic acid-based, 23
Drug PKs, 21–22
Drug-polymer interactions, 191
Drug release mechanisms, 21, 82, 113, 172
Drug solubility, 38, 43, 112
Drug susceptibility, 24
Dry eye disease, 40
Dynamic light scattering (DLS), 13, 193

E

Elastomers, 96
Electroporation, **7**, 71, 132
Electrostatic interaction, 190
ELISA, *See* Enzyme-linked immunosorbent assay (ELISA)
Emulsification-solvent evaporation, 191
Encapsulation method, 191
Endogenous enzymes, localized, 35
Energy-dispersive X-ray spectroscopy (EDS), 176
Entrapment efficiency (EE)
  of ART-loaded SLNs, 141–142
  of SLNs, 147, 149
Environmental modifications, 28
Enzymatic degradation, protection from, 192–193
Enzyme-linked immunosorbent assay (ELISA), 16
Enzyme-mediated drug release, 53
Enzyme-sensitive systems, 53
Enzyme sensitivity, 191
Epidermal growth factor receptor (EGFR) expression, 174
Epigenetic modulators, 38
Ethanol in lipid bilayers, 10
Ethosomal drug delivery systems, 11
Ethosomes, 10, *10*
  advantages and disadvantages of, 11
  applications of, 11
  morphological characterization of, 15
  structure and composition of, 10–11, 15
  and transferases, 15
Eudragit® S, 83, 108
Exosomes, 81, 132
Exparel (liposomal bupivacaine), 40
Eye disease, 40

F

FA, *See* Folic acid (FA)
Fabrication methods of nano-systems, **157**
Fatty acids, stearic and/or palmitic, 5
Fibrosis, cardiac, 39
Fluorescence imaging, 45
Fluorescence molecular tomography (FMT), 45
Fluorescence resonance energy transfer (FRET), 45
5-Fluorouracil (5-FU), 98
FMT, *See* Fluorescence molecular tomography (FMT)
Focused ultrasound-mediated drug delivery, 69, *70*
Folate-modified self-microemulsifying drug delivery systems (FSMEDDS), 115
Folic acid (FA), 100, 115
Folic acid-loaded dendrimer-functionalized selenium nanoparticles (FA-PAMAM-SeNPs), 47
Fourier transform infrared (FTIR) spectroscopy, 14
Franz cell diffusion assay, 16
Franz diffusion cell technique, 150
FRET, *See* Fluorescence resonance energy transfer (FRET)

G

GalNAc–siRNA conjugate, 26
Gastrointestinal diseases, 108
Gastrointestinal mucosal barrier, 110
Gastrointestinal tract (GIT), 110
GDIs, *See* Glaucoma drainage implants (GDIs)
Gene delivery
  nanoparticle-based, 39
  viral, 88
Generally recognized as safe (GRAS), 113
Gene therapy, 80
  targeted, 56
GIT, *See* Gastrointestinal tract (GIT)
Glaucoma drainage devices, 85
Glaucoma drainage implants (GDIs), 85, *87*
Glycerophospholipids, 5
Glycodendrimers, **6**
Glycol chitosan (GC), 48
Glycoprotein transferrin (Tf), 49
GNPs, *See* Gold nanoparticles (GNPs)
GOD-loaded nanoreactors, 47
Gold nanoparticles (GNPs/AuNPs), 38, 43, 81, 171–174, *174*, 185–186
Gradient copolymer, 95
Grafted copolymer, 95
Green synthesis of metal and metal oxide nanoparticles, 171, *173*, **173**

H

HA, *See* Hyaluronic acid (HA)
Hausner's ratio
  of lyophilized ART-loaded SLNs, 143
  of lyophilized SLNs, 150
Health Research Authority (HRA), 35
HeLa cells, 48
Hematite, 175, *175*
Herpes simplex viruses, 129
Hexagonal boron nitride nanosheet drug delivery system, 23
High-density polyethylene (HDPE), 94
High-intensity focused ultrasound (HIFU), 51
High-performance liquid chromatography (HPLC), 15
Homopolymer, 95
Human immunodeficiency virus (HIV), 39
Human reproductive tract-related functions, 40
Human serum albumin (HSA), 99
Hyaluronic acid (HA), 49–50, 115–116

Hyaluronic acid-based drug nanocarriers drug delivery systems, 23
Hyaluronidase 1 (Hyal-1), 50
Hybrid hydrogels, 11
Hydrogels, 11, 16, 43, 188–190
advantages and disadvantages of, 11
in vivo study for, 16
nanosized, 48
nanotechnology-based drug delivery systems, 66
structural characterization of, 16
synthesis of, 11
Hydrophilic-lipophilic balance (HLB), 9
Hydrophobic drugs, 8
Hydrophobic/nonpolar molecules, 4
Hydrothermal and solvothermal synthesis, MONP, 169, *169*

I

Imaging techniques in drug delivery, 43–46
computed tomography, 44–45
fluorescence imaging, 45
intravital microscopy, 45
MRI (magnetic resonance imaging), 44
nuclear imaging, 45
Immediate release (IR), 21
Immunogenicity, 103
Inflammatory bowel disease (IBD), 81, 108
Innovative drug delivery systems, 1–2, 168
Innovative nanomaterials, 163
Inorganic nanoparticles, **3**, 43, 163, 188
In silico for studying, 117
In situ gel drug delivery system, 23–24
Integrated Research Application System (IRAS), 35
Intellectual property (IP), 55
Interleukin-4 receptor (IL-4Receptor)-binding peptide-1 (IL4RPep-1), 50
Intracranial implants, 87
Intranasal drug delivery systems, 67–68
Intravital microscopy, 45
In vitro for studying, 116–117
In vitro permeation studies, TDDS, 14–15
In vivo animal models for studying, 117
In vivo behavior of nanomaterials, 43
In vivo study for hydrogels, 16
IONPs, 176
Iontophoresis, **7**
Iron oxide nanoparticles, *175*, 175–177

K

Ktoprofen–cyclodextrin complexes, 15

L

Lactoferrin receptor, 67
Laser ablation, 169, 174
Laser ablation-based nanoparticle production, *170*
Laser light scattering (LLS), 14
Lasers, 71
Lentivirus, 130
Ligand-receptor-mediated drug delivery system, 115–116
Ligands, 38
attachment, 192
for nanoparticle drug delivery, 49–50
Light-triggered drug delivery, 51–52
Linear polymer, 94
Lipid-based delivery system, 115
Lipid-based nanocarriers, 42, 43, 163
Lipid-based nanoparticles, 3, **3**, 40
Lipid-based vesicular systems, 168
Lipid bilayers, ethanol in, 10
Lipid nanoparticles (LNP), 28, 80, 130, 185, 188
Lipids
natural, 5
synthetic, 4
Lipinski's rule of five, 25
Lipophilic medication pinacidil monohydrate, 14
Liposomal doxorubicin, PEGylated, 22
Liposomal drug delivery systems, 168
Liposome-based DDS, 12
Liposome nanoparticles (LNPs), 80
Liposomes, 3–4, *4*, 14
advantages of, 4
applications of, 5
^{64}Cu-radiolabeled, 46
formation of, 4
nanotechnology-based drug delivery systems, 66
PEGylated, 34
structure and composition of, 4–5
thermosensitive, 51
Liposomes/SLNs, 13
morphological characterization of, 12–13
structural characterization of, 13
Live-cell therapy, 25, **26**, **27**
Liver-cell-therapies, **27**
LLS, *See* Laser light scattering (LLS)
LNP, *See* Lipid nanoparticles (LNP)
Localized brain drug delivery, 74
Localized endogenous enzymes, 35
Low-density polyethylene (LDPE), 95
Lung disease, 38
Lupron Depot (LD), 189
Lymphatic system, 111
Lyophilization, 192
Lyophilized ART-loaded SLNs, 142–143
Lyophilized solid lipid nanoparticles
angle of repose of, 150
bulk density of, 150
Carr's Index of, 150
Hausner's ratio of, 150
in vitro drug release of, 150
particle size distribution of, 149
shape of, 150
tapped density of, 150

M

Machine learning (ML), 117
Maghemite, 175, *175*
Magnetic nanoparticles (MNPs), **156**, 176
antibody-loaded, 41
drug delivery system, 72
protein/peptide drug delivery systems, 188
Magnetic resonance-guided focused ultrasound (MRgFUS), 69
Magnetic resonance imaging (MRI), 44
Magnetite, 175, *175*
Magnetophoresis, **7**, 71–72
MDR, *See* Multidrug resistance (MDR)
Membrane-coated erythrocyte nano-formulations, 23
Mertansine, 25
Mesenteric lymph nodes (MLNs), 111
Mesoporous block copolymers, 48
Mesoporous silica nanoparticles (MSNPs), 52, 186
Messenger RNA (mRNA) vaccine development, 124
Metal-based nanoparticles, 3, **3**
challenges and limitations, 178
green synthesis of, 171, *173*
types, 171–178
Metal oxide nanoparticles (MONPs), 168
challenges and limitations, 178
comparative analysis of different, 178
green synthesis of, 171, *173*
synthesis methods of, 168–171
types, 171–178
Micelles, polymeric, 47
Microbial ecosystem, colonic, 110
Microbiota, colonic assessment, 112–113
Microbiota-triggered systems, pH-triggered with, 113–114
Microenvironmental tactics, 28
Microfluidic liquid manipulation techniques, 161
Microfluidics technology, 160
Microfluidizer technology, 160
Microneedles (MNs), **7**, 69–71
Microparticle system, polymer nanoparticle and, 114–115
Microspheres, *8*, 9, 15, 189
advantages and disadvantages of, 9
applications of, 10
morphological characterization of, 15
structure and types of, 9
vaccine-containing, 10
Microwave-assisted synthesis, MONP, 171, *172*
MLNs, *See* Mesenteric lymph nodes (MLNs)
MMP-cleavable peptide-linked triblock copolymer, 47
MNs, *See* Microneedles (MNs)
Molecular dynamics (MD), 117
Molecular-level functional systems engineering, 155
Moloney murine leukemia virus, 89
Monoclonal antibody treatment, 56
Monomers, 94, *94*, 95
Mononuclear phagocytic system (MPS), 97
MONPs, *See* Metal oxide nanoparticles (MONPs)
Mouth-dissolving tablets, 11
MPS, *See* Mononuclear phagocytic system (MPS)
MRI, *See* Magnetic resonance imaging (MRI)
MSNs, *See* Mesoporous silica nanoparticles (MSNs)
Mucosal barrier, gastrointestinal, 110
Multidrug resistance (MDR), 6
Multifunctional nanocarriers, 163–164
Mycapssa®, 113

N

Nananoparticles, polymeric, 46
Nanocapsules, 97
Nanocarriers
drugs, **158**
lipid-based, 42, 43
multifunctional, 163–164
Nanoceria, *See* Cerium oxide nanoparticles
Nanodrug carriers, biological ligands for targeting, *50*
Nano drugs, 3
Nanoemulsions, perfluorocarbon, 51

Nanoformulations, 14, 39
Nanomaterials
and biomedical applications, **156**
innovative, 163
in vivo behavior of, 43
Nanomedicine (NNM), 38, 40, 54
advanced preclinical models and imaging techniques in, 46
advancement of, 45
approved for market, **56–57**
in cardiology, 164
to clinical settings, 54–55
clinical transition of, 56
commercialization of, 55
in dermatology, 165
development through translation approach, 55–56
feasibility and challenges in scale-up technology of, 162–164
formulation, intricacy, 55
future perspective on, 164–165
imaging techniques for development of, **45**
nanoparticles technology and, 155–156
in neurology, 164–165
production, translational imaging and scale-up, 46
regulatory guidelines for, 162
regulatory supervision of, 55
safety concerns regarding, 165
scale-up technologies of, 156
from small to large preclinical models, 45–46
Nanoparticles (NPs), 2–3, 12
carbon-based, 2
delivery systems, clinical applications, 193–194
dispersion, 141
drug delivery
ligands, 49–50
recent advancements, 3
formulations of fentanyl citrate, 40
gene delivery, 39
gold, 38
inorganic, 43
lipid-based, 3
metal-based, 3
metal oxide, 168
morphological characterization of, 12–13
as nanomedicine, 155
polymeric, 3, 43
protein/peptide drug delivery systems by, 78, *79*
semiconductor, 3
technology and nanomedicine, 155–156
types, characteristics, and advantages, **3**
Nanoparticulate Drug Delivery Systems (NDDSs), 160
Nanoparticulate system, surface charge of, 104
Nanoplatforms, 38
Nano precipitation, 192
Nanoreactors (NRs), 47
Nanoscale metal–organic frameworks (nmofs), **156**
Nanosized hydrogels, 48
Nanospheres, 97
Nano-systems, fabrication methods of, **157**
Nanotechnology, 2, 35, 85, 155
in drug delivery systems, 35, **36–37**, *41*
cancer, 38
cardiovascular, 39
COVID-19, 38–39
dementia, 39
eye disease, 40
human immunodeficiency virus, 39
lung disease, 38
obesity, 40
orthopedic and prosthetic, 40
pain control, 40
reproductive disorders, 40
urology, 40
in ophthalmology, 165
in regenerative medicine, 165
Nanotechnology-based drug delivery systems, 65, 184
hydrogels, 66
liposomes, 66
polymeric micelles, 66
polymeric nanoparticles, 65–66
Nanotechnology-mediated cell therapy, 39
Nasal anatomy physiology, 83–84
NASAL drug delivery, 84
Nasopulmonary drug delivery system, 83
National Centre for Advancing Translational Sciences (NCATS), 35
National Institute for Health Research (NIHR), 35
Natural lipids, 5
Natural phenotypic shift, 28
Natural polymeric materials, 9
Natural polymers, 96
in controlled drug delivery systems, 98–100
NCEs, *See* New chemical entities (NCEs)
NDDS, *See* Novel drug delivery systems (NDDS)
Near-infrared fluorescence (NIRF) imaging, 45
Network polymer, 95
Neural tissue engineering, 72–73
Neurodegenerative diseases, 39
Neurology, nanomedicine in, 164–165
Neurovascular unit (NVU), 65
New chemical entities (NCEs), 53
Niosomes (NS), *8*, 8–9, 15
advantages and disadvantages of, 9
applications of, 9
composition of, 9
morphological characterization of, 15
structural characterization of, 15
structure of, 9
NNM, *See* Nanomedicine (NNM)
Nonionic surfactants, 9
Nonviral gene therapy, 80
Non-viral vectors, 130–132
Nose-to-brain (N2B) drug delivery systems, 71
Noval intranasal drug delivery to target CNS, 84–88
microemulsion, 84
microsphere, 84
nanoparticles, 84
nasal in situ gel, 84
ocular drug delivery, 84–85, *85*, *86*
Novel drug delivery systems (NDDS)
advancements in, 1
advantages, 2
biochemical mechanism, 1
characterization of, 11–16, *12*
ideal characteristics of, 2
limitations, 2
mechanisms employed in, 1
need for, 1–2
physical mechanism, 1
recent developments in, 2–6
nanoparticles, 2–3
types, *2*
NPs, *See* Nanoparticles (NPs)
NRs, *See* Nanoreactors (NRs)
Nuclear imaging, 45
Nucleic acid, **26**, **27**
Nucleic acid-based therapeutics, 124–125
clinical studies using, **126–127**
delivery challenges, 125
delivery system, 78–81, 125
non-viral vectors, 130–132
physical methods, 132
viral vectors, 125, 128–130
DNA-based therapeutics, 125
RNA-based therapeutics, 125
Nucleic acid delivery, viral and non-viral vectors for, *129*
Nucleic acids, 25
Nucleic acid therapies, FDA and EMA approved, **133–134**
Nucleoside modifications, 28

O

Obesity, 40
OCDS, *See* Oral colonic drug delivery systems (OCDS)
Ocular drug delivery, 84–85, *85*, *86*
Oligomers, 94
Ophthalmic drug delivery, 10
Ophthalmology, nanotechnology in, 165
Opticore, 83
OPTICORE™ system, 113–114
Oral colonic drug delivery systems (OCDS), 109
Organic nanoparticles, **3**
Orthopedic and prosthetic, 40
Osmotic pressure-driven release, 190
Outlining quality target product profile (QTPP), 140
and CQAs of SLN, 144

P

Pain control, 40
Palbociclib (PAL), 101
Palmitic fatty acids, 5
PALS, *See* Phase analysis light scattering (PALS)
PAMAM dendrimers, **6**, 131
Parkinson's disease (PD), 67, 74
Particle characteristics, optimization of, 192
PAT, *See* Process analytical technologies (PAT)
PDI, *See* Polydispersity index (PDI)
PDMS (Polydimethylsiloxane), 160–161
PEGylated liposomal doxorubicin, 22
PEGylated liposomes, 34
PEGylation, 192
PEI, *See* Polyethyleneimine (PEI)
Peptides, 14, 24, 50, 116
dendrimer, **6**
and proteins, **26**, **27**
Perfluorocarbon nanodroplets (PFCnDs), 71
Perfluorocarbon (PFC) nanoemulsions, 51
Periodic copolymer, 95
Peripheral nervous system (PNS) damage, 72
Periprosthetic joint infection (PJI), 40
PFCnDs, *See* Perfluorocarbon nanodroplets (PFCnDs)

PGA, *See* Polyglycolic acid (PGA)
pH and microbiota-triggered combination system, 82–83
Pharmaceutical drug delivery system, 1
Pharmacodynamics, 34, 55, 109
Pharmacokinetics (PKs), 21, 34, 113
Phase analysis light scattering (PALS), 13
pH Buffers, 192
Phloral®, 82–83, 113
pH modifiers, 28
Phosphatidylcholine, 6
Phospholipids, 5
  synthetic, 5
Phosphorus dendrimers, **6**
Photodynamic therapy (PDT), 5
Photoresponsive systems, 51
Photothermal therapy (PTT), 5
pH-responsive DDSs, 52
pH-sensitive liposomes, 53
pH sensitivity, 52, 191
pH-triggered and time-triggered systems, 114
pH-triggered with microbiota-triggered systems, 113–114
Phytosomal vesicle (PDE), entrapment efficiency of, 13
Phytosomes, *4*, 5, 13
  applications of, 5–6
  morphological characterization of, 13
  structure and composition of, 5
PJI, *See* Periprosthetic joint infection (PJI)
PKs, *See* Pharmacokinetics (PKs)
PLA, *See* Polylactic acid (PLA)
Plasmonic nanoparticles, **156**
PLGA, *See* Polylactic-co-glycolic acid (PLGA)
Pluripotent stem cells, 25
PMs, *See* Polymeric micelles (PMs)
Poly (β-amino ester) nanoparticles, 52
Polycaprolactone (PCL), 102
Poly (D, l-lactide-co-glycolide) microspheres, 15
Polydispersity index (PDI), 101, 192
Polyethyleneimine (PEI), 131
Polyglycolic acid (PGA), 101
Polylactic acid (PLA), 100–101
Polylactic-co-glycolic acid (PLGA), 101–102
Poly-L-lactide, 100
Polymer, 42, 94
  biodegradable and biocompatible, 46
  chain, 94, *94*
  and lipid nanocarriers, **156**
  natural, 9
  synthetic, 103
Polymeric cubosomes, 48
Polymeric drug delivery system, 46–48, *97*, 98, 103
  dendrimers, 47–48
  nanosized hydrogels, 48
  polymer-based nanoparticles, 46–47
  polymeric cubosomes, 48
  polymeric micelles, 47
  polymersomes, 47
Polymeric micelles (PMs), 47
  nanotechnology-based drug delivery systems, 66
Polymeric nanocarrier, for drug delivery, **48–49**
Polymeric nanoparticles, 3, **3**, 43, 46, 80, 97, 163, 188
  classification of, 97
  future of polymer and, 105–106
  nanotechnology-based drug delivery systems, 65–66
  physicochemical properties of, 97
  protein/peptide drug delivery systems, 185
  surface of, 97
Polymerization, 94
Polymer-lipid hybrid nanoparticles drug delivery system, 23, *23*
Polymer nanoparticle and microparticle systems, 114–115
Polymers classification, 94–97
  arrangement of monomers, 95, *95*
  monomer source, 96–97
  structure of polymer chains, 94–95
  thermal behavior, 95–96, *96*
Polymersomes, 47
Poly(N-isopropyl acrylamide) (PNIPAM), 51
Polysaccharide, 49–50, 98–99, 131–132
Positron emission tomography (PET), 45
PPI dendrimers, **6**
Precision radiotherapy, 38
Process analytical technologies (PAT), 53
Protein, 14, 24, 49–50, 99–100
  nanoparticles, 188
  and peptides, stability, 191
Protein/peptide drug delivery systems, 78, 184
  challenges in delivering, 185
  importance of, 184–185
  nanoparticles in, 78, *79*, 185–188
Protein-peptide nanocarrier systems-I, *186*
Proteins and peptides nanoparticles
  administration routes and targeting strategies, 193
  bioavailability improvement, 192
  clinical applications, 193–194, *194*
  controlled-release mechanisms, 192
  design for stability and bioavailability, 191–192
  evaluation of nanoparticle systems, 193
  future perspectives and challenges, 195
  in vivo stability, 192
  optimization of particle characteristics, 192
  regulatory and manufacturing considerations, 195
  stabilization strategies, 192
  surface modification, 192
  techniques for encapsulating, 188–190
Protein transduction domains (PTDs), 132

Q

Quality-by-Design (QbD) approach, 66, 140, *140*, 144
Quality target product profile (QTPP), elements and CQAs for SLNs-based formulations, **141**, **145**
Quantum dots (QDs), 3, 38

R

Radio-frequency microchannels, **7**
RAGE (receptor for advanced glycation end-products), 186
Random copolymer, 95
Rare-earth-doped upconversion nanoparticles, **156**
Receptor-mediated targeting systems, 66–67
Receptor-mediated transcytosis (RMT), 66
Rectal drug delivery, 116
Red blood cell membrane-camouflaged nanoparticles drug delivery system, 23
Red blood cell (RBC) membrane coating, 38
Regenerative medicine, 41, 165
Regorafenib (REG), 101
Regulatory considerations in drug delivery system, 53–54
Release mechanisms, 190
Release profiles, factors influencing, 190–191
Reproductive disorders, 40
Reticulo endothelial system (RES), 8, 139
Retroviral systems, 89–90
Retroviruses, 130
Risk assessment matrix (RAM), 140
RMT, *See* Receptor-mediated transcytosis (RMT)
RNA-induced silencing complex (RISC), 80
Routes of drug administration, 83, 168
Rybelsus®, 113

S

scCD98-functionalized nanoparticles, 115
Self-microemulsifying drug delivery system (SMEDDS), 23, 115
Semiconductor nanoparticles, 3, **3**, **156**
Short-chain fatty acids (SCFAs), 112
Silica nanoparticles, protein/peptide drug delivery systems, **156**, 186–187
Silicon and glass, 160
Silver nanoparticles (AgNPs), 3, 174–175
Simvastatin-loaded HDL (S-HDL), 46
Single photon emission computed tomography (SPECT), 45
Single-stranded RNA (ssRNA), 130
SLNs, *See* Solid lipid nanoparticles (SLNs)
Small interfering RNA (siRNA), 28, 51
Small-molecule controlled-release technologies, 28
Small molecules, **26**, **27**
SMEDDS, *See* Self-microemulsifying drug delivery system (SMEDDS)
Sol-gel process, MONP synthesis, 169, *169*
Solid lipid nanoparticles (SLNs), 3, 7, *8*, 13, 139
  advantages and disadvantages of, 8
  analysis of SLNs using QbD approach, 144
  angle of repose of lyophilized, 150
  ANOVA-based statistical parameters for, **146**
  applications of, 8
  bulk density of lyophilized, 150
  Carr's Index of lyophilized, 150
  central composite design, 141
  components of, 7–8
  design of experiments for, 145
  DoE-based optimization of selected factors of, 145–146
  effect of variable on particle size of, 147
  effect of variables on entrapment efficiency of, 147
  entrapment efficiency of, 147, 149
  in vitro drug release of lyophilized, 150
  Hausner's ratio of lyophilized, 150
  lyophilized, 150
  morphological characterization of, 13
  optimization of, 141
  overlay plot and validation of model, 147
  particle size distribution of lyophilized, 149
  permeability studies, 150–151

Solid lipid nanoparticles (cont.)
pharmacokinetic studies, 151
physicochemical characterization of ART-loaded, 141–144
preparation of, 144, 149
PXRD spectra of prepared, 150
QTPP and CQAs of, 144
risk assessment studies of, 140, 144–145
selection of lipids of, 144
statistical analysis of data of, 146
structural characterization of, 13
surface morphology of, 149
tapped density of lyophilized, 150
XRD of, 149
Solid lipid nanoparticles-based formulations, QTPP elements and CQAs for, **141**
Sonoporation, 132
SPION, *See* Superparamagnetic iron oxide nanoparticles (SPION)
SPIONPAMAM (ID-NPs), 14
Spray drying, 192
Stearic fatty acids, 5
Stem cells, pluripotent, 25
Stimuli-responsive drug delivery, 51–53
Stimuli-responsive release, 190, 192
Stratum corneum, 6
Suitable delivery system, virus infection and, 88–90
Sulfasalazine, 81
Supercritical fluid technology (SCF)
pilot-scale studies, 158–162
fluid purification and recycle, 159
membrane extrusion, 161–162, *162*
microfluidics technology, 160
microfluidizer technology, 160
particle collection, 159
particle generation, 159
process optimization for scale-up, 162
residual solvent stripping, 159
scale-up considerations, 159–161
scale-up results, 159–160
principles and applications of, 157, *158*
process optimization for scale-up, 157–158
Superoxide dismutase (SOD), 177
Superparamagnetic iron oxide nanoparticles (SPION), 14
Surface engineering, 164
Surfactants, 5
nonionic, 9
Swelling-controlled release, 190
Synthetic lipids, 4
Synthetic phospholipids, 5
Synthetic polymers, 96–97, 103
in controlled drug delivery systems, 100–101

T

Tablets, mouth-dissolving, 11
t-AML, *See* Therapy-related acute myeloid leukemia (t-AML)
Target CNS, noval intranasal drug delivery to, 84–88
Targeted drug delivery system, 24, 49–53, 66–67
Targeted drug release, 53
Targeted gene therapy, 56
Targeting nanodrug carriers, biological ligands for, *50*
T cells with chimeric antigen receptors, 25
TDDS, *See* Transdermal drug delivery system (TDDS)
Temperature sensitivity, 191
Therapeutic delivery system, nucleic acid-based, 78–81
Therapeutic nucleic acids, *128*
Therapy-related acute myeloid leukemia (t-AML), 54
Thermal analysis techniques, 176
Thermodynamically stable niosomes, 9
Thermoplastic polymers, 95–96
Thermoresponsive DDSs, 51
Thermosensitive liposomes (TSLs), 51
Thermoset polymers, 96
Tight junctions (TJs), 87
Time-triggered systems, 114
pH-triggered, microbiota-triggered, and, 114
Titanium dioxide nanoparticles, 177–178
Titanium metals, 41
Topical non-invasive controlled-release delivery techniques, 22
Total joint arthroplasty (TJA), 40
Total joint replacements (TJRs) fail, 40
Toxicity, 3, 6–8, 22, 30
TPP, 80
Traditional drug delivery approaches, 1
Transdermal drug delivery system (TDDS), 6–7, 10, 14
advantages, disadvantages and applications, **7**
factors affecting, 7
in vitro permeation studies, 14–15
Transferases, ethosomes and, 15
Transferosomes, 14
Transferrin (Tf), 49
Transferrin receptor, 67
Translational drug delivery, clinical success in, 56
Translational medicine, 49
Transnational respiratory research (TRR), 38
Transvascular medication delivery, 87
Trastuzumab, 56
Triazine dendrimers, **6**
Tumor microenvironment (TME), 38, 164
Tumor necrosis factor alpha (TNFα), 52

U

Ulcerative colitis (UC), 108
Ultra-deformable vesicles (UDV), 15
Ultrasound, **7**
Ultrasound-mediated drug delivery, focused, 69, *70*
United States Food and Drug Administration (USFDA), 46
Urology, 40

V

Vaccine-containing microspheres, 10
Vaccine development, 80
Vasaka-loaded phytosome DDS, 14
Vascular endothelial growth factor (VEGF), 56
Vascular system, 111
lipid-based, 168
Viral and nonviral gene therapy, *80*
Viral gene delivery system, 88
Viral vectors, 125, 128–130
Virus infection and suitable delivery system, 88–90
Viscosity, colonic luminal, 82

W

Water soluble drugs, 3

X

X-ray diffraction (XRD) of solid lipid nanoparticles, 149

Z

Zeta potential, 104
Zinc oxide nanoparticles, *170*, 177

For Product Safety Concerns and Information please contact our
EU representative GPSR@taylorandfrancis.com Taylor & Francis
Verlag GmbH, Kaufingerstraße 24, 80331 München, Germany